Mechanisms and Machine Science

Volume 34

Series editor

Marco Ceccarelli, Cassino, Italy

More information about this series at http://www.springer.com/series/8779

Veniamin Goldfarb · Natalya Barmina
Editors

Theory and Practice of Gearing and Transmissions

In Honor of Professor Faydor L. Litvin

Springer

Editor
Veniamin Goldfarb
Institute of Mechanics
Kalashnikov Izhevsk State Technical University
Izhevsk
Russia

Assistant Editor
Natalya Barmina
Institute of Mechanics
Kalashnikov Izhevsk State Technical University
Izhevsk
Russia

ISSN 2211-0984 ISSN 2211-0992 (electronic)
Mechanisms and Machine Science
ISBN 978-3-319-19739-5 ISBN 978-3-319-19740-1 (eBook)
DOI 10.1007/978-3-319-19740-1

Library of Congress Control Number: 2015944439

Springer Cham Heidelberg New York Dordrecht London

Printed on acid-free paper

Springer International Publishing AG Switzerland is part of Springer Science+Business Media (www.springer.com)

Preface

The present book is published to honor the outstanding scientist, the founder of the modern theory of gearing, Prof. Faydor L. Litvin, who celebrated his 100th birthday in 2014. The International Symposium "Theory and Practice of Gearing" was dedicated to this event. It was held during January 21–23, 2014 in Izhevsk, Russia. The idea of organizing this Symposium was supported by the Executive Council of International Federation for the Promotion of Mechanism and Machine Science (IFToMM). The book comprises results of scientific research which were discussed at the Symposium and supplemented later with new information. The contributors to this book are scientists in the field of the theory and practice of gearing, many of whom were students of Prof. F. Litvin.

The book is prefaced by two contributions dedicated to the work of Prof. Litvin in different periods of his life in Russia and USA. These two papers comprise an almost complete list of his publications within these periods. The greatest part of the contributions touches upon issues of development of the theory of gearing and practice of design, research and contact quality control for different types of gears—helical and asymmetrical cylindrical, worm and spiroid, bevel and hypoid, planetary and composite gears. A number of papers present the development of gear generating methods and methods of designing the applied machine tools. One paper is devoted to the description of a new concept of design process for gearboxes and gearbox systems.

The book will be of interest and advantage to scientists and practical engineers dealing with gear design, research and production and also to young researchers and students interested in new perspective directions of creating and applying the progressive gears and methods of their synthesis and analysis.

The authors are thankful to IFToMM for the support of our activity and to Springer, which is one of the most authoritative publishing houses of scientific literature, for their assistance in publishing this book.

Veniamin Goldfarb

Contents

Prof. Faydor L. Litvin: A Life Dedicated to the Development of the Modern Theory of Gearing

A. Fuentes

Abstract The American period in the development of the theory of gearing by Prof. Faydor L. Litvin is described. Professor Litvin has dedicated his life to the development of the theory of gearing, and has made significant contributions to the development of the theory of mechanisms. He has developed two professional careers, one in Russia and another in the United States, having become a respected and renowned authority in kinematics and the theory of gearing, and having trained and supervised more than 85 Ph. D. students and visiting scholars from all over the world, all of whom went on to receive prominent research and teaching positions in the USA, Japan, Australia, Bulgaria, China, Taiwan, Italy, Spain, and Russia.

Keywords Faydor L. Litvin · Theory of gearing

Faydor L. Litvin - Year 2002

A. Fuentes (✉)
Department of Mechanical Engineering, Polytechnic University of Cartagena (UPCT), Cartagena, Spain
e-mail: alfonso.fuentes@upct.es

V. Goldfarb and N. Barmina (eds.), *Theory and Practice of Gearing and Transmissions*, Mechanisms and Machine Science 34,
DOI 10.1007/978-3-319-19740-1_1

1 Introduction

Professor Faydor L. Litvin was born in 1914 in Russia. In 1937, he received a Master's degree from the Saint Petersburg Technical Institute. He taught and performed research initially at Saint Petersburg State Polytechnic University (1947–64) and then at Saint Petersburg Institute of Precision Mechanics and Optics (1964–78), where he was professor and head of the Department of Mechanisms Theory and Machine Elements. After his retirement in 1978, he emigrated to the United Stated with his wife, Shifra Litvin, to join their daughter and son. He left behind a distinguished career in Russia to start from scratch in the United States, with only a few dollars in his pocket and no job. Meanwhile, in the 1970s, the US Army established drive system technology as part of its research to support its ever increasing use of helicopters. In 1977, NASA started component work. However, there were no experts in gear theory working at NASA at that time. In 1978, NASA received a call from Prof. Litvin. He had just arrived in US and was living with his son in New Jersey. He was looking for a job and offered to give a lecture about his pioneering ideas on the theory of gearing and its applications. Without entering into details on the many circumstances that merged at that time, with the invaluable support and sponsorship of NASA, Dr. Litvin joined the University of Illinois in Chicago in 1979, at the age of 65, 4 months after his arrival in the US, starting yet another productive career and further cementing this reputation as a respected and renowned authority in kinematics and the theory of gearing.

2 Teaching, Research, and Awards

Dr. Litvin's teaching experience includes analysis and kinematics of mechanisms, theory of gearing and applications, advanced theory of gearing, dynamics of machinery, analysis and design of manipulators, and special topics in advanced kinematics and dynamics of mechanisms. He has performed research in the area of mechanisms, manipulators, theory of gearing, and computerized design, generation and simulation of meshing of gear drives. He has proposed and developed advanced techniques for the design and analysis of gear drives, including spur, helical, worm, spiral bevel, and face gear drives. These techniques have been adapted by many US aerospace companies in service of their products. Dr. Litvin's methods and theories have been the catalyst for changing the design and manufacture of gears to achieve major operational improvements in helicopter gear systems [1, 2]. The application of his innovative technology has resulted in design power increased by 100 % with same volume of gear transmissions.

He achieved those improvements by applying principles based on differential geometry and the theory of gearing to correct many problems associated with alignment and manufacturing errors that shift the bearing contact and cause transmission errors. One of the most significant achievements was the design of optimized face-milled spiral bevel gears with decreased levels of noise, vibration, and stresses solely as the results of changes in machine-tool settings computed by application of his unique algorithm of local synthesis that allows for the desired conditions of meshing directly, without application of the ever common trial-and-error procedure of experienced gear designers. The improved designs allowed for transmission of 25 % more power by the same gear set. Also, the application of formate-cut spiral bevel gears was extended to gear sets with gear ratio close to one, with improved contact conditions, low noise and stresses, and lower manufacturing costs. Litvin has been a pioneer in considering transmission errors in the design of gear drives, and proposing methods to absorb them when errors of alignment or manufacturing occur. Nowadays, it is considered one of the key factors when designing low noise and vibration gear drives.

Not only Prof. Litvin contributed to the development of the theory of gearing, but he has also proposed pioneering ideas in the field of the theory of mechanisms. Some of these are now universally accepted, but they were not when they were proposed in the early '80s, a clear indication of his status as a visionary. He was probably the first to propose the idea of using singular configurations as a tool for finding all solutions for position analysis in the theory of mechanisms. He understood the potential of the theorem of implicit functions as a tool for investigating the number of possible solutions of a non-linear system. Indeed, the theorem states that the determinant of the Jacobian matrix of a square system becomes zero when a singular configuration occurs and the input-output relationship is no longer one-to-one. Few authors recognized the importance of singularities as a key factor for the position analysis. His student and friend, Prof. Parenti-Castelli, states "It has been a golden mine in which many researchers have worked for years". Most relevant papers gave credit to the pioneering work of Prof. Litvin, and nowadays, these ideas/concepts are reported as principles for the topic. Prof. Litvin has supervised the research of a total of 85 Ph.D. students, 42 of whom graduated from UIC, making the professor with the most students to have graduated in the history of that university's Department of Mechanical Engineering. His students have received prominent research and teaching positions in the USA, Japan, Australia, Bulgaria, China, Taiwan, Italy, and Russia. He has also trained and supervised many visiting scholars from Italy and Spain, definitively changing their professional lives in the process. Prof. Litvin is author of nearly 300 publications in referred journals and prestigious gear conferences. His publications include the following books published in the USA:

1. Theory of Gearing, NASA Reference Publication 1212, 1989.
2. Gear Geometry and Applied Theory, Prentice Hall, 1994, ISBN: 0-13-211095.
3. Development of Gear Technology and Theory of Gearing, NASA Reference Publication 1406, 1998.
4. Gear Geometry and Applied Theory, 2nd Edition, Cambridge University Press, 2004.
5. Noncircular Gears: Design and Generation, Cambridge University Press, 2009.

He is holder of 26 inventions (3 of them are U.S. Patents). On of his patents, *Apparatus and Method for Precision Grinding Face Gear*, has singularly provided a way to reduce the weight of helicopter transmissions by 40 %, promoting fuel savings, reduced emissions and lower seat prices. The grinding method invented by Prof. Litvin has, for the first time, enabled a way to grind hardened face gears that will provide the same level of safety as spiral bevel gears. His efforts have opened the way for the use of these low cost, high capacity gears in applications that will significantly benefit the aerospace, automotive and shipping industries. In his American period of research work, Prof. Litvin has been granted the following awards:

- 12 NASA tech-brief awards in the period from 1983 to 2001.
- Best Paper Award, 1990 ASME Mechanism Conference.
- The Allan S. Hall Award of the Applied Mechanism and Robotics Conference, 1991, for contribution to the advancement of the Science of Mechanisms and Robotics.
- Honorable Dr. Degree from Miskolc University, Hungary, 1999.
- UIC Inventor of the Year, 2001.
- The Thomas Bernard Hall Prize, granted by the Institute of Mechanical Engineers, UK, 2001.
- ASME Thomas A. Edison Patent Award, 2004.
- The University of Bologna Sigillum Magnum Award for the pioneering studies and outstanding achievements of his research work in the theory of mechanism and advanced gear technology, not to mention his dedication to teaching and training generations of students and scholars, 2009.
- IFToMM Award of Merits, granted by the International Federation for the Promotion of Mechanism and Machine Science, 2014.

He has been an ASME Fellow since 1992 and member of the editorial advisory board of the Journal of Mechanism and Machine Theory, as well as associate editor of Computer Methods in Applied Mechanics and Engineering, 1995–2000.

3 The Gear Research Center at UIC

Professor Litvin founded the Gear Research Laboratory, the name of which was later changed to the Gear Research Center in 2000. The research performed at the Gear Research Center has been centered around computerized design, generation, and simulation of meshing and contact of aligned and misaligned gear drives with the primary goal of localization and stabilization of bearing contact and reduction of transmission errors. His research has encompassed the study of all types of gear drive applications for transformation of rotation between parallel, intersected, and crossed axes. In addition, cycloidal gearing applied in screw compressors, Wankel engines, Root's blowers, and pumps were studied. Since the foundation of the Gear Research Laboratory, it has been completely funded by memberships, grants, and contracts. Prof. Litvin has worked with a large number of companies in the theory and application of gearing, gear design and manufacturing, and computerized simulation of meshing and contact analysis. He has also become the most prolific contributor to NASA, having written 28 NASA contractor reports and technical manuscripts in the period 1982–2003. Professor Litvin is a believer in creativity and a mentor of youth. In [3], he states: "I can offer new ideas to advance the modern theory of gearing and develop manufacturing procedures, but if I try to develop this research by myself it is not productive". He has always shared his ideas with members of his research team, making them co-authors and co-inventors of the developed ideas. Actually, he has stated, referring to his students and scholars at the Gear Research Center [3]: "Here, there are many brains. We work together and enrich my ideas". Professor Litvin has worked enthusiastically, with love to his job, during his whole life. After his retirement, and supported by the funds remaining from the activity of the Gear Research Center, the University of Illinois at Chicago offered the Graduate Student Faydor Litvin Award, which provided funds for recruiting the best students from around the world, and the Faydor L. Litvin Scholarship in Mechanical Engineering, providing scholarships to undergraduate students in the Department of Mechanical Engineering at the University of Illinois at Chicago, with preference going to students with a GPA of at least 4.0 on a 5.0 scale.

4 The Gallery of Fame

One of Prof. Litvin's goals was to pay tribute to those who, in one way or another, have contributed to the development of the theory of gearing. In [1], a comprehensive history of the development of the theory of gearing through biographies of major contributors to the field was published. Prof. Litvin's unique historical perspective was achieved through assiduous research into the lives of courageous, talented, and creative men who made significant contributions to the field of gearing. Very often, they came from humble backgrounds, seeking education and facing obstacles, making personal sacrifices to attain goals, and working hard for many years to fulfill their creative aspirations. The task of accumulating information about those men was extremely difficult, because many were deceased and facts existed only in family records and the archives of both libraries and the companies for which they worked. The meeting room of the Gear Research Center has been decorated with the portraits of those that contributed to the theory of gearing as a way of making them participants in the discussions that were held in that room. Visitors of the Gear Research Center have had the opportunity to listen to Prof. Litvin talk about their biographies and tell anecdotes about his efforts to assemble their portraits, many of which were obtained from family collections held by descendants many generations later.

5 Final Remarks

I was given the opportunity to be trained by and learn so much from Prof. Litvin, beginning with my visit to the Gear Research Center at UIC in 1999 as a postdoc visiting scholar. I have had him as a reference scientist, teacher, and friend for more than 15 years. He shared with me many of his achievements during the last stage of his active research work. My professional career would certainly have been different had I not met Prof. Litvin in 1999. His advice is always taken into account in my day-by-day professional activity. He is a great example in my life. Thanks Professor!

Appendix

List of Scientific Works and Inventions During the Period 1978–2010 of Distinguished Emeritus Professor Dr. Tech. Sc. Faydor L. Litvin

1. Litvin, F.L., *Analysis of Undercutting Conditions and of Appearance of an Envelope to Contact Lines on Gear Tooth Surfaces*, ASME Journal of Mechanical Design, Vol. 100, pp. 423–432, 1978.
2. Litvin, F.L., *Application of Theorem of Implicit Function System Existence for Analysis and Synthesis of Linkages*, Mechanism and Machine Theory, Vol. 15, pp. 115–125, 1980.
3. Litvin, F.L., *Criteria of Force Transmission for Linkages and Their Application for Synthesis*, ASME Journal of Mechanical Design, Vol. 102, pp. 38–44, 1980.
4. Litvin, F.L., Gutman, Y., *Methods of Synthesis and Analysis for Hypoid Gear Drives of Formate and Helixform. Parts 1, 2, and 3*. ASME Journal of Mechanical Design, Vol. 103, No. 1, pp. 83–113, 1981.
5. Litvin, F.L., Gutman, Y., *Method of Local Synthesis of Gears Based on the Connection between the Principal and Geodesic Curvatures of Surfaces*, ASME Journal of Mechanical Design, Vol. 103, No. 1, pp. 114–125, 1981.
6. Litvin, F.L., Rahman, P., Goldrich, R.N., *Mathematical Models for the Synthesis and Optimization of Spiral Bevel Gears Tooth Surfaces*, NASA CR-3553, 1982.
7. Litvin, F.L., Kukvari, I.S., Krueger, J.E., *Synthesis of Conjugated Surfaces for Sealing Systems*, ASME Journal of Mechanisms, Transmissions and Automation in Design, Vol. 105, pp. 113–121, 1983.
8. Litvin, F.L., Goldrich, R.H., Coy, J.J., Zaretsky, E.V., *Precision of Spiral-Bevel Gears*, ASME J. Mech., Trans. and Auto. Design, Vol. 105, pp. 310–316, 1983.
9. Litvin, F.L., Goldrich, R.H., Coy, J.J., Zaretsky, E.V., *Kinematic Precision of Gear Trains*, ASME J. Mech., Trans. and Auto. Design, Vol. 105, pp. 317–326, 1983.

10. Litvin, F.L., *The Theory of Gear Meshing*, Shangai Science and Technology Press (in Chinese), Shangai, 1984.
11. Litvin, F.L., Coy, J.J., *Special Cases of Friction and Applications*, ASME J. Mech., Trans. and Auto. Design, Vol. 106,pp. 256–260, 1984.
12. Jamalow, R.I., Litvin, F.L., Roth, B., *Analysis and Design of RCCC Linkages*, Mechanism and Machine Design, Vol. 19, pp. 397–407, 1984.
13. Litvin, F.L., Tsay, C.-B., *Helical Gears with Circular Arc Teeth: Simulation of Conditions of Meshing and Bearing Contact*, ASME J. Mech., Trans. and Auto. Design,, Vol. 107, pp. 556–564, 1985.
14. Litvin, F.L., Rahman, P., Tsay, C.B., Tsung, W.J., Coy, J., *Computer-Aided Simulation of Kinematical Errors and Bearing Contact for Vehicle Gears, and their Adjustment to Errors of Manufacturing and Assembly*, International Journal of Vehicle Design, Vol. 6, pp. 630–633, 1985.
15. Litvin, F.L., Castelli, V.P., *Configurations of Robots Manipulators and Their Identification, and the Execution of Prescribed Trajectories. Part 1: Basic Concepts*, ASME J. Mech., Trans. and Auto. Design, Vol. 107, pp. 170–178, 1985.
16. Litvin, F.L., Costopoulos, T., Castelli, V.P., Shaheen, M., Yukishige, Y., *Configurations of Robots Manipulators and Their Identification, and the Execution of Prescribed Trajectories. Part 2: Investigations of Manipulators Having 5-degrees, 7-degrees, and 8-degrees of Freedom*, ASME J. Mech., Trans. and Auto. Design, Vol. 107, pp. 179–188, 1985.
17. Litvin, F.L., Fanghella, P., Tan, J., Zhang, Y., *Singularities in Motion and Displacement Functions of Spatial Linkages*, ASME J. Mech., Trans. and Auto. Design, Vol. 108, pp. 516–523, 1986.
18. Litvin, F.L., Yi, Z., Castelli, V.P., Innocenti, C., *Singularities, Configurations, and Displacement Functions for Manipulators*, International Journal of Robotics Research, Vol. 5, pp. 52–65, 1986.
19. Litvin, F.L., Yi, Z., *Robotic Bevel-Gear Differential Train*, International Journal of Robotics Research, Vol. 5, pp. 75–81, 1986.
20. Litvin, F.L., Castelli, V.P., Phillips, R.H., *Manipulators—Execution of Prescribed Trajectories Special Link Positions and Versions of Assembly*, Mechanism and Machine Theory, Vol. 21, pp. 173–185, 1986.
21. Litvin, F.L., Tsung, W.-J., Coy, J., Heine, C., *Method for Generation of Spiral Bevel Gears with Conjugate Gear Tooth Surfaces*, ASME J. Mech., Trans. and Auto. Design, Vol. 109, pp. 163–170, 1987.
22. Litvin, F.L., Castelli, V.P., Shaheen, M., *Singularities in Motion and Displacement for a 7-degree-of-freedom Manipulator*, Journal of Robotic Systems, Vol. 4, pp. 397–421, 1987.
23. Litvin, F.L., Zhang, Y., Lundy, M., Heine, C., *Determination of Settings of a Tilted Head Cutter for Generation of Hypoid and Spiral Bevel Gears*, ASME J. Mech., Trans. and Auto. Design,, Vol. 110, pp. 495–500, 1988

24. Litvin, F.L., Zhang, J., Handschuh, R.F., *Crowned Spur Gears: Methods for Generation and Tooth Contact Analysis—Part I: Basic Concepts, Generation of the Pinion Tooth Surface by a Plane*, ASME J. Mech., Trans. and Auto. Design, Vol. 110, pp. 337–342, 1988.
25. Litvin, F.L., Zhang, J., Handschuh, R.F., *Crowned Spur Gears: Methods for Generation and Tooth Contact Analysis—Part 2: Generation of the Pinion Tooth Surface by a Surface of Revolution*, ASME J. Mech., Trans. and Auto. Design, Vol. 110, pp. 343–347, 1988.
26. Litvin, F.L., Zhang, J., *Spur Gears. Optimal Geometry, Methods for Generation and Tooth Contact Analysis (TCA) Program*, NASA Contractor Report 4135, AVSCOM Technical Report 88-C-002, 1988.
27. Litvin, F.L., Tan, J., *Singularities in Motion and Displacement Functions for the RCRCR Linkage*, ASME J. Mech., Trans. and Auto. Design, Vol. 110, pp. 373–377, 1988.
28. Litvin, F.L., *Theory of Gearing*, NASA Reference Publication 1212 (AVSCOM 88-C-C035), Washington, D. C., 1989.
29. Litvin, F.L., Zhang, Y., Krenzer, T.J., Goldrich, R.N., *Hypoid Gear Drive with Face-Milled Teeth: Condition of Pinion Non-Undercutting and Fillet Generation*, AGMA paper 89FTM7, 1989.
30. Litvin, F.L., Shaheen, M.A., Tan, J., *Toward the Inverse Kinematics of a 7 Degree-of-freedom Manipulator*, Robotersysteme, Vol. 5, pp. 238–246, 1989.
31. Litvin, F.L., Tan, J., *Singularities in Motion and Displacement Functions of Constrained Mechanical Systems*, International Journal of Robotics Research, Vol. 8, pp. 30–43, 1989.
32. Litvin, F.L., Tan, J., *Determination of Bearing Reactions of Spatial Linkages and Manipulators*, ASME Journal of Mechanical Design, Vol. 112, pp. 168–174,1990.
33. Litvin, F.L., Kin, V., Zhang, Y., *Limitations of Conjugate Gear Tooth Surfaces*, ASME Journal of Mechanical Design, Vol. 112, pp. 230–236, 1990.
34. Litvin, F.L., Chaing, W.-S., Lundy, M., Tsung, W.-J., *Design of Pitch Cones for Face-hobbed Hypoid Gears*, ASME Journal of Mechanical Design, Vol. 112, pp. 413–418, 1990.
35. Litvin, F.L., Yi, Z., Kieffer, J., Handschuh, R.F., *Identification and Minimization of Deviations of Real Tooth Surfaces*, ASME Journal of Mechanical Design, Vol. 113, pp. 55–62, 1991.
36. Litvin, F.L., Kuan, C., Kieffer, J., Bossler, R., Handschuh, R.F., *Straddle Design of Spiral Bevel and Hypoid Pinions and Gears*, ASME Journal of Mechanical Design, Vol. 113, pp. 422–426, 1991.
37. Litvin, F.L., Zhang, Y., *Local Synthesis and Tooth Contact Analysis of Face-Milled Spiral Bevel Gears*, NASA Contractor Report 4342, AVSCOM Technical Report 90-C-028, 1991.

38. Litvin, F.L., *Applied Theory of Gearing*, Chapter in Dudley's Gear Handbook, McGraw-Hill, NY, 2nd Ed., 1991.
39. Litvin, F.L., Chaing, W.S., Kuan, C., Lundy, M., Tsung, W. J., *Generation and Geometry of Hypoid Gear-Member With Face-Hobbed Teeth of Uniform Depth*, International Journal of Machine Tools and Manufacture, Vol. 31, pp. 167–181, 1991.
40. Handschuh, R.F., Litvin, F.L., *A method for Determining Spiral-Bevel Gear Tooth Geometry for Finite Element Analysis*, NASA Technical Paper 3096, AVSCOM Technical Report 91-C-020, 1991.
41. Kieffer, J., Litvin, F.L., *Swept Volume Determination and Interference Detection for Moving 3-D Solids*, ASME Journal of Mechanical Design, Vol. 113, pp. 456–463, 1991.
42. Kieffer, J., Litvin, F.L., *Local Parametric Representation of Displacement Functions for Linkages and Manipulators*, Mechanism and Machine Theory, Vol. 26, pp. 41–53, 1991.
43. Litvin, F.L., Wang, J.-C., Chen, Y.-J., Bossler, R.D., Heath, G., Lewicki, D. G., *Face Gear Drives: Design, Analysis and Testing for Helicopter Transmission Applications*, AGMA Paper 92FTM2, 1992.
44. Litvin, F.L., Kin, V., *Computerized Simulation of Meshing and Bearing Contact for Single-Enveloping Worm-Gear Drives*, ASME Journal of Mechanical Design, Vol. 114, pp. 313–316, 1992.
45. Litvin, F.L., Zhang, Y., Wang, J.-C., Bossler, R. B., Chen, Y.-J. D., *Design and Geometry of Face-Gear Drives*, ASME Journal of Mechanical Design, Vol. 114, pp. 642–647, 1992.
46. Litvin, F.L., Zhang, Y., Kuan, C., Handschuh, R.F., *Computerized Inspection of Real Surface and Minimization of Their Deviations*, International Journal of Machine Tools and Manufacture, Vol. 32, pp. 141–145, 1992.
47. Litvin, F.L.., Kuan, C.P., Zhang, Y., Masseth, J., *Computerized Inspection of Hypoid Pinion Face-Milled Tooth Surfaces*, International Journal of Machine Tools and Manufacture, Vol. 32, pp. 869–884, 1992.
48. Litvin, F.L., Kuan, C., Wang, C.C., Handschuh, R.F., Masseth, J., Maruyama, N., *Minimization of Deviations of Gear Real-Tooth Surfaces*, ASME Journal of Mechanical Design, Vol. 115, No. 4, pp. 995–1001, 1993.
49. Litvin, F.L., Hsiao, C.-L., *Computerized Simulation of Meshing and Bearing Contact of Enveloping Gear Tooth Surfaces*, Computer Methods in Applied Mechanics and Engineering, Vol. 102, pp. 337–366, 1993.
50. Litvin, F.L., Chen, N.X., Zhang, Y., Krenzer, T.J., Handschuh, R.F., *Computerized Generation of Surfaces with Optimal Approximation to Ideal Surfaces*, Computer Methods in Applied Mechanics and Engineering, Vol. 110, pp. 39–55, 1993.
51. Litvin, F.L., Lu, J., *Computerized Simulation of Generation, Meshing and Contact of Double Circular-Arc Helical Gears*, Mathematical and Computer Modeling, Vol. 18, No. 5, pp. 31–47, 1993.

52. Litvin, F.L., *Gear Geometry and Applied Theory*, Prentice Hall, Inc., Englewood Cliffs, New Jersey, 1994.
53. Litvin, F.L., Hsiao, C.L., Wang, J.-C., Zhou, X., *Computerized Simulation of Generation of Internal Involute Gears and Their Assembly*, ASME Journal of Mechanical Design, Vol. 116, No. 3, pp. 683–689, 1994.
54. Litvin, F.L., Wang, J.-C., Bossler, R.B., Chen, Y.-J., Heath, G., Lewicki, D. G., *Application of Face-Gear Drives in Helicopter Transmissions*, ASME Journal of Mechanical Design, Vol. 116, No. 3, pp. 672–676, 1994.
55. Litvin, F.L., Hsiao, C.-L., *Direct Relations between Curvatures of Surfaces Being in Continuous Tangency*, Zeitschrift fur Angewandte Mathematik und Mechanik, Vol. 74, No. 7, pp. 225–234, 1994.
56. Litvin, F.L., Zhao, X., *Computerized Design and Analysis of Face-Milled Uniform Tooth Height*, Low Noise Spiral Bevel Gear Drives, NASA Contractor Report 4704, 1994.
57. Lewicki, D.G., Handschuh, R.F., Henry, Z.S., Litvin, F.L., *Low-Noise, High Strength Spiral Bevel Gears for Helicopter Transmission*, J. Propul. Power, Vol. 10, No. 3, pp. 356–361, 1994.
58. Zhang, Y., Litvin, F.L., Maruyama, N., Takeda, R., Sugimoto, M., *Computerized Analysis of Meshing and Contact of Gear Real Tooth Surfaces*, ASME Journal of Mechanical Design, Vol. 116, No. 3, 1994.
59. Litvin, F.L., Chen, N.X., Lu, J., Handschuh, R.F., *Computerized Design and Generation of Low-Noise Helical Gears with Modified Surface Topology*, ASME Journal of Mechanical Design, Vol. 117, No. 2, pp. 254–261, 1995.
60. Litvin, F.L., *Applied Theory of Gearing: State of the Art*, ASME Journal of Mechanical Design, Special 50th Anniversary Issue, Vol. 117, pp. 129–134, 1995.
61. Litvin, F.L., Chen, N.X., Chen, J.-S., *Computerized Determination of Curvature Relations and Contact Ellipse for Conjugate Surfaces*, Computer Methods in Applied Mechanics and Engineering, Vol. 125 of 1–4, September, 1995.
62. Litvin, F.L., Lu, J., *Computerized Design and Generation of Double Circular-Arc Helical Gears With Low Transmission Errors*, ASME Journal of Mechanical Design, Vol. 127, No. 1–4, pp. 57–86, 1995.
63. Litvin, F L., Chen, J.-S., Sep, T.M., Whang, J.-C., *Computerized Simulation of Transmission Errors and Shift of Bearing Contact for Face-Milled Hypoid Gear Drives*, ASME Journal of Mechanical Design, Vol. 117, No. 2, pp. 262–268, 1995.
64. Zhang, Y., Litvin, F.L., Handschuh, R.F., *Computerized Design of Low-Noise Face-Milled Spiral Bevel Gears*, Mechanism and Machine Theory, Vol. 30, No. 8, pp. 1171–1178, 1995.
65. Lu, J., Litvin, F.L., Chen, J.S., *Load Share and Finite Element Stress Analysis for Double Circular-Arc Helical Gears*, Mathematical and Computer Modeling, Vol. 21, No. 10, pp. 13–30, 1995.

66. Litvin, F.L., Wang, A.G., Handschuh, R.F., *Computerized Design and Analysis of Face-Milled Uniform Tooth Height Spiral Bevel Gear Drives*, ASME Journal of Mechanical Design, Vol. 118, No. 4, pp. 573–579, 1996.
67. Litvin, F.L., Seol, I.H., *Computerized Determination of Gear Tooth Surface as Envelope to Two Parameter Family of Surfaces*, Computer Methods in Applied Mechanics and Engineering, Vol. 138, No. 1–4, pp. 213–225, 1996.
68. Litvin, F.L., Feng, P.-H., *Computerized Design and Generation of Cycloidal Gearings*, Mechanism and Machine Theory, Vol. 31, No. 7, pp. 891–991, 1996.
69. Litvin, F.L., Seol, I.H., Kim, D., Lu, J., Wang, A.G., Egelja, A., Zhao, X., Handschuh, R.F., *Kinematic and Geometric Models of Gear Drives*, ASME Journal of Mechanical Design, Vol. 118, No. 4, pp. 544–550, 1996.
70. Litvin, F.L., Chen, J.-S., Lu, J., Handschuh, R.F., *Application of Finite Element Analysis for Determination of Load Share, Real Contact Ratio, Precision of Motion, and Stress Analysis*, ASME Journal of Mechanical Design, Vol. 118, No. 4, pp. 561–567, 1996.
71. Litvin, F.L. et al., *Computerized Design and Generation of Gear Drives with Localized Bearing Contact and Low Level of Transmission Errors*, Proceedings of International Conference on Gears, VDI-Beritchte, No. 1230, pp. 63–82, 1996.
72. Seol, I.H., Litvin, F.L., *Computerized Design, Generation and Simulation of Meshing and Contact of Worm-Gear Drives with Improved Geometry*, Computer Methods in Applied Mechanics and Engineering, Vol. 138, No. 1–4, pp. 73–103, 1996.
73. Seol, I.H., Litvin, F.L., *Computerized Design, Generation and Simulation of Meshing and Contact of Modified Involute, Klingelnberg and Flender Type Worm-Gear Drives*, ASME Journal of Mechanical Design, Vol. 118, No. 4, pp. 551–555, 1996.
74. Litvin, F.L., Kim, D.H., *Computerized Design, Generation and Simulation of Meshing of Modified Involute Spur Gear With Localized Bearing Contact and Reduced Level of Transmission Errors*, ASME Journal of Mechanical De-sign, Vol. 119, pp. 96–100, 1997.
75. Litvin, F.L., Feng, P.-H., *Computerized Design, Generation and Simulation of Meshing of Rotors of Screw Compressor*, Mechanism and Machine Theory, Vol. 32, No. 2, pp. 137–160, 1997.
76. Litvin, F.L., Egelja, A., Tan, J., Heath, G., *Computerized Design, Generation and Simulation of Meshing of Orthogonal Offset Face-Gear Drive with a Spur Involute Pinion with Localized Bearing Contact*, Mechanism and Machine Theory, Vol. 33, pp. 87–102, 1998.
77. Litvin, F.L., Egelja, A.M., De Donno, M., *Computerized Determination of Singularities and Envelopes to Family of Contact Lines on Gear Tooth Surface*, Computer Methods in Applied Mechanics and Engineering, Vol. 158, No. 1–2, pp. 23–34, 1998.
78. Litvin, F.L., De Donno, M., Lian, Q., Lagutin, S.A., *Alternative Approach for Determination of Singularity of Envelope to a Family of Parametric*

Surfaces, Computer Methods in Applied Mechanics and Engineering, Vol. 167, pp. 153–165, 1998.

79. Litvin, F.L., Wang, A.G., Handschuh, R.F., *Computerized Generation and Simulation of Meshing and Contact of Spiral Bevel Gears with Improved Geometry*, Computer Methods in Applied Mechanics and Engineering, Vol. 158, No. 1–2, pp. 35–64, 1998.
80. Litvin, F.L., Hsiao, C.-L., Ziskind, M.D., *Computerized Overwire (Ball) Measurement of Tooth Thickness of Worms*, Screws and Gears, Mechanism and Machine Theory, Vol. 33, No. 6, pp. 851–877, 1998.
81. Litvin, F.L., Zhang, J., Chaing, W.-S., Coy, J.J., Handschuh, R.F., *Crowned Spur Gears: Optimal Geometry and Generation*, Gear Technology, The Journal of Gear Manufacturing, Vol. 5, No. 5, pp. 9–15, 1998.
82. Argyris, J., Litvin, F.L., Peng, A., Stadtfeld, H.J., *Axes of Meshing and Their Application in Theory of Gearing*, Computer Methods in Applied Mechanics and Engineering, Vol. 163, No. 1–4, pp. 293–310, 1998.
83. Litvin, F.L. De Donno, M., *Computerized Design and Generation of Modified Spiroid Worm-Gear Drive with Low Transmission Errors and Stabilized Bearing Contact*, Computer Methods in Applied Mechanics and Engineering, Vol. 162, No. 1–4, pp. 187–201, 1998.
84. Litvin, F.L., *Development of Gear Technology and Theory of Gearing*, NASA Reference Publication 1406, ARL-TR-1500, 1998.
85. Argyris, J., Litvin, F.L., Lian, Q., Lagutin, S.A., *Determination of Envelope to Family of Planar Parametric Curves and Envelope Singularities*, Computer Methods in Applied Mechanics and Engineering, Vol. 175, No. 1–2, pp. 175–187, 1999.
86. Feng, P.-H., Litvin, F.L., Townsend, D.P., Handschuh, R.F., *Determination of Principal Curvatures and Contact Ellipses for Profile Crowned Helical Gears*, ASME Journal of Mechanical Design, Vol. 121, No. 1, pp. 107–111, 1999.
87. Litvin, F.L., Peng, A., Wang, A.G., *Limitation of Gear Tooth Surfaces by Envelopes to Contact Lines and Edge of Regression*, Mechanism and Machine Theory, Vol. 34, No. 6, pp. 889–902, 1999.
88. Litvin, F.L., Lu, J., Townsend, D.P., Howkins, M., *Computerized Simulation of Meshing of Conventional Helical Involute Gears and Modification of Geometry*, Mechanism and Machine Theory, Vol. 34, No. 1, pp. 123–147, 1999.
89. De Donno, M., Litvin, F.L., *Computerized Design and Generation of Worm Gear Drives with Stable Bearing Contact and Low Transmission Errors*, ASME Journal of Mechanical Design, Vol. 121, No. 4, pp. 573–578, 1999.
90. De Donno, M., Litvin, F.L., *Computerized Design, Generation, and Simulation of Meshing of a Spiroid Worm-Gear Drive with a Ground Double-Crowned Worm*, ASME Journal of Mechanical Design, Vol. 131, No. 2, pp. 264–273, 1999.
91. Litvin, F.L., Argentieri, G., De Donno, M., Howkins, M., *Computerized Design, Generation and Simulation of Meshing and Contact of Face Worm-*

Gear Drives, Computer Methods in Applied Mechanics and Engineering, Vol. 189, No. 3, pp. 785–801, 2000.

92. Litvin, F.L., De Donno, M., Peng, A., Vorontsov, A., Handschuh, R.F., *Integrated Computer Program for Simulation of Meshing and Contact of Gear Drives*, Computer Methods in Applied Mechanics and Engineering, Vol. 181, No. 1–3, pp. 71–85, 2000.
93. Litvin, F.L., Lian, Q., Kapelevich, A.L., *Asymmetric Modified Spur Gear Drives: Reduction of Noise, Localization of Contact, Simulation of Meshing and Stress Analysis*, Computer Methods in Applied Mechanics and Engineering, Vol. 188, No. 1–3, pp. 363–390, 2000.
94. Litvin, F.L., Feng, P.-H., Lagutin, S.A., *Computerized Generation and Simulation of Meshing and Contact of New Type of Novikov-Wildhaber Helical Gear*, NASA Contractor Report CR 2000-209415, ARL-CR-428, 2000.
95. Argyris, J., De Donno, M., Litvin, F.L., *Computer Program in Visual Basic Language for Simulation of Meshing and Contact of Gear Drives and its Application for Design of Worm Gear Drives*, Computer Methods in Applied Mechanics and Engineering, Vol. 189, No. 2, pp. 595–612, 2000.
96. Litvin, F.L., Egelja, A., Tan, J., Chen, D.Y.-D., Heath, G., *Handbook on Face Gear Drives with a Spur Involute Pinion*, NASA/CR-2000-209909, ARL-CR-447, 2000.
97. Litvin, F.L., Demenego, A., Vecchiato, D., *Formation by Branches Of Envelope to Parametric Families of Surfaces and Curves*, Computer Methods in Applied Mechanics and Engineering, Vol. 190, pp. 4587–4608, 2001.
98. Litvin, F.L., Fuentes, A., Howkins, J.M., Handschuh, R.F., *Design, Generation and Tooth Contact Analysis (TCA) of Asymmetric Face Gear Drive with Modified Geometry*, NASA/TM 2001-210614, ARL-TR-2373, 2001.
99. Litvin, F.L., Fuentes, A., Demenego, A., Vecchiato, D., Fan, Q., *New Developments in the Design and Generation of Gear Drives*, Inst. Mech. Eng., Mechanical Engineering Science, Part C., Vol. 215, No. C7, pp. 747–758, 2001.
100. Litvin, F.L., Fuentes, A., Howkins, M., *Design, Generation and TCA of New Type of Asymmetric Face-Gear*, Computer Methods in Applied Mechanics and Engineering, Vol. 190, pp. 5837–5865, 2001.
101. Litvin, F.L., Fan, Q., Vecchiato, D., Demenego, A., Handschuh, R.F., Sep, T.M., *Computerized Generation and Simulation of Meshing of Modified Spur and Helical Gears Manufactured by Shaving*, Computer Methods in Applied Mechanics and Engineering, Vol. 190, pp. 5037–5055, 2001.
102. Litvin, F.L., Fan, Q., Fuentes, A., *Computerized Design, Generation and Simulation of Meshing and Contact of Face-Milled Formate Cut Spiral Bevel Gears*, NASA Glenn Research Center, Technical Report, NASA/CR-2001-210894, ARL-CR-467, May, 2001.

103. Vecchiato, D., Demenego, A., Argyris, J., Litvin, F.L., *Geometry of a Cycloidal Pump*, Computer Methods in Applied Mechanics and Engineering, Vol. 190, No. 18–19, pp. 2309–2330, 2001.
104. Demenego, A., Vecchiato, D., Litvin, F.L., Nervegna, N., Manco, S., *Design and Simulation of Meshing of a Cycloidal Pump*, Mechanism and Machine Theory, Vol. 37, pp. 311–332, 2002.
105. Litvin, F.L., Fuentes, A., Zanzi, C., Pontiggia, M., *Face Gear Drive with Spur Involute Pinion: Geometry, Generation by a Worm, Stress Analysis*, NASA CR-2002-211362, ARL-CR-491, pp. 1–44, 2002
106. Argyris, J., Fuentes, A., Litvin, F.L., *Computerized Integrated Approach for Design and Stress Analysis of Spiral Bevel Gears*, Computer Methods In Applied Mechanics and Engineering, Vol. 191, pp. 1057–1095, 2002.
107. Litvin, F.L., Fuentes, A., Fan, Q., Handschuh, R.F., *Computerized Design, Simulation of Meshing and Contact, and Stress Analysis of Face-Milled Formate Generated Spiral Bevel Gears*, Mechanism and Machine Theory, Vol. 37, No. 5, pp. 441–459, 2002.
108. Fuentes, A., Litvin, F.L., Mullins, B.R., Woods, R., Handschuh, R.F., Lewicki, D.G., *Design, Stress Analysis and Experimental Tests of Low-Noise Adjusted Bearing Contact Spiral Bevel Gears*, Proc. International Conference on Gears, VDI-Beritchte, No. 1665, Vol. 1, pp. 327–340, 2002.
109. Litvin, F.L., Nava, A., Fan, Q., Fuentes, A., *New Geometry of Face-Worm Gear Drives with Conical Cylindrical Worms: Generation, Simulation of Meshing, and Stress Analysis*, Comput. Methods Appl. Mech. Engrg, Vol. 191, No. 27, pp. 3035–3054, 2002.
110. Litvin, F.L., Fuentes, A., Zanzi, C., Pontiggia M., Handschuh R.F., *Face Gear Drive with Spur Involute Pinion: Geometry, Generation by a Worm, Stress Analysis*, Computer Methods in Applied Mechanics and Engineering, Vol. 191, No. 25, pp. 2785–2813, 2002.
111. Litvin, F.L., Vecchiato, D., Demenego, A., Karedes, E., Hansen, B., Handschuh, R.F., *One Stage Planetary Gear Train with Improved Conditions of Load Distribution and Reduced Transmission Errors*, ASME Journal of Mech. Design, 124, pp. 745–752, 2002.
112. Argyris J., Fuentes A., Litvin F.L., *Computerized integrated approach for design and stress analysis of spiral bevel gears*, Comput. Methods Appl. Mech. Engrg, Vol. 191, pp 1057–1095, 2002.
113. Litvin F.L., Vecchiato D., Fuentes A., Gonzalez-Perez I., 2002, *Automatic determination of guess values for simulation of meshing of gear drives*, Comput. Methods Appl. Mech. Engrg., Vol. 193, pp 3745–3758, 2004.
114. Litvin F.L., Fuentes A., *Gear Geometry and Applied Theory (2-nd edition)*, Cambridge University Press, New York, 2004.
115. Litvin F.L., Sheveleva G.I., Vecchiato D., Gonzalez-Perez I., Fuentes A., *Modified approach for tooth contact analysis of gear drives and automatic determination of guess values*, Comput. Methods Appl. Mech. Engrg., Vol. 194, pp. 2927–2946, 2005.

116. Litvin F.L., Gonzalez-Perez I., Fuentes A., Hayasaka K., Yukishima K, *Topology of modified surfaces of involute helical gears with line contact developed for improvement of bearing contact, reduction of transmission errors, and stress analysis*, Math. and Computer Modelling, Vol. 42, No. 9–10, pp. 1063–1078, 2005.
117. Litvin F.L., Gonzalez-Perez I., Fuentes A., Vecchiato D, Sep T.M., *Generalized concept of meshing and contact of involute crossed helical gears and its application*, Comput. Methods Appl. Mech. Engrg., Vol. 194, No. 34, pp. 3710–3745, 2005.
118. Litvin F.L., Gonzalez-Perez I., Fuentes A., Vecchiato D, Bruce D., Hansen B.D., Binney D., *Design, generation and stress analysis of face-gear drive with helical pinion*, Comput. Methods Appl. Mech. Engrg., Vol. 194, No. 36, pp. 3870–3901, 2005.
119. Litvin, F.L., Vecchiato, D., Gurovich, E., Fuentes, A., Gonzalez-Perez, I., Hayasaka, K., Yukishima, K., *Computerized developments in design, generation, simulation of meshing, and stress analysis of gear drives*, Meccanica, 40 (3), pp. 291–323, 2005.
120. Litvin F.L., Fuentes A., Hayasaka K., *Design, manufacture, stress analysis, and experimental tests of low-noise high endurance spiral bevel gears*, Mechanism and Machine Theory, Vol. 41, No. 1, pp. 83–118, 2006.
121. Litvin, F.L., Vecchiato, D., Yukishima, K., Fuentes, A., Gonzalez-Perez, I., Hayasaka, K., *Reduction of noise of loaded and unloaded misaligned gear drives*, Computer Methods in Applied Mechanics and Engineering, 195 (41–43), pp. 5523–5536, 2006.
122. Litvin F.L., Gonzalez-Perez I., Yukishima K, Fuentes A., Hayasaka K., *Design, simulation of meshing, and contact stresses for an improved worm gear drive*, Mechanism and Machine Theory, Vol. 42, No. 8, pp. 940–959, 2007.
123. Litvin, F.L., Gonzalez-Perez, I., Yukishima, K., Fuentes, A., Hayasaka, K., *Generation of planar and helical elliptical gears by application of rack-cutter, hob, and shaper*, Computer Methods in Applied Mechanics and Engineering, 196 (41–44), pp. 4321–4336, 2007.
124. Gonzalez-Perez, I., Fuentes, A., Litvin, F.L., Hayasaka, K., Yukishima, K., *Application and investigation of modified helical gears with several types of geometry*, 2007 Proceedings of the ASME International Design Engineering Technical Conferences and Computers and Information in Engineering Conference, DETC2007, 7, pp. 19–28., 2007.
125. Lewicki, D.G., Litvin, F.L., Woods, R.L., Fuentes, A., *Evaluation of a low-noise formate spiral-bevel gear set*, 2007 Proceedings of the ASME International Design Engineering Technical Conferences and Computers and Information in Engineering Conference, DETC2007, 7, pp. 305–325, 2007.
126. Fuentes, A., Gonzalez-Perez, I., Litvin, F.L., Hayasaka, K., Yukishima, K., *Determination of basic machine-tool settings for generation of spiral bevel gears from blank data*, 2007 Proceedings of the ASME International Design Engineering Technical Conferences and Computers and Information in Engineering Conference, DETC2007, 7, pp. 57–68, 2007.

127. Lewicki, D.G., Litvin, F.L., Woods, R.L., Fuentes, A., *Evaluation of a low-noise, formate spiral bevel gear set*, Gear Technology, 25 (1), pp. 46–58, 2008.
128. Litvin, F.L., Gonzalez-Perez, I., Fuentes, A., Hayasaka, K., *Design and investigation of gear drives with non-circular gears applied for speed variation and generation of functions*, Computer Methods in Applied Mechanics and Engineering, 197 (45–48), pp. 3783–3802, 2008.
129. Litvin, F.L., Gonzalez-Perez, I., Fuentes, A., Hayasaka, K., *Tandem design of mechanisms for function generation and output speed variation*, Computer Methods in Applied Mechanics and Engineering, 198 (5–8), pp. 860–876, 2009.
130. Litvin F.L., Fuentes-Aznar A., Gonzalez-Perez I., Hayasaka K., *Noncircular Gears: Design and Generation*, Cambridge University Press., 2009.
131. Gonzalez-Perez, I., Fuentes, A., Litvin, F.L., Hayasaka, K., *Computerized design of multi-gear drives for function generation*, Proceedings of the ASME International Design Engineering Technical Conferences and Computers and Information in Engineering Conference 2009, DETC2009, 6, pp. 29–38., 2009.
132. Fuentes, A., Gonzalez-Perez, I., Litvin, F.L., Hayasaka, K., *Computerized design of gear drives with modified elliptical centrodes*, Proceedings of the ASME Design Engineering Technical Conference, 6, pp. 13–19, 2009.
133. Gonzalez-Perez, I., Fuentes, A., Litvin, F.L., Hayasaka, K., *Computerized design of multi-gear drives for function generation*, Proceedings of the ASME Design Engineering Technical Conference, 6, pp. 29–38, 2009.
134. Fuentes, A., Gonzalez-Perez, I., Litvin, F.L., Hayasaka, K., *Computerized design of gear drives with modified elliptical centrodes*, Proceedings of the ASME International Design Engineering Technical Conferences and Computers and Information in Engineering Conference 2009, DETC2009, 6, pp. 13–19, 2009.
135. Fuentes, A., Nagamoto, H., Litvin, F.L., Gonzalez-Perez, I., Hayasaka, K., *Computerized design of modified helical gears finished by plunge shaving*, Computer Methods in Applied Mechanics and Engineering, 199 (25–28), pp. 1677–1690, 2010.

US Patents:

- US Patent No. 6,128,969; "*Hybrid Gear Drive for Transformation of Power and Rotation Between Crossed and Intersected Axes*"; October 10, 2000. Inventors: Litvin, F.L., Egelja, A., and Seol, I.H.
- US Patent No. 6,146,253; "*Apparatus and Method for Precision Grinding Face Gear*", November 14, 2000. Inventors: Litvin, F.L., Chen, Y.-J., Heath, G.F., Sheth, V.J., and Chen, N.
- US Patent No. 6,205,879; "*Helical and Spur Gear Drive with Double Crowned Pinion Tooth Surfaces and Conjugated Gear Tooth Surfaces*", March 27, 2001. Inventors: Litvin, F.L., Feng, P.-H., Lagutin, S.A., Townsend, D.P., and Sep, T.M.

References

1. Litvin FL (1998) Development of gear technology and theory of gearing, NASA reference publication 1406, ARL-TR-1500
2. Ravani B (1994) A dedication to Professor Faydor L. Litvin in celebration of his eightieth birthday. J Mech Des 116(3):671
3. Engineering news, Colleague of engineering, University of Illinois at Chicago [UIC] (2002)

Prof. F.L. Litvin: Contribution to the Formation of the Russian School of the Theory of Gearing

S. Lagutin and N. Barmina

Abstract A brief overview of selected scientific works by Prof. F.L. Litvin performed up to 1980 is presented. His role in the development of the Russian school of the theory of gearing is shown.

Keywords Generating of gearing · Kinematic method of investigation · Envelope surfaces · Analysis and synthesis of gears

1 Introduction

The end of 1950s was marked by two outstanding events in the formation of the Russian school of the theory of gearing. First, in 1959, the engineer Colonel M.L. Novikov was granted an afterlife with the Lenin award for the development of gears with a new system of gearing, which caused an outbreak of interest in gears. Second, in 1960, the first edition of a monograph by Prof. F.L. Litvin was published, though at the time it was noticed only by a very limited number of experts.

The fundamentals of gearing research were laid in the first half of the 20th century through works by outstanding scientists, namely Ch.I. Gochman, Ch.F. Ketov, N.I. Kolchin, V.N. Kudryavtsev, and V.A. Gavrilenko. But publication of the monograph by Litvin [1] specifically marked the identification of the theory of gearing as a separate independent scientific subject. Beginning exactly at its inception, a "Sturm und Drang" period in the development of this theory started.

S. Lagutin
JSC "EZTM", Electrostal, Russia
e-mail: sergeylagutin@gmail.com

N. Barmina (✉)
Institute of Mechanics, Kalashnikov Izhevsk State Technical University, Izhevsk, Russia
e-mail: barmina-nat@mail.ru

V. Goldfarb and N. Barmina (eds.), *Theory and Practice of Gearing and Transmissions*, Mechanisms and Machine Science 34,
DOI 10.1007/978-3-319-19740-1_2

At that time a whole constellation of bright researchers gathered around F.L. Litvin. Among them were: Ya.S. Davydov, L.V. Korostelev, N.N. Krylov, K.M. Pismanik, K.I. Gulyaev, G.I. Sheveleva, M.L. Erikhov et al. Cooperating and competing with each other, within a few years they had created a marvelous building of a new science at the junction of differential geometry and machine mechanics.

While we can talk about each of these brainstorming participants as contributing some part to this new science, F.L. Litvin laid the essential primary foundation most completely represented in monograph [2].

An attempt is made below to enumerate, even if not in great detail, the separate, closely related blocks of this foundation.

2 A Little Prehistory. Principles of Generation of Spatial Gearing

Fundamentals of the theory of spatial gearing were laid out by a French geometrician Theodore Olivier [3]. He was the first to consider the generation of conjugated surfaces as the enveloping process and introduce the concept of an auxiliary generating surface. Considering tooth flanks as enveloping to the auxiliary surface, he found the condition of their linear and point contact. Olivier suggested then that the theory of gearing is a particular subject of projective geometry.

Restrictions imposed by projective geometry were withdrawn by Gochman [4]. Having developed analytic investigation methods, he projected the theory of gearing into the area of analytical and differential geometry.

Olivier, Theodore
(1793-1853)

Gohman, Chaim I.
(1851-1916)

Principles of Olivier-Gochman, generalized by Ya.S. Davydov for the case of applying non-rigid and non-congruent pairs of generating surfaces [5], have been and remain now the foundation of any method for the generating of gears cut by enveloping. While developing a system of gearing named after him, M.L. Novikov proposed a principle of generating of gearing that differed from the principles of Olivier-Gochman [6]; however, industrial implementation of Novikov gears became possible only on the basis of applying the basic rack profile of a pair of non-congruent generating racks. Even the recently outlined trend of cutting gears by pin tools with NC machines will not save the design engineers from the necessity of virtual tooth cutting by Olivier-Gochman principles.

3 Kinematic Method for Investigation of Spatial Gearing

Principles of generating of gearing require an adequate mathematical apparatus for their investigation. While developing the analytical methods of Ch.I. Gochman, N.I. Kolchin brought them into a state for solving the important practical tasks, such as: investigation of spur bevel gears, non-orthogonal worm gears and globoid gears of Cone type, profiling the cutting tool and analysis of errors in gearing [7].

Kolchin, Nikolai I.
(1894-1975)

Davydov, Yakov S.
(1914-2003)

However, the methods of analysis by Gochman and Kolchin were very bulky. The next step in the development of the theory of gearing was creation of a kinematic method of investigation, based on the simple and obvious fact that at any contact point of mutually enveloping surfaces their common contact-normal **n** should be perpendicular to the vector of their relative velocity $V^{(1,2)}$. The kinematic approach to gearing investigation was proposed at the beginning of the 1950s by three researchers independent of each other. But while Shishkov [8] described such an approach verbally, and Davydov [9] expressed it through projections of the above-mentioned vectors, it was F.L. Litvin who presented the main law of gearing in a perfect and concise way [10]:

$$nV^{(1,2)} = 0.$$

This law allows to describe the enveloping surface and to investigate the conditions of surface contact analytically. For 60 years already this exact law has been the basis of a majority of investigations both in the area of analysis and the synthesis of gearing.

However, to be fair, in a number of cases, a non-differential approach to solving the enveloping problem, i.e., unrelated to finding normal vectors to surfaces, is sometimes of greater advantage. Fundamentals of such an approach were proposed in [11, 12].

4 Matrix Method of Coordinate Transformation in the Theory of Gearing

In any case, the investigation of gearing is related to a repeated transformation of equations of lines and surfaces from one coordinate system to another: from the system related to a tool to a fixed coordinate system, from a fixed one to a system related to the gearwheel to be cut, etc. In order to describe these transformations, F.L. Litvin proposed applying a 4 × 4 matrix, where the 3 × 3 determinant is for cosines of angles between axes of the old and new coordinate systems and the 4th column is for coordinates of the vector connecting the origins of these systems. This method is so vital that Litvin describes it in the first chapters of all his monographs up to the last edition [13], and all his further investigations were performed based on application of this method.

The versatility of this method becomes even more important when one considers that, according to the reasonable note of M.L. Erikhov, having mastered this mathematical apparatus, any average post-graduate is promptly capable of solving the task of investigating any type of new gearing. This statement is proved by the fact that 48 Ph.D. theses were presented under the direct supervision of F.L. Litvin in Leningrad. And the number of Litvin's absentee followers, from both Russia and abroad, who applied his methods, including the method of matrix transformation, is uncountable.

5 General Problems of the Theory of Spatial Gearing

Among the most important theoretical tasks successfully solved by F.L. Litvin in his works and described in the monograph [2], the following should be mentioned:

(i) Applying the kinematic method of investigation, he obtained equations of a surface enveloping a family of generating surfaces both for one and, what is more complicated, two parameters of enveloping. He revealed the conditions sufficient for the existence of the envelope to the family of generating surfaces represented by equations both in parametrical and implicit forms.

(ii) Investigating curvatures in the gearing, he determined the main and normal curvatures of the enveloping surface and its torsion. He found direct relations between curvatures of the enveloped and enveloping surfaces in one- and two-parametrical enveloping processes. He determined the dimensions and orientation of the Dupin indicatrix as the prototype of the instant contact ellipse.

(iii) He developed the method for finding singular points on the enveloping surface and methods of their prevention at its active segment or, in engineering language, methods of preventing tooth undercutting.

(iv) The concept (well-known for gears with parallel and intersected axes) of the axis of meshing as the straight line intersected by all contact normals of the gear was extended by him to gears with crossed axes. He proved the existence of a pair of such equivalent axes in the case in which one of engaged surfaces is helical with a constant pitch, and he showed how application of these axes allows for simplifying the investigation of gearing, in particular, worm gears.
(v) He considered and described mathematically seven different ways of synthesis of conjugated gearing, including by means of a rigid or non-rigid non-congruent pair of generating surfaces as well as by that of generating lines.
(vi) Litvin proposed methods of synthesis of approximated gearing with the optimal conditions of gearing in the vicinity of the predesigned contact point.
(vii) He investigated the effect of errors in gear manufacturing and assembly on the function of transmission errors in gearing with linear and point contact.

6 Geometry of Gearing for Specific Types of Gears

Statement of the above enumerated general problems of the theory of gearing takes up about one third of the monograph [2]. The remaining two thirds are devoted to investigation of gearing geometry for specific types of gears. Among them are:

(i) Non-circular gears with closed centrodes, intended to transmit the rotation with variable gear ratio. The first monograph by the author [14] was devoted to investigation of these gears.
(ii) Helical gears with parallel axes, both involute and Novikov-Wildhaber gears cut by a non-congruent pair of generating racks.
(iii) Helical gears with crossed axes, including involute ones and a new type of crossed gears cut by a congruent pair of concave and concise generating racks, proposed together with V.V. Shultz.
(iv) Worm gear drives, orthogonal and non-orthogonal, with cylindrical worms, both with ruled surfaces ZA (Archimedes), ZN (convolute), and ZI (involute), and non-ruled surfaces, such as surface ZK, reproduced by a grinding cone and, finally, with two new versions of ground worms with a concave profile, later included into Russian standards as types ZT1 and ZT2.
(v) Face-gear drives with orthogonal and non-orthogonal, intersecting and crossed axes, for which engineering techniques of geometrical design were developed taking into account conditions preventing tooth undercut and sharpening and providing contact localization.
(vi) Face-milled spiral bevel gears with tapered and uniform teeth, Helixform bevel gears with helicoid tooth surface and cyclo-palloid bevel gears with cycloid tooth profile of the generating wheel. In all cases, investigation is brought up to obtaining the equations of conjugate surfaces and the design of the set-up parameters of cutter heads to provide contact localization in gears.

7 Investigations Made in the 1970s

During this time, F.L. Litvin's main attention was focused on scientific supervision of his numerous post-graduates. For instance, at the 6th All-Union Conference on ToMM in 1973, he presented his survey paper with seven young co-authors [15]. The following issues were discussed in this paper, amongst others: analysis of pitch surfaces of hypoid and spiroid gears, synthesis of involute spiroid gearing, synthesis of double-enveloping worm gears with a grounded worm, synthesis of double-enveloping worm gears with reduced vibration activity, synthesis of bevel gears with circular teeth based on uniting the local synthesis through solving the so-called reverse task of the theory of gearing.

Among other practically important works of that period, the authors should mention the work [16], in which fundamentals of the synthesis of cylindrical worm gear drives with localized contact have been presented.

Simultaneous to working through the issues of specific gear geometry, F.L. Litvin continued investigating general issues of the theory of gearing.

In particular, he generalized the formula of Euler-Savary for determining the relation between the main curvatures of tooth flanks in spatial gearing [17].

Together with M.L. Erikhov, he created the vector field of normals for ordinary nodes of contact of the enveloped surface [18].

Working through the concept of the node line of contact as the envelope to the family of instant contact lines, Litvin revealed the necessary and sufficient conditions of such a line existence on the generating surface, having thus determined the boundary of this surface beyond which the area not participating in the enveloping process is located [19].

He revealed the sufficient conditions of existence for the similar node line of contact and the edge of regression on the enveloping surface, i.e., determined its theoretical boundaries [20].

The full list of scientific publications and patents that Doctor of Technical Sciences Prof. Litvin published in Russia within the period from 1939 until 1978 comprises 121 papers and 22 inventor's certificates. This list was made by F.L. Litvin himself and is presented in the Appendix.

The 1970s were also the beginning of Litvin's publication abroad. The first of these papers was devoted to investigation of the relation between tooth curvatures in spatial gearing [21]. Another notable publication of this period is the fundamental generalization of methods of gearing synthesis by two-parameter enveloping, written together with N.N. Krylov and M.L. Erikhov [22]. It is difficult for a modern internet user to understand the psychological effect these breakthrough Russian articles, coming from behind the Iron Curtain to the international podium, had on his contemporaries.

In 1979, Faydor L'vovich immigrated to the USA with his family. There, he continued his active scientific work. He organized the Gear Research Centre at the University of Illinois in Chicago, and his credibility lends this Centre the support of the leading mechanical engineering enterprises in the country. This period of activity is better described by his post-graduate students in Chicago.

We note only that, working in the United States, F.L. Litvin didn't interrupt his personal and scientific relations with his Russian colleagues. In particular, in 2003, he visited Moscow, and at the ensuing workshop, which brought together all contemporary Russian experts, delivered a brilliant lecture on the latest scientific achievements of his school.

8 Conclusion

Thankfully, Faydor L'vovich is still with us, healthy and continuing his activity. He works 364 days a year. On the 365th, Judgment Day, he reports to God and begins a new work. Thus, he does not allow us, his students and followers, to overestimate our small successes and rest on our laurels.

He does not let us forget that, if we want to live in a civilized world, someone has to design and cut those gears without which we can drive neither car wheels, nor power plant turbines, nor gear pumping units for oil extraction, nor mill rolls.

Let us hope that supreme forces will maintain his health and buoyancy until the age of 120, as given to a man in one old and wise book.

Appendix

Full List of Scientific Publications and Inventions of Professor, Doctor of Technical Sciences F.L. Litvin for the Period of 1939–1978

Publications

1. Basic statements of shaving process. Proc. of Committee on tooth-cutting engineering. VNITOMASH-LONITOMASH, N1, 1939 (in Russian).
2. To calculation of rectilinear guidance mechanism of the type of Evans mechanism. "Journal of Engineers and Technicians", N9, 1939 (in Russian).
3. Influence of eccentricity of spur pinions of involute profile on reading accuracy gears. Jnl. "Optical and Mechanical Industry", N4, 1940 (in Russian).
4. Control of spur pinions by means of wires. Jnl. "Optical and Mechanical Industry", N6, 1941 (in Russian).
5. Spur and helical involute pinions with crossed axes; influence of errors on accuracy of operation. PhD thesis, maintained in Tomsk polytechnic institute in 1944 (in Russian).
6. Crossed helical gears. Proc. "Gearing", N1, Mashgiz, L., 1946. (Awarded by LONITOMASH as the best scientific research work) (in Russian).

7. Main problems of the theory of non-circular gearwheels. In Proc. of LONITOMASH, N6 "Theory and Analysis of Gearwheels", Mashgiz, L.,1949 (in Russian).
8. Some problems of the theory of gearing for non-circular spur and helical gearwheels. "Proceedings of Seminar on Machine and Mechanism Theory", N29, USSR AS, 1949 (in Russian).
9. Non-circular gearwheels (monograph, the first edition). Mashgiz, 1950 (in Russian).
10. Basic theorems of planar gearing and their application to profiling the flat gearing. "Proceedings of LPI", N7, Mashgiz, 1950 (in Russian)
11. Non-circular gearwheels for fluid meter. "Journal of Engineers and Technicians", N3, 1950 (in Russian).
12. To the question of noncircular gearwheel tooth cutting by generating method "Proceedings of Seminar on Machine and Mechanism Theory". N 34, USSR AS, 1950 (in Russian).
13. Determination of tooth thickness of worms and helical gearwheels by rollers and balls. "Proceedings of Seminar on Machine and Mechanism Theory", N39, USSR AS, 1951 (in Russian).
14. Application of bevel gearing analytical theory to determination of tooth thickness of straight bevel gearwheels by balls. "Proceedings of Seminar on Machine and Mechanism Theory", N44, USSR AS, 1952 (in Russian).
15. Calculation methods in production and control of gear parts (coauthor: N.I. Kolchin), Mashgiz, 1952 (in Russian).
16. Cutting non-circular gearwheels by division method. "Report of technical information" N1, Glavpribor, 1952 (in Russian).
17. Toward the Investigation of Spatial Gearing with Line Contact of Surfaces. "Proceedings of Seminar on Machine and Mechanism Theory", Vol. 49, USSR AS, 1953 (in Russian)
18. Tool profiling for helical surface machining according to pitch. "Proceedings of LPI", N4, 1953 (in Russian).
19. Improvement of gear hobbing machine for cutting gearwheels (coauthor: G.G. Pavlov). "Machines and tools", N4, 1953 (in Russian).
20. Theory of gearing, design and production of non-circular gearwheels. "Gears in Mechanical Engineering. Proceedings of the first Moscow conference on gears in mechanical engineering", USSR AS, 1953 (in Russian).
21. Adjustment of stroke, force calculation and balancing of a piston machine (coauthors: N.I. Kolchin, A.A. Valkovskiy). Study guide. LPI Publ., 1954 (in Russian).
22. Production of multi-thread worm gears with new geometry (coauthors: S.G. Printsental, G.F. Shigorin). Mashgiz, 1953 (in Russian).
23. Theory and methods of production of gears with variable gear ratio (Doctor of technical sciences thesis, maintained in Leningrad polytechnic institute in 1954), (in Russian).
24. Device for cutting non-circular gearwheels (coauthor: G.G. Pavlov). "Machines and tools", N2, 1954 (in Russian).

25. Analytical method of investigating the worm gearing with variable gear ratio and variable center distance. "Proceedings of LPI", N3, 1954 (in Russian).
26. Analysis of cylindrical screw manifolds (coauthor: D.L. Faynshtein). "Machines and tools", N7, 1955 (in Russian).
27. Designing non-circular gearwheels, applied in mechanical engineering. "Proceedings of IMASH. Seminar on TMM", N55, USSR AS, 1954 (in Russian).
28. Application of matrices and theory of screws to investigation of spatial gearing. "Proceedings of LPI", N182, 1955 (in Russian).
29. Reproducing the function of one independent variable by a simple row of non-circular gearwheels. "Proceedings of LPI", N182, 1955 (in Russian).
30. Non-circular gears: design, theory of gearing and manufacture, 2nd ed, Mashgiz, 1956. 320p. (in Russian).
31. Non-circular gears. "Machines and tools", N9, 1956 (in Russian).
32. Addendum correction of spur gears and drawing of their meshing (coauthor: N.I. Kolchin). Study guide. LPI Publ., 1957 (in Russian).
33. On design and cutting teeth of multithread non-circular gearwheels (coauthor: N.S. Yablonskiy). "Priborostroeniye", N6, 1957 (in Russian).
34. To investigation of steady motion of machine aggregate with elastic driving shaft for forces dependent on velocity and position of links. "Proceedings of IMASH. Seminar on TMM", N68, USSR AS, 1958 (in Russian).
35. Simplified method of determining the contact lines and contact area of a worm gearing with a cylindrical worm. "Vestnik mashinostroyeniya", N2, 1959 (in Russian).
36. To calculation of a functional shunted potentiometer (coauthor: S.G. Kislitsin). "Automatics and telemechanics", N11, 1959 (in Russian).
37. Simplified method of determining the position of a ball or roller in the root of a screw or worm. "Izvestiya Vuzov. Mashinostroyeniye", N1, 1960 (in Russian).
38. Simplified method of investigating the worm gearing with a cylindrical worm. "Proceedings of the 2nd All-Union Conference on main problems of TMM", USSR AS, 1960 (in Russian).
39. Problems of geometry and manufacture of worm gears with the increased load-carrying capacity (coauthor: I.K. Koryshev). "Proceedings of the conference on technique of gears", Mashgiz, 1960 (in Russian).
40. Analytical methods of spatial gearing investigation. "Proceedings of IMASH. Seminar on TMM", N84, USSR AS, 1961 (in Russian).
41. Engagement of bevel gears with epi- hypotrochoid longitudinal profile (coauthor: N.B. Firun). "Proceedings of IMASH. Seminar on TMM", N86, USSR AS, 1961 (in Russian).
42. Theory of gearing (monograph, the 1st ed.) FIZMATGIZ, 1960 (in Russian).
43. Grinding method and geometry of thread surfaces of cylindrical worms with concave profile. "Izvestiya OTN USSR AS. Mechanics and mechanical engineering", N4, 1961 (in Russian).

44. Dynamic balancing of rotors (coauthors: S.G. Smirnov, N.F. Utekhin). LPI Publ., 1961 (in Russian).
45. Geometry of worm gear drives with cylindrical worms with concave profile, ground by a toroidal wheel. “Proceedings of Odessa conference”, Mashgiz, 1961 (in Russian).
46. Geometry of surfaces and efficiency of screw pairs with rolling friction (coauthor: B.A. Konstantinov). “Machines and tools”, N7, 1962 (in Russian).
47. Geometrical investigation of a variety of Novikov gearing. “Proceedings of mechanical institute”, N24, 1962 (in Russian).
48. Undercutting of bevel gearwheels with circular teeth (coauthor: Guo-Kai). “Machines and tools”, N11, 1962 (in Russian).
49. New types of worm gear drives with cylindrical worms (monograph). Mashgiz, 1962 (in Russian).
50. Investigation of meshing of bevel gears with tapered teeth. “Theory of machines and mechanisms” (coauthor: Guo-Kai). “Theory of machines and mechanisms”, N92-93, USSR AS, 1962 (in Russian).
51. Method of determination of contact lines and limiting points of tooth flanks of a spatial gearing mechanism. “Izvestiya OTN USSR AS. Mechanics and mechanical engineering”, N6, 1962 (in Russian).
52. Matrix method of determination of relation between curvatures of tooth flanks in a gearing (coauthor: Chzhan Tsy-syan). “Izvestiya OTN USSR AS. Mechanics and mechanical engineering”, N3, 1963 (in Russian).
53. Tooth undercutting of planar and spatial gearing. “Proceedings of the 3rd all-Russian meeting on TMM”, Vol. 4, Mashgiz, 1963 (in Russian).
54. Improvement of engagement conditions of bevel gearwheels with arch teeth (coauthor: Guo-Kai). “Proceedings of Seminar on Machine and Mechanism Theory”, N98-99, USSR AS, 1963 (in Russian).
55. Investigation of engagement conditions in vicinity of the tooth flanks contact point. “Proceedings of Seminar on Machine and Mechanism Theory”, N100, USSR AS, 1963 (in Russian).
56. Position function and transmission ratio of the driving arm mechanism. “Analysis and Synthesis of Mechanisms and Gear Theory”, Nauka, 1965 (in Russian).
57. Application of the kinematic method to determine the relation between curvatures of mutually enveloping surfaces, and conditions for the absence of teeth undercutting.. “Proceedings of Seminar on Machine and Mechanism Theory”, N103, USSR AS, 1964 (in Russian).
58. Main problems of the theory of spatial gearing and methods of their solution. Proceedings of Miskolc-Universitat, Hungary, 1964 (in Hungarian).
59. Geometry and methods of manufacture of worm gear drives with concave worm profile. Proceedings of Miskolc-Universitat, Hungary, 1964 (in Hungarian).
60. Reference book of designer of precision instrument industry. (Chief editing and author of several sections). “Mashinostroyeniye”, 1964 (in Russian).

61. Functional cam mechanisms (monograph). Publ. house of Leningrad institute of precise mechanics and optics, 1965 (in Russian).
62. Synthesis of approximated bevel and hypoid gearing with improved parameters. "Analysis and synthesis of mechanisms and theory of gearing", M. Nauka, 1965 (in Russian).
63. Determination of initial parameters, their correction when cutting bevel gearwheels (coauthor: A.A. Zaostrovskiy). "Theory of gears in machines", Mashinostroyeniye, 1966 (in Russian).
64. Helical gearwheels with new geometry, providing the increased load-carrying capacity (coauthor: V.V. Schultz). "Theory of gears in machines", Mashinostroenie, 1966 (in Russian).
65. Improving the engagement conditions in Helixform hypoid gears (coauthor: Yu.I. Shurygin). "Theory of gears in machines", Mashinostroenie, 1966 (in Russian).
66. Determination of the error of the dividing circuit for tooth hobbing machines by electro-magnetic method and error influence on the accuracy of gearwheels to be cut (coauthor: N.G. Lindtrop). "Proc. of technical economic conference of Neva plant n.a. Lenin", Mashinostroyeniye, 1966 (in Russian).
67. Determination of main parameters of gearwheels of Helixform bevel gears with helicoid tooth surface (coauthor: B.D. Zilberman). "Mechanics of machines", N1-2, Nauka, 1966 (in Russian).
68. Application of kinematic method to determine curvatures and conditions of undercutting elimination for the envelope of two-parameter set of tool surfaces (coauthor: M.L. Erikhov). "Mechanics of machines", N3-4, Nauka, 1966 (in Russian).
69. Engagement and cutting the bevel gearwheels with cycloidal tooth lines (coauthors: V.N. Kedrinskiy, K.I. Gulyaev, N.B. Firun). "Machines and tools", N2, 1967 (in Russian).
70. Synthesis of approximated gearing according to local conditions (Yu.A. Sinichkin, V.S. Smirnov). "Mashinovedeniye", N2, 1967 (in Russian).
71. Worm gear drives with the concave profile worm (coauthors: V.N. Komkov, I. P. Bernatskiy). "Vestnik mashinostroeniya", N9, 1967 (in Russian).
72. To generation of non-circular bevel gearwheels (coauthor: G.A. Livshits). "Mashinovedeniye", N1, 1968 (in Russian).
73. Theory of gearing (monograph, 2nd ed.). "Nauka" Publ., 1968 (in Russian; it was translated into Hungarian and published in Hungary in 1973).
74. Transmission of motion by eccentrically fixed helical gearwheels (coauthor: N.G. Lindtrop). "Gears and worm gear drives", Mashinostroenie, 1968 (in Russian).
75. Contact conditions and radii of curvatures in the design point for straight bevel gears cut by continuous circular drawing (coauthors: Yu.A. Sinichkin, V.S. Smirnov). "Theory of mechanisms and machines", N5, Kharkov, 1969 (in Russian).

76. Determination of main curvatures of tooth flanks and parameters of cutting heads for spiral bevel gears with tapered teeth (coauthor: B.P. Timofeyev). "Mashinovedeniye", N1, 1969 (in Russian).
77. Curvature determination for tooth flanks of straight bevel gearwheels at synthesis according to local conditions (coauthors: L.Ya. Liburkin, B.D. Zilberman, Yu.A. Sinichkin). "Mashinovedeniye", N5, 1969 (in Russian).
78. Cutting of non-circular bevel gearwheels by intermittent face milling (coauthor: R.Sh. Varsimashvili). "Machines and tools", N5, 1970 (in Russian).
79. Generalization of the Euler-Savary formula and determination of relations between main curvatures of tooth flanks in spatial gearing. "Mashinovedeniye", N6, 1970 (in Russian).
80. Determination of function of the position in a spatial rod mechanism by a conditional open circuit. "Mashinovedeniye" N3, 1970 (in Russian).
81. Finding the line of boundary points on the tool surface (coauthors: V.D. Britskiy, B.P Timofeyev). "Izvestiya VUZov. Mashinostroyeniye, N3, 1970 (in Russian).
82. "Theory of gears in machines" (Chief editing). Mashinostroyeniye, 1970. (in Russian).
83. Development of geometry and theory of gearing. "Theory of gears in machines", Mashinostroyeniye, 1970 (in Russian).
84. Vector field of normals in ordinary node contact points of the enveloped surface (coauthor: M.L. Erikhov). "Theory of gears in machines". Mashinostroyeniye, 1970 (in Russian).
85. Analysis of settings to cut Helixform bevel gears with tapered teeth (coauthor V.N. Rubtsov). "Machines and tools", N12, 1970 (in Russian).
86. Worm link stopper of rotation (coauthors: V.D. Britskiy, I.P. Tyrsa). "Izvestiya VUZov. Priborostroeniye", N8, 1970 (in Russian).
87. Peculiarities of generating the contact pattern of bevel gears with circular teeth (coauthors: V.N. Rubtsov, Kh.I. Marinov, B.P. Timofeyev). "Machines and tools", N6, 1971 (in Russian).
88. Mathematical modeling of approximated gearing and their optimization. "Proceedings of international congress on theory of mechanisms and machines", 1971, Belgrad (in Russian).
89. Synthesis of face-milled and Helixform bevel gearwheels with uniform teeth (coauthors: V.N. Rubtsov, Kh.I. Marinov). "Izvestiya Vuzov. Mashinostroyeniye", N1, 1971 (in Russian).
90. Kinematic way of determining the curvature of centrodes (coauthor: E.E. Peisakh). "Izvestiya Vuzov. Mashinostroyeniye", N1, 1971 (in Russian).
91. Elimination of node contact points at synthesis of involute spiroid gears (coauthor: V.A. Ganshin). "Theory of gears in machines", Nauka, 1971 (in Russian).
92. Probabilistic estimation of kinematic error of eccentric gearwheels (coauthors: Y.I. Gutman, V.A. Kutsokon). "Izvestiya VUZov. Priborostroeniye", N10, 1972 (in Russian).

93. Synthesis of straight bevel gears cut by circular broaching (coauthors: L.Ya. Liburkin, B.D. Zilberman, Yu. A. Sinichkin). "Mechanics of machines", N 31–32, 1972 (in Russian).
94. Transmission of forces and friction losses in spur involute gears with spring mechanism to adjust the dead stroke (coauthors: Yu.Ya. Grossmann, M.A. Nozdrin. "Izvestiya VUZov. Priborostroeniye", N5, 1972 (in Russian).
95. Transmission of forces in zoom-lens (coauthors: E.E. Peisakh, Ya.M. Kruger). "Izvestiya VUZov. Priborostroeniye", N12, 1972 (in Russian).
96. Synthesis of face-milled and Helixform bevel gearwheels according to local conditions (coauthors: B.P. Timofeyev, V.N. Rubtsov). "Mechanics of machines", N31-32, 1972 (in Russian).
97. To definition of some geometrical parameters of gear bearing capacity when cutting the teeth by enveloping method (coauthor: V.D. Britskiy, B. P. Timofeyev). "Mashinovedeniye", N5, 1972 (in Russian).
98. Kinematic error of a planetary wave gear (coauthor: I.I. Vasilyeva). "Izvestiya VUZov. Priborostroeniye", N7, 1973 (in Russian).
99. Some problems of spatial gearing synthesis (coauthors: B.D. Britskiy, V.A. Ganshin et al.). "Theory of gears in machines", Nauka, 1973 (in Russian).
100. Designing the mechanisms and parts of devices (monograph), "Mashinostroyeniye", 1973 (in Russian).
101. Probabilistic method of determining the program for automatic amateur photo cameras. Jnl. "Optical and Mechanical Industry", N9, 1973 (in Russian).
102. Investigating the influence of assembly errors on accuracy of spiral bevel gears (coauthor: D.I. Ivanova). "Izvestiya Vuzov. Mashinostroyeniye", N11, 1973 (in Russian).
103. Guidelines on analysis of kinematic error of spur and helical gears (coauthor: Y.I. Gutman). LITMO Publ., 1974 (in Russian).
104. Guidelines on analysis of cams for scanning systems (coauthor: L.L. Alperin). LITMO Publ., 1974 (in Russian).
105. Determining the relation between main curvatures at arbitrary value of gear ratio derivative (coauthor: Y.I. Gutman). "Mechanics of machines", N45, Nauka, 1974 (in Russian).
106. Combining the local synthesis and solution of the reverse task to optimize the approximated gearing as applied to hypoid gears (coauthor: Y.I. Gutman). "Mechanics of machines", N 45, Nauka, 1974 (in Russian).
107. Localization of bearing contact in worm gear drives with cylindrical worms (coauthor: V.I. Rybakov). "Izvestiya Vuzov. Mashinostroyeniye", N8. 1974 (in Russian).
108. Applying the method of conditional cutting the contour to determine the motion parameters of seven-, six-, five-link spatial level mechanisms (coauthor: S.E. Karpovich. "Mashinovedeniye". N6, 1975 (in Russian).
109. Analysis and design of mechanisms and parts of devices. Chief editor and author of several chapters, "Mashinostroyeniye", 1975 (in Russian).
110. Determination of Envelope of Characteristics of Mutually Enveloping Surfaces. "Izvestiya Vuzov. Matematika", N10, 1975 (in Russian)

111. Analysis of versions of clock gearing by solving the reverse task (coauthor: S. P. Ganchev). “Izvestiya VUZov. Priborostroeniye”, N8, 1975 (in Russian).
112. Determining the dimensions of blank plates, generating the solid limited by spatial polyhedron (coauthor: S.E. Karpovich). “Izvestiya Vuzov. Mashinostroyeniye”, N10, 1975 (in Russian).
113. Structural synthesis of geared kinematic circuits and designing the bearing block to mount the cylindrical gear in the circuit (coauthor: Yu.A. Grossman). LITMO Publ., 1975 (in Russian).
114. Geometric and analytic determination of existence domain for a spatial four-link mechanism of RSSR type. “Mashinovedeniye”, N4, 1976 (in Russian).
115. Conditions of existence of one or two cranks in mechanism of RCCC type (coauthor: S.E. Karpovich). “Izvestiya Vuzov. Mashinostroyeniye”, N8, 1976 (in Russian).
116. Cutting the non-circular spur and helical gears by discontinuous face milling method with numerical control. “Reports of AS of Georgia SSR”, N3, 1976 (in Russian).
117. Probabilistic calculation of the backlash for spur involute gears (coauthors: F.I. Tutma, E.V. Shalobayev, Yu.A. Yuryev). “Izvestiya VUZov. Priborostroeniye”, N9, 1976 (in Russian).
118. Determination and elimination of unfavorable zones of engagement in worm gear drives with cylindrical worms (coauthor: I.P. Bernatskiy). “Vestnik mashinostroyeniya”, N12, 1976 (in Russian).
119. Calculation of gearing in two-screw pump, consisting of a convolute and conjugated to it screws (coauthor: S.E. Karpovich). “Vestnik mashinostroyeniya”, N11, 1976 (in Russian).
120. Method of analysis of rod mechanisms with abundant links and its applications (coauthors: R.I. Jamalow, Y.I. Gutman). “Mashinovedeniye”, N4, 1978 (in Russian).
121. Existence domain of one- and two-crank rod mechanisms of RCCC type with account of force transmission (coauthors: R.I. Dzhamalov, A.B. Kikin). “Izvestiya Vuzov. Mashinostroenie”, N11 1977 (in Russian).
122. Synthesis of rod mechanisms with osculating cylinders (coauthors: T.S. Bil, Yu.Z. Tenenbaum). “Izvestiya VUZov. Priborostroeniye”, N9, 1978 (in Russian).
123. Litvin, F.L., Relation Between Curvatures of Teeth Surfaces in Spatial Gearing, *Zeitschrift fur Angewandte Mathematik und Mechanik* (in German), Vol. 49, p. 685–690, 1969.
124. Litvin, F.L., Synthesis of Approximate Meshing for Spatial Gears, *Journal of Mechanisms*, Vol. 4, Pergamon press, 1969.
125. Litvin, F.L., Petrov, K.M., and Ganshin, V. A., The Effect of Geometrical Parameters of Hypoid and Spiroid Gears on Their Quality Characteristics, *ASME Journal of Engineering for Industry*, Vol. 96, pp. 330–334, 1974.

126. Litvin, F.L., Krylov, N.N., and Erikhov, M. L., Generation of Tooth Surfaces by Two-Parameter Enveloping, *Mechanism and Machine Theory*, Vol. 10, No. 5, pp. 365–373, 1975.
127. Litvin, F.L., Simplification of the Matrix Method of Linkage Analysis by Division of a Mechanism into Unclosed Kinematic Chains, *Mechanism and Machine Theory*, Vol. 10, pp. 315–326, 1975.
128. Litvin, F.L., Determination and Analysis of Links Position Functions of a Spatial Linkage, *Theor. and Appl. Mechanics*, Bulgarian Academy of Sciences, year VII, No 1, 1976.
129. Litvin, F.L., Jamalow, R.I. and Petrov, K.M., Analysis of Overconstrained Spatial Mechanisms, *Theor. and Appl. Mechanics*, Bulg. Acad. of Sciences, Vol. VIII, No 3, 1977.
130. Litvin, F.L., Analysis of Undercutting Conditions and of Appearance of an Envelope to Contact Lines on Gear Tooth Surfaces, *ASME Journal of Mechanical Design*, Vol. 100, pp. 423–432, 1978.

Inventor's Certificates (Granted in Russia)

1. Planetary gearbox. Certificate of authorship N 83854, priority dated January 19, 1949 (coauthor: Ch. F. Ketov).
2. Method of checking the centrodes of non-circular gearwheels. Certificate of authorship N 88033 priority dated April 25, 1949 (coauthors: Ch.F. Ketov and B.S. Kogan).
3. Device to gear-hobbing machine to cut convex-concave non-circular gearwheels. Certificate of authorship N 91505 priority dated January 24, 1950.
4. Device to gear-hobbing machine. Certificate of authorship N 91506 priority dated November 16, 1949.
5. Caliper pig to check the profile of gearwheels. Certificate of authorship N93866 priority dated November 27, 1950 (coauthor: N.A. Gayevskiy).
6. Gear-hobbing machine for cutting the non-circular gearwheels with concave centrodes. Certificate of authorship N97454 priority dated May 11, 1950 (coauthor: G.G. Pavlov).
7. Method of generating the helical surface of the worm. Certificate of authorship N 139531 priority dated January 26, 1961.
8. Gear-hobbing machine for cutting the non-circular gearwheels. Certificate of authorship N10005 priority dated January 19, 1954.
9. Gear-hobbing machine for cutting the non-circular gearwheels. Certificate of authorship N154700 priority dated June 19, 1961 (coauthors: G.G. Pavlov, N.S. Yablonskiy, V.A. Ziskindovich).
10. Device for cutting the non-circular gearwheels at gear-hobbing machines. Certificate of authorship N 162016 priority dated April 11, 1949 (coauthors: Kh.F. Ketov, B.S. Kogan, K.Kh. Klyamikin, M.P. Makarov).
11. Device for cutting the non-circular gearwheels at gear-hobbing machines. Certificate of authorship N162017 priority dated April 11, 1949.

12. Low-frequency vibration stand. Certificate of authorship N169845 priority dated February 29, 1964 (coauthors: G.G. Pavlov, A.V. Frolova, G.A. Smirnov).
13. Method of machining the rotational flanks of centrifugal pump impeller. Certificate of authorship N 192601 priority dated July 9, 1963 (coauthor: V.A. Ziskindovich).
14. Method of machining the worm gear with contact localization. Certificate of authorship N 263394 priority dated June 19, 1967 (coauthors: V.D. Britskiy, V.I. Vasylyev, V.F. Gruzdev, N.I. Vyushkin, V.I. Rybakov).
15. Method of cutting the conjugated screws. Certificate of authorship N 272007 priority dated December 19, 1967 (coauthor: A.M. Politavkin).
16. Spiroid involute gear. Certificate of authorship N364783 priority dated February 22, 1971 (coauthors: L.S. Borovich, V.A. Ganshin).
17. Device for controlling the kinematic error of gears. Certificate of authorship N435448 priority dated October 29, 1971 (coauthors: B.Ya. Karasik, R.S. Lipkovich, N.A. Anikeyeva, A.A. Zaostrovskiy). This invention was awarded by "USSR inventor" badge on November 29, 1978.
18. Device for demonstrating the reduction of friction losses in bearing supports. Certificate of authorship N 43638 priority dated March 21, 1972 (coauthors: V.I. Rybakov, V.A. Udalov).
19. Demonstrator prototype of wave gear. Certificate of authorship N477446 priority dated March 20, 1972 (coauthors: V.I. Rybakov, E.A. Bugrov).
20. Gear with crossed axes. Certificate of authorship N 530135 priority dated May 24, 1974 (coauthors: I.A. Bazavluk, L.S. Borovich, V.A. Ganshin, Y.I. Gutman).
21. Lever mechanism. Certificate of authorship N561040 priority dated April 5, 1976. (coauthors: T.S. Bil, Yu.Z. Tenenbaum).
22. Synchronous lock between SON-2 and PUAZO-3. 1943. Technical improvement was done during WWII.

References

1. Litvin FL (1960) Theory of gearing. M.: Physmatgiz, 444 pp (in Russian)
2. Litvin FL (1968) Theory of gearing, 2nd ed. M.: Nauka, 584 pp (in Russian)
3. Olivier T (1842) Theorie Geometrique des Engrenages. Bachelier, Paris
4. Gochman ChI (1886) Theory of gearing generalized and developed analytically, Odessa (in Russian)
5. Davydov YaS (1963) About one generalization of olivier method to generate conjugated surfaces in gears // «Theory of gears in machines». M.: Mashgis (in Russian)
6. Novikov ML (1958) Gears with new engagement. M.: Publ. VVIA n.a. Zhukovsky, 186 pp (in Russian)
7. Kolchin NI (1949) Analytical investigation of planar and spatial gearing. M.-L.: Mashgiz, 210 pp (in Russian)

8. Shishkov VA (1951) Forming the surfaces by cutting according to generating method. M.: Mashgiz, 150 pp (in Russian)
9. Davydov YaS (1950) Non-involute gearing. M.: Mashgiz. 189 pp (in Russian)
10. Litvin FL (1953) Toward the investigation of spatial gearing with linear contact of surfaces. // Proc. of Seminar on TMM, no. 49. USSR AS, pp 16–55 (in Russian)
11. Sheveleva GI (1969) Algorithm of numerical calculation of the machined surface. // Machine-tools and tooling, no. 8. pp 17–20 (in Russian)
12. Nesmelov IP, Goldfarb VI (1983) Non-differential approach to solving the enveloping task. // Mechanics of machines, N61. M.: Nauka, pp 3–10 (in Russian)
13. Litvin FL, Fuentes A (2004) Gear geometry and applied theory, 2nd edn. Cambridge University Press, 800 pp
14. Litvin FL (1956) Non-circular gears: design, theory of gearing and manufacture, 2nd ed. Mashgiz, 320 pp (in Russian)
15. Litvin FL, Britsky VD, Ganshin VA, Marinov KhI, Petrov-Minkov M, Rubtsov VN, Sorkin GI, Timofeev BP (1973) Selected issues of synthesis of spatial gearing. // Theory of gears in machines. M.: Nauka, pp 20–27 (in Russian)
16. Litvin FL, Rybakov VI (1974) Localization of bearing contact in cylindrical worm gears. // Izv. VUZov. Mashinostroenie, no. 8. pp 57–61 (in Russian)
17. Litvin FL (1970) Generalization of Euler-Savary formula and determining of dependencies between main curvatures of tooth flanks in spatial gearing. "Mashinovedenie", no. 6, pp 6–14 (in Russian)
18. Litvin FL, Erikhov ML (1970) Vector field of normals in ordinary nodes of contact of the enveloping surface // Theory of transmissions in machines. M.: Mashinostroenie, pp 27–38 (in Russian)
19. Litvin FL (1975) Determining the envelopes to contact lines on mutually enveloping surfaces. // Izv. VUZov. Mathematics 10(161):47–50 (in Russian)
20. Zalgaller VA, Litvin FL (1977) Sufficient feature of existence of the envelope characteristics and edge of regression on the enveloping surface. // Izv. VUZov. Mathematics 3(178):20–23 (in Russian)
21. Litvin F L (1969) Relation between curvatures of teeth surfaces in spatial gearing // Zeitschrift fur Angewandte Mathematik und Mechanik (in German), vol 49, pp 685–690
22. Litvin FL, Krylov NN, Erikhov ML (1975) Generation of tooth surfaces by two-parameter enveloping // Mech Mach Theor 10(5):365–373

Considerations on History of Mechanism and Machine Science with an IFToMM Role for Future Developments

M. Ceccarelli

Abstract Mechanism Science (MS) has been the core of mechanical engineering and industrial engineering since the beginning of engineering practice in modern times. A short survey is presented to outline the main characteristics of MS and its evolution, with the additional aim of identifying challenges and the role of MS in future developments of technology for the benefit of society. The community of IFToMM working in MS is presented, identifying various members and their respective roles. Modern systems with mechatronic features still need careful attention from a mechanism design viewpoint to properly achieve the goals of forwarding technological developments in helping or substituting for human operators in their activities.

Keywords Mechanism science · History of iftomm · Trends in MMS · Mechatronic systems

1 Introduction

New systems today are constantly expected to perform according to the latest updates, as well as accommodating multidisciplinary integration as mechatronic solutions more and more.

Two primary facts can be observed as supporting the claim that MMS (Mechanism and Machine Science), which is related to mechanical systems, is still a fundamental aspect of teaching, research, and practice, namely:

M. Ceccarelli (✉)
University of Cassino and South Latium, Cassino, Italy
e-mail: ceccarelli@unicas.it

V. Goldfarb and N. Barmina (eds.), *Theory and Practice of Gearing and Transmissions*, Mechanisms and Machine Science 34,
DOI 10.1007/978-3-319-19740-1_3

- Human beings operate and interact with their environments using many systems based in actions of a mechanical nature; therefore, mechanical systems will always be an essential part of systems that assist or substitute for human beings in their actions and other tasks.
- There is a continuing need to update problems and solutions in Technology since Society's needs are continually evolving; thus, mechanical systems are similarly expected to evolve so as to deal with new and/or updated problems, a process that must be based in ever-growing knowledge, means, and operation for the successful application of mechanical systems.

Historical insight can be useful both to outline past developments for understanding evolution towards the future and to indicate trends and open problems as determined by new/updated conditions and needs in Technology and Society. Historical backgrounds have been outlined from several technical viewpoints (also as a part of surveys of the History of Science) in several works on the historical evolution of Technology and Engineering in recognizing the original paternity of machine achievements, such as in [7–9, 17, 20, 22–28, 38–40], just to cite a few relevant sources in reasonably accessible literature within the context of mechanical engineering. A specific conference forum has been established within IFToMM (The International Federation for the Promotion of MMS) as the HMM (History of Machines and Mechanisms) Symposium during which various views and studies are discussed [5, 7–9, 29, 42].

In this paper, a survey is presented for describing the significant role of MMS in technological developments, along with recent IFToMM activity, as outlined in [13, 14, 16], by using historical outlines and general considerations based on the author's experience.

2 A Short History of MS

Over time, the needs of Society and Technology and the tasks we carry out to fulfill them have required continuous evolution of mechanical systems and their uses, with or without a rational technical awareness [2].

Mechanisms and machines have attracted attention since the beginning of Technology, and have thus been studied and designed with successful activity and specific results. But TMM (Theory of Machines and Mechanisms) only reached maturity as an independent discipline in the 19th century, [4]. Today, we refer to TMM as MMS (Mechanism and Machine Science), since a wider engineering area can be identified as being related to mechanical systems.

Historical developments of mechanisms and machines can be divided into periods as functions of specific technical events, according to the author's personal opinion, such as:

- Utensils in Prehistory
- Antiquity: 5th cent. BCE (Mechanos in Greek theatre plays)

- Middle Ages: 275 CE (sack of School of Alexandria with destruction of Library and Academy)
- Early design of machines: 1420 CE (the book Zibaldone with designs by Filippo Brunelleschi)
- Early discipline of mechanisms: 1577 CE (book Mechanicorum Liber by Guidobaldo Del Monte)
- Early Kinematics of mechanisms: 1706 CE (book Traitè des Roulettes by Philippe De La Hire)
- Beginning of TMM: 1794 CE (Foundation of Ecole Polytechnique)
- Golden Age of TMM: 1841 CE (the book Principles of Mechanism by Robert Willis)
- World War I Period: 1917 CE (the book Getriebelehre by Martin Grübler)
- Modern TMM: 1959 CE (journal paper Synthesis of Mechanisms by means of a Programmable Digital Computer by Ferdinand Freudenstein and Gabor N. Sandor)
- MMS Age: 2000 CE (re-denomination of TMM to MMS by IFToMM).

The historical evolution towards the current MMS can be briefly outlined by looking at developments that have occurred since the Renaissance period. Mechanisms and machines were used and designed as a means to achieve and improve solutions in various fields of human activity. Specific fields for mechanisms design grew in results and awareness, and their first personalities were recognized as brilliant experts, like Francesco Di Giorgio Martini and Leonardo Da Vinci, amongst many others, individuals who enjoyed considerable social reputations as well as professional [11]. At the end of the Renaissance period, Mechanics of Machinery also began to attract great attention in the Academic world, starting from the first classes given by Galileo Galilei in 1593–1598, [10]. In the 18th century, the designer figure evolved with a professional status and strong theoretical bases, finalizing a process that in the Renaissance saw the activity of small, closed communities of pupils/co-workers following 'mastros' and 'maestros', as stressed in [3]. Academic activity increased basic knowledge of the rational design and operation of mechanical systems. The first mathematizations were attempted and fundamentals on mechanism kinematics were proposed by the pioneering investigators, who were specifically dedicated to mechanism issues, for example, Philippe De la Hire, amongst many others. The successful practice of mechanisms was fundamental during the Industrial Revolution, when many practitioners and researchers implemented the evolving theoretical knowledge in practical applications and new powered machines. The 19th century can be considered the Golden Age of machines and its science TMM, since relevant novelties were proposed both in theoretical and practical fields. Mechanisms formed the core elements of any machinery and any technological advance at that time. A community of professionals was identified and a specific academic formation was established worldwide. TMM played an important role in the development of Technology and Society and several personalities expressed the fecundity of the field with their activity and results. The first half of the 20th century saw the prominence of TMM in mechanical (industrial) engineering, but also more and more of its integration into other technologies. A further evolution was experienced when the advent of Electronics made it possible to handle contemporaneously several motors in multi-d.o.f. applications of mechanical systems in 3D tasks. The increase of performance (not only in terms of speed and accuracy) required more sophisticated and accurate calculations that became possible with the advent of Informatics (involving both computers and programming).

Technically, MMS can be understood as an evolution of TMM with a broader content and vision of Science, including new disciplines. Historically, TMM has included as its main disciplines: History of TMM; Mechanism Analysis and Synthesis; Theoretical Kinematics; Mechanics of Rigid Bodies; Mechanics of Machinery; Machine Design; Gearing Systems; Experimental Mechanics; Teaching of TMM; Mechanical Systems for Automation; Transportation Machinery, Control and Regulation of Mechanical Systems; Rotordynamics; Human-Machine Interfaces; and Biomechanics. The modernity of MMS has augmented TMM with new vision and means but also with many new disciplines, the most significant of which can be recognized in Robotics; Mechatronics; Computational Kinematics; Computer Graphics; Computer Simulation; CAD/CAM for TMM; Tribology; Multibody Dynamics, Medical Devices, Micro and Nano Machines, Energy Sustainable Systems, and Service Systems.

In 2000, the evolution of the name from TMM to MMS included a change in the denomination of the IFToMM Federation from "IFToMM: the International Federation for TMM" to "IFToMM, the International Federation for the Promotion of MMS", [6]. This also corresponded with an increase in the engineering community worldwide.

Today, a modern machine is a combination of systems of different natures, and this integration has led to the modern concept of Mechatronics (Fig. 1). Thus, most of the recent advances in machinery are sometimes considered to be in fields other than MMS. But Mechanism Design can still be recognized as a fundamental aspect for developing successful systems that operate in the mechanical world of human beings. Tasks and systems for human beings must generally have a mechanical nature and a careful Mechanism Design is still fundamental for obtaining systems that assist or substitute for human beings in their operations. Most of those tasks are already performed with mechanical solutions that can be seen as traditionally successful but that nevertheless will benefit from further update or re-consideration because of new operational strategies and/or materials and components. Therefore, MMS can still be considered of current interest in research for developing mechanical systems of proper modern performance.

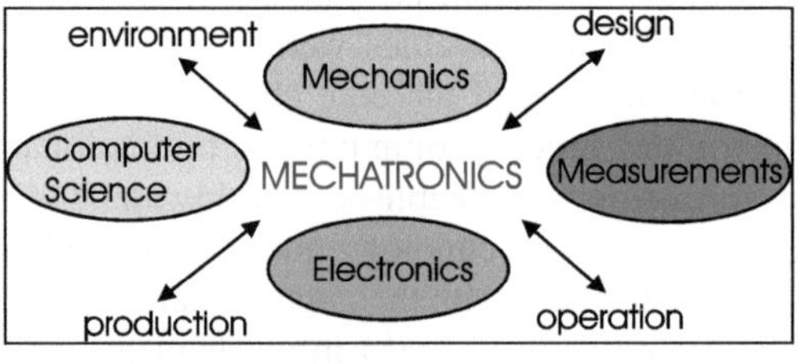

Fig. 1 A scheme for the concept of mechatronics

3 Challenges in MMS

The most popular trends and challenges in current MMS research lie mainly in:

- 3D Kinematics
- Modeling and mathematization for MS
- Multi-d.o.f. multibody systems
- Spatial mechanisms and manipulators
- Unconventional mechanisms (particularly compliant, under- or over-constrained types, among others)
- Scaled mechanisms
- Tribology issues
- Creative design
- Mechatronic designs
- Human-machine interactions for user-oriented systems
- Reconsideration and reformulation of theories and mechanism solutions.

These topics and many others in MMS are also motivated by needs of formation and the activity of professionals, who will be able to conceive and transmit innovation into both i production and service frames.

Teaching in MMS requires attention to modern methodologies that make efficient use of computers and software, which are still evolving rapidly. Thus, there is also a need to update teaching in order to make use of simulations and computer-oriented formulations. In addition, mechatronic layouts of modern mechanical systems suggest that mechanisms should be taught from the very beginning as being integrated with other components like actuators and sensors.

The mission of academia needs to be revitalized and better understood as a result of the high expertise of teachers that, with intense research activity and links to the professional and industrial world, is available to their students as well. This requires more attention and vision not only from academia but even more so from society as a whole, which, through government, should give more and more support to the educational system.

Professional activity requires novel applications and high performance machines since there is a continuous need to evolve/update systems and engineering tasks. In addition, there is a need to make new methodologies understandable to professionals for practical implementation both in their use and results. New solutions and innovations are continuously required not only for technical needs but also for the political/strategic goals of company success, including pure economic prominence.

In general, MMS activity will be directed towards further development by searching for:

- information and understanding of the functionality and impact of systems
- algorithms for design, operation, and evaluation of systems within user/task - oriented performance
- operation and application for full tasks, as constrained by environmental limits
- performance evaluation and economic merit of systems

- transfer of innovation
- human-machine interface and interaction.

Thus, the role of mechanical systems in Mechatronics can be understood according to such primary aspects as:

- Human-machine interaction and user/task-oriented performance
- Mechanical tasks in motion operations
- Structural design for sizing dimensions.

Therefore, the 'hot' topics for MMS for Mechatronics development would be:

- analysis and investigation of the operation and performance of mechatronic systems
- analysis and investigation of actions against the environment and within the mechatronic system
- focus on safety and security issues for both the system and its human operators
- consideration of the Mechanics of interactions
- evaluation of situations with mechanical contacts and force transmissions
- ranking of system actions according to task/user oriented requirements
- achievement of desired goals and proper working of the overall system
- consideration of complex motions, such as spatial movement at high acceleration
- investigation of integrated systems via suitable modeling of components other than those of a mechanical nature.

Trends in system composition can be summarized as in the examples in Table 1. In the future more than at present, mechanical components will be reduced percentage-wise but nevertheless will still be necessary and indeed fundamental for the use and operation of systems.

Figure 2 shows an example of a new machine design relating to fully autonomous intelligent tracks for agricultural purposes. This involves challenges for MMS since the primary machine task is still focused on the motion through:

- Path-planning
- Power transmission
- Terrain interaction.

Although most of the design improvements are in autonomous intelligent operation, the core of the machine operation still depends on:

Table 1 Examples of evolution of system composition

1960–2000	Mechanics (%)	Electronics/Informatics (%)
CARS	90–50	10–50
CALCULATORS	100–10	0–90
CAMERAS	100–30	0–70

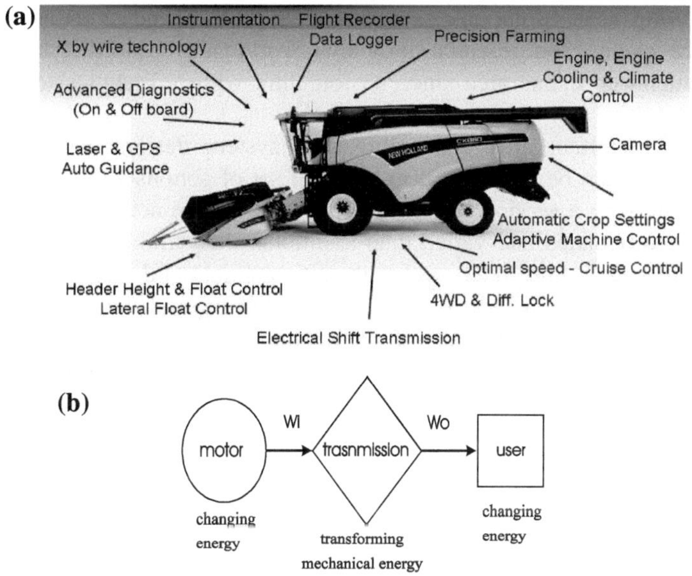

Fig. 2 An autonomous field track as: **a** example of intelligent mechatronic machine; **b** mechanics-dependency from mechanical transmissions

- the Gear box
- the Suspension mechanisms
- the Steering mechanisms.

But intelligent solutions require integration of these mechanical components into a well-balanced mechatronic design. The relevance of aspects of MMS in such modern mechatronic systems can be summarized as shown in the scheme in Fig. 2b, in which MMS features can be recognized mainly (but not only) in the transmission block for machine motion, along with strong relationships with other components of different natures and goals.

From Fig. 2b, emphasis can continue to be on gearing systems, even in high performance intelligent mechatronic systems, as fundamental components for the successful transmission and adaptation of mechanical energy during operation. In fact, Gear Technology in particular is one of the key aspects of MMS that, despite the high level of performance achieved, is still a subject of research for improvements and innovations. Gear Technology still attracts a tremendous amount of interest in the search for more efficient solutions, both in design procedures and design products with advanced operations, through the updating and innovating of theory, material, and manufacturing issues, even when integrated with other disciplines, such as Control Theory, Tribology, Energy Sustainable Systems.

The evolution of gear systems has produced considerable improvements over the last two centuries, as outlined in [32]. But challenges for relevant research activities still remain in regard to problems related to design algorithm precision manufacturing, scaled solutions, energy efficient profiles, and so on, as discussed in conference events and research programs worldwide. IFToMM has dedicated specific attention to this area since its founding with a Technical committee on gearing systems.

Another important area demanding new system designs would be service operations that can be understood in terms of set of actions and behaviors geared towards achieving a service task [13, 14]. Those service actions and tasks can be much more articulated and varied than traditional industrial applications.

A service task can be understood as the ultimate goal of the design and operation of a service machine, a term that can be applied beyond the notion of a robot. A service task may be identified with well-defined properties and by a large variety of situations. This is the main aspect that makes service systems (robots or machines) challenging in regard to design in practical applications where they need to be efficient and successful so as to provide a desired service at proper user capability.

The multi-disciplinarity for a service system is much wider than in any other field of engineering, since it includes technical aspects, human attitudes (of operators and/or users), human-machine interaction, and issues of environment. Indeed, in developing and operating service robots, other than technical expertise, it is increasingly necessary that competences from other fields of human life and environmental considerations be incorporated. For example, psychologists and biologists (and many others) are welcomed on the R&D teams for designing service robots.

Referring to technical aspects, Fig. 3 summarizes the mechatronic character of a service robot, as a traditional robot, but with specific emphasis on those above-mentioned peculiarities in terms of interaction with the environment and human beings, and in terms of a careful consideration of the environment. Those interactions should be understood not only in terms of engineering issues (mainly mechanical ones) but by looking at more general aspects, for example, psychological attitudes and social impacts. A consideration of the environment should also

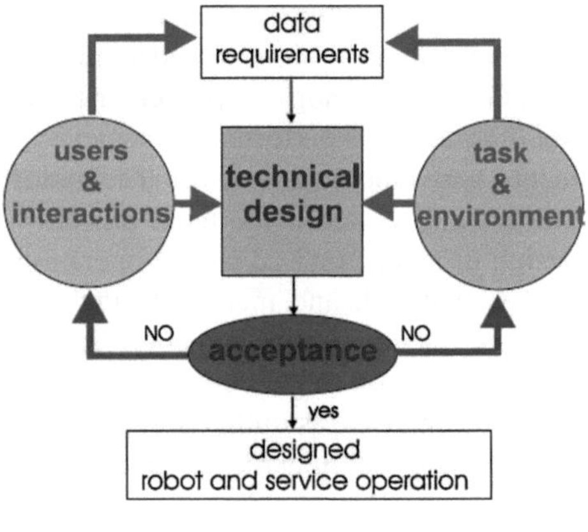

Fig. 3 A general flowchart with new peculiarities for designing and operating service systems

address the problem of how a service robot affects or is affected by that environment, by analyzing and designing for the variety of conditions and situations, including contamination and pollution.

In particular, the main flow of technical design activity is indicated in the central streamline as referring to data identification in both technical and non-technical aspects, considerations of technical constraints/issues, analysis of service operation and goal, and design activity and system programming, with final checks by operators and users. The care in technical design activity is indicated in the blocks for system design and operation planning, since they are strongly influenced by aspects and activities grouped in the two lateral blocks concerning interactions with human beings and the environment, respectively. Each block will refer to aspects that, even with non-technical concerns, must be included in the machine development with proper models and problem formulation as synthetically indicated in the lateral boxes for features and constraints on task and environment and users and interactions.

In addition, service systems can be considered efficient and successful when the ultimate cost, both in design and operation, can be properly sized as a function of the service task and mainly as a function of the affordable budget of users and operators. Thus, indeed, economic evaluation and management will be included both in the R&D and in the design of service robots, even from an outset with a strong influence on technical issues.

All the above considerations can be considered as also pertaining to the process of building and transferring innovation, which will be understood not only as a technical advance but more widely as an enhancement of the quality of life in all its aspects with the help and support of technical means.

4 IFToMM and Its Role in MMS

The terms IFToMM, TMM, and MMS are related to the fields of Mechanical Engineering concerned with Mechanisms in a broad sense. TMM is often misunderstood, even within the IFToMM Community, although it is recognized as the specific discipline of Mechanical Engineering related to mechanisms and machines, as commented upon in [19] in the announcement of the birth of IFToMM. The meaning of TMM, now MMS, can be clarified by looking at IFToMM terminology [30]:

- Machine: a mechanical system that performs a specific task, such as the forming of material, and the transference and transformation of motion and force.
- Mechanism: a system of bodies designed to convert motions of, and forces on, one or several bodies into constrained motions of, and forces on, other bodies.

The meaning of the word 'theory' needs further explanation. The Greek word for 'theory' (θεωρία) comes from the corresponding verb, whose main semantic meaning is related both to examination and observation of existing phenomena. But even in the classic Greek language, the word includes such practical aspects of observation as experiencing the reality of phenomena, so that 'theory' also means practice with analytical results. In fact, this latter aspect is what was included in the discipline of modern TMM when Gaspard Monge (1746–1818) established it at the Ecole Polytechnique at the beginning of 19th century, [18], (see, for example, the book by Lanz and Betancourt [31], whose text includes early synthesis procedures and hints for practical applications). Later (see, for example [33]) and even today (see, for example [41]), many textbooks have been entitled "Theory of Mechanisms", since they describe both the fundamentals and the applications of mechanisms in machinery.

The term MMS was adopted within the IFToMM Community in the year 2000, after a long discussion (see Ref. [4] in the IFToMM Newsletter), with the aim of giving a better identification of the modern enlarged technical content and broader view of knowledge and practice in regard to mechanisms. Indeed, the use of the term MMS also stimulated an in-depth revision in the IFToMM terminology, since the definition of MMS has been given as [30]:

- Mechanism and Machine Science: The branch of science which deals with the theory and practice of geometry, motion, dynamics, and control of machines, mechanisms, and elements and systems thereof, together with their application in industry and other contexts, e.g., in Biomechanics and the environment. Related processes, such as the conversion and transfer of energy and information, also pertain to this field.

The developments in TMM have stimulated cooperation around the world at various levels. One of the most relevant results was the foundation of IFToMM in 1969 (Fig. 4). IFToMM was founded as a Federation of territorial organizations, but was based on the activity of individuals within a family frame with the aim of facilitating co-operation and exchange of opinions and research results in all the fields of TMM. As stressed in [19]. Many individuals have contributed and continue to contribute to the success of IFToMM and related activity (see IFToMM webpage: www.iftomm.org) under the coordination of various IFToMM Presidents over time.

IFToMM was founded as the International Federation for the Theory of Mechanisms and Machines in Zakopane, Poland on September 27, 1969, during the Second World Congress on TMM (Theory of Mechanisms and Machines). The main promoters of the IFToMM World Federation were Academician Ivan I. Artobolevski (USSR) and Prof. Erskine F.R. Crossley (USA), whose principal aim was to bypass the obstacles at the time of the Cold War in developing international collaboration in TMM science for the benefit of the world society. IFToMM started as a family of TMM scientists, among whom we may identify the IFToMM founding fathers who signed or contributed to the foundation act with the initial 13 Member Organizations (Fig. 5): Academician Ivan I. Artobolevski (USSR), Prof. Erskine F.R. Crossley (USA), Prof. Michael S. Konstantinov (Bulgaria), Dr. Werner Thomas (GFR), Prof. B.M. Belgaumkar (India), Prof. Kenneth H. Hunt (Australia), Prof. Jan Oderfeld (Poland), Prof. Jack Phillips (Australia), Prof.

We, the undersigned chief delegates at the Inaugural Assembly of the International Federation for the Theory of Machines and Mechanisms (IFTOMM) here at Zakopane Poland on 27th September 1969, declare that we have founded the above-mentioned Federation and that we have adopted its Constitution which is attached hereto and decided to the following categories (see Article 8.4 of the Constitution).

Territory	Chief delegate	Proposed Category	Signature
Australia	JACK PHILLIPS	IV *	
Bulgaria	Georgi Rusanov	IV	
German Democratic Republic *	Wolfgang Rössner	III *	
German Federal * Republic	Werner Thomas	III *	
Hungary *	Zénó TERPLÁN	IV *	
India *	J. S. RAO	V *	
Italy *	GIOVANNI BIANCHI	IV *	
Poland	Adam Morecki	IV	
Rumania	Nicolae I. Manolescu	IV (6)	
United Kingdom *	L. Maunder	III *	
U.S.A.	DOUGLAS MUSTER	I	
U.S.S.R.	[illegible]	I	
Yougoslavia	Ilić Branisky	IV (x)	

Fig. 4 The foundation act of IFToMM, the International Federation for the Theory of Machines and Mechanisms, in Zakopane (Poland) on 27 September 1969, (Courtesy of IFToMM Archive)

George Rusanov (Bulgaria), Prof. Wolfgang Rössner (GDR), Prof. Zènò Terplàn (Hungary), Prof. Jammi S. Rao (India), Prof. Giovanni Bianchi (Italy), Prof. Adam Morecki (Poland), Prof. Nicolae I. Manolescu (Romania), Prof. Leonard Maunder (UK), Prof. Douglas Muster (USA), and Prof. Ilic Branisky (Yugoslavia).

The foundation of IFToMM was the result of intense activity in the interest of stimulating and promoting international collaboration, more than had been done previously, a process begun in the late 1950s, as documented by several letters that are stored in the IFToMM Archive at CISM in Udine, Italy. A first World Congress on TMM (Theory of Mechanisms and Machines) was held in 1965 in Varna, Bulgaria, during which the foundation of IFToMM was planned as it would later be agreed to during the Second World Congress on TMM in Zakopane, Poland. This series of Congresses was immediately recognized as the IFToMM World Congresses, and in 2011, we celebrated the 13th event with the participation of delegates from 46 Member Organizations representing more than 50 countries.

IFToMM activity has grown in many respects, as, for example, concerning the number of member organizations (growing from the 13 founding members to the current 46 members), the size and scale of conference events (with many other

Fig. 5 First IFToMM President Artobolevskii speaking at the foundation of IFToMM, Fig. 4, (Courtesy of IFToMM Archive): *1* Prof. Ivan Ivanovic Artobolevskii (USSR); *2* Prof. Adam Morecki (Poland); *3* Prof. Kurt Luck (Germany); *4* Prof. Michale S. Konstantinov (Bulgaria); *5* Prof. Nicolae I. Manolescu (Romania); *6* Prof. Erskine F. Crossley (USA); *7* Prof. Giovanni Bianchi (Italy); *8* Prof. Aron E. Kobrinskii (USSR); *9* Prof. Werner Thomas (Germany); *10* Prof. Jan Oderfeld (Poland)

conferences, even on specific topics, at national and international levels, in addition to the MMS World Congress), and the number and focus of technical committees working on specific discipline areas of MMS. IFToMM was founded in 1969, and today, a fourth generation of IFToMMists is becoming active, as they distinguish themselves among those working within the IFToMM community. Knowing the History of IFToMM and how we arrived at today's modus operandi gives a greater awareness of the community's identity and significance [13, 14, 16].

The IFToMM community evolved in character from that of a family of a few enthusiastic pioneers/visionaries and founders into a worldwide scientific community across the following generations:

- 1950s–1979 First Generation: founding fathers and their friends/colleagues; lasts through the 4th IFToMM World Congress in Newcastle-upon-Tyne in 1975, with Prof. Leonard Maunder as Congress Chair
- 1980–1995 Second Generation: students and others educated by the founding fathers and their friends/colleagues; lasts through the 9th World Congress in Milan in 1995, with Prof. Alberto Rovetta as Congress Chair
- 1996–2011 Third Generation: educated people within the framework of IFToMM and IFToMM activity, with 48 national organizations as IFToMM members, with Prof. Carlos Lopez-Cajùn as General Chair for the 2011 Congress.
- Today- Fourth Generation: educated people in local organizations linked to IFToMM and within IFToMM activity, with 46 organizations as IFToMM members.

IFToMM officers (who are the Chairs of IFToMM Member Organizations, the Chairs of TCs and PCs, and the members of the Executive Council) have contributed and continue to lead in contributing to the mission of IFToMM, stated in the 1st article of the Constitution as: '…the promotion of Mechanism and Machine Science'. A complete list of IFToMM officers over time is available in the Proceedings of the second International Symposium on the History of Machines and Mechanisms HMM2004, published in 2004 by Kluwer/Springer and now available at the IFToMM webpage.

In particular, Presidents and Secretaries General have had significant roles in guiding the growth and success of IFToMM. Their personalities are also representative of the IFToMM community in terms of reputation and visibility worldwide. The list of Presidents is as follows: Ivan I. Artobolevsky (1969–1971 and 1972–1975) (USSR), Leonard Maunder (1976–1979) (UK), Bernard Roth (1980–1983) (USA), Giovanni Bianchi (1984–1987 and 1988–1991) (Italy), Adam Morecki (1992–1995) (Poland), Jorge Angeles (1996–1999) (Canada), Kenneth J. Waldron (2000–2003 and 2004–2007) (USA), Marco Ceccarelli (2008–2011) (Italy), and Yoshihiko Nakamura (2012–2015) (Japan), Fig. 6. The Secretaries General were Michael S. Konstantinov (Bulgaria), Emil Stanchev (Bulgaria), Adam Morecki (Poland), Elizabeth Filemon (Hungary), L. Pust (CSSR), Tatu Leinonen (Finland), Marco Ceccarelli (Italy), and Teresa Zielinska (Poland).

Details of the History of IFToMM can be found in the first Chapter of the Proceedings of the first International Symposium on the History of Machines and Mechanisms HMM2000 (published by Kluwer) in which all the past IFToMM Presidents have outlined their historical perspectives on IFToMM in contributed papers with references [1]. Additional references can be found in [9, 12–15, 20, 21, 34–37]. More information on IFToMM and its activity can be found at the website: http://www.iftomm.org.

The structure of the IFToMM is summarized in Fig. 7 in which the actions and activities of the IFToMM Bodies are indicated as flowing from the IFToMM constitution. According to the IFToMM mission as stated in the constitution, IFToMM activity is specifically designed to provide leadership for cooperation and

(a) (b)

Fig. 6 **a** IFToMM Presidents at HMM 2000 in Cassino (Italy); **b** the current President

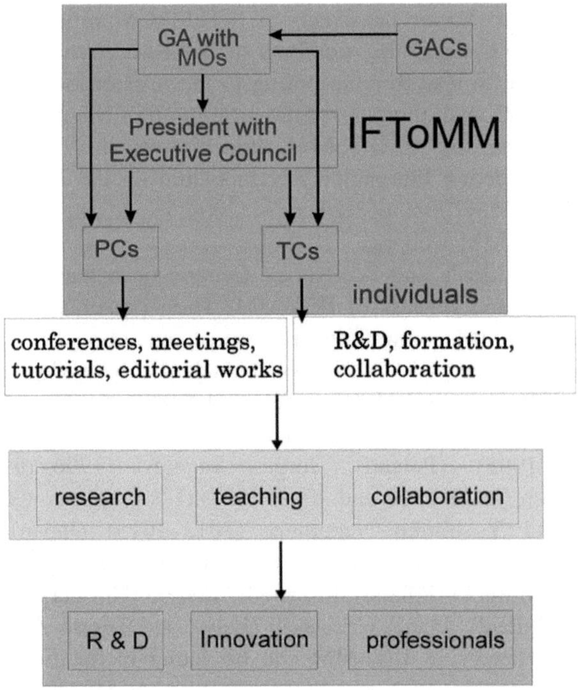

Fig. 7 A scheme for structure and activity of IFToMM

development of modern results in the Mechanism and Machine Sciences by assisting and enhancing international collaboration.

The bodies of IFToMM can be described synthetically as:

- General Assembly: the supreme body of the Federation that determines its policy. It is composed of the Chief Delegates of IFToMM Organization members (46 as of 2015) and members of the Executive Council.
- Executive Council: group that manages the affairs of the Federation between the sessions of the General Assembly. It is elected every 4 years, meets annually, and is composed of the President, Vice-President, Secretary-General, Treasurer, and six ordinary members.
- Permanent Commissions (PCs) and Technical Committees (TCs): Each PC and TC is composed of a Chairperson, appointed by the Executive Council, a Secretary, and members nominated by the Chairperson and appointed by the Executive Council. A Chairperson shall not serve for more than two terms consecutively. The general goals for the PCs and TCs are aimed at promoting their fields of interest by attracting researchers and practitioners, including young individuals, in order to:
- define new directions in research and development within their technical areas;
- establish contacts between researchers and engineers;

- initiate and develop bases and procedures for modern problems;
- promote the exchange of information;
- organize national and international symposia, conferences, summer schools, and meetings.

As of 2015, 13 TCs are active in the fields of: Biomechanical Engineering, Computational Kinematics, Gearing and Transmissions, Linkages and Mechanical Controls, Micromachines, Multibody Dynamics, Reliability, Robotics and Mechatronics, Rotordynamics, Sustainable Energy Systems, Transportation Machinery, Tribology, and Vibrations. Additional TCs are under consideration for hot topics within the greater IFToMM community. There are currently active PCs on the subjects of: Communications, Publications and Archiving, Education, History of MMS, and Standardization of Terminology.

The main aspects of IFToMM activity are emphasized in Fig. 7 as related to Research, Teaching, and Collaboration for final goals in R&D frames, Innovation, and Professional Formation. Indeed, they are closely interrelated since the education of our modern professional communities must be based on the sort of successful research activity that can only be achieved today through the collaboration of teams from all parts of the globe sharing approaches and results.

IFToMM activity can be summarized mainly as:

- conference events, on both wide-form and specific subjects not only linked to TCs, at local and international levels, some in series and some in one-shot events. A World Congress is organized every 4 years, with the 14th IFToMM World Congress scheduled to be held on 25–30 October, 2015, in Taipei (http://iftomm2015.yohoz.com)
- meetings and visit exchanges, including joint teaching
- collaborative projects in research programs and teaching plans
- organization of fields of interests for aggregation of an international community and planning actions as those mentioned above
- publications for dissemination of research results and knowledge transfer. This is achieved at the moment through 6 affiliated journals and two book series specifically dedicated to MMS and the IFToMM community. The journals are: Mechanism and Machine Theory (http://www.elsevier.com), Problems of Mechanics (http://pam.edu.ge), Open-access Mechanical Sciences (http://www.mech-sci.net), Chinese Journal of Mechanical Engineering (http://www.cjmenet.com), Advances in Vibration Engineering (http://www.tvi-in.com/index.asp), Mechanics Based Design of Structures and Machines (http://www.tandf.co.uk/journals/titles/15397734.asp). Springer has published book series on MMS (http://www.springer.com/series/8779) and on the History of MMS (http://www.springer.com/series/7481).

The significance of the IFToMM can be summarized as residing within its status as the world's unique federation in MMS spurred by the following motivations:

- Yesterday: To initiate and facilitate international collaboration between Eastern and Western countries

- Today: To help and enhance international collaboration and modern results on MMS
- Tomorrow: To foster cooperation and development in MMS Mechanism and Machine Science.

5 Conclusions

Not everything is new or recently developed in MMS, although innovation seems to be a priority today. But this does not mean that there is no interest in MMS, nor that there is no need to work on developing and enhancing knowledge and application of MMS. New challenges have and will be determined for MMS through the new needs of Technology and Society, both in terms of developing new solutions and updating those from the past. An awareness of the historical background can give not only a conscious understanding of past efforts and solutions, including their paternity, but even more importantly, it can help to find/develop ideas for new and updated problems to be solved. But the rapidly evolving needs of Technology and Society will require a continuous re-thinking and re-conceiving of methodologies and solutions in suitably updated applications. Thus, the main challenges for future success in MMS may be recognized in the community's capability to continue updating the field and, therefore, in being ready to solve new and updated problems with new ideas or by refreshing past solutions, as has been done successfully in the past.

References

1. Angeles J, Bianchi G, Bessonov AP, Maunder L, Morecki A, Roth B (2004) A history of IFToMM, Chapter 2. In: Proceedings of HMM2004—the second IFToMM international symposium on history of machines and mechanisms, Springer, Dordrecht, 2004, pp 25–125
2. Bautista Paz E, Ceccarelli M, Echavarri Otero J, Munoz Sanz JJ (2010) A brief illustrated history of machines and mechanisms, Springer, Dordrecht
3. Ceccarelli M (1998) Mechanism schemes in teaching: a historical overview. ASME J Mech Design 120:533–541
4. Ceccarelli M (1999) O the meaning of TMM over time, Bulletin IFToMM Newsletter, vol 8, no 1. (http://www.iftomm.org)
5. Ceccarelli M (ed) (2000) International symposium on history of machines and mechanisms—proceedings of HMM2000, Kluwer, Dordrecht
6. Ceccarelli M (2001) From TMM to MMS: a vision of IFToMM, Bulletin IFToMM Newsletter, vol 10, nr 1. (http://www.iftomm.org)
7. Ceccarelli M (2004) Evolution of TMM to MMS: an illustration survey, keynote lecture. In: 11th IFToMM world congress in mechanism and machine science, vol 1. Tianjin, pp 13–24
8. Ceccarelli M (2004) IFToMM activity and its visibility, Bulletin IFToMM Newsletter, vol 13, nr 1. (http://www.iftomm.org)

9. Ceccarelli M (ed) (2004) International symposium on history of machines and mechanisms—proceedings of HMM2004, Kluwer, Dordrecht
10. Ceccarelli M (2006) 2006, Early TMM in Le Mecaniche by Galileo Galilei in 1593. Mech Mach Theor 41(12):1401–1406
11. Ceccarelli M (2008) Renaissance of machines in Italy: from Brunelleschi to Galilei through Francesco di Giorgio and Leonardo. Mech Mach Theor 43:1530–1452
12. Ceccarelli M (2010) IFToMM celebration for 40-th year celebration. Mech Mach Theor 45 (2010):119–127
13. Ceccarelli M (2011) President report for IFToMM general assembly. IFToMM Archive, Udine
14. Ceccarelli M (2011) Activity and trends in MMS from IFToMM community. In: Role of MMS and IFToMM in technology development, Book series on machines and machine science, vol 1, Springer, Dordrecht, pp 3–24
15. Ceccarelli M (2013) Twenty-five years of activity in IFToMM. J Theor Mech Mach 11(2): 3–14. http://tmm.spbstu.ru. (St Petersburg State University)
16. Ceccarelli M (2015) A short account of history of IFToMM and its role in MMS. Mech Mach Theor (in print). 10.1016/j.mechmachtheory.2014.09.007
17. Chasles M (1837) Apercu historique sur l'origin et le développement des méthodes en géométrie …, Mémoires couronnés par l'Académie de Bruxelles, Vol.11. (2nd ed, Paris, 1875)
18. Chasles M (1886) Exposé historique concernant le cours de machines dans l'enseignement de l'Ecole Polytechinique. Gauthier-Villars, Paris
19. Crossley FRE (1970) The international federation for the theory of machines and mechanisms. J Mech 5:133–145
20. Crossley FRE (1988) Recollections from forty years of teaching mechanisms. ASME J Mech, Trans Autom Des 110:232–242
21. Crossley FRE (1991) The early days of IFToMM. In: Proceedings of 8-the IFToMM world congress, vol 1. Prague, pp 4–9
22. De Groot J (1970) Bibliography on kinematics. Eindhoven University, Eindhoven
23. De Jonge AER (1943) A brief account of modern kinematics, Transactions of the ASME, pp 663–683
24. Dimarogonas AD (1993) The origins of the theory of machines and mechanisms. Modern kinematics—developments in the last forty years. Wiley, New York, pp 3–18
25. Ferguson ES (1962) Kinematics of mechanisms from the time of watt. Contri Mus His Technol Wash 27:186–230
26. Hain K (1967) Applied kinematics. McGraw-Hill, New York
27. Hartenberg RS, Denavit J (1956) Men and machines… an informal history. Mach Des 3 May 1956, pp 75–82; 14 June 1956, pp 101–109; 12 July 1956, pp 84–93
28. Koetsier T (2000) Mechanism and machine science: its history and its identity. In: Proceedings of HMM 2000—the first IFToMM international symposium on history of machines and mechanisms, Springer, Dordrecht, pp 5–24
29. Koetsier T, Ceccarelli M (eds) (2012) Explorations in the history of machines and mechanisms—proceedings of HMM2012, Springer, Dordrecht
30. IFToMM (2003) Special issue 'standardization and terminology'. Mech Mach Theor 38:7–10
31. Lanz JM, Betancourt A (1808) "Essai sur la composition des machines", Paris
32. Litvin FL (1998) Development of gear technology and theory of gearing, NASA Reference Publication 1406, ARL-TR-1500
33. Masi F (1897) Teoria dei meccanismi, Bologna
34. Maunder L (1980) The progress of IFToMM. Mech Mach Theory 15:415–417
35. Maunder L (1988) Report: the scientific activity of IFToMM. Mech Mach Theory 23:329–332
36. Morecki A (1995) Past present and future of IFToMM. Mech Mach Theory 30:1–9
37. Morecki A (1999) International friendly thinkers organization (who likes) machines and mechanisms (IFToMM)—where are we going ?. In: Proceedings of 10-the IFToMM world congress, Oulu
38. Nolle H (1974) Linkage coupler curve synthesis: a historical review—I and II. Mech Mach Theory 9(2):147–168, 325–348

39. Reuleaux F (1875) Theoretische kinematic, Chapter 1, Braunschweig
40. Roth B (2000) The search for the fundamental principles of mechanism design. Proceedings of HMM2000—the first IFToMM International symposium on history of machines and mechanisms, Springer, Dordrecht pp 187–195
41. Uicker JJ, Pennock GR, Shigley JE (2003) Theory of machines and mechanisms. Oxford University Press, New York
42. Yan HS, Ceccarelli M (eds) (2008) International symposium on history of machines and mechanism—proceedings of HMM2008, Springer, Dordrecht

Development of the Theory and Practice of Spiroid Gears

V. Goldfarb

Abstract The paper presents a short history of the development of the theory of spiroid gears and gearboxes design in the Institute of Mechanics of Kalashnikov ISTU, and aspects of implementation and development of their production in the "Mechanic" Ltd.

Keywords Spiroid gears and gearboxes · Theory of design · Implementation of production

1 Introduction

The history of the existence of gears spans thousands of years and includes a great number of wonderful events and outstanding engineers and scientists, thanks to whom this field of engineering became essential. Over the years, hundreds of gear types and varieties have appeared, a well-composed theory of gearing has been created, one of the outstanding authors of which is Prof. F.L. Litvin [1, 2, 3 and other], and effective techniques of gear manufacturing, equipment and tools for their implementation have been developed. This process was actively developed in the 19th and 20th centuries and continues nowadays with greater intensity, finding new forms and development trends.

One such event that attracted the attention of scientists was the invention in 1954 of a spiroid gear by a talented engineer named Saari [4], a gear which took a respectable position among other gears with skew axes thanks to peculiarities of meshing geometry and kinematics and, as a result, high loading capacity, wear resistance, resistance to impact and vibration loads, technological effectiveness of production and mounting and other advantages [5]. The first and rather detailed description of the gear and its design procedure was given in works by Nelson [6],

V. Goldfarb (✉)
Institute of Mechanics, Kalashnikov Izhevsk State Technical University, Izhevsk, Russia
e-mail: veniamingoldfarb@yahoo.com

V. Goldfarb and N. Barmina (eds.), *Theory and Practice of Gearing and Transmissions*, Mechanisms and Machine Science 34,
DOI 10.1007/978-3-319-19740-1_4

and the path-breaker in organizing the production of spiroid gears and gearboxes and their wide implementation into different branches of engineering was the company Illinois Tool Works (USA), which patented the terms SPIROID and HELICON (a variety of spiroid gear with a cylindrical worm, also invented by Saari [7]) (Fig. 1) and published thorough catalogues advertising successful applications of the gears in the machine-tool industry, weight-handling and military equipment, precise devices and other fields, and demonstrating their great potential, including the possibility of obtaining the gear ratios from 8 to 360 in one gear pair.

Young scientists B.D. Zotov and N.S. Golubkov were the first researchers in Russia (USSR at that time) to begin studying spiroid gears at the Izhevsk Mechanical Institute (since 1993—Izhevsk State Technical University). The first samples of spiroid gearboxes were produced and tested (one of them had already been operating at the Izhevsk Heat Electric Generation Plant for a long time) and an attempt was made to develop their analytical procedure. Further development of these works was proceeding under the aegis of of Georgiev [8, 9], who organized the Laboratory of Spiroid Gears in 1965, from which a whole generation of engineers and researchers would appear: V.I. Goldfarb, V.A. Shubin, V.A. Long, S.V. Ezerskaya, A.S. Kuniver, S.D. Man'shin, V.A. Ivaykin, V.A. Modzelevsky, L.N. Saburov, E.K. Shibanov, V.I. Matveyev and others. Research and development works went off in various directions: the creation of the geometrical theory of spiroid gears and the engineering method of their analysis, the development of new varieties, two of which were patented in a number of countries [10, 11] (Fig. 2), the development of gearbox designs for general and specific engineering, the carrying out of testing, and the development of terminology [12] and a number of industry standards.

In the 1980s, the center of influence of research and development activity of spiroid gears was displaced to a Special Design Department of Gears organized by V.I. Goldfarb and rearranged in 1993 into the Institute of Mechanics (IM) of ISTU, which became the main scientific center in this area. This paper will describe the activity in the development of the theory of spiroid gears and gearbox design, the carrying out of the analytical, numerical and experimental research works, the development of production technology and the mastering of their manufacture realized at the IM and its subsequent basis for the production enterprise "Mechanic Ltd".

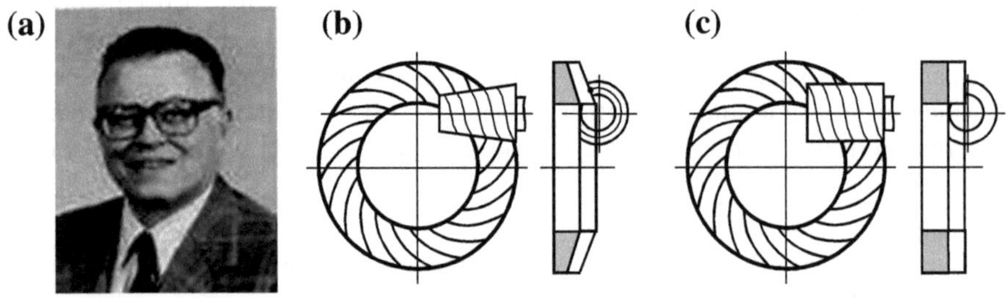

Fig. 1 Oliver Saari (1924–2003) [4] **a** and Spiroid **b** and Helicon **c** gears invented by him

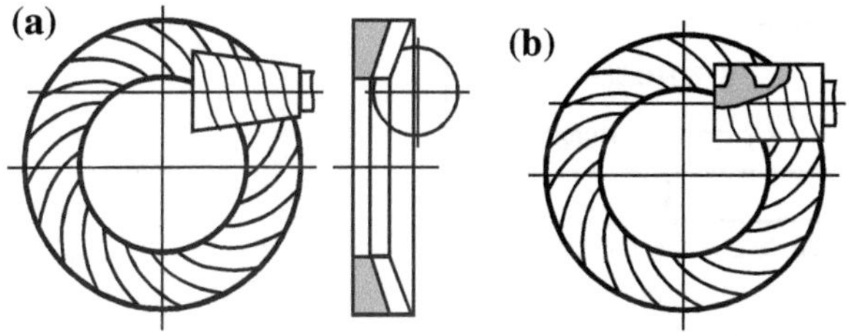

Fig. 2 Types of spiroid gears invented in Russia by A.K. Georgiev **a** and patented in USSR, USA, and Germany, and by A.K. Georgiev with V.I. Goldfarb **b** and patented in USSR, USA, Germany, UK, France, Japan, Italy, Sweden, and Canada

Note. The number of works published in different countries in journals and proceedings of many international conferences devoted to spiroid gears exceeds 400, more than half of which were written by IM staff. Here and hereafter, only a small number of them are listed.

2 Development of the Theory and Practice of Spiroid Gearbox Design

Solving this problem is the main scientific focus of the Institute of Mechanics. The activity surrounding the search for this solution has been and is carried out through the following activities.

1. Development of the theory of conjugate gearing of a spiroid gear for a general (non-orthogonal) case of axes location, development of principles, structure (Fig. 3), mathematical support and tools of computer-aided design and research of the gear. Results of this activity provided the basis for a thesis for the Doctoral degree in engineering of Goldfarb [13] (the first of four theses for the Doctoral degree dealing with spiroid gears; the remaining three [14–16] were also carried out at the IM) and CAD system «SPDIAL» [17], which was the basic tool for gear design, research of gearing geometry and kinematics and calculation of their force characteristics.

This activity included the following research works:

- analysis and synthesis of gear schemes [18, 19] that resulted in the patenting of the new varieties [20–22 and other] (Fig. 4);
- non-orthogonal spiroid gears [23];
- spiroid gears with worms, having variable thread pitch [24];
- combined double-stage gears and gearboxes [25];
- accuracy of spiroid gears [26, 27].

On the basis of the approach created, general engineering gearboxes and gear-motors, gears of machine-tools, automatic lines and a number of precise devices were developed and implemented in serial production (Fig. 5).

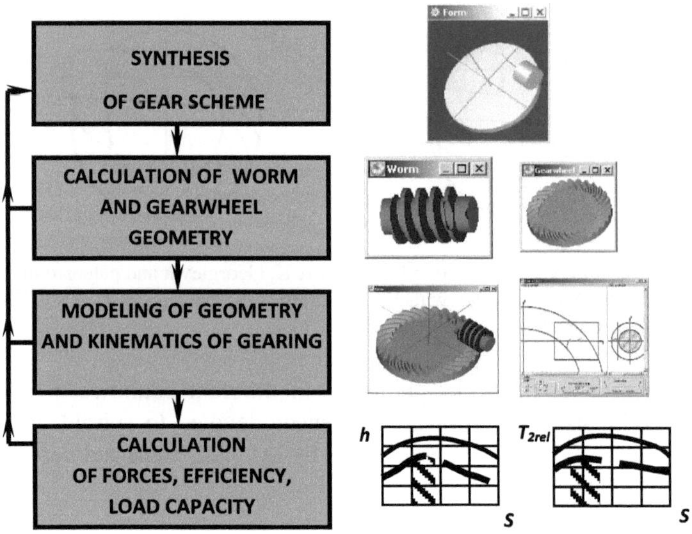

Fig. 3 Structure of ideal spiroid gearing design

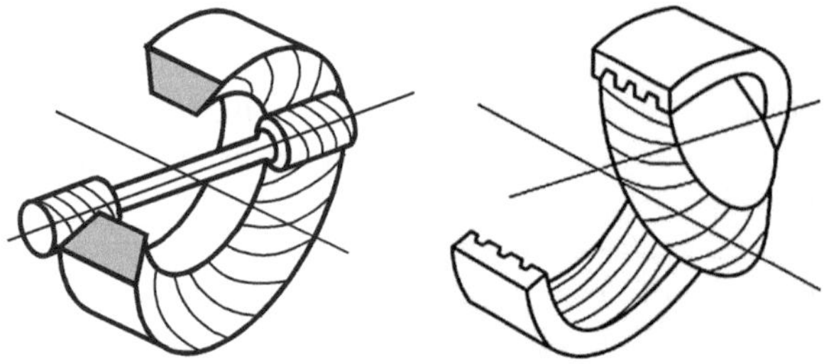

Fig. 4 Schemes of some new non-standard type of skew-axis gears synthesized and patented by V.I. Goldfarb with coauthors

Fig. 5 Spiroid gearboxes for general-purpose engineering

2. Development of the theory of real spiroid gearing, taking into account possible errors in manufacturing, mounting and deformation of both gear elements and gearbox units as a whole. This activity began with a Thesis for the DSc in the

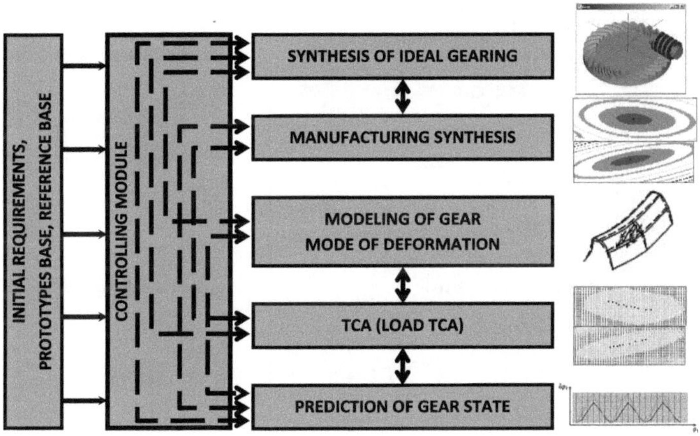

Fig. 6 Structure of real spiroid gearing design

Engineering degree of Trubachev [16]. Figure 6 shows the structure of a real spiroid gearing design.

At present time, the works in this area are the main scientific toolkit allowing the IM to solve (by means of reconstruction of the base of the concepts developed around the CAD system "SPDIAL+" [28]), in particular, the following fundamental problems of great practical importance:

- the development of principles, structure and design methods of real spiroid gears and gearboxes;
- determination of the point of tangency of the meshing surfaces when there are factors breaking conjugate (ideal) contact (solution of the inverse problem of the theory of gearing), that, in turn, gives considerable possibility to the solution of the whole class of problems connected with influence research of practically any error, including those artificially created during the tooth modification to achieve the desired contact localization, on the meshing quality and gear accuracy such as development of the accuracy regulation system;
- the search for position, dimensions, shape of instant contact areas and summary contact pattern in real meshing;
- control of the position of the contact pattern through changing the parameters of the set-up and the geometry of the cutting tool for cutting the gearwheel and the worm; this problem is of great practical importance for decreasing the gear sensitivity to different errors.

It should be noted that the developed concepts, mathematical support, and system "SPDIAL+" are: (a) the basic tools for solving the design problem, as well as investigating the majority of the works at the IM; (b) universal tools for the research and design not only of spiroid, but also worm gear drives with cylindrical worms as verified by numerous practical calculations (Fig. 7).

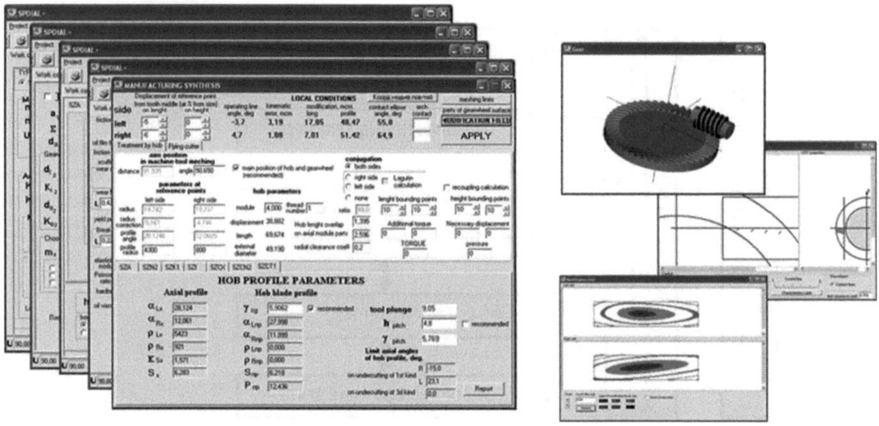

Fig. 7 Fragment of CAD "SPDIAL+" functioning

3. The previous works gave active development to a trend connected with the evaluation of loaded state, deformability, strength of gears on the basis of the proposed models of loaded state [29], force distribution in real gearing with application of FEM and the results of numerous experimental investigations (Fig. 8).

One important scientific and practical aim of this trend is an adequate evaluation of spiroid gear strength at different modes and conditions of loading. This trend is closely connected with the works of gear wear resistance evaluation [15] initiated by V.N. Anferov at the Sverdlovsk Scientific Research Institute of Chemical Engineering [30] and which have been actively developed at the Siberian Transport University (Novosibirsk) in cooperation with the IM.

4. Manufacturing design is also an actively developing scientific trend carried out at the IM and without which it would be impossible to create effective production.

In addition to solving the typical problems of manufacturing design, a new principal approach was developed according to which gear design is subjected to two conditions:

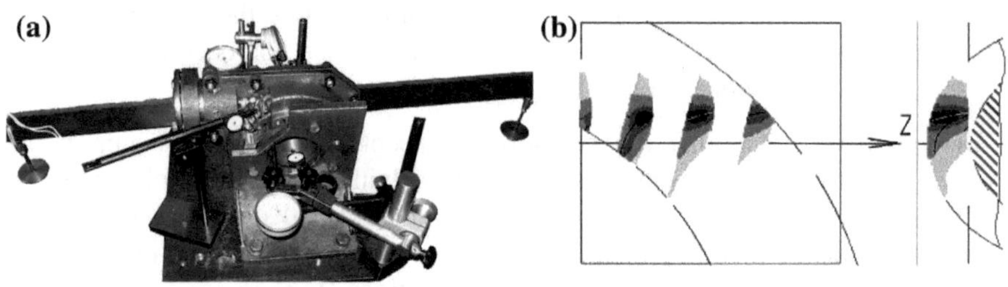

Fig. 8 Experimental unit **a** and results of investigation **b** of load distribution along the contact area

- application of one of the hobs from a preliminarily known and limited row for a gearwheel hobbing;
- ensuring necessary contact localization which guarantees small sensitivity of the gear to errors and deformations.

Solution of the first problem allows for significant reduction of the nomenclature of the tools for producing gearwheels of a rather large nomenclature, which, in turn, leads to an appreciable reduction of cost required for the hobs' manufacturing. The problem of the hobs' unification is solved on the basis of the developed approach. This approach is invariant relative to worm-type gears. The development prospect of this trend is the creation of new effective schemes and equipment and tools for generating the gearwheel teeth and the worm threads.

5. Formalization of the gearbox design process, which will ensure maximum automation of this process. This trend is new and is currently in the stage of development.

6. Application of polymer materials for spiroid gearwheels. The first works created in this activity [31] turned out to be perspective. Such works could be connected with the application of powder materials and nanotechnologies.

7. Research and development of design concepts of new varieties of gear and gears with non-traditional combination of parameters, for example, gears with small gear ratio (<10).

8. Experimental research of gears and gearboxes (Fig. 9), elaboration of control methods, and development of concepts and equipment for testing. The works generated from this activity are of great importance for checking the validity of the developed mathematical models and software, accumulation of information concerning the state of gears and gearboxes and their possibilities in different combinations of parameters and loading modes, and evaluation of operating ability of the developed and produced gears and gearboxes. In addition to giving answers to the described questions, the testing results also provide a basis for constructing the design and experimental models for evaluating the most important characteristics of gears and gearboxes.

Besides the indicated works, the following themes were elaborated at the IM related to:

- development of a new design approach for involute spur and helical gears on the basis of a method of blocking contours [32, 33] that allows for predicting the gear quality at initial design stages and solving the optimized design problem;
- creation of a theory of building-up the systems of computer-aided synthesis of objects and processes on the basis of characterization analysis (under the supervision of Malina [34]).

Carrying out research and development works in the described trends, IM assistants are working hard to develop spiroid gearboxes and gear-motor design for mastering their further production and implementation into different industry branches, technological planning and support of this production through the scientific production company "Mechanic, Ltd".

Fig. 9 Equipment for spiroid gearboxes testing

3 Mastering the Production of Spiroid Gears and Gearboxes

Industrial implementation of the developed gears and gearboxes has always been and still remains one of the most important tasks. Starting from the first works, a large amount of pilot, industrial samples and small lots of spiroid gearboxes have been manufactured. The first serial production of spiroid gear-motors of transport units of flexible manufacturing systems, mechanisms for chip removing of metal-cutting machine tools, and pivot mechanisms of vacuum units, developed in the Special Design Department of Gears and also at the IM, was organized in the beginning of the 1980s at the Plant of Precision Engineering (the city of Chaikovsky). It is fair to say that, in those years, S.A. Lagutin designed and mastered the serial production of the spiroid gearboxes of metallurgical equipment at the Plant of Heavy Mechanical Engineering (the city of Electrostal).

At the beginning of the 1990s, many manufacturing enterprises were closed because of the well-known political and economic situation in Russia. In order to support institutional scientific schools, which were mainly financed by the industry, the Ministry of Higher Education of the Russian Federation initiated state-financed programs of innovation; one of the obligatory conditions of their realization was compulsory implementation of the projects into the manufacturing process. The Institute of Mechanics began participating in these programs in 1991 and, in 1994, it oversaw the creation of the Federal Special-purpose Program "Progressive Gears", becoming its chief participant and uniting many leading scientific schools within Russian higher education in the field of gears. At that moment, an idea appeared and was rapidly realized of the independent manufacturing of science-intensive spiroid gearboxes and gear-motors within the IM of the ISTU on the basis of existing machine-tools, soon to be known as the scientific production company "Mechanic". At first, the company only mastered the production of spiroid gear-motors of small dimensions (loading torque up to 100Nm), but in the middle of the 1990s, the specialized manufacturing of spiroid gearboxes for drives of pipeline valves began at the request of the enterprise "Samaravolgomash, Ltd", producing ball valves for pipelines. At the present time, the serial nomenclature counts for 10 items of quarter-turn gearboxes (loading torques from 500 to 64000 Nm and gear ratios from 6 to 80 for single-stage modifications and from 100 to 2000 and more—for double-stage ones) and 3 items of multi-turn gearboxes.

Peculiarities of design and manufacturing of these gearboxes are described in [35], as applied at large valve-producing plants. It is interesting to note that this unique enterprise ("Mechanic" is the only producer of spiroid gearboxes for pipeline valves) became the first special independent gearbox enterprise in the field of valve-producing in Russia (Fig. 10).

It is obvious that all works at the enterprise are carried out on the basis of the developed techniques and with the direct involvement of the IM. Together with the ISTU Department "Design and Manufacture Staging of Mechanical Engineering Production", it forms the scientific educational manufacturing association, as well as functioning as a manufacturing base for carrying out laboratory works, practical manufacture education, implementation of scientific techniques into real science-based production in graduation and thesis works.

Fig. 10 Fragments of dimensional rows of quarter-turn **a** and multi-turn **b** spiroid gearboxes of pipeline valves

4 Conclusion

The paper gives a short historical review and describes the current state of scientific investigations and manufacturing development of spiroid gears and gearboxes at the Institute of Mechanics of ISTU and the company "Mechanic". The advanced knowledge on spiroid gears, the method for their design, the development of new varieties, constructions, production technologies, searching for new materials, lubricants, and new perspective areas for their application are a permanent focus of attention for the IM and "Mechanic, Ltd" workers. Due to the works performed, and judging from the known publications, the IM is the current worldwide leader in this specified area. Cooperation of the IM and "Mechanic" in preparing workers, engineers and scientific staff ensures permanent participation of the youth in scientific and manufacturing activity.

In conclusion, I would gratefully like to mention my colleagues who took part in the specified works and those participating now: Dr. Sc. E.S. Trubachev, Dr. Sc. V.N. Anferov, Dr. Sc. A.S. Kuniver, Dr. Sc. O.V. Malina, Ph.D. Yu. V. Puzanov, Ph.D. A.S. Kuznetsov, Ph.D. E.I. Popova, Ph.D. N.A. Barmina, Ph.D. A.A. Tkachev, Ph.D. D.V. Koshkin, Ph.D. V.Yu. Puzanov, Ph.D. E.V. Lukin, V.A. Ivaykin, M.P. Nadeina, D.V. Ivanov, T.V. Savelyeva, A.V. Beresneva, A.M. Sannikov, A.A. Kornilov, M.O. Vereshchagin, L.A. Korovkina, D.V. Pushkarev, A.Yu. Ovchinnikov, S.V. Kosolapov, enterprise workers.

We gratefully and warmly remember the unforgettable N.S. Golubkov, D.P. Gromov, E.K. Shibanov, V.I. Matveyev, S.A. Lenkova, and V.M. Spiridonov.

References

1. Litvin FL (1968) Theory of gearing. Nauka Publishers, Moscow, 584 pp
2. Litvin FL (1994) Gear geometry and applied theory. Prentice Hall, NJ
3. Litvin FL (1997) Development of gear technology and theory of gearing. NASA Reference Publication 1406, ARL-TR-150
4. Saari O (1954) Speed reduction gearing. Pat USA №26996125
5. Goldfarb VI (2006) What we know about spiroid gearing. In: Proceedings of the international conference on mechanical transmissions. China, vol 1. Science Press, p 19–26
6. Nelson WD (1961) Spiroid gearing. Machine design, № 3, p 136–144, №4, p 93–100, №5, p 165–171
7. Saari O (1961) Pat USA №29547046
8. Georgiev AK (1965) Elements of geometrical theory and some aspects of designing and manufacturing of hypoid-worm gears. Ph.D. thesis in Engineering, Izhevsk, 263 pp (scientific supervisor Prof. Vorobyov NV)
9. Georgiev AK, Goldfarb VI (1972) Aspects of geometrical theory and results of investigation of spiroid gears with cylindrical worms. Mech Mach:31:70–80 (Nauka Publishers, Moscow)
10. Georgiev AK (1967) Orthogonal worm-bevel gear. Inventor's certificate of USSR №201864, published in bulletin of inventions №18, patented in USA and Federal Republic of Germany

11. Georgiev AK, Goldfarb VI (1968) Skew-axis gear. Inventor's certificate of USSR № 208396, published in bulletin of inventions №3, patented in USA, Germany, England, Canada, France, Italy, Japan, Sweden
12. State Standard GOST 22850-77 (1978) Spiroid gears. Terms, definitions and symbols. Put in January 1 1979. Standards Publishers, Moscow, 65 pp
13. Goldfarb VI (1985) Fundamentals of theory of computer-aided geometrical analysis and synthesis of general-type worm gears: thesis for the Doctor's degree in Engineering, Izhevsk, 415 pp
14. Kuniver AS (2001) Theoretical fundamentals of meshing synthesis of modified spiroid cylindrical gears: thesis for the Doctor's degree in Engineering, Izhevsk, 342 pp (scientific supervisor Prof. Goldfarb VI)
15. Anferov VN (2002) Constructing of drives for handling equipment based on spiroid gears: thesis for the Doctor's degree in Engineering, Novosibirsk, 262 pp (scientific supervisor Prof. Goldfarb VI)
16. Trubachev ES (2004) Fundamentals of analysis and synthesis of real spiroid gears: thesis for the Doctor's degree in Engineering, Izhevsk, 347 pp (scientific supervisor Prof. Goldfarb VI)
17. Goldfarb VI (1983) Experience of spiroid gears design with application of interactive CAD. In: Digest: development and implementation of CAD systems in machine-building engineering, Izhevsk, pp 78–79
18. Goldfarb VI (1994) The synthesis of nontraditional kind of skew axis gearing. In: Proceedings of the International Gearing Conference, BGA transmission technology. Newcastle, MEP, London, pp 513–516
19. Russkikh AG (1977) Computer-aided synthesis of skew-axes gear schemes: Ph.D. thesis in Engineering, Izhevsk, 169 pp (scientific supervisor Prof. Goldfarb VI)
20. Inventor's certificate of USSR №806935. Non-orthogonal skew-axes gear. Goldfarb VI, Nesmelov IP, published in bulletin of inventions №7, 1981
21. Inventor's certificate of USSR №875133. Double-stage gearbox. Goldfarb VI, Nikitin AS, Nesmelov IP, published in bulletin of inventions №39, 1981
22. Inventor's certificate of USSR №1059325. Double-rim non-orthogonal skew-axes gear, Goldfarb VI, Nesmelov IP, Teterin AN, published in bulletin of inventions №45, 1983
23. Trubachev ES (1999) Research of parameters space of non-orthogonal spiroid gears: Ph.D. thesis in Engineering, Izhevsk, 170 pp (scientific supervisor Prof. Goldfarb VI)
24. Isakova NV (1999) Development and research of spiroid gear with ideal thread screw parameter: Ph.D. thesis in Engineering, Izhevsk, 145 pp (scientific supervisor Prof. Goldfarb VI)
25. Barmina NA (2002) Structural and parametric analysis of double-stage gearboxes with spiroid and cylindrical gears: Ph.D. thesis in Engineering, Izhevsk, 183 pp (scientific supervisor Prof. Goldfarb VI)
26. Abramov AI (1996) Theoretical and experimental research of kinematical accuracy and vibration activity of spiroid gears: Ph.D. thesis in Engineering, Izhevsk, 148 pp (scientific supervisor Prof. Goldfarb VI)
27. Koshkin DV (1999) Research of errors influence and geometrical modeling of localized contact in spiroid gear: Ph.D. thesis in Engineering, Izhevsk, 159 pp (scientific supervisor Prof. Goldfarb VI)
28. Trubachev ES, Oreshin AV (2003) CAD of spiroid gears. Informational mathematics, №1 (3), Moscow: Physical and Mathematical Literature Publishers, pp 159–165
29. Kuznetsov AS (2005) Theoretical and experimental research of static loading of spiroid gears. Ph.D. thesis in Engineering, Izhevsk, 174 pp (scientific supervisor Prof. Trubachev ES)
30. Anferov VN (1982) Research of wear resistance of spiroid gears. Ph.D. thesis in Engineering, Kurgan, 162 pp (scientific supervisor Prof. Drozdov Yu.N, scientific adviser Prof. Goldfarb VI)
31. Popova EI (2004) Development of tools and manufacturing technology of metal-polymer gearwheels of spiroid gears. Ph.D. thesis in Engineering, Izhevsk, 164 pp (scientific supervisor Prof. Goldfarb VI)
32. Tkachev AA (2001) Development of system for interactive design of involute cylindrical gears. Ph.D. thesis in Engineering, Izhevsk, 173 pp (scientific supervisor Prof. Goldfarb VI)

33. Goldfarb VI, Tkachev AA (2004) Designing the involute spur and helical gears. In: New approach. ISTU Publishers, Izhevsk, 94 pp
34. Malina OV (2002) Theory and practice of structural synthesis automation of objects and processes with application of characteristic analysis methods, thesis for the Doctor's degree in Engineering, Izhevsk, 392 pp (scientific supervisor Prof. Goldfarb VI)
35. Goldfarb VI, Glavatskikh DV, Trubachev ES, Kuznetsov AS, Lukin EV, Ivanov DE, Puzanov V Yu (2011) Spiroid gearboxes of pipeline valves. Veche Publishers, Moscow, 222 pp

Cause of Failure Beyond Conjugate Theory of Gear Meshing

A. Kubo

Abstract It is obvious that the basics of gear technology is the theory of geometry for conjugate action of mating gears for transmission of motion through contacting tooth flanks. Teeth of actual gears though contact not only on tooth flank to tooth flank, but also at edges of tooth tip and of tooth sides. Such edge contact is usually out of the conjugate meshing theory of gearing. Many causes of failure of highly loaded gear teeth are initiated due to the contact at such tooth edges. Contact of tooth edge produces considerable amount of wear debris. Meshing teeth crush intruded wear debris and damage themselves. With run of gear operation, gear failure develops as a phenomenon of such a positive feedback system.

Keywords Edge contact · Trochoidal interference · Pitting · Spalling · Case crush · Crush of wear debris · Positive feedback · Surface durability

1 Introduction

The theory of conjugate action of gears is the fundament of mechanical transmission of motion and power. Cylindrical involute gears, for instance, realize the conjugate action for motion and power transmission with perfect geometrical theory, but in practice, gear failure occurs. In reality, actual gearboxes and teeth have some amount of manufacturing and alignment errors. Gear shafts, bearings and gear boxes deform through transmitting loads and as a result of both internal heating due to friction and power loss and external heating. Fundament of installation also deforms. Alignment of shafts in actual usage of a gearbox is usually somewhat different from that of the designed state. This causes uneven load distribution on the tooth flank and is major cause of frequent gear failure. When a side edge of a gear tooth of smaller width comes into contact with a wider tooth flank, very severe local

A. Kubo (✉)
Research Institute for Applied Sciences, Kyoto 606-8022, Japan
e-mail: aizokubo@hera.eonet.ne.jp

V. Goldfarb and N. Barmina (eds.), *Theory and Practice of Gearing and Transmissions*, Mechanisms and Machine Science 34,
DOI 10.1007/978-3-319-19740-1_5

slip occurs between the contacting side edge and the tooth flank in the tooth lead direction under high contact pressure. Such contact repeats with tooth meshing. This very local severe slip in tooth lead direction induces pitting and micro-cracks on the wider tooth flank. In the long run, such failure develops into fatal tooth breakage. With gear operation, gear failure itself develops as a phenomenon of a positive feedback system of gear failure development. That means gear engineers must be aware of the fact that many additional factors beyond the consideration of the conjugate geometry of gears, and also beyond the conventional methods of calculation of load carrying capacity, are necessary to raise the performance of gears.

2 Examples of Failure

So-called spalling or case-crushing or flaking, as shown in Fig. 1, is an example of the result of such tooth flank failure at the middle of tooth width, where the lead form modification of crowning makes the 3D position of the tooth flank most projected.

Figure 2 shows another example of the same kind of failure, but in this case, the damaged addendum remains on the tooth. A subsurface crack initiated from the apex of the triangular spalled off part on the dedendum proceeds between case and core inside the carburized and hardened gear tooth and reaches up to the back flank of the tooth, cf. Fig. 3. Such a failed part of a tooth will fall off during operation and partial tooth breakage, as shown in Fig. 1, will result.

From the perspective of Figs. 1 or 2, it is not so clear as to what position near the root of the tooth the tooth tip edge of a mating driven gear comes into contact. Figure 4 shows a replica of the failure shown in Fig. 2. The change in surface texture of the tooth flank is far more easily seen on the acetylcellulose replica than

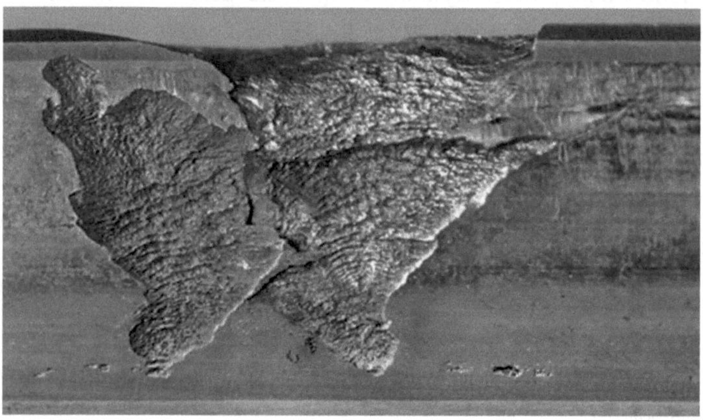

Fig. 1 Flaking failure of tooth flank

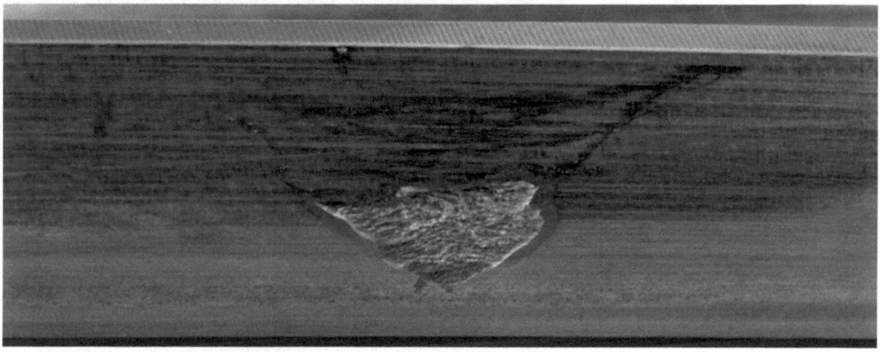

Fig. 2 Middle stage of flaking progress

Fig. 3 Crack reaching the back flank of a tooth

Fig. 4 Replica figure of Fig. 2

Fig. 5 Region of trochoidal interference on the dedendum

in the photo of the actual tooth flank. Almost half of the depth of the dedendum has been rubbed strongly by the tooth tip edge of the mating driven gear, due to so-called trochoidal interference. The depth of trochoidal interference is usually believed to be the narrow scratched band near the contact limit at the tooth root, but that is the cut-in region inside the trochoidal interference zone and the actual depth of trochoidal interference is far greater than that. Utilizing a photo with more sophisticated lighting, you can observe a clear state of trochoidal interference in full-length on an actual dedendum tooth flank, as shown in Fig. 5. Its length spans almost half of the dedendum.

3 Contact of Tooth Tip Edge and Stress Distribution Over Tooth Form

Figure 6 shows the beginning state of tooth meshing of involute gears. A meshing tooth pair exists near the middle position of distance AB, though it is not shown in this figure. Due to the elastic deformation of a meshing tooth pair under load transmission, the rotational position of driven gear delays and the position of the tooth form of the driven gear shifts from the theoretically-defined position through the conjugate action theory of gear meshing. Also, due to lead crowning or the existence of pitch error, etc., the position of the tooth form of the driven gear shifts from the theoretically-defined position by a corresponding amount. The theoretical beginning point of meshing is B. That means that when the tooth tip edge of a driven gear comes to point B, the involute curve of the tooth form of the driving gear passes that point, where the pinion and gear are assumed to be error-free, rigid bodies. In case that rotational delay, etc., between driving and driven gears exists, the beginning point of meshing moves: when the tooth tip edge of the driven gear comes to point C, the involute curve of the tooth form of the driving gear passes this point. That means point C is the actual beginning point of contact. The tooth tip

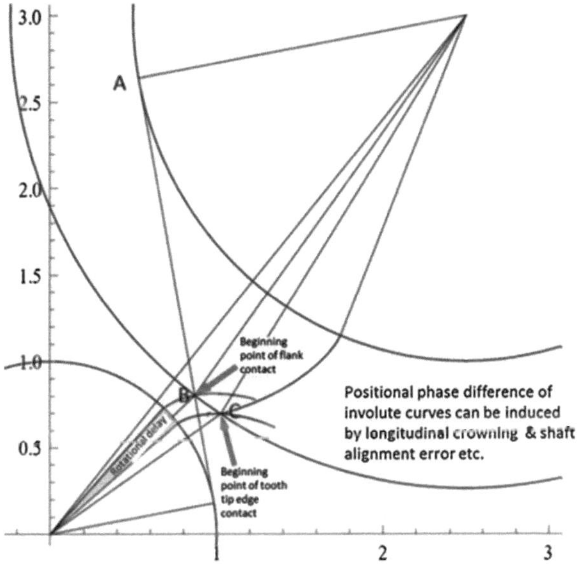

Fig. 6 Beginning of tooth contact outside of LoA

edge of the driven gear rubs the dedendum of the driving tooth from point C to point B. When the position of contact reaches point B, the tooth flank of the involute helicoid of the driven gear first comes into contact with the tooth flank of the driving tooth. That means the "involute-helicoid to involute-helicoid" contact of tooth forms first begins here and the contact point moves from B toward A on the line of action, as taught in books on gear theory.

One example of contact stress as a function of gear rotation during the trochoidal interference (from C to B) is shown in Fig. 7. The Hertzian stress during edge contact is always far higher than that of contact between involute helicoids.

When the state of contact stress is expressed on a gear tooth, schematically, it looks as shown in Fig. 8. The stress during "tooth flank to tooth flank" contact is not so high in comparison with that of tooth tip to tooth dedendum contact, where trochoidal interference occurs. All the prediction methods today for load carrying capacity of tooth flank are valid only for "tooth flank to tooth flank" contact. They are not valid in the region of trochoidal interference, though many actual causes of gear failure are triggered by the surface failure occurring inside such a rather wide region of trochoidal interference.

Figure 9 shows an example of the state of slipping velocity during tooth tip edge contact. Slipping velocity increases with gear rotational position from B to C and is larger than the maximum value of the tooth flank contact at B. The result of Figs. 7 and 9 means the state of stress and frictional heat generation during trochoidal interference is more severe than the state during tooth flank contact. Because material fails at the position where the induced stress, including temperature and lubricating condition, becomes highest, trochoidal interference should be eliminated.

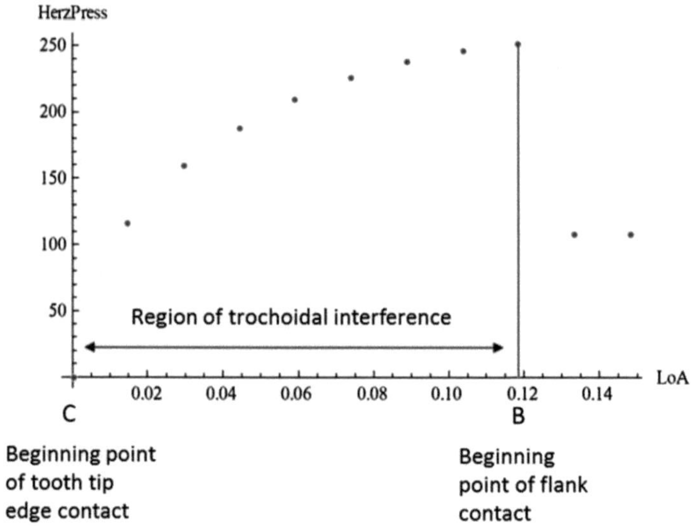

Fig. 7 Contact stress during trochoidal interference

Fig. 8 Contact stress over tooth form

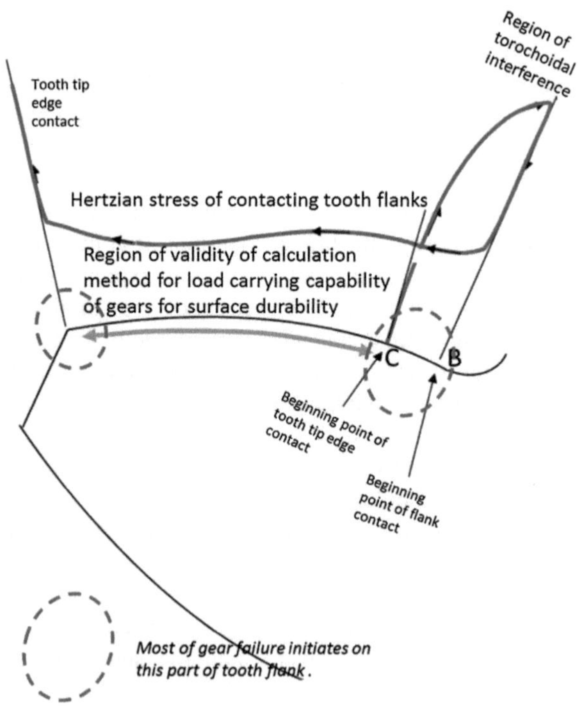

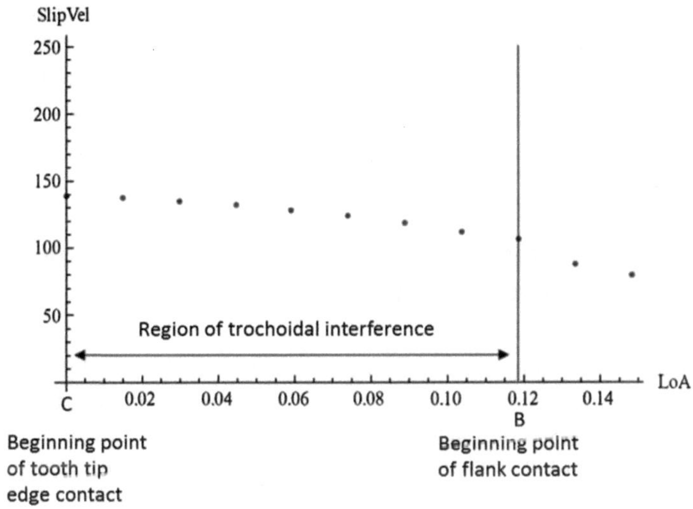

Fig. 9 Slip velocity during trochoidal interference

Fig. 10 Surface state inside trochoidal interference zone

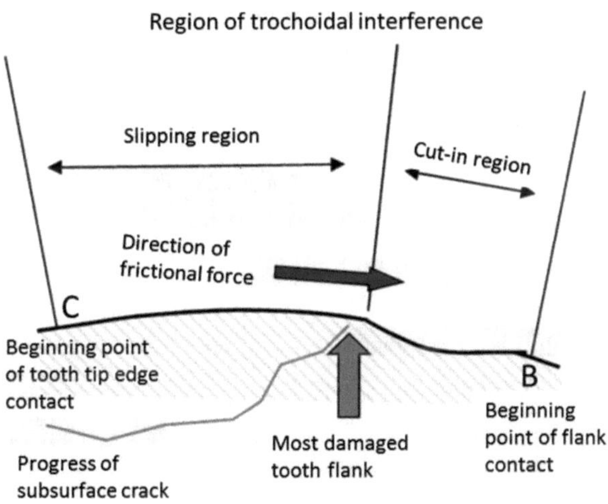

Figure 10 shows the schematic state of the dedendum surface inside trochoidal interference. The pitch point exists on the outer left side of the figure. At point C, the tooth tip edge of the mating gear comes into contact. The dedendum flank of the driving gear carries strong contact stress and frictional force. High heat generation occurs there. The flank, however, does not fail badly and a "slipping region" is formed. But at some position, the surface could not bear further stress and the tooth tip edge of the driven gear cuts into the dedendum surface of the driving pinion to make the "cut-in region". Then, an inflection point in the tooth form curve is created at the position at which the "slipping region" and the "cut-in region" are joined. Subsequently, contact stress there becomes high and a foreign object, such as wear debris, will often crash there. As a result, a subsurface crack initiates from that point. Forthcoming "Involute helicoid to Involute helicoid" contact squeezes that crack so that it opens wide and develops upward in the direction of the addendum. This is the reason why the failure shown, e.g., in Figs. 1 and 2, etc., shows an apex as the initiating point of flaking and the flaking development progressing in fan-form.

4 Increase of Rotational Delay Due to Lead Crowning and Tooth Form Correction

Rotational delay between driving and driven gears, that is, the shifting amount of the rotational position of a tooth form, is the major cause of trochoidal interference. One of the main causes of rotational delay is elastic deflection of meshing tooth pairs due to transmitting load. In the field of power transmitting gears, almost all gears have a tooth flank form correction for obtaining good tooth flank bearing. On the other hand, many researchers investigated the stiffness of gear teeth and the result is incorporated into the gear design standard. These tooth stiffness values, however, are always the stiffness of an error-free gear tooth that realizes 100 % of the tooth flank contact bearing. Figure 11 shows an example to indicate the tremendous influence of tooth flank form correction on the value of tooth stiffness. For

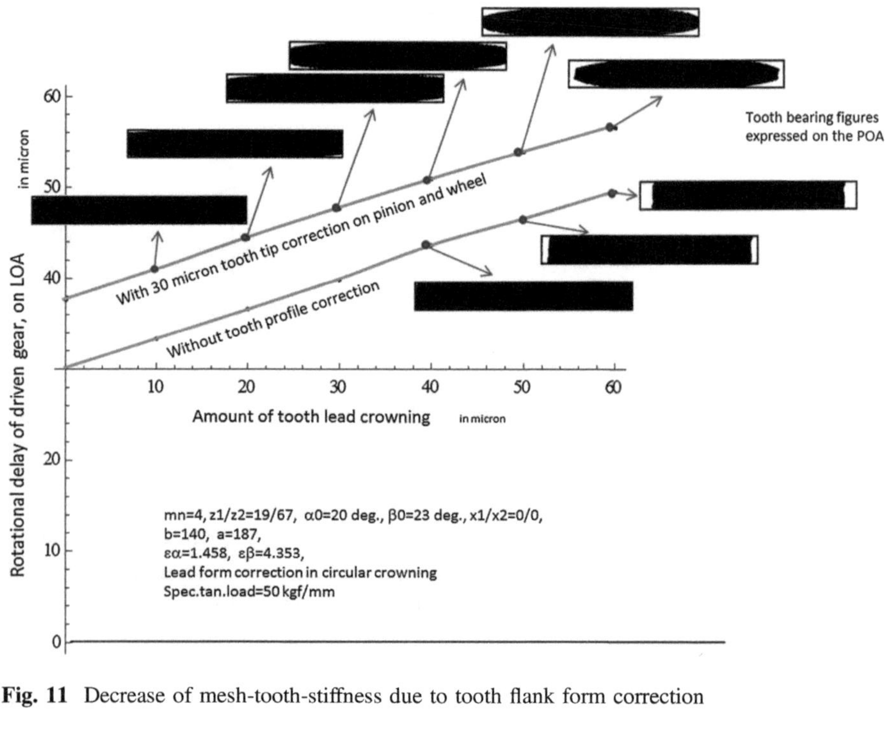

Fig. 11 Decrease of mesh-tooth-stiffness due to tooth flank form correction

instance, the average stiffness of a meshing tooth pair with 0.06 mm lead crowning and 0.03 mm tooth form correction is ca.1.8 times softer than that of ideal correction-free gears. When the lead crowning of this gear is designed without this knowledge, it is a rational result that this gear would have gear failure such as shown Figs. 1 or 2, also in such a case as most of gear designers usually might think that gear is safe enough, because the gear has a fairly large margin of safety factor for tooth flank durability calculated by using the standard calculating method for the load carrying capacity of gears. The reason for this is that the initiating point of such failure is not covered by the region of validity of the standardized calculating methods.

5 Trochoidal Interference and Mal-Effect on Surface Durability

Figure 12 shows an example of gear failure initiated from a triggering failure inside the trochoidal interference zone. The tooth-bearing pattern in the tooth width direction is OK, but on the dedendum, you can observe a strong trace of trochoidal interference reflecting the form of crowning of longitudinal tooth lead correction. This is because (a) the position of the involute tooth form at the center of the tooth width and that at the sides is different and (b) tooth deformation due to transmitting torque has become significant. Accordingly, the flaking failure initiated from that position of the most projected tooth form position, i.e., at the middle of the tooth width.

Figure 13 shows an example of a tooth bearing of a heavily loaded spur gear with a rather large lead crowning on a driving planet of vehicle driving planetary gear box for land construction. At the tooth tip edge, a worn figure reflecting the form of the tooth lead crowning is observed. This wear is the result of trochoidal interference at the recessing contact. At the inflection line between this worn tip part and the addendum flank, in the middle of the tooth width, pitting occurs as a result of high contact stress. On the dedendum flank, the normally worn tooth flank is seen as the lighter reflected part. Beneath this dedendum, a black narrow band is observed. This is the region of cut-in region inside trochoidal interference in approaching contact. It is very important to know that the width of this trochoidal interference is different from the width of tooth flank contact on the dedendum.

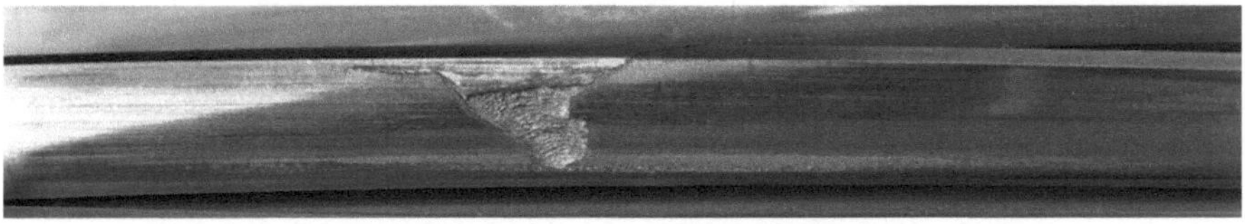

Fig. 12 Trochoidal interference due to lead crowning and flaking failure

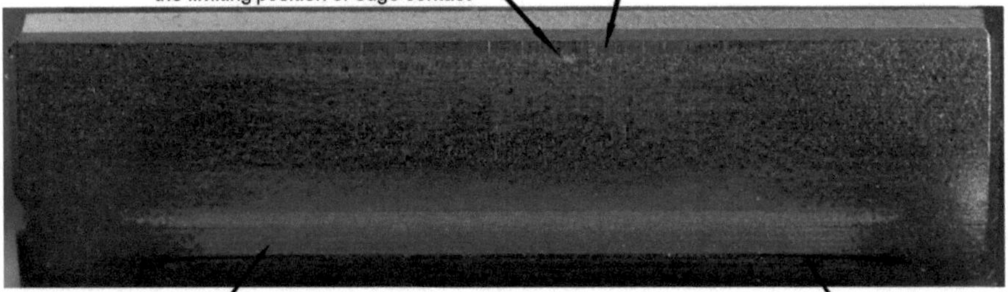

Fig. 13 Tooth bearing figure, trochoidal interference and tooth tip wear

Fig. 14 Chipping of tooth tip edge of driving pinion

That means the region of tooth tip edge contact and that of tooth flank contact is discontinuous. The stress change there is quite sudden, as seen in Fig. 8. This is the reason why the tooth tip edge makes such strong contact, though the addendum tooth flank near the tooth tip carries little shared value of transmitting load.

As the result of trochoidal interference, not only do the stress state and thermal state of the interfered dedendum become severe, but so do those at the mating tooth tip edge and the nearby addendum, as seen in Fig. 13. It is a usual case that the stress situation at trochoidal interference becomes more severe to produce tip edge chipping: the material of the tip edge falls off the tooth as wear debris. An example of tooth tip edge chipping is shown in Fig. 14. The tendency of over-carburizing of the material there plays, of course, some important role in such chipping.

Comparing the trochoidal interference at the approaching contact with that at the recessing contact, heat generation due to friction is greater at the recessing contact: At the approaching trochoidal interference, the cut-in region inside the trochoidal interference zone softens the contact stress and consequently the heat generation there. The main cause of heat generation by the approaching trochoidal interference would be material deformation at the cut-in action by the driven tooth tip edge into the dedendum flank. The temperature of the dedendum flank becomes, therefore, high, and the hardness of the material there becomes softer. In such a thermal and lubricating condition, the tooth tip edge of the driven gear moves in the cut-in direction and tooth edge tip chipping barely occurs. Contrastingly, at the recessing trochoidal interference, the tooth tip of driving pinion moves in contra cut-in direction and no cut-in region inside trochoidal interference zone appears. Addendum tooth flank very near to the tooth tip edge of the driving pinion rubs the dedendum of the driven gear strongly and significant frictional heat is generated. The tooth tip temperature of the driving pinion becomes high and chipping of the tooth tip occurs rather easily, becoming more severe than the state of the driven tooth tip.

6 Contact of Tooth Side Edge

In practical gear design, it is usual to make the tooth width of the driving pinion different from that of the driven gear. In many cases, wide tooth width of the pinion meshes with the small tooth width of the wheel. That means that the side edge of the wheel tooth comes into contact with the wide tooth flank of the pinion. Figure 15 shows how the pinion tooth deforms in such a case. In the figure, the displacement of tooth deformation is enlarged by 100 times according to the tooth's geometry.

The wheel tooth plunges into the pinion tooth flank and the tooth side end deforms, i.e., the tooth side wall bulges toward the outside of the tooth. On the other hand, the pinion tooth bends like an arc in its width direction. It is therefore obvious from that deformed figure that the side edges of the wheel teeth make far stronger contact than we might normally imagine. The blue and brown colors in the figure indicate the magnitude of displacement of the tooth flank material in plus and minus directions of the tooth lead, respectively. These two colors near both sides of the tooth contact area mean that the surface material inside the contact zone and a little outside of that position of the wider tooth flank moves into the middle of the contact length, i.e., in the direction towards the center of the tooth width.

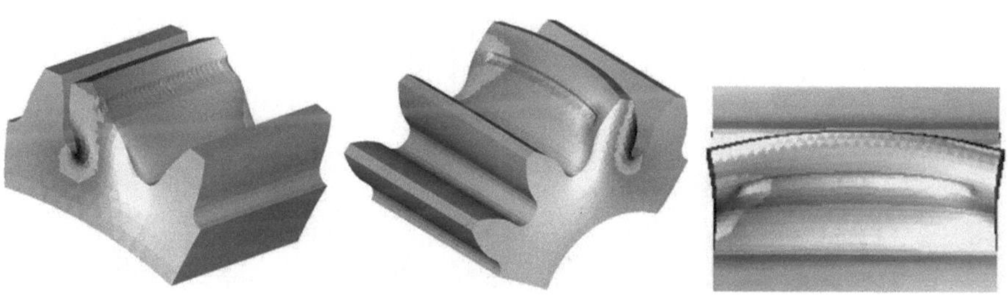

Fig. 15 Deformation of tooth and tooth flanks at contact of narrower and wider tooth width (courtesy of Dr. A. Ueda, AMTEC Inc.)

The surface material at the side edge of the narrower tooth width bulges in the direction outside of the tooth width as shown in Fig. 16. The tooth edge of the narrower wheel makes contact with the wider pinion tooth flank under such a figure of deformation. The side edge of the wheel tooth in the figure (upper part) moves right, whereas the wider tooth flank (lower part) material of the pinion moves left under very high contacting pressure, changing its magnitude as the tooth contacting progresses. This situation repeats every tooth meshing cycle. As a result, the tooth flank of the wider pinion at which the tooth side edge of the narrower wheel makes contact is strongly rubbed and polished: we can easily observe a clear polished line in the tooth form direction after a rather short time of on-load gear operation. As shown in Fig. 17, the contact there generates considerable frictional heat. The wider

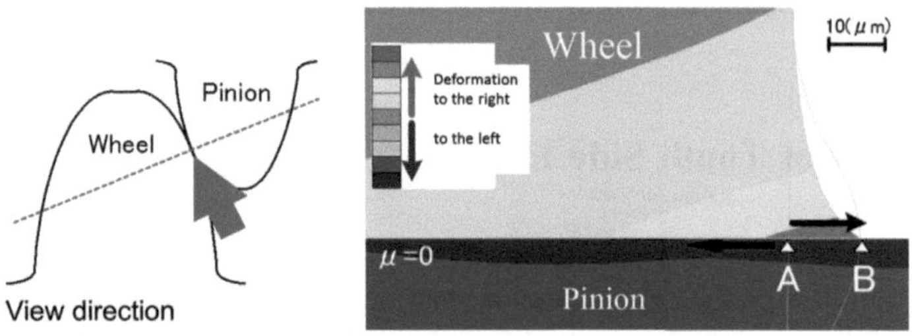

Fig. 16 Relative displacement at the tooth side edge of contact with the wider tooth flank

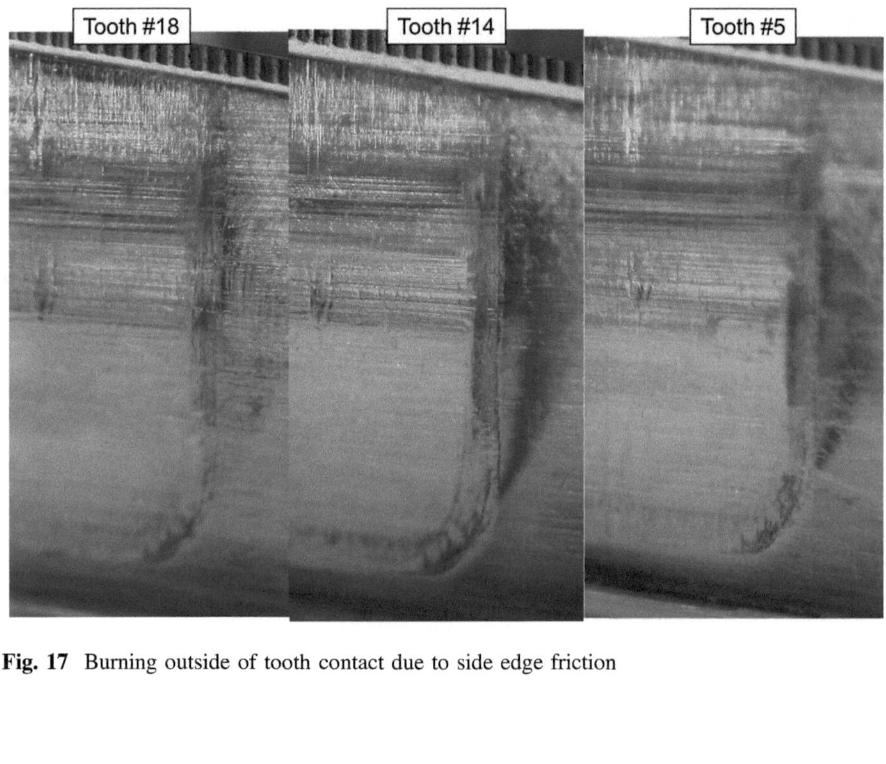

Fig. 17 Burning outside of tooth contact due to side edge friction

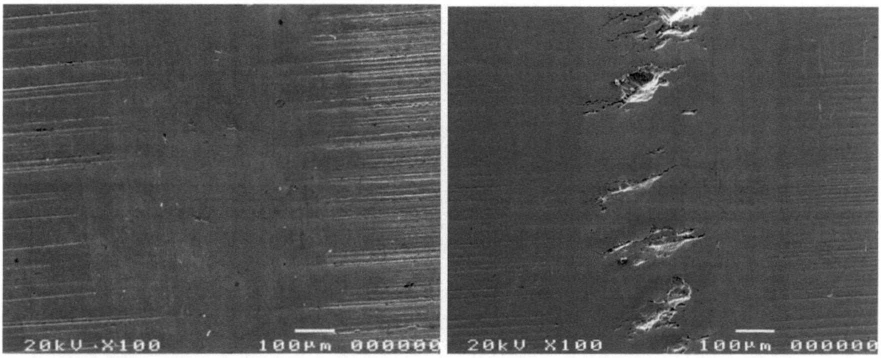

Fig. 18 Micro-crack initiation in the polished band due to tooth side edge contact

tooth flank in such side edge contact is damaged during much repetition of the loading cycles and many micro cracks appear there, as shown in Fig. 18.

Figure 18 shows an SEM picture of the surface state in this polished band on the wider tooth flank making contact against the tooth side edge of the narrower tooth. The left figure shows the state at 7×10^5 loading cycles on the lightly contacting side of the tooth. The left half from the polished band in this picture shows the state of the loaded flank and the right half shows the non-contacting flank. Ground marks, ridges of surface asperities, on the loaded tooth flank are plastically deformed or worn, but the ground marks still remain. In the polished band, surface roughness asperities are completely worn and small micro-cracks have already appeared. The right figure of Fig. 18 shows the state of the same tooth flank position at 1.7×10^7 loading cycles. The micro-cracks inside the polished band have developed in large size. Material inside the cracks falls off from the crack-wall due to repeating contacts. These micro-cracks are the cause of the small pit inside the polished band and the forthcoming development of flaking, cf. Fig. 19.

Figure 19, from left to right, shows the initiation and growth of the surface failure, triggered by Fig. 18 like micro-pitting. The pictures are taken on the opposite side end of the tooth in Fig. 18: this side of the tooth makes stronger contact than is the case in Fig. 18.

After 6.6×10^5 load repetition, a very small pit was born inside the polished band, as seen in the leftmost figure of Fig. 19, and that became the trigger of the spalling that will come. At 1.48×10^7 loading cycles, the state of spalling develops as shown in the upper figure of Fig. 20. It is commonly understood that the contact stress at the spalled-off part of the tooth flank was excessive. The end stage of this failure development is as shown in the lower figure of Fig. 20 and in Fig. 21, a case of partial tooth breakage. Most gear engineers who observed this state of failure noted, "This gear was operated under strongly uneven contact state in tooth width direction and the contact pressure at severer tooth side became too high. That induced severe progressive pitting and it developed to spalling. From the basin of spalled off tooth flank, fatal crack initiated and it developed to destroy the tooth partially".

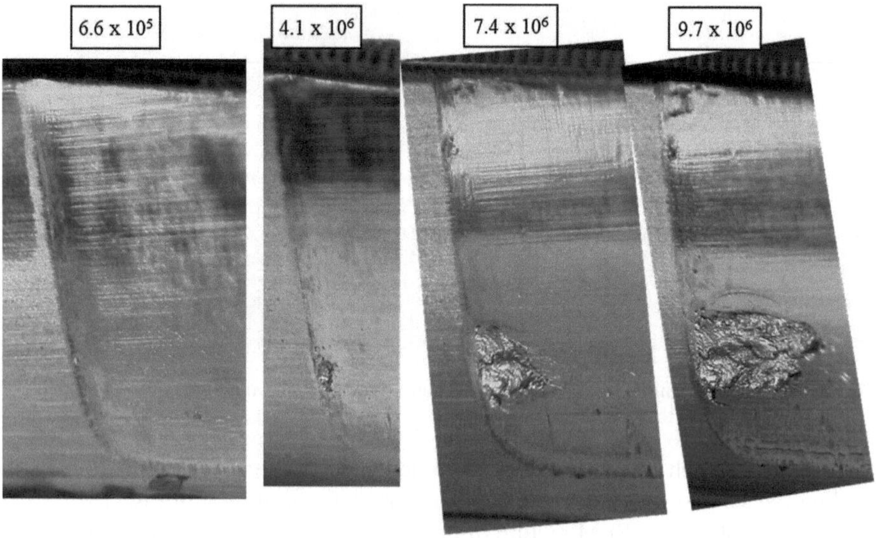

Fig. 19 Initiation of the tooth breakage in Fig. 21 [1]

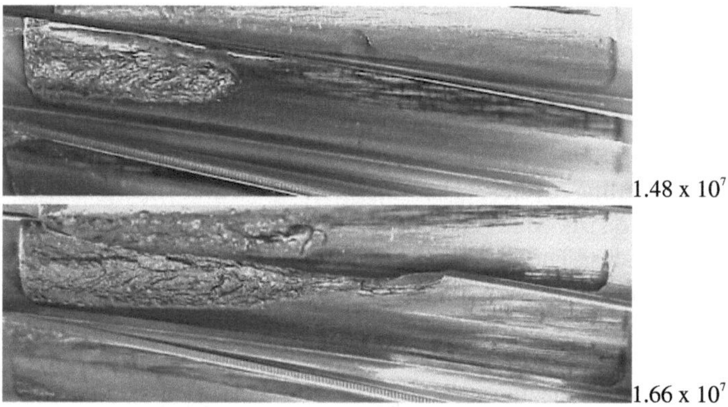

Fig. 20 Severe spalling and partial tooth breakage [1]

Fig. 21 Tooth breakage the cause of which is neither the excess tooth bending stress nor the excess Herzian stress between tooth flanks, 1.66×10^7 loading cycles [1]

However, the true cause of the failure was the micro-crack and the small pit inside the polished band at the point of contact on the tooth side edge, where the nominal contact stress according to the gear design formula was not especially high. This failure, though small, developed progressively and became the fatal failure for this gear set. Today's theory for the prediction of tooth flank durability cannot predict such initiation and progress of surface failure. The case of such a mode of failure trouble has not yet been solved during the many cases of accidents in practical gear usage.

As you can see from Fig. 16, the state of subsurface stress inside the contacting side edge of the narrower wheel tooth is also very severe. Figure 22 shows the state of shear stress in the cross section of the tooth in the tooth lead over a half of the contact length. Usually, the tooth edge of a carburized and hardened gear is over carburized and becomes brittle. If the material there is not tough enough, the edge will chip off rather easily. At that moment of edge rupture, the crack often runs inside the tooth, and develops into a fatal crack that causes the tooth to break, cf. Figs. 23 and 24. This crack usually runs along the most severely bent stressed section at tooth fillet, i.e., Hofer's 30 deg. position, in the tooth width direction.

Figure 25 shows a tooth breakage experienced with case carburized helical gears: A tooth is broken off whole and when you see the middle or largest part of

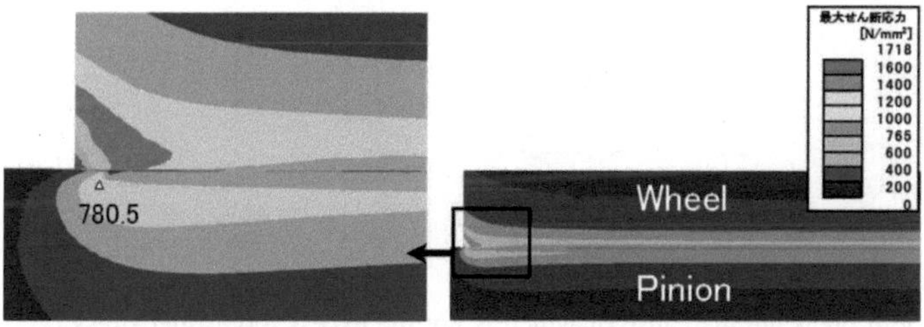

Fig. 22 Maximum shear stress induced subsurface of tooth side edge contact, cf. Fig. 16

Fig. 23 Chipping at tooth side edge and induced crack for the case of Fig. 25 [1]

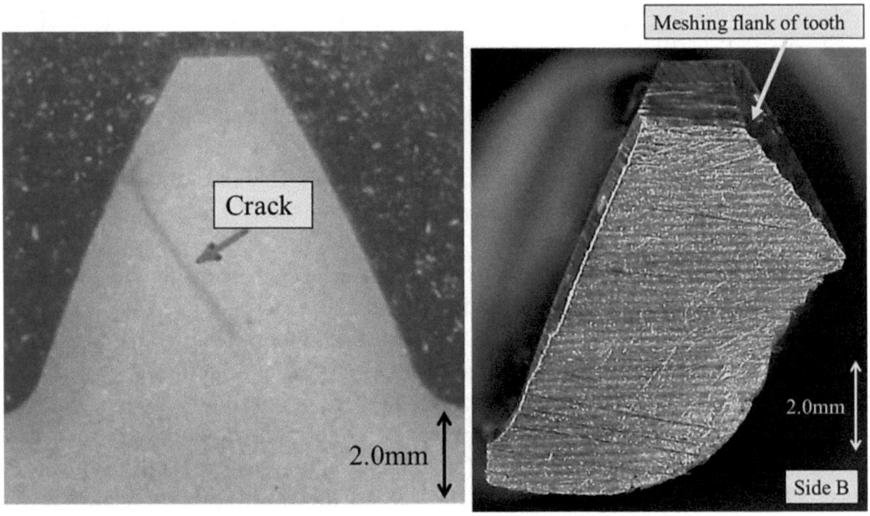

Fig. 24 Crack propagation from the tooth side edge chipping and tooth breakage. Cause of tooth breakage for the case of Fig. 25 [1]

Fig. 25 Tooth breakage the cause of which is not the excess tooth fillet stress

the tooth, the section of tooth breakage appears to exist at the Hofer's 30 deg. position, i.e., the most severely bent stressed part at the tooth fillet. Many gear engineers think, therefore, that this tooth breakage is bending fatigue failure of the gear tooth as a canti-lever model, but this is not true. Thorough investigation of this failure has been carried out and the real cause and process of failure development, from the beginning to the tooth breakage, was clarified: The endurance test was carried out under very heavy loading conditions. The over-carburized and brittle tooth side edge made strong contact with the mating tooth flank. During the endurance test, such subsurface stress as shown in Fig. 22 has been induced in every tooth meshing, and as a result, the edge has chipped off, cf. Fig. 23. A micro-crack run deeply inside tooth, and is propagated to break off the tooth, cf. Fig. 24. From this history, it is a matter of course that the conventional method for predicting tooth bending strength cannot be valid to predict this failure.

The subsurface stress concentration at the tooth side edge of the shorter tooth width, as shown in Fig. 22, induces another kind of tooth failure. At the side edge of

Fig. 26 Crack initiation inside the tooth side chamfer and the side edge rupture of the driven gear

Fig. 27 Heat damage of the dedendum due to trochoidal interference

the helical gear tooth on the acute angle side, fatigue crack can initiate inside the side-chamfer of the tooth, as shown in the leftmost picture of Fig. 26. This crack usually develops and reaches considerable depth, which is why it has such a bad effect in regard to tooth breakage.

This part of the tooth side edge of the strong helical gear tooth on the acute angle side is usually over-carburized and becomes very brittle. Often, this tooth side edge ruptures at the point of trochoidal interference, as shown in the central and rightmost pictures of Fig. 26. Beneath the lower limit of tooth contact, bad oil sludge covers the tooth fillet. This indicates the surface temperature there has become very high during gear operation. Inside this recessing trochoidal interference zone, severe rag wear* occurs. The material hardness and strength there decreases and a crack initiates, propagating easily.

Figure 27 shows a damaged tooth dedendum. Inside the recessing trochoidal interference zone, severe rag wear occurs and severe edge contact generates significant heat. The surface temperature of the dedendum tooth flank in and near the trochoidal interference zone becomes very hot, and the tooth flank hardness decreases. The capable durability allowance of the material there becomes

Fig. 28 Outlook of tooth side rupture

considerably lower from the value of virgin gears, i.e., from the value taken at the gear design stage. The surface crack initiates there and then acts as a notch, going deeper and becoming a macro crack that breaks the tooth. According to the endurance test run soon after this photo was taken, this tooth was broken due to the crack at this section.

Figure 28 shows what a tooth side edge rupture on the acute angle side ends of teeth looks like. Figure 29 shows the resulting tooth breakage. Most mechanical engineers never imagined that the real cause of such bad tooth breakage could be such a small tooth side edge rupture.

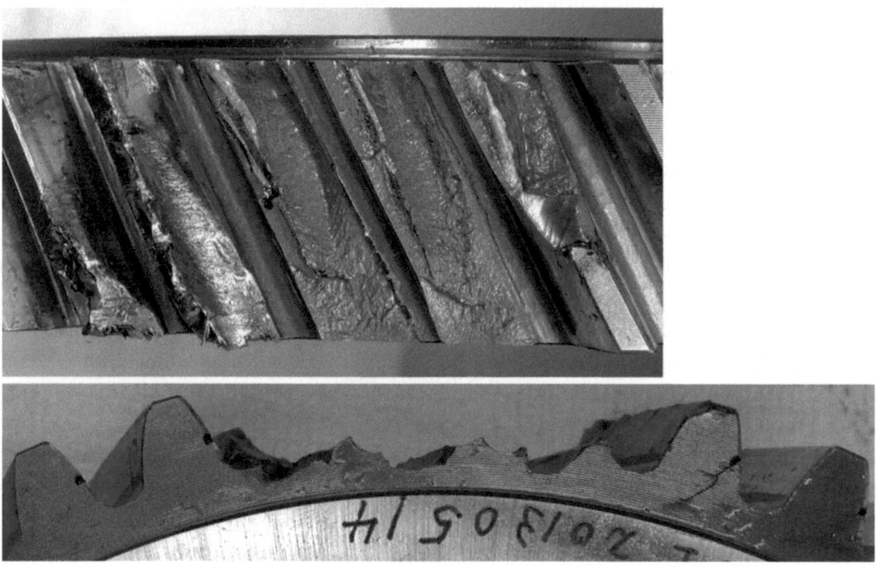

Fig. 29 Total tooth breakage induced by a tooth side edge rupture

7 Merging of Subsurface Cracks from Side and Root

The subsurface crack that initiates from the pitting inside the trochoidal interference zone at the tooth root and the subsurface crack inside the polished band at the side edge contact develops during repetition of severe contact stress by the power transmitting tooth meshing. Both the cracks then merge together to become one large spall, as shown, for example, in Fig. 30. The large spalled-off part of the tooth flank then develops to cover a considerable portion of the tooth flank, leading to the fatal state, i.e., a bad case-crushing and partial tooth breakage.

Figure 31 shows another example: The subsurface crack that initiates from the pitting inside the approaching trochoidal interference zone at the tooth root and the

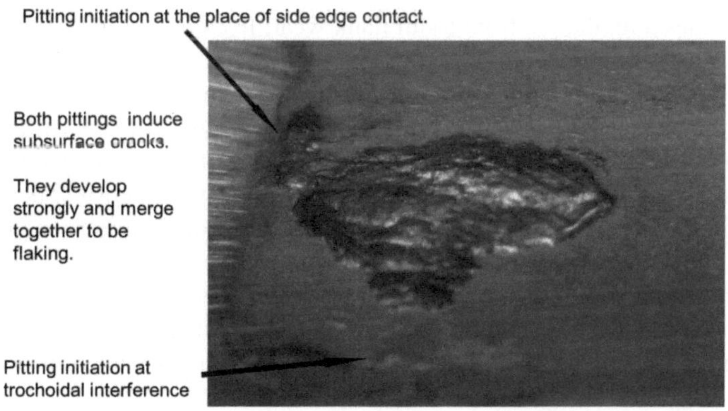

Fig. 30 Merging of subsurface cracks from approaching trochoidal interference and from surface failure by side edge contact

Fig. 31 Merging of subsurface cracks from approaching trochoidal interference and from side edge chipping

crack induced by tooth side edge chipping merge together. With repetition of tooth flank contact of gear meshing, the merged macro-crack proceeds inside subsurface of the tooth between the core and the case to break the gear tooth.

High strength steel is commonly brittle and crack propagation rate is high. The result of this nature of gear material is frequent outbreaks of such gear failure. It is agreed that many cases of failure of highly loaded gear teeth are triggered by such phenomena as “tooth flank to edge” contact.

8 Crush of Wear Debris

The lubricating oil in the gear box becomes dark after some operating time has passed. This is because of wear debris suspended in the oil. Most wear debris comes from gear meshing, that is, from tooth flank wear, from pitting and chipping due to edge contact of the corners of tooth tips and tooth side edges. Wear debris produced in such a fashion, from particles of considerable size to very fine, intrudes between the engaging tooth flanks of gears, and is subsequently crushed. Crushing of wear debris by engaging tooth flanks not only causes considerable damage on the tooth flank, but also produces a considerable amount of new wear debris. The probability of intrusion of wear debris suspended in lubricating oil between engaging tooth flanks is not so high, but some are crushed. In the case that trochoidal interference occurs, the chipped-off edge material and wear particles produced probably stay on and near the tooth flank surface. Forthcoming “involute helicoid to involute helicoid” tooth flank contact then crushes those foreign objects with considerably high probability.

When we examine the tooth flanks of heavily operated gears under sophisticated lighting, we generally find many traces of crushed foreign objects, usually in the form of scratching, cf. Figs. 37 and 39. Figure 32 shows another such example of severe crushing of an intruding foreign object: this is observed on the tooth flank of the planet of the final drive of a land construction machine. Though plastic indentation and surface irregularity of the tooth flank is emphasized by the lighting technique in the photograph, we find there occurred a very strong intrusion of foreign objects between the meshing teeth and the engaging tooth flanks crushed them during machine operation. This is not a special case, but is common in the normal operation of gears in actual heavy usage of machines. Even the tooth flank is case hardened, it deformed plastically and scratched through the crushing of wear debris.

Figure 33 shows another example: The gear is case carburized, hardened and ground. The gear accuracy is good and the tooth lead has a considerable amount of form modification of crowning. That means the load distribution near the side end of the tooth is very weak, but in this case, a bad spall occurred on that portion of the tooth flank. When we think of the reliability of the calculating method of the load carrying capability of gears concerning tooth flank durability, this failure is then contradictory to the conventional theory or to the calculating standards. By exact

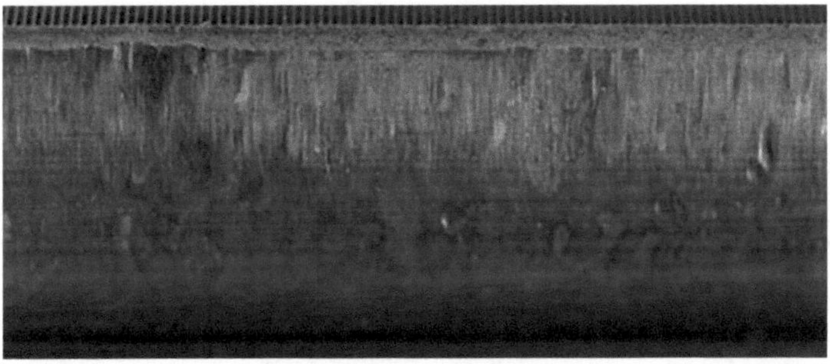

Fig. 32 Tooth flank that crushes foreign objects (Final drive gears of bulldozer)

Fig. 33 Drag and crush of a foreign object to make a spall

observation, adjusting the lighting of the damaged tooth flank, the trace of the dragging of a foreign object is observed from outside of the tooth contact region at the tooth fillet to the apex of the spalled-off surface failure of the tooth flank. Surely the intruding foreign object was crushed at the apex position of the flaking to be able to destroy the tooth flank. In this example, intrusion of foreign objects and their subsequent crushing is the suspect in this failure.

Figure 34 shows an assumed model of the crushing of foreign objects intruding between the meshing tooth flanks, especially at the state inside the approaching trochoidal interference zone. Intruded wear debris between meshing tooth flanks collides (1) against the inflection point of the tooth flank curvature between the sliding region and the cut-in region inside the approaching trochoidal interference zone, and (2) against the top of ridge of surface roughness, e.g., of the grinding mark, cf. Figs. 34, 35 and 39. Upon the crushing of foreign objects by tooth flanks at the inflection point of the tooth flank curvature, a subsurface crack often initiates, cf. Fig. 36, and proceeds upward toward the tooth tip inside the tooth, usually between the core and the case of the carburized and hardened gear tooth, cf. Figs. 2 and 3.

Figure 35 shows an observed pitting initiation resulting from these two cases of collision of foreign objects on the tooth flank to indicate the validity of the model in

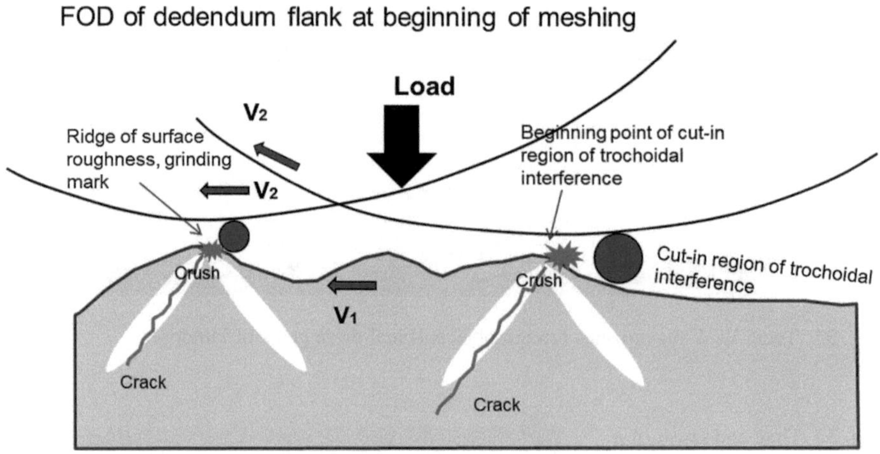

Fig. 34 Crushing of foreign object and damage of tooth flank

Fig. 34. Note the obscure figure of pitting aligned along the line of the inflection point of the tooth flank curvature between the sliding region and the cut-in region inside the approaching trochoidal interference zone. This in-line aligned state of

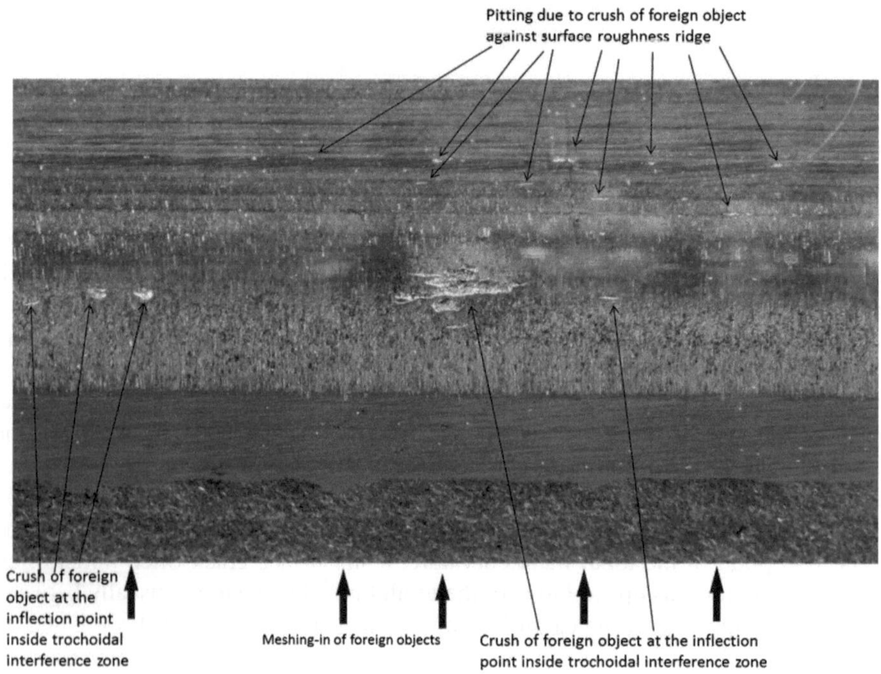

Fig. 35 Different kind of pitting initiated due to the crushing of wear debris

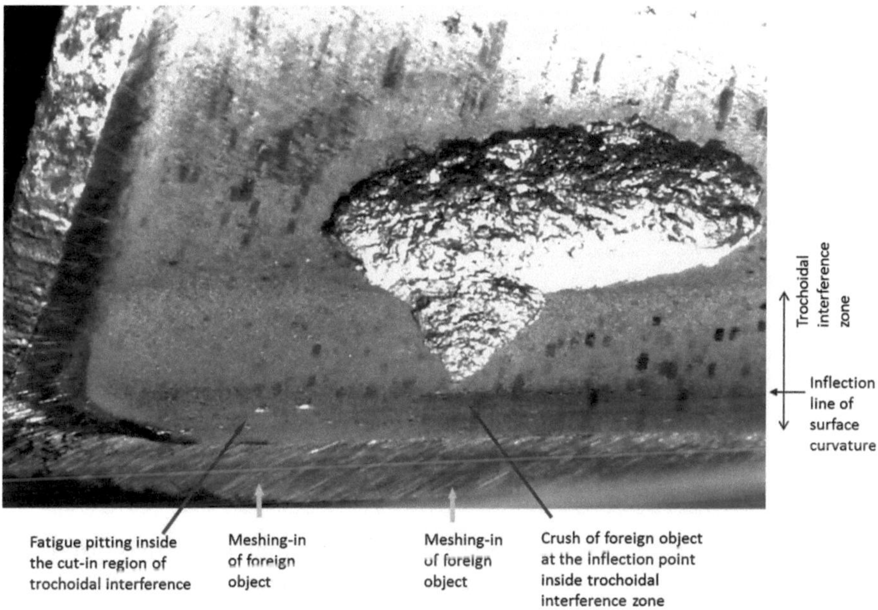

Fig. 36 State of pitting and spalling initiated due to the crushing of wear debris

pitting and the obscure figure of individual pitting suggests that all of this pitting must initiate through the collision of wear particles and is never fatigue pitting. Note also many small instances of pitting aligned along the top ridges of surface roughness of the grinding mark: Most of them are surely made by the collision of fine foreign objects. Of course, surface fatigue along some tops of surface roughness asperity could be the cause of micro-pitting.

When we see the tooth flank of power transmission gears after an endurance test, we often judge easily, giving it only a glance without exact detailed observation, and say it looks OK. But when the lighting is well adjusted to emphasize the state of surface irregularity, we can usually see considerable scratching, traces of the collision of fine wear debris over whole tooth flank. Figure 36 shows a typical failure observed on the tooth flank at the beginning part of tooth engagement of a strong helical driving pinion for a personal car drive. Black scratching (in reality, this is glare) on the tooth flank is are traces of wear debris having been dragged, rubbed, crushed and polished during the endurance test of this gear set.

Inside the cut-in region in the trochoidal interference zone, micro-pitting due to material fatigue is observed. The initiation of much spalling is due to the collision and crushing of intruding foreign objects at the inflection point of the tooth flank, and at the joining point of the cut-in and slipping regions inside the approaching trochoidal interference zone. You can observe the trace of dragging and rubbing of foreign objects from the outside of the lower limit of tooth contact to the apex of that spalling.

Fig. 37 Tooth tip chipping due to the crushing of wear debris

In Fig. 37, the pitting-like failure at the tooth tip on the upper right part of the photograph is perhaps not failure due to material fatigue. Downward from that failure, you can observe a trace of the rubbing and dragging of a foreign object on the tooth flank. Surely a foreign object of considerable size intruded between the engaging tooth flanks and rubbed the tooth flank at high pressure. At the moment that the foreign object passed over the tooth tip edge, the edge ruptured and has resulted in chipping.

Figure 38 shows another, clearer example of such edge rupture and chipping failure: Considerable chipping failure was observed on the tooth tip edge of the driving pinion of heavy duty spiral bevel gears for small ship propulsion. Downward from that chipping, a trace of the rubbing and dragging of a foreign object on the tooth flank is recognized. Without question, at the moment that the foreign object passed over the tooth tip edge, the edge ruptured and resulted in chipping. The forthcoming tip edge contact by recessing trochoidal interference accelerates the chipping failure.

Figure 39 shows the state of scratching after an endurance test [1] by an FZG rig using ground gears from 16MnCr5 steel and gear oil of VG320 with EP-additive of S: rich 0.53 %, P: rich 0.04 %. The state of the tooth flank in this picture looks very rough, but it is the usual state of a ground tooth flank of case carburized and hardened gears of common industrial production procedure: There remains considerable surface roughness, i.e., a ridge of grinding marks. During the endurance test, much wear debris, from fine to of a considerable size, intrudes between the engaging tooth flanks and the tooth flanks crush it. The state of the tooth flank shows that the fine wear debris looks to destroy the top of the ridges of the grinding mark, as if it had been bombarded. That is, most of the micro-pitting-like failure occurring at the projecting part of the surface roughness, the white ruptured part at

Fig. 38 Tip rupture, chipping, due to passing of a foreign object (wear debris)

Fig. 39 Scratching by soft wear debris against surface roughness asperities. Some black balls, perhaps melted steel, are also found there

Fig. 40 Craters, scratching and melted steel due to the crushing of intruding wear debris

the top ridge of the grinding marks, looks as if it were destroyed by the collision of wear debris and is perhaps not a fracture due to material fatigue. So-called micro-pitting perhaps includes both of those failures from different causes, i.e., those due to material fatigue and those due to the collision of fine wear debris against the ridge of the grinding mark. Near the grooves of rather strong scratching, some black balls, perhaps melted steel, are found. This means that the instantaneous temperature of the crushing surface on the tooth flank or of the wear debris at the collision reaches a temperature above that of the melting temperature of steel.

Figures 40 and 41 show very different outlooks of damaged tooth flanks [2, 3], one lubricated by ATF with EP-additive of S: rich 0.12 %, Ca: poor 0.005 %, P: const. 0.03 %, and one lubricated by ATF with EP-additive of S: poor 0.06 %, Ca: rich 0.01 %, P: const. 0.03 %. In this endurance test, we have put the same wear debris into the test gear box of the FZG rig. The wear debris used is taken from the contaminated lubricating oil of another heavily tested gear box. The amount and nature of wear debris added in the lubricant of these two tests was therefore the same. On the tested tooth flank, considerable scratching is recognized. At some position, such scratching destroys the ridge of the grinding mark through the collision of fine foreign objects.

The resulting outlook of the damaged tooth flank is quite different between the two cases shown in Figs. 40 and 41: in the case of Fig. 40, the crushing of large

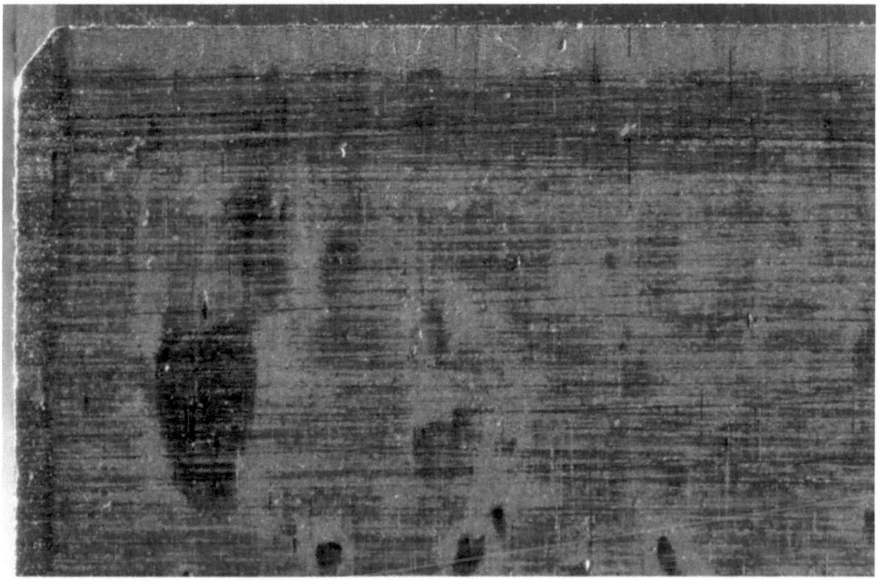

Fig. 41 Worn craters and evidence of melted steels on the basin of craters

foreign objects induces plastic deformation of the tooth flank in the forms of a crater and a horseshoe. It also induces cracks inside the tooth flank. We can see black oxidized steel at the side of deep scratching grooves and on the crater mountain. It shows steel was melted very locally.

In the case of Fig. 41, the size of the intruding foreign object seems to be very large, although the size of the wear debris put in the lubricant at the beginning of test is the same as in the case of Fig. 40. The abrasive wear of the crater mountain and the projecting part of the tooth flank, i.e., wear of surface roughness asperity, is very fine. On the bottom of the craters, very local melting of the tooth flank or melting of crushed wear debris is observed.

9 Positive Feedback System of Failure Development

Summing up much of the observation of tooth flanks of heavily operated gears, we feel the following impression becomes apparent: the progress of tooth flank wear as a function of operating time is described by the positive feedback system, as shown in Figs. 42 and 43.

Figure 42 describes the positive feedback system of gear failure development using names of commonly specified failure. During the life time of gears, the situation of tooth flanks is one of being under "progressive failure" or

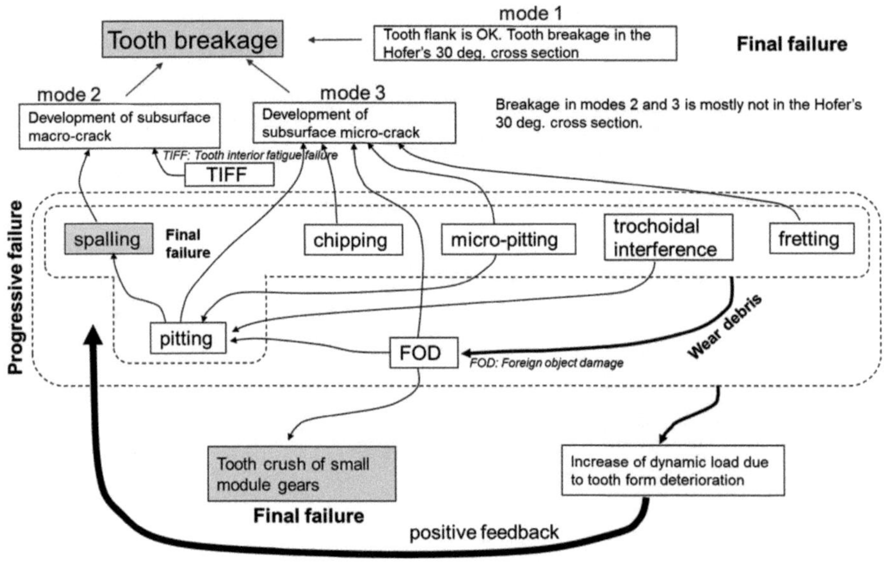

Fig. 42 Positive feedback system for the development of gear failure, progressive failure to final failure

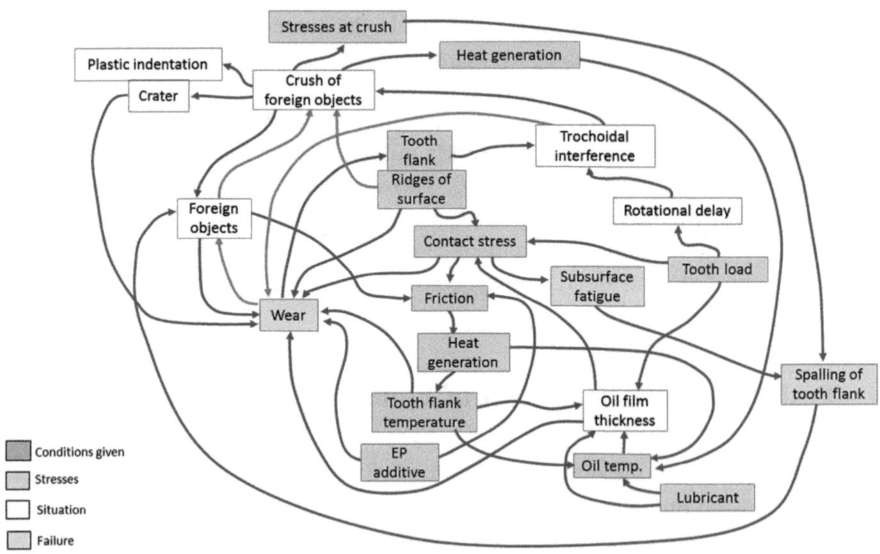

Fig. 43 Relation between the surface failure of a gear tooth flank and the influencing factors acting in a positive feedback system

"failure-in-progress", and this situation becomes worse due to the feedbacked dynamic loading and deterioration of gear accuracy. The final state of this gear life is tooth breakage, heavy spalling or the wearing out of teeth. Most gear life carries on under a state of failure-in-progress, and different kinds of failure modes end up integrating and changing their own magnitude of failure modes.

Figure 43 describes the positive feedback system of such gear failure development as the relation between failure names and influencing factors. During the life time of gears, this complex feedback system works to develop failure of the gear tooth flank. According to each practical case of gearbox usage in various fields, the strength of the feedback route is different. The fashion of failure development is the result of this system behavior, and it is different in practice on "case by case" basis.

In this complex feedback system of failure development, the intruding and crushing of foreign objects by engaging tooth flanks is a serious trigger of forthcoming fatal failure. Gear design and operation executed under the recognition of the importance of not producing wear debris is crucial for realizing long life and reliability in a gear unit. Lubricant, especially its containing EP additive, e.g., phosphorus, sulphur, calcium etc., has considerable influence on the generation and formation of characteristics of wear debris.

10 Conclusions

The geometrical analysis of the conjugate action of gears is the first and most fundamental step for realizing the gears' ability to transmit motion and power from one shaft to another. To infuse the operation of that gearing with reliability is the next step in making that gearing useful. That is the work beyond the conjugate theory of gear meshing. Much work has been carried out for predicting the load carrying capacity of gears, but the roles of tooth edge contact and the unavoidable crushing of wear debris in the development of gear failure have not been taken in consideration until now. Tooth edge contact at the tip and side ends plays a big role in considering the reliability of gearing, especially when high strength steel is employed as gear material. All gear failure develops as a behavior of the positive feedback system. The crushing of wear debris, mainly brought about by edge contact, also has significant influence on that. Gear tooth design considering those complex influences is essential for realizing a reliable gear unit.

Appendix

* Wording: "Rag-wear"

For convenience of discussion, I defined here the concept of following gear tooth flank damage as "Rag-wear":

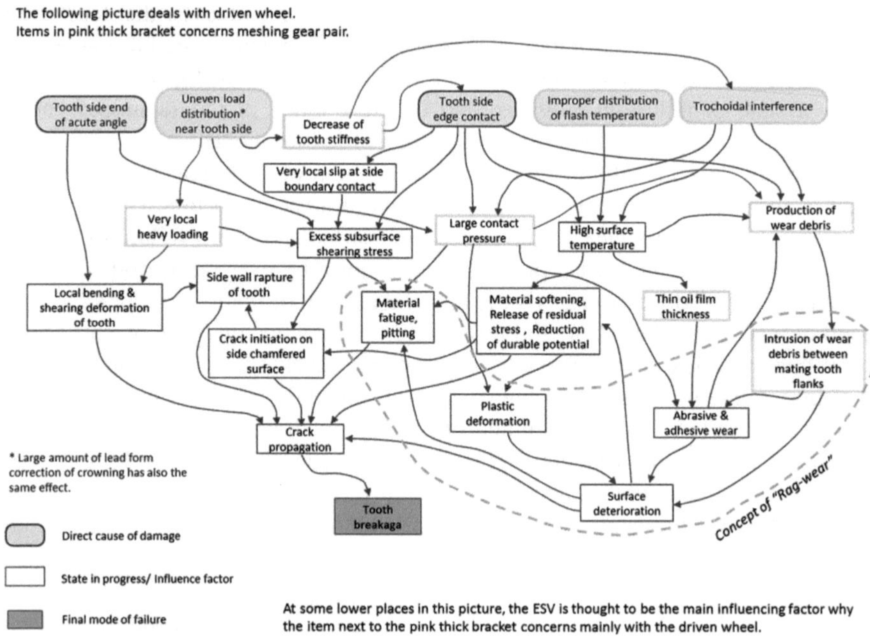

Fig. 44 One typical mode of tooth flank failure that depends on run of operational time

Under very high surface temperature, the thickness of lub.oil film becomes very thin and granular falling-off of surface material occurs, adhesive wear occurs, pitting occurs, crush of wear debris occurs, plastic deformation of that tooth flank occurs simultaneously. Such complex situation of gear tooth flank wear develops itself with run of operating time as positive feedback system of failure development. During the damage development, tooth flank wear and plastic deformation often covers pitted area and eliminates pitting.

The expression "Rag wear" indicates such state of gear tooth flank wear with one word (Fig. 44).

References

1. Kubo A (2013) Theory and practice of gearing. In: Proceedings of International Symposium, Izhevsk, Russia, January 2014, pp 50–55
2. Kubo A (2014) In: Proceedings of International Gear Conference, Lyon, France, August 2014, vol II, p 825
3. Ichihashi T, Fujita H, Matsumoto S (2013) Proceedings of International Conference Gears, Garching, Germany, October 2013, VDI-Berichte Nr.2199, p 841

Several Issues of Tooth Generating Process by Two-Parametric Families of Generating Lines

E. Trubachev

Abstract The paper describes methods of tooth generating by two-parametric families of generating lines (cutting tool edges) proposed for spiroid gears. Several general issues of tooth generating process by means of these families are considered.

Keywords Gears · Tooth generating process · Generating line

1 Introduction

It has become rather common to apply generating lines in the production of gearwheel teeth for spatial gears. Demonstrative examples here are the cases of cutting spiral teeth of bevel gearwheels by a *Cyclo palloid system* [1] or worm machining by a cutter. At the same time, less attention has been paid to its theoretical aspect as compared to application of generating surfaces. Synthesis of conjugated gearing based on applying the generating lines with one parameter of motion is considered in the works of Davydov [2] and Litvin [3]. As for investigation of surfaces formed by the two-parametric motion of generating lines, according to M. L. Erikhov, "…one can note their insufficient development which is probably related to the limited application of such surfaces as conjugated ones" [4]. Litvin [3] and (under his direction) Gulyaev [5] reduced generation of spiral type teeth of bevel gearwheels by a generating line (cutting edge of the cutter of the face cutter head) to an enveloping process of an artificially introduced generating surface. However, according to Erikhov [3], the possibility of applying such a technique requires clarification. Many experts have investigated the formation of fillet surfaces of teeth through a generating line, i.e., an edge of the generating surface. This question has been thoroughly studied by Babichev [6].

E. Trubachev (✉)
Kalashnikov Izhevsk State Technical University, Izhevsk, Russia
e-mail: truba@istu.ru

V. Goldfarb and N. Barmina (eds.), *Theory and Practice of Gearing and Transmissions*, Mechanisms and Machine Science 34,
DOI 10.1007/978-3-319-19740-1_6

The reason for the new comprehension of applying the generating lines for the author of the present paper was his proposal of three versions of applying the generating lines to produce teeth in spiroid gears within a short period of time. First of all, let us briefly consider the essence of these methods.

1.1 Formation of Spiroid Gearwheel Teeth by a Generating Line—Cutting Edge of the Fly Cutter

Modification of the tooth cutting method by a fly cutter, as proposed in [7], provides greater opportunities to control the localized contact pattern in a spiroid gear thanks to the additional variable parameter, while at the same time, it maintains the advantage of this method—tool simplicity. The very fact of a method of cutting spiroid and worm gearwheel teeth by fly cutters is not novel. It is applied mostly in single-piece production, and sometimes in short-run production, though there was a proposal to apply such a method for large-scale production of automobile gears that found no practical implementation. The classical version of this method (Fig. 1a) implies that the fly cutter rotating in coordination with the blank (parameter—angle of rotation $\varphi_0^{(1)}$, rotation velocity—$\omega_0^{(1)}$) is fed along the axis of its rotation (parameter—coordinate $z_0^{(2)}$, velocity—$\boldsymbol{v}_0^{(2)}$). Concurrently, the fly cutter makes an additional turning around its axis (to be more exact, the correspondent additional turning—$\varphi_2^{(2)}$—is imparted to the gearwheel). The result of the generating process does not differ much from the result of enveloping the helicoid generating surface. While being fed, the fly cutter is engaged in some kind of motion inside this surface and cutting edges are per se its generators (the helical line can be considered the directional one). Variation of the feed motion (Fig. 1b) by introduction of radial or vertical components or their combination (depending on the flat or bevel shape of the gearwheel to be cut) changes the situation: in this case, formation by generating lines is not equivalent to formation by the generating helicoid, at least due to the fact that the axis of tool rotation changes its position relative to the axis of the gearwheel to be cut.

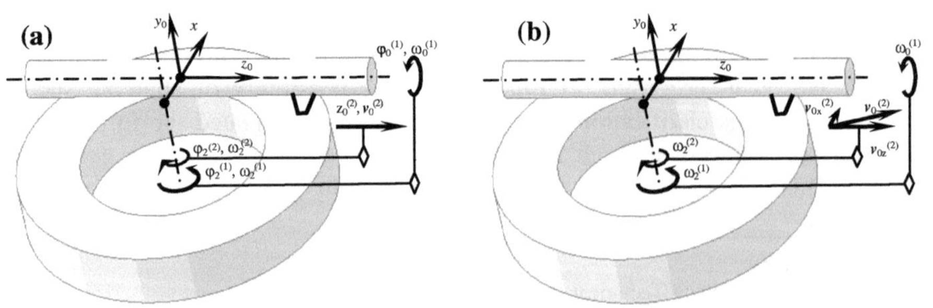

Fig. 1 Formation of spiroid gearwheel teeth by a generating line—cutting edge of the fly cutter

In this case, formation by a generating line is not reduced to formation by the unchangeable generating surface. The generated tooth surface should be considered to be formed by a two-parametric family of generating lines. Two independent parameters of the family are the angle $\varphi_0^{(1)}$ and the coordinate $z_0^{(2)}$. The corresponding change of coordinates x_0, y_0 within radial and vertical feeds and the additional turning $\varphi_2^{(2)}$ of the gearwheel to be cut are dependent on $z_0^{(2)}$.

For the stopped feed (fixed value of parameter $z_0^{(2)}$) during each turning of the cutter around the tool axis (within variation of the first parameter of the family of generating lines $\varphi_0^{(1)}$), there is the trace of cut in the gearwheel blank (Fig. 2a). The surface of each trace is the result of relative motion of the generatrix—the generating line (cutting edge of the cutter). Two families of distinct lines can be found on the surface of the trace, the lines being considered parametric—the family of lines l_{fi} coinciding with the generating line in its various positions (tangent vector $\boldsymbol{f}_2$) and the family of lines $l_{\varphi(1)j}$—trajectories of relative motion of points of generating lines (tangent vector $\boldsymbol{v}_{s2}^{(1)} = \boldsymbol{v}_{f2}^{(1)}—\boldsymbol{v}_2^{(1)}$ is the vector of relative velocity of the first of two independent motions) (Fig. 2b). The normal to the trace surface can be determined as the vector product of the pointed vectors:

$$\boldsymbol{n} = \boldsymbol{f}_2 \times \boldsymbol{v}_{s2}^{(1)} \tag{1}$$

When the second parameter of motion (($z_0^{(2)}$)—cutter feed along and across the tool axis) is changed, traces of cuts are shifted with respect to the blank, changing their shape a little. The surface generated by a fly cutter can be represented as the enveloping line of the family of such artificially introduced generating surfaces (Fig. 2c). In general, when changing the second parameter, the surface of the family is not moving as a whole but as each point separately, being thus deformed, unlike the usual unchangeable generating surfaces. Nevertheless, the kinematic condition to seek for the enveloping line of the family of artificially introduced and deformed surfaces is the same as for non-deformed generating surfaces, i.e., orthogonality of the normal $\boldsymbol{n}$ and vector $\boldsymbol{v}_{s2}^{(2)}$ of the relative velocity within the second of two independent motions:

$$\boldsymbol{n}\boldsymbol{v}_{s2}^{(2)} = 0 \tag{2}$$

Substitution of (1) into (2) gives the condition to seek for points of the formed tooth surface without artificial introduction of the generating surface:

$$\left[\boldsymbol{f}_2 \boldsymbol{v}_{s2}^{(1)} \boldsymbol{v}_{s2}^{(2)}\right] = 0 \tag{3}$$

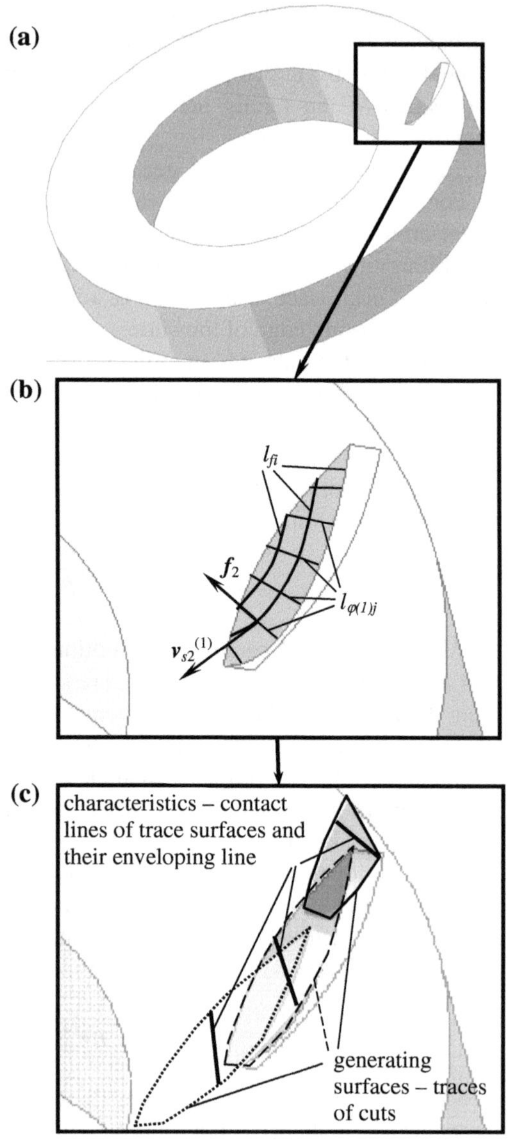

Fig. 2 Formation of spiroid gearwheel tooth root surfaces by a fly cutter

1.2 Formation of Spiroid Gearwheel Teeth by a Generating Line—Cutting Edge of the Cutter of the Flat Cutter Head

The drawback of the conventional technique of spiroid gearwheel machining is comparatively low production efficiency (by the way, this drawback becomes even greater when tooth-cutting by a fly cutter). First of all, it is because of low tool

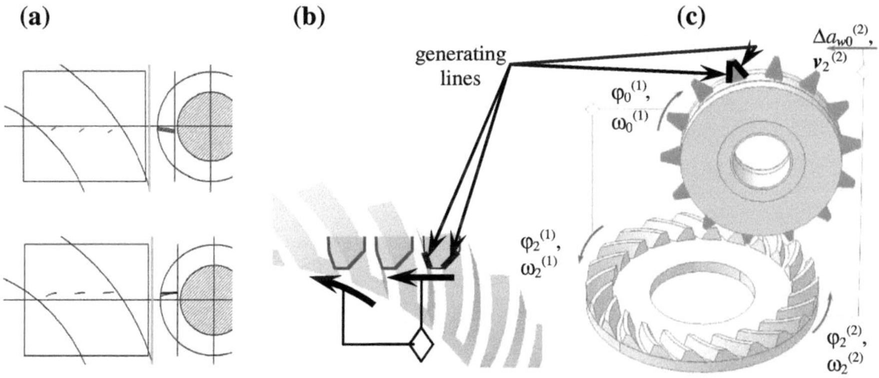

Fig. 3 Formation of spiroid gearwheel teeth by a generating line—cutting edge of the cutter of the face cutter head

rigidity, hard conditions of heat removal from the cutting zone and rather unfavorable conditions of chip flow. The desire to overcome this drawback motivated the development of a new scheme of gear machining [8] with application of a flat cutter head (Fig. 3). Initial premises of proposing this scheme are the elongated zone of spiroid mesh along the worm axis (Fig. 3a) and the constant axial pitch of the worm. Taking this into account, one can obtain a good approximation of the conjugated surface of gearwheel teeth when the generating line is moving uniformly, for example, the cutting edge of the cutter coinciding with one of the lines of conjugated contact is moving along the worm axis and the gearwheel to be cut is rotated proportionally (Fig. 3b). To implement this scheme, it is necessary to provide reciprocal motion of the tool and single indexing per tooth. The pointed tool motion in the space related to the blank provides one-parametric family of generating lines, solely generating the tooth flank.

The reciprocal motion of one cutter inconvenient to be implemented can be replaced by a rotary motion of several cutters mounted in a cutter head, changing each other and thus providing the continuous tooth indexing. Relation of velocities of points when rotating around axes of the cut gearwheel and cutter head is that the vector of relative velocity becomes almost parallel to the head axis within cutting, thus allowing for placement of the cutters on its face surface. Propagation of cutting along the whole tooth length is carried out by tool feed along the rotation plane at the concurrent additional rotation of the gearwheel to be cut (Fig. 3c).

Similar to the case of a fly cutter, the tooth surface generated by a flat cutter head should be considered to be formed within the two-parametric relative motion of the generating line. Two independent parameters of motion are the angle $\varphi_0^{(1)}$ of cutter head turning and the centre distance $a_{w0}^{(2)}$ in machine-tool gearing. In [9], modification of this method has been described, when the cutter head is fed not along the straight line but along the circumference arc (Fig. 4) for better contact localization. In this case, the second independent parameter should be the angle $\gamma_0^{(2)}$

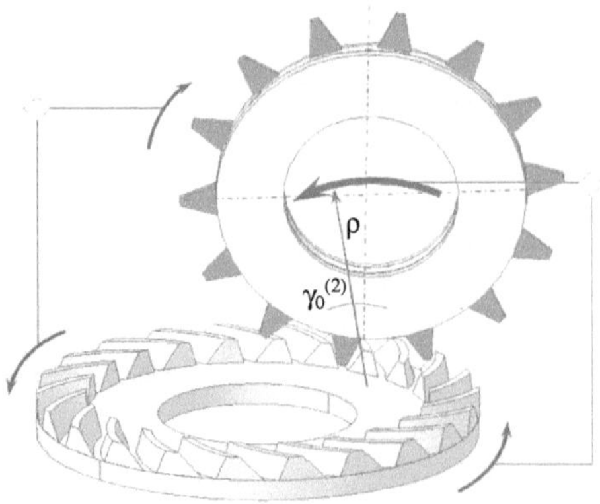

Fig. 4 Modification of the cutting scheme—feed along the circumference arc

characterizing the position of the cutter head on the pointed arc. All discussions of the sequence of tooth generation given above for the case of generation by a fly cutter are also true for generation by a cutter head, which is why points of the surface to be formed can also be determined by the relation (3).

1.3 Formation of Multithread Worms by a Generating Line—Cutting Edge of the Face Cutter Head

Formation of helical surfaces is the area that has been developed rather thoroughly and deeply (evidently due to its prevalence in manufacturing tasks). However, we dare to think that not all reserves for advancement have been revealed here, especially with regard to the formation of multi-thread worms. In order to provide high productive efficiency of cutting, the manufacturers propose formation by a disk tool (hob or whirling cutter head) at a single indexing per thread (see, for example, [10]). The scheme of formation by a face cutter head proposed in [11] possesses a number of advantages, in our opinion:

- continuous indexing per thread is provided;
- widespread equipment (comparatively low cost CNC lathe) and tool (face cutter head) are applied;
- rather simple machine-tool setting is provided;
- simple re-adjustment to the formation of worms with a low number of threads by means of a generating cone or torus is practically possible with the same manufacturing equipment.

The cutting scheme is shown in Fig. 5. The cutter head and the worm to be generated are rotated consistently with angular velocities $\omega_0^{(1)}$ and $\omega_1^{(1)}$, correspondingly, with the gear ratio i_{01} equal to the ratio of the number of threads to be cut $z_{(1)}$ to the number of cutters $z_{(0)}$ (or pairs of cutter, where one cutter is for the left surfaces of threads, and the other is for the right surfaces) of the cutter head. When

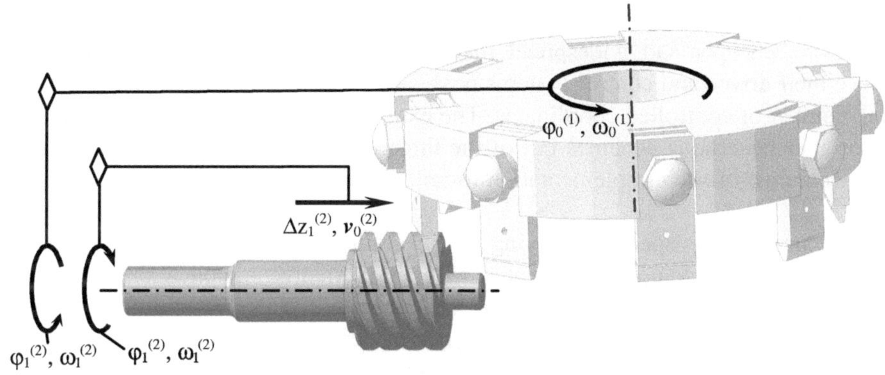

Fig. 5 Formation of multithread worms by a generating line—cutting edge of the face cutter head

this requirement is met, each next cutter (or the next pair) enters each next root between threads. Moreover, the feed of the cutter head is carried out at the velocity $\boldsymbol{v}_0^{(2)}$ along the worm axis, which agrees with the additional proportional revolution (angular velocity $\omega_1^{(2)}$) of the latter.

In this case, thread flanks are also formed within the two-parametric motion of generating lines—cutting edges of the head cutters. The first parameter is the angle $\varphi_0^{(1)}$ of the cutter head turning, and the second one is the coordinate $z_1^{(2)}$ characterizing its arrangement along the worm axis. In order to calculate coordinates of the generated surface points, the relation similar to (3) can also be used in this case (subscripts are enumerated in the formula according to worm generation—the first element of the gear):

$$\left[f_{01} \boldsymbol{v}_{s1}^{(1)} \boldsymbol{v}_{s1}^{(2)} \right] = 0 \tag{4}$$

1.4 Practical Implementation and Prospects

As for all described schemes, one of independent motions (the main cutting motion) is rotation of the tool consistent with rotation of the gearwheel blank, and the other is the tool feed also consistent with the additional rotation of the blank. Such a combination is worth considering mainly due to the simplicity of providing rotational motions and continuous indexing per tooth. Let us agree that, from herein, we are discussing exactly this method of formation of two-parametric families of generating lines within tooth generation.

The pointed combination of motions is normally implemented in gear-cutting equipment: universal gear hobbers and machine-tools for cutting bevel and hypoid gearwheels. One consequence of a breakthrough increase in CNC machine-tool possibilities was implementation of gear machining functions at lathe type machining centers. However, it is also possible for comparatively simple and, therefore, cheaper and widespread machine-tools after their slight modification, since their drives and control systems possess necessary possibilities for increasing the number of controlled coordinates. The example is a practical implementation of two latter generating schemes out of the three enumerated above. Figure 6 shows the technique of such implementation for an NC thread-cutting lathe with a straight frame.

The practical prospect of applying the generating line to form gearwheel teeth is almost always evaluated as compared with an alternative—application of generating surfaces. Advantages of applying the generating lines under different manufacturing conditions and for different types of gear can be revealed separately or altogether. The main advantages are as follows:

- greater possibilities for controlling the localized contact;
- better features of the localized contact;
- higher production efficiency and lower dynamics due to continuous indexing, increased tool rigidity and more favorable conditions for heat removal from the cutting zone, chip flow and removal;
- simplicity of tool manufacture and operation and efficiency within a wide range of manufacturing conditions—from single-piece to mass production;
- simple implementation for widespread and comparatively low cost equipment.

One of the technique advancement reserves within the considered area is the proposal and development of competitive generating schemes. These schemes can involve both well-known and widely applied face cutter heads (Fig. 7a) and cutter heads which have not yet found wide application: flat (Fig. 7b), disk (Fig. 7c), and cylindrical (Fig. 7d).

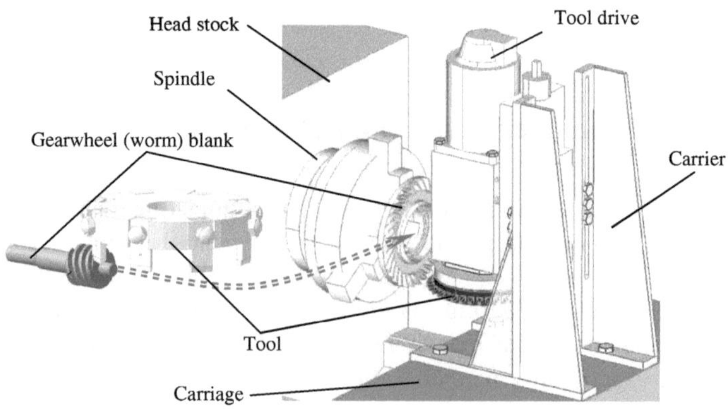

Fig. 6 Implementation of generating schemes for gearwheel teeth and worm threads for an NC thread-cutting lathe

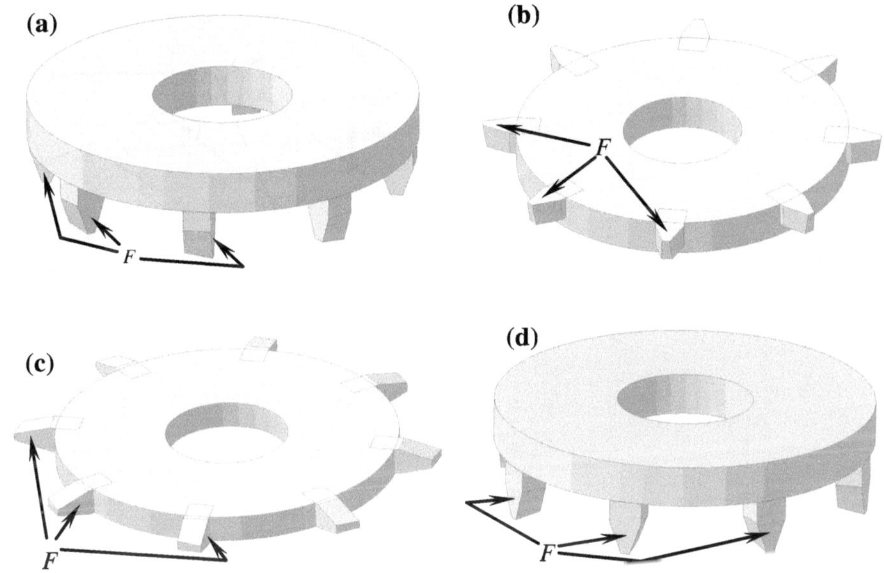

Fig. 7 Versions of cutter heads implementing the rotation of the generating line (F is the front face of the cutter) **a** face, **b** falt, **c** disk, **d** cylindrical

1.5 Choice of Generating Parameters

This issue can be related to the synthesis process of both conjugated gearing and gearing with localized contact. In many cases, the design of a conjugated gear precedes the synthesis of localized contact; the synthesis can be carried out here by changing the parameters of generating the conjugated tooth surfaces. When designing a spiroid gear as the object of the above-described generating schemes, we applied the following sequence of geometry analysis [12]:

- choice of gear scheme—axes arrangement, gear ratio and pitch surfaces (Fig. 8);
- analysis of conjugated geometry of worm threads and gearwheel teeth;
- choice of generating parameters according to local conditions in designed points [13];
- non-local evaluations of localized contact [14, 15].

Worm thread generating parameters are determined at step *ii*. As for the above-described new generating scheme, it is done according to the condition of collinearity of vectors $\boldsymbol{v}_{s1}^{(1)}$ and $\boldsymbol{v}_{s1}^{(2)}$ and the vector $\boldsymbol{w}_1$ (tangent to the worm helical line) at the point of tangency of the worm reference cylinder and cutter head reference plane:

Fig. 8 Gear scheme

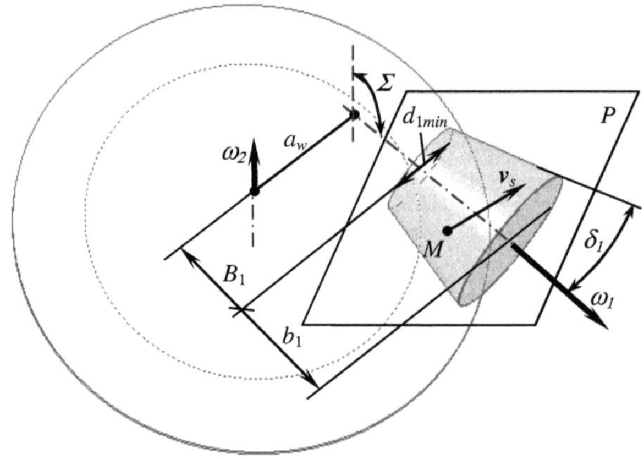

$$\boldsymbol{w}_1 \times \boldsymbol{v}_{s1}^{(1)} = 0, \tag{5}$$

$$\boldsymbol{w}_1 \times \boldsymbol{v}_{s1}^{(2)} = 0. \tag{6}$$

In many aspects, the key parameter for the gearwheel that allows us to look into the chosen generating parameters is step *iii*. In turn, the main local conditions at this step are first-order conjugation conditions [16] providing tangency of modified surfaces and nominal gear ratio at designed points. Within the sequence *i*...*iv*, the conjugated surfaces are determined completely at step *ii*. This allows for choosing the designed points at these surfaces and stating the following conditions of choosing the generating parameters:

- orthogonality of the contact normal $\boldsymbol{n}$ in the conjugated gearing and vectors $\boldsymbol{v}_{s2}^{(1)}$ and $\boldsymbol{v}_{s2}^{(2)}$:

$$\boldsymbol{n}\boldsymbol{v}_{s2}^{(1)} = 0; \tag{7}$$

$$\boldsymbol{n}\boldsymbol{v}_{s2}^{(2)} = 0. \tag{8}$$

- orthogonality of the normal $\boldsymbol{n}$ and the vector $\boldsymbol{f}_{02}$ tangent to the generating line:

$$\boldsymbol{n}\boldsymbol{f}_{02} = 0 \tag{9}$$

The given calculation sequence is related to the spiroid gear design process, and one can say that it is natural for a worm-type gear generated according to the second Olivier principle. However, conditions are used in calculation which are invariant with respect to the type of gear, allowing us to apply this sequence for other types of gear by assigning the generating parameters step by step. For example, parameters for the pinion are first assigned, then for the gearwheel, thus adjusting its geometry to the conjugated surfaces of the gearwheel [14]. One can state an even more common condition for the choice of parameters without account of, first, the preliminary assignment of at least one conjugated surface and, second, specific gear type and generating scheme. Here we accept that:

- gear scheme is known at the beginning of synthesis (the step i is made—see above)—gear ratio, parameters of axes arrangement and pitch surfaces are known;
- we consider the point of conjugated contact;
- we take aside the possible tooth interference, that is, we consider the first-order conjugation conditions [16].

The tangent plane to each conjugated surface is determined by any pair from a group of three vectors, non-collinear in general—$\boldsymbol{f}_{1(2)}$, $\boldsymbol{v}_{s1(2)}^{(1)}$, $\boldsymbol{v}_{s1(2)}^{(2)}$. Tangent planes of both surfaces engaged in a gear must coincide. Also, the common tangent plane Π should involve vectors $\boldsymbol{f}_{1(2)}$, $\boldsymbol{v}_{s1(2)}^{(1)}$, $\boldsymbol{v}_{s1(2)}^{(2)}$ and, additionally, the vector $\boldsymbol{v}_s$ of the relative velocity in the gear:

$$\left(\boldsymbol{f}_1, \boldsymbol{v}_{s1}^{(1)}, \boldsymbol{v}_{s1}^{(2)}, \boldsymbol{f}_2, \boldsymbol{v}_{s2}^{(1)}, \boldsymbol{v}_{s2}^{(2)}, \boldsymbol{v}_s\right) \in \Pi \tag{10}$$

The condition (10) is principal to provide the conjugation of surfaces, each of them being formed by a two-parametric family of generating lines. If it is fulfilled along the whole length of the operating segment of the line of meshing, the conjugated gearing is generated. As for localized contact synthesis, its fulfillment is required within the local vicinity of the point M taken as the designed one.

The condition (10) implies much freedom in the choice of direction of its component vectors. Let us consider additional conditions, as was agreed, not breaking the generality: they are unrelated both to the sequence of generated surface calculation and to specific gear types and generating schemes. As usual, it is reasonable to choose the point M of the pitch surface tangency (Fig. 8) as the one designed for the synthesis of machine-tool gearing and at the first approximation to place vectors $\boldsymbol{v}_{s1(2)}^{(1)}$, $\boldsymbol{v}_{s1(2)}^{(2)}$ of the point M of generating line velocities with respect to each of the elements on the designed plane P which is tangent to reference surfaces and has the normal vector $\boldsymbol{N}$. Therefore, the first additional condition can be written as follows:

$$\begin{aligned} \boldsymbol{N}\boldsymbol{v}_{s1(2)}^{(1)} &= 0, \\ \boldsymbol{N}\boldsymbol{v}_{s1(2)}^{(2)} &= 0. \end{aligned} \tag{11}$$

In the widespread case when each of the elements performs rotary motion around its axis while generating, relation (11) can be written in the other form, that is, in terms of absolute velocities $\boldsymbol{v}_{f1(f2)}^{(1)}$ and $\boldsymbol{v}_{f1(f2)}^{(2)}$ of the point M with regard to independent motions of generating lines. The pointed velocities will automatically be located on this plane (vectors $\boldsymbol{v}_1$ and $\boldsymbol{v}_2$ of elements velocities in rotary motions are located on plane P):

$$\begin{aligned} \boldsymbol{N}\boldsymbol{v}_{f1(f2)}^{(1)} &= 0, \\ \boldsymbol{N}\boldsymbol{v}_{f1(f2)}^{(2)} &= 0. \end{aligned} \tag{12}$$

Even more rigid requirements can be stated according to the mutual tangency of tooth lines at reference surfaces at the designed point M. This is fulfilled by matching the direction components of (10)—four vectors of relative velocities within independent motions of generating lines—$\boldsymbol{v}_{s1}^{(1)}$, $\boldsymbol{v}_{s1}^{(2)}$ $\boldsymbol{v}_{s2}^{(1)}$, $\boldsymbol{v}_{s2}^{(2)}$. For gears with parallel and crossed axes, the choice of tooth line direction is arbitrary to a great extent (at the point of reference surface tangency, the relative velocity in the gear is equal to 0). For hyperboloid gears, the vector tangent to tooth lines (and vectors $\boldsymbol{v}_{s1}^{(1)}$, $\boldsymbol{v}_{s1}^{(2)}$ $\boldsymbol{v}_{s2}^{(1)}$, $\boldsymbol{v}_{s2}^{(2)}$) must be collinear to the vector $\boldsymbol{v}_s$ of the relative velocity in the gear. Therefore:

$$\boldsymbol{v}_{s1}^{(1)} // \boldsymbol{v}_{s1}^{(2)} // \boldsymbol{v}_{s2}^{(1)} // \boldsymbol{v}_{s2}^{(2)} // \boldsymbol{v}_s \tag{13}$$

This requirement can be fulfilled at point M by changing the direction of absolute velocities $\boldsymbol{v}_{f1(f2)}^{(1)}$ and $\boldsymbol{v}_{f1(f2)}^{(2)}$ on plane P and changing the relation between values of these velocities and velocities of elements $\boldsymbol{v}_{1(2)}^{(1)}$ $\boldsymbol{v}_{1(2)}^{(2)}$ within tooth generation.

Meeting requirement (13) at point M allows for freedom in the choice of direction of one of the generating lines (one of vectors $\boldsymbol{f}_1$ or $\boldsymbol{f}_2$), and the direction of the second of these vectors must be chosen according to condition (10).

1.6 Pairs of Remarkable Straight Lines

In order to decide how to choose the contact zone (position of the designed point with respect to elements axes) and the arrangement of generating lines of rotation axes with respect to gearwheels axes, one can use pairs of remarkable straight lines of stationary space.

First of all, let us digress from the subject of the present paper and turn to the concept of axes of meshing in order to draw the analogy. As is known [3, 17, 18], for a hyperboloid gear with one of its elements representing a helicoid, a pair of axes of meshing exists (can exist, strictly speaking) in a stationary space, being straight lines through which all normal contact of the gear are passing. One of the important features of these straight lines is the fact that they comprise points where the vector of the relative velocity in a gear is tangent to helical lines of the helicoid element. This feature is used in problems of gearing synthesis, both conjugated [18, 19] and with localized contact [20]. This feature implies in its essence that axes of meshing represent a locus of stationary space where vectors of two helical vector fields of this space become collinear. The two fields are the field of relative velocities and the field of tangent vectors of helical lines of a helicoid generated in rotation of the latter. Let us give our attention to two more circumstances [3, 17]:

- axes of these fields (the axis of the helicoid element and the axis of instant helical motion in the gear) are crossed;
- axes of meshing belong to the plane perpendicular to the interaxial line of two helical fields, intersecting this line.

Turning back to the subject of the present paper, let us consider the gearing where gear elements and generating lines are involved simultaneously. Physically, simultaneous engagement is certainly impossible, but in each phase of engagement, it can be imagined for the corresponding contact point (including, of course, the designed point *M*) where the engagement is therefore taking place in the gear simultaneously with two types of machine-tool gearing to generate the pinion and gearwheel teeth. Figure 9a shows the arrangement of axes for this engagement at

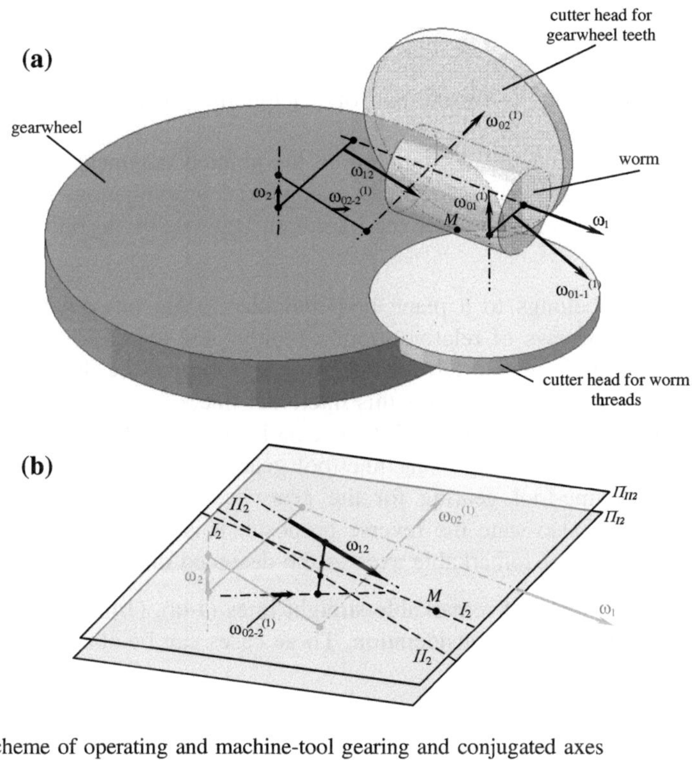

Fig. 9 Scheme of operating and machine-tool gearing and conjugated axes

the designed point M for the above-considered second and third generating method, for the gearwheel and worm, correspondingly.

Let us fix the values of one of the parameters in each machine-tool gearing, namely that parameter which, in practice, corresponds to the tool feed. Therefore, three pairs of matching rotations are performed in a stationary space: both gearwheels with respect to each other and each gearwheel with respect to the generating lines. In the general case, axes of instant relative motions in these engagements are crossed, thus generating three helical vector fields of stationary space. Figure 9 shows the directing vectors of these axes: $\boldsymbol{\omega}_{12}$, $\boldsymbol{\omega}^{(1)}_{01\text{-}1}$ and $\boldsymbol{\omega}^{(1)}_{02\text{-}2}$.

Regarding the above-described similar situation with straight lines—axes of meshing, one can state that, in each pair composed of these fields and, correspondingly, in each pair of gearing ("operating gearing—machine-tool gearing for the pinion", "operating gearing—machine-tool gearing for the gearwheel" and, finally, the pair of machine-tool gearing), pairs of remarkable straight lines exist (can exist) with the points at which vectors of relative velocities are collinear, namely:

$$\boldsymbol{v}^{(1)}_{s2}//\boldsymbol{v}_s\text{—for the pair "operating gearing—machine-tool gearing for the pinion";} \tag{14a}$$

$$\boldsymbol{v}^{(1)}_{s2}//\boldsymbol{v}_s\text{—for the pair "operating gearing—machine-tool gearing for the gearwheel";} \tag{14b}$$

$$\boldsymbol{v}^{(1)}_{s1}//\boldsymbol{v}^{(1)}_{s2}\text{–for the pair of machine-tool gearing.} \tag{14c}$$

Let us put aside the detailed grounding of the pointed statement with derivation and analysis of corresponding formulas as a special investigation. In addition to (14a, 14b and 14c), let us note here the important features, in our opinion, of these straight lines:

- each of them belongs to a plane perpendicular to the interaxial line of two corresponding screws of relative motion (conjugated axes I_2-I_2 and II_2-II_2 on planes $\Pi_{\text{II}2}$ and $\Pi_{\text{I}2}$ in Fig. 9b for operating gearing and machine-tool gearing for the gearwheel) and intersects this interaxial line;
- when conditions (13) are fulfilled, two remarkable straight lines corresponding to pairs "operating gearing—machine-tool gearing for the pinion", "operating gearing—machine-tool gearing for the gearwheel" intersect at the designed point M (let us also state the reverse proposal: it is convenient to accept the intersection point of remarkable axes as the designed one).

Cases of coincidence of remarkable straight lines (14a), (14b) and (14c) are of interest and require special consideration. These cases can be obtained by varying gear ratios in the considered gearing and the arrangement of rotation axes, including the arrangement of generating lines with respect to the tool axis (Fig. 7).

In order to avoid misunderstanding, let us note once again both the common and uncommon features of axes of meshing and considered remarkable straight lines. Along both of them, there is conjugation of helical vector fields with skew axes, therefore it is reasonable to call them the conjugation lines of screws. However, one can talk about conjugation of helical fields of relative velocities in operating and machine-tool gearing independently of the shape of tooth surface (helicoid or non-helicoid, and only when considering at least two types of gearing simultaneously and in common). If both lines of conjugation of relative velocities (remarkable straight lines under discussion) and axes of meshing are present in this consideration, generally speaking, they do not coincide. Note also that the pointed remarkable straight lines can be determined and their features also used for the case of tooth formation by means of rotational generating surfaces.

1.7 Are Parameters Independent While Tooth Machining? Can the Relation Between Them Be Changed Arbitrarily in Real Tooth Cutting?

This issue has two aspects. On the one hand, the ratio of velocities at which the two-parametric family of generating lines was formed is not significant. To demonstrate, let us change the main motion and feed motion for any of the examples given above. In the first example, the main (more rapid) motion is the fly cutter drawing along the tool axis (variation of parameter $z_0^{(2)}$), concurrent with the additional turning of the gearwheel blank (variation of $\varphi_2^{(2)}$)—Fig. 1. This motion will be carried out at the stopped rotation of the cutter (fixed value of parameter $\varphi_0^{(1)}$) for better visualization. As a result, the spiral type trace of cut will appear in

Fig. 10 Alternative version of generating the surfaces of the spiroid gearwheel tooth root by a fly cutter

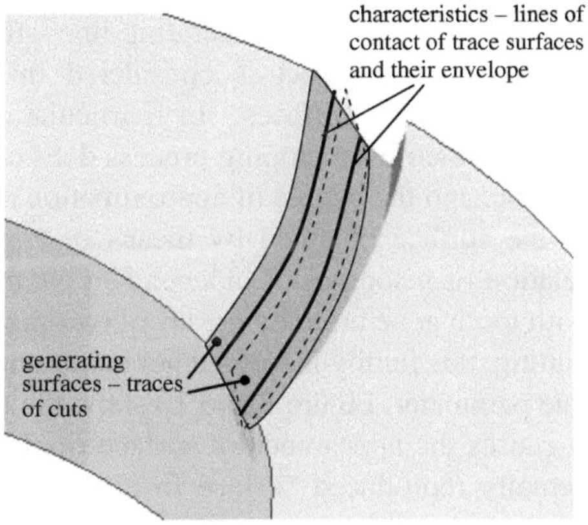

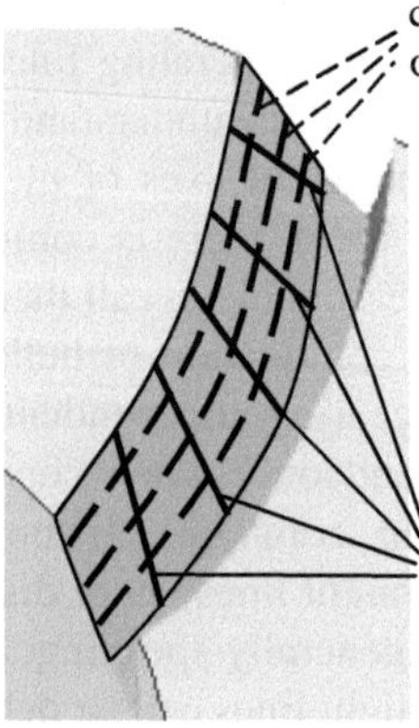

Fig. 11 Net of parametric characteristic lines on the tooth flank

the blank solid (longitudinal lines obtained at face sections are elongated or shortened involutes)—compare Fig. 2 and Fig. 10.

A similar (somewhat different) trace of cut is formed if the cutter is rotated at some angle ($\varphi_0^{(1)}$), while the cut gearwheel is concurrently rotated (at the angle $\varphi_2^{(1)}$) and the drawing is repeated. If this procedure is repeated with a small pitch of variation of parameter $\varphi_0^{(1)}$, the family of spiral type traces generated in space of the blank will have an envelope—it will actually be formed by a two-parametric family of generating lines. Statements made above for the usual relation of velocities of independent motion remains the same. As a result, the same relation (3) will be obtained to determine the coordinates of points of the generated surface. The same effect can be obtained by introducing any other relation between velocities of variation of parameters $\varphi_0^{(1)}$ and $z_0^{(2)}$, implying the independence of these parameters. The ability to change the sequence of tooth generation without detriment to the geometry allows for determining the net of lines on the generated surface which are, generally speaking, not tangent and not coincident with each other. These lines are the characteristic line for artificially introduced generating surfaces (traces of cut) and their one-parameter envelopes. Figure 11 shows the arrangement of characteristic lines for two considered sequences of generation of spiroid gearwheel tooth surface by means of a generating line—the cutting edge of the fly cutter.

The second aspect is considered in detail by Erikhov [4] for the case of two-parametric surfaces. In particular, he stated that, at real cutting, the two-parametric enveloping process does not exist. Both parameters are the function of time, and the degree of approximation of a surface actually reproduced in cutting to the surface obtained by means of two-parametric enveloping depends on the relation of velocities of independent parameter variation. A similar situation exists with tooth generation by means of two-parametric family of generating lines. In real cutting, this family is formed not continuously, but discretely with respect to at least one parameter. Figure 2 and 10 show the exaggerated process of family generating. It causes the appearance of surface rippling (faceting), that is, the difference of an actually reproduced surface from the designed one according to (3) or (4). The height of this rippling depends on the value of tool feed $\Delta z_0^{(2)}$ per one turning of the blank as for the sequence of generation shown in Fig. 2, or the value of transposition $\Delta\varphi_0^{(1)}$ carried out per one double pass of the tool, as for the sequence shown in Fig. 10. Turning back to the question stated in the heading of this section, the tooth geometry actually reproduced by two-parametric motion of the generating line in cutting depends again on the relation of velocities of independent parameter variation.

1.8 Is It Possible to Generate Teeth by Means of the Family of Generating Lines with a Number of Independent Parameters Greater Than 2?

It was stated above that variation of one of the parameters of the family of generating lines (to be more specific, the family related to the main cutting motion) results in some surface in the space of a blank, and variation of the second one results in the single-parametric family of such surfaces, possessing the envelope. For such a representation, it is not forbidden to form the two-parametric family of the pointed surfaces, that is, for the generating line, it is their three-parametric family. The example here can be a practically implementable scheme of cutting a usual involute spur gearwheel by means of the fly cutter (Fig. 12). Generating lines (cutting edges of the fly cutter) are accepted to slide along the surface of the involute helicoid while drawing along the tool axis (parameter $z_0^{(2)}$). If the gearwheel rotation is concurrent with the rotation of the mandrel with the cutter (rotation of the pointed imaginary helicoid—parameter $\varphi_0^{(1)}$), the two-parametric family of generating lines is formed to generate the tooth flank of the corresponding

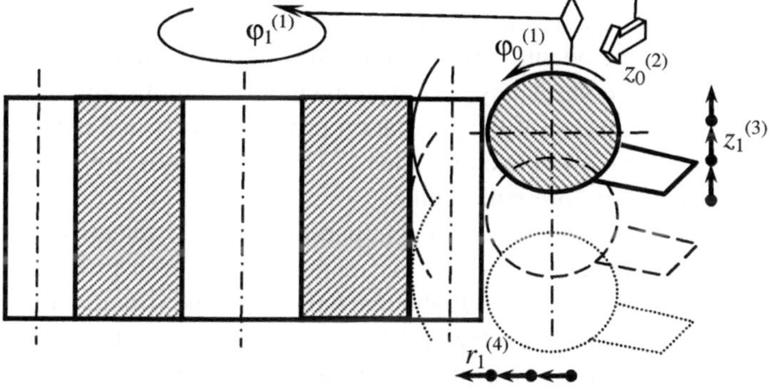

Fig. 12 Generation scheme for involute spur gearwheel teeth by means of four-parametric family of generating lines—cutting edges of the fly cutter

worm gearwheel. After each drawing, the mandrel cutter can be rearranged along the axes of the gearwheel to be cut (varying the parameter $z_1^{(3)}$) and the three-parametric family of generating lines can be thus obtained to form the involute surface. If we follow Babichev [21] and cut in the blank not to the full height of the tooth, but consequently by some height parts increased from one step to the other (varying the parameter $r_1^{(4)}$), then we obtain the four-parametric family of generating lines. Therefore, the answer to the section heading is yes, it is possible. However, this question has its practical aspect: in practice, the number of independent parameters does not usually exceed two, one of them corresponding to the main cutting motion and the other to the tool feed. It is possible to organize the additional feed as in the given example (by tool re-adjustment), but is there solid argumentation for this? The negative effect is obvious in this case, being a considerable increase in machining time. Therefore, the necessity of practical implementation of application of the family of generating lines with three or more independent parameters for tooth generation is of great doubt, except perhaps for certain specific manufacturing cases.

1.9 About Terminology

1. How should the surface generated by the two-parametric family of generating lines be called? Application of the term "envelope of generating lines" is doubtful here: the envelope of the family of lines is the line itself. The single-parametric family of spatial curves generates, generally speaking, the surface rather than the envelope. Lines of the family do that by themselves and the surface is per se comprised of these lines. In special cases, such families may have envelopes [22]. For example, a single-parametric family of characteristic lines on the surface of the envelope to the family of surfaces may have an enveloping line—the edge of regression. Information about two-parametric families of lines in the three-dimensional space is rather rare in commonly available references. Favar [22] and Erikhov [4] after him called the surface generated by such a family (to be more exact, the surface tangent to all curves of the family in some vicinity of the chosen values of parameters) the "focal surface of curves congruency". Is it convenient to apply this term? Does it reflect the representation of the generation process clearly and comprehensively for experts as the term "envelope" does? These questions are, in our opinion, ambiguous, and therefore still open.
2. The second terminological question is particular and it relates to the gear for which the above-described generation methods have been proposed. Can this gear be considered "spiroid"? On the one hand, both the structure of design and, in many respects, the geometry and kinematic features of the gear have been kept. On the other hand, application of general techniques of calculation of both the conjugated gearing and the gearing with localized contact, modification of motion schemes carried out in tooth machining, change the properties of gears

and make the hyperboloid gear comprising the pinion with helical teeth (spiroid gear) the particular, though very important, case. Therefore, there are reasons to re-name the hyperboloid gear with elements generated by a two-parametric family of generating lines. Practice of development of new generation methods must provide the final answer to this still unsolved problem.

2 Conclusion

Only several issues of application of two-parametric families of generating lines for tooth generation have been considered in this paper. These issues have not been completely developed. Some questions remained beyond the scope of this paper: undercut in machine-tool gearing, gearwheel teeth conjugation, one of the gearwheels being formed by a single-parametric family of generating lines or by a generating surface, the other being formed by a two-parametric family of lines, and many others. Nevertheless, the author of this paper would be glad to have aroused the discussion and further development of the considered theme.

References

1. Stadtfeld HJ (1995) Gleason Bevel gear technology. In: Manufacturing, inspections and optimization: collected publications 1994/95. The Gleason Works, Rochester, 202 p
2. Davydov Ya S (1965) Generation of conjugation surfaces in gears by means of two curves. In: Analysis and synthesis of mechanisms and theory of gearing. Nauka, Moscow, pp 12–23 (in Russian)
3. Litvin FL (1968) Theory of gearing. Nauka, Moscow, 584 p (in Russian)
4. Erikhov ML (1972) Principles of systematization, analysis methods and questions of synthesis of gear schemes DSc (Engineering) Thesis. Khabarovsk, 373 p (in Russian)
5. Gulyaev KI (1964) Methods of generating the conjugated gearing of bevels gearwheels with cycloidal longitudinal tooth profile. Izvestiya vuzov. N12, Mashinostroyeniye, Moscow (in Russian)
6. Babichev DT Development of kinematic method of theory of gearing to determine areas of tooth flanks produced by jogs of generating solids, In the present volume
7. Trubachev ES, Semakin NN (2002) Development of the method of spiroid gearwheel machining by generating cutter. Innovative techniques in mechanical engineering and device manufacturing. In: Proceedings of international scientific technical conference dedicated to 50 anniversary of ISTU, part 2, Izhevsk, pp 327–336 (in Russian)
8. Trubachev ES, Zlobina PA, Loginov SE (2011) New scheme of spiroid gearwheel tooth generation. In: Intelligent systems in manufacturing, N3. ISTU Publications, Izhevsk, pp 178–184 (in Russian)
9. Trubachev ES, Loginov SE (2014) New method of spiroid gearwheel tooth cutting. In: Theory and practice of gearing: Proceedings of international symposium (January 21–23, Russia, Izhevsk). ISTU Publications, Izhevsk, 2013. 580 p, ISBN 978-5-7526-0629-8. pp 232–237 (in Russian)
10. http://www.holroyd.com/holroyd-precision/machines/rotor-milling-machines.php

11. Trubachev ES, Zubkova AI (2014) Method of cutting the threads of multi-thread cylindrical worms. In: Theory and practice of gearing: Proceedings of international symposium (January 21–23, Russia, Izhevsk). Izhevsk, ISTU Publications, 2013. 580 p, ISBN 978-5-7526-0629-8. pp 238–243 (in Russian)
12. Goldfarb VI, Trubachev ES (2002) Development and application of computer-aided design and tooth contact analysis of spiral-type gears with cylindrical worms. Technical Paper for AGMA FTM, USA, p 17–21
13. Litvin FL, Timofeyev BP, Rubtsov VN (1971) Synthesis of face-milled and Helixform bevel gears with circular teeth according to local conditions. Mechanics of machines, N31–32. Nauka, Moscow, pp 40–49 (in Russian)
14. Segal MG (1972) To determination of boundaries of tooth contact pattern for bevel and hypoid gears, N4. Mashinovedeniye, Moscow, pp 61–68 (in Russian)
15. Trubachev ES (2014) Synthesis of multi-pair loaded spiroid gear. In: Theory and practice of gearing: Proceedings of international symposium (January 21–23, 2014, Russia, Izhevsk). ISTU Publications, Izhevsk, 2013, 580 p. ISBN 978-5-7526-0629-8, pp 359–366 (in Russian)
16. Sheveleva GI (1999) Theory of generation and contact of moving solids. Stankin, Moscow, 494 p (in Russian)
17. Litvin FL, Fuentes A (2004) Gear geometry and applied theory, 2nd edn, 800 p, Cambridge University Press, New York
18. Lagutin SA Analogs of axes of meshing in general type worm gearing. In the present volume
19. Goldfarb VI, Trubachev ES, Puzanov V Yu (2009) New possibilities of non-orthogonal worm gears. In: Proceedings of the 3rd International Conference Power Transmissions 09. Greece, Chalkidiki, pp 139–145 (in Russian)
20. Lagutin SA (1998) Synthesis and applications of general type worm gears with localized contact, Dresden (Germany): Preprint IOCOMO-06-1998, 16 p
21. Babichev DT (2004) To application of multi-parametric enveloping in computer-aided simulation of generation processes in operating and machine-tool types of gearing. In: Theory and practice of gearing: proceedings of scientific-technical conference with international participation, Izhevsk, pp 302–315 (in Russian)
22. Favar J (1960) Course of local differential geometry. Publishing of foreign literature, 559 p (in Russian)

Direct Gear Design for Asymmetric Tooth Gears

A. Kapelevich

Abstract In many gear drives, the tooth load on one flank is significantly higher and is applied for longer periods of time than for the opposite one. An asymmetric tooth shape reflects this functional difference. Design intent of asymmetric gear teeth is to improve performance of the primary drive profiles at the expense of the performance off the opposite coast profiles. The coast profiles are unloaded or lightly loaded during a relatively short work period. Asymmetric tooth profiles also make it possible simultaneously to increase the contact ratio and operating pressure angle beyond the conventional gear limits. The main advantage of asymmetric gears is contact stress reduction on the drive flanks, resulting in higher torque density (load capacity per gear size). The paper presents an application of the Direct Gear Design® method to asymmetric tooth gears. This is an alternative approach to traditional design of involute gears, separating gear geometry definition from the generating (tooling) rack to maximize gear drive performance. The paper describes asymmetric tooth and gear mesh characteristics, limits of asymmetric tooth gearing, tooth geometry optimization, analytical and experimental comparison of symmetric and asymmetric tooth gears, and implementation of gears with asymmetric teeth.

Keywords Direct gear design · Asymmetric tooth gears · Gear geometry optimization · Performance maximization

1 Introduction

The flanks of a gear tooth are functionally different for many gear drives. Tooth load on one flank is significantly higher and is applied for longer periods of time than that on the opposite one. An asymmetric tooth shape reflects this functional difference (Fig. 1). A design objective of asymmetric gear teeth is to improve the

A. Kapelevich (✉)
AKGears, LLC, Shoreview, USA
e-mail: ak@akgears.com

V. Goldfarb and N. Barmina (eds.), *Theory and Practice of Gearing and Transmissions*, Mechanisms and Machine Science 34,
DOI 10.1007/978-3-319-19740-1_7

Fig. 1 Gears with asymmetric teeth

performance of primary drive profiles at the expense of the performance of opposite coast profiles. The coast flanks are unloaded or lightly loaded during a relatively short work period.

The first asymmetric gears had a buttress tooth shape with a low pressure angle at the drive tooth flanks, while supporting coast flanks had a high pressure angle. According Woodbury [1]: "In Leonardo da Vinci we find some drawings of tooth form—one very like a buttress tooth." In 1841, Willis [2] had shown the asymmetric buttress gear teeth with the following explanation: "If a machine be of such a nature that the wheels are only required to turn in one direction, the strength of the teeth may be doubled by an alteration of form". Leutwiler [3] applied involute profiles for both drive and coast flanks of the buttress or, as he called them, "hook-tooth" gears. He suggested the 15° pressure angle for drive flanks and the 35° pressure angle for coast flanks.

Many gear researchers [4–11] have defined asymmetric gear geometry traditionally by the pre-selected asymmetric generating gear rack parameters, which are typically modified from the standard symmetric rack by increasing the pressure angle of one flank. The opposite flank and other rack tooth proportions remain unchanged.

It is well known that gear transmission load capacity and power density depend mainly upon the tooth flank surface durability, which is defined by the contact stress level and scuffing resistance. From this point, the application of a higher pressure angle for drive tooth flanks and a lower pressure angle for coast tooth flanks is more promising than the buttress tooth shape. In addition, this tooth form provides lower stiffness and better gear mesh impact dampening.

Prof. E.B. Vulgakov had applied his theory of generalized parameters to asymmetric gears [12, 13], defining their geometry without using rack generation parameters.

Publications [14–17] suggested an asymmetric tooth formed with two involutes of two different base circles. Figure 2 shows that the symmetric tooth with identical drive pressure angle and tooth thicknesses at the reference and tip diameters has a much shorter involute active involute flank than the drive flank of an asymmetric tooth.

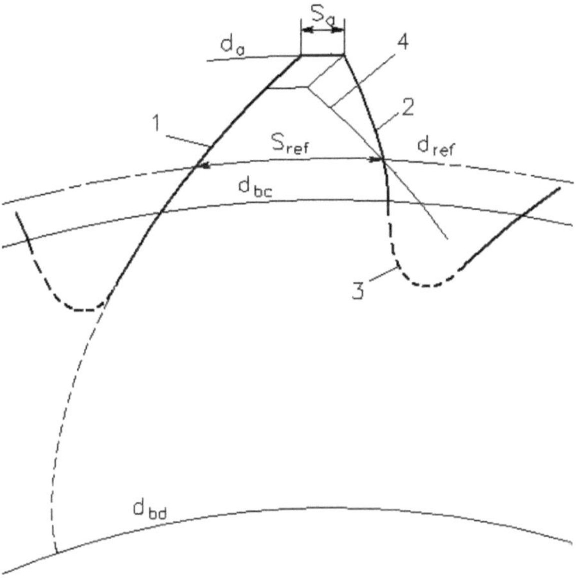

Fig. 2 Asymmetric tooth constructed with two involutes; *1* drive flank from base diameter d_{bd}, *2* coast flank from base diameter d_{bc}, *3* root fillet, *4* symmetric tooth profile with the same drive flank, and tooth thicknesses S_{ref} and S_a at the reference diameter d_{ref} and the tooth tip diameter d_a

The Direct Gear Design approach [18], based on Prof. E.B. Vulgakov's theory of generalized parameters, allowed for maximizing the performance of asymmetric tooth gears. Asymmetric tooth profiles make it possible simultaneously to increase the contact ratio and operating pressure angle beyond those limits achievable with conventional symmetric gears. The main advantage of asymmetric gears is contact stress reduction on the drive flanks that results in higher power transmission density (load capacity per-gear size). Another important advantage is the possibility of designing the coast tooth flanks independently from the drive tooth flanks, i.e., managing tooth stiffness while keeping a desirable pressure angle and contact ratio of drive flanks. This allows for an increase in tooth tip deflection, thus damping tooth mesh impact and resulting in a reduction of gear noise and vibration.

2 Geometry of Asymmetric Tooth Gears

Two involute flanks of the asymmetric tooth (see Fig. 3) are unwound from two different base diameters d_{bd} and d_{bc}. The symbol "*d*" is used for the drive flank and the symbol "*c*" is used for the coast flank of an asymmetric tooth. A diameter dx at the drive flank point *X* can be defined from (1)

$$d_x = d_{bd}/\cos\alpha_{xd} = d_{bc}/\cos\alpha_{xc}. \quad (1)$$

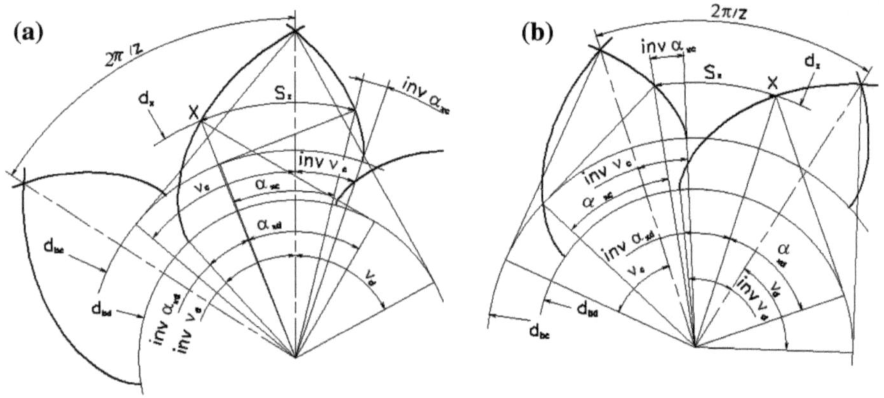

Fig. 3 Involute flanks of external (**a**) and internal (**b**) asymmetric gear teeth, *z* number of teeth

Then, the tooth asymmetry factor K is

$$K = d_{bc}/d_{bd} = \cos\alpha_{xc}/\cos\alpha_{xd}. \tag{2}$$

For many applications, the drive flank profile angle α_{xd} is greater than the coast flank profile angle α_{xc}. This means $d_{bd} < d_{bc}$ and the asymmetry factor $K > 1.0$.

At the coast flank base circle d_{bc}, the coast flank profile angle $\alpha_{xc} = 0$ and the drive flank profile angle α_{xd} from (2) is

$$\alpha_{xd} = \arccos\left(1/K\right). \tag{3}$$

The base tooth thickness of the asymmetric tooth can be defined only at the coast flank base circle d_{bc}:

- for external tooth

$$S_b = \frac{d_{bc}}{2} \times (inv(v_d) + inv(v_c) - inv(\arccos(1/K)), \tag{4}$$

- for internal tooth

$$S_b = \frac{d_{bc}}{2} \times (2\pi/z - inv(v_d) - inv(v_c) + inv(\arccos(1/K)). \tag{5}$$

The tooth thickness at the diameter d_x is

- for external tooth

$$S_x = \frac{d_x}{2} \times (inv(v_d) + inv(v_c) - inv(\alpha_{xd}) - inv(\alpha_{xc})) \tag{6}$$

or

$$S_x = \frac{d_{bd}}{2\cos\alpha_{xd}} \times (inv(v_d) + inv(v_c) - inv(\alpha_{xd}) - inv(\alpha_{xc})); \tag{7}$$

- for internal tooth

$$S_x = \frac{d_x}{2} \times (\frac{2\pi}{z} - inv(v_d) - inv(v_c) + inv(\alpha_{xd}) + inv(\alpha_{xc})) \tag{8}$$

or

$$S_x = \frac{d_{bd}}{2\cos\alpha_{xd}} \times (\frac{2\pi}{z} - inv(v_d) - inv(v_c) + inv(\alpha_{xd}) + inv(\alpha_{xc})). \tag{9}$$

An asymmetric tooth profile also includes the tip land and the root fillet between teeth (see Figs. 4 and 5). The tooth tip diameter d_a can be defined from (1)

$$d_a = d_{bd}/\cos\alpha_{ad} = d_{bc}/\cos\alpha_{ac}, \tag{10}$$

where α_{ad} and α_{ac} are the drive and coast profile angles at the diameter d_a.

The tooth tip land S_a of the external gear tooth is defined considering tooth tip radii equal to zero from Eq. (6) or (8)

- for external tooth (Fig. 4)

$$S_a = \frac{d_a}{2} \times (inv(v_d) + inv(v_c) - inv(\alpha_{ad}) - inv(\alpha_{ac})) \tag{11}$$

or

$$S_a = \frac{d_{bd}}{2\cos\alpha_{ad}} \times (inv(v_d) + inv(v_c) - inv(\alpha_{ad}) - inv(\alpha_{ac})); \tag{12}$$

- for internal tooth (Fig. 5)

$$S_a = \frac{d_a}{2} \times (2\pi/z - inv(v_d) - inv(v_c) + inv(\alpha_{ad}) + inv(\alpha_{ac})) \tag{13}$$

or

$$S_a = \frac{d_{bd}}{2\cos\alpha_{ad}} \times (2\pi/z - inv(v_d) - inv(v_c) + inv(\alpha_{ad}) + inv(\alpha_{ac})). \tag{14}$$

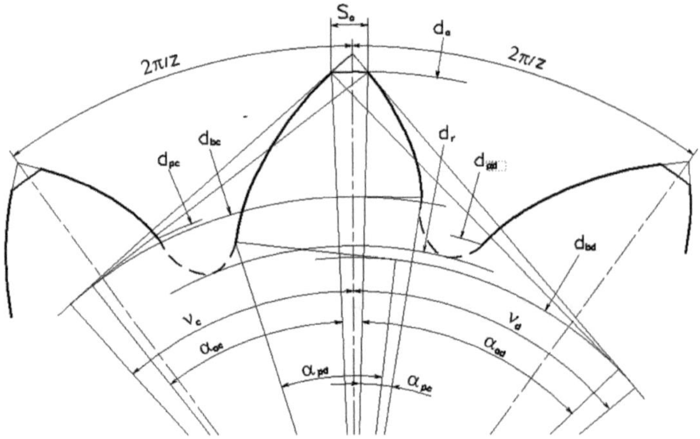

Fig. 4 External asymmetric gear tooth

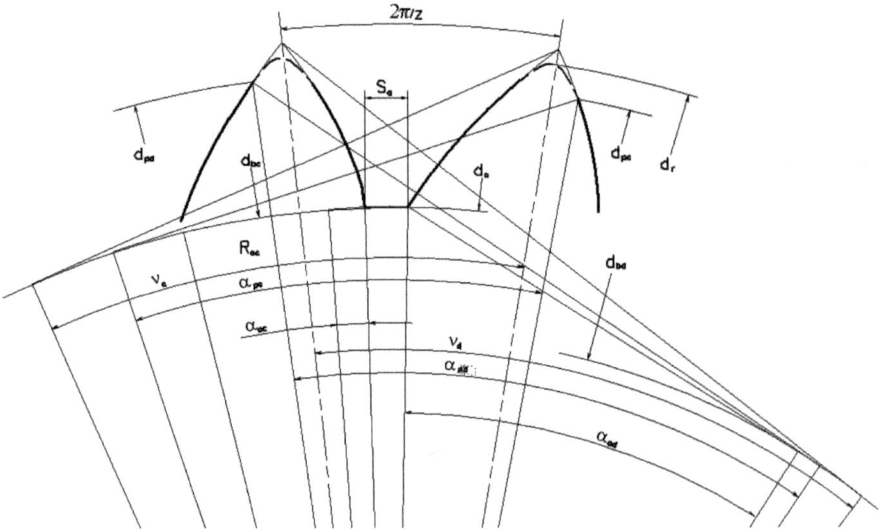

Fig. 5 Internal asymmetric gear tooth

The diameter d_{pd} and d_{pc}, at the last points of contact near the root fillet and related profile angles α_{pd} and α_{pc} are defined considering a mesh with the mating gear. The root diameter d_r is defined as a result of the root fillet profile optimization.

3 Gear Mesh Characteristics

Asymmetric external and internal gear meshes are shown in Fig. 6.

In all figures and equations describing gears with asymmetric teeth, indexes "d" and "c" are for parameters related to the drive and coast tooth flanks accordingly.

The pinion and gear tooth thicknesses S_{w1} and S_{w2} at the operating pitch diameters $d_{w1,2}$ are defined by Eqs. (7) and (9) as

$$S_{w1} = \frac{d_{bd1}}{2\cos\alpha_{wd}}(inv(v_{d1}) + inv(v_{c1}) - inv(\alpha_{wd}) - inv(\alpha_{wc})), \qquad (15)$$

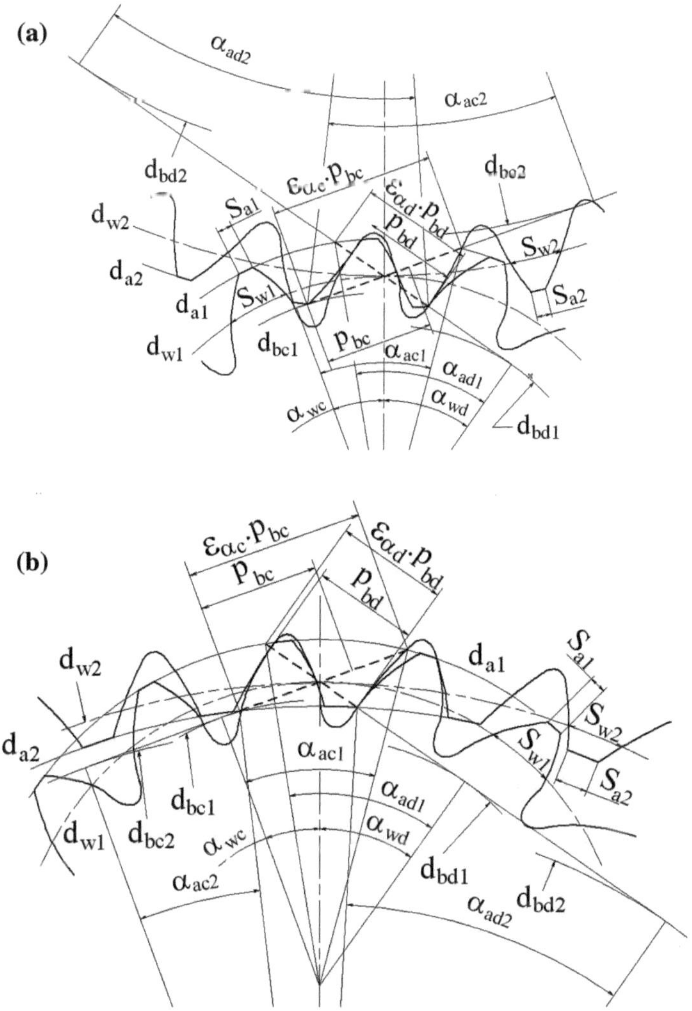

Fig. 6 Asymmetric gear mesh; **a** external, **b** internal

- for external gearing

$$S_{w2} = \frac{d_{bd2}}{2\cos\alpha_{wd}}(inv(\nu_{d2}) + inv(\nu_{c2}) - inv(\alpha_{wd}) - inv(\alpha_{wc})), \tag{16}$$

- for internal gearing

$$S_{w2} = \frac{d_{bd2}}{2\cos\alpha_{wd}}(\frac{2\pi}{z_2} - inv(\nu_{d2}) - inv(\nu_{c2}) + inv(\alpha_{wd}) + inv(\alpha_{wc})). \tag{17}$$

for the Metric System

$$S_{w2} = \pi m - \frac{d_{bd1}}{2\cos\alpha_d}(inv(\nu_{d1}) + inv(\nu_{c1}) - inv(\alpha_d) - inv(\alpha_c)), \tag{18}$$

for the English System

$$S_{w2} = \frac{\pi}{DP} - \frac{d_{bd1}}{2\cos\alpha_d}(inv(\nu_{d1}) + inv(\nu_{c1}) - inv(\alpha_d) - inv(\alpha_c)). \tag{19}$$

The operating pressure angles for the drive flanks α_{wd} and for the coast flanks α_{wc} are defined by substitution of the S_{w1} and S_{w2} from Eqs. (15) and (16) or (17) into the operating circular pitch equation

$$p_w = S_{w1} + S_{w2}. \tag{20}$$

Then, for external gear

$$inv(\alpha_{wd}) + inv(\alpha_{wc}) = \frac{1}{1+u}(inv(\nu_{d1}) + inv(\nu_{c1}) + u(inv(\nu_{d2}) + inv(\nu_{c2})) - \frac{2\pi}{z_1}), \tag{21}$$

- for internal gear

$$inv(\alpha_{wd}) + inv(\alpha_{wc}) = \frac{1}{u-1}(u(inv(\nu_{d2}) + inv(\nu_{c2})) - inv(\nu_{d1}) + inv(\nu_{c1})). \tag{22}$$

The relation between pressure angles for the drive flanks α_{wd} and for the coast flanks α_{wc} is defined from (2) as

$$\cos \alpha_{wc} = K \cos \alpha_{wd}. \tag{23}$$

The profile angles at the last points of contact near the fillet are

- for external gear mesh, drive flanks

$$\alpha_{pd1} = \arctan((1+u)\tan \alpha_{wd} - u \tan \alpha_{ad2}), \tag{24}$$

$$\alpha_{pd2} = \arctan(\frac{1+u}{u}\tan \alpha_{wd} - \frac{1}{u}\tan \alpha_{ad1}), \tag{25}$$

- for external gear mesh, coast flanks

$$\alpha_{pc1} = \arctan((1+u)\tan \alpha_{wc} - u \tan \alpha_{ac2}), \tag{26}$$

$$\alpha_{pc2} = \arctan(\frac{1+u}{u}\tan \alpha_{wc} - \frac{1}{u}\tan \alpha_{ac1}), \tag{27}$$

- for internal gear mesh, drive flanks

$$\alpha_{pd1} = \arctan(u \tan \alpha_{ad2} - (u-1)\tan \alpha_{wd}), \tag{28}$$

$$\alpha_{pd2} = \arctan(\frac{u-1}{u}\tan \alpha_{wd} + \frac{1}{u}\tan \alpha_{ad1}), \tag{29}$$

- for internal gear mesh, coast flanks

$$\alpha_{pc1} = \arctan(u \tan \alpha_{ac2} - (u-1)\tan \alpha_{wc}), \tag{30}$$

$$\alpha_{pc2} = \arctan(\frac{u-1}{u}\tan \alpha_{wc} + \frac{1}{u}\tan \alpha_{ac1}). \tag{31}$$

If asymmetry factor $K > 0$, interference occurs first for the coast involute flanks, which are unwound from the larger base circle. If the profile angles α_{pc1} or α_{pc2} in the external mesh or angle α_{pc1} in the internal mesh are less than zero, then their involute flanks close to the base diameters are interfering with the mating tooth tips. This leads to the involute flank profile undercut.

The transverse contact ratios:

- for external gear mesh, drive flanks

$$\varepsilon_{\alpha d} = \frac{z_1}{2\pi}(\tan \alpha_{ad1} + u \tan \alpha_{ad2} - (1 + u) \tan \alpha_{wd}), \tag{32}$$

- for external gear mesh, coast flanks

$$\varepsilon_{\alpha c} = \frac{z_1}{2\pi}(\tan \alpha_{ac1} + u \tan \alpha_{ac2} - (1 + u)\tan \alpha_{wc}), \tag{33}$$

- for internal gear mesh, drive flanks

$$\varepsilon_{\alpha d} = \frac{z_1}{2\pi}(\tan \alpha_{ad1} - u \tan \alpha_{ad2} + (u - 1) \tan \alpha_{wd}), \tag{34}$$

- for internal gear mesh, coast flanks

$$\varepsilon_{\alpha c} = \frac{z_1}{2\pi}(\tan \alpha_{ac1} - u \tan \alpha_{ac2} + (u - 1) \tan \alpha_{wc}). \tag{35}$$

4 Asymmetric Tooth Gearing Limits

Figure 7 presents ranges of the drive pressure angles for a different number of teeth and asymmetry factors *K*.

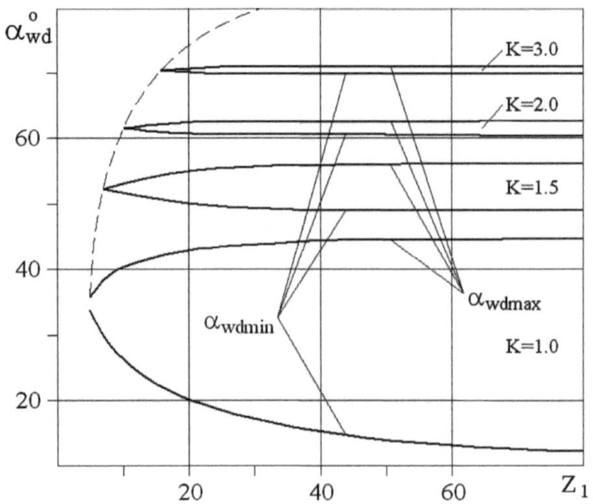

Fig. 7 Minimum and maximum pressure angles for external spur gears with gear ratio $u = 1$ and various asymmetry factors *K*

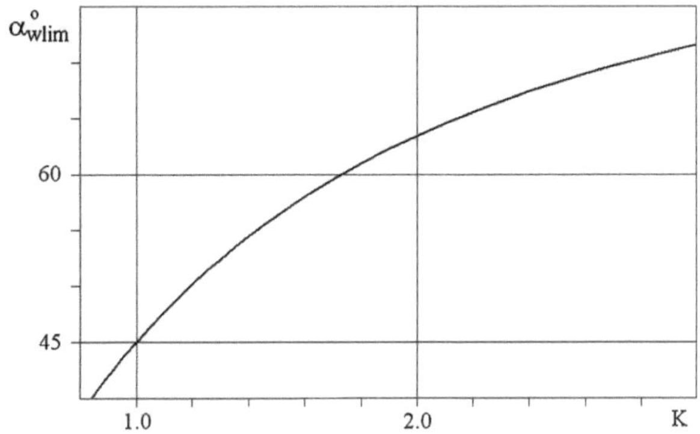

Fig. 8 Pressure angle limits external spur gears

The pressure angle limit for the external spur gears with $z_1 + z_2 \to \infty$ is [14]

$$\alpha_{\text{wlim}} = \arctan K. \quad (36)$$

A chart of the pressure angle limit α_{wlim} as a function of the asymmetry factor K is shown in Fig. 8.

Direct Gear Design greatly expands the transverse contact ratio range. Its maximum value depends on type of gearing (external or internal), tooth profile (symmetric or asymmetric), number of teeth, and relative tooth tip thicknesses.

Maximum transverse contact ratios for spur external reversible asymmetric gears with the relative tooth tip thicknesses $m_{a1,2} = 0$ are shown in Table 1.

The Table 1 data indicated that the maximum contact ratio of the reversible asymmetric gears is just slightly greater than it is for the symmetric gears, and asymmetry of such gears is very low.

Practical maximum pressure angle and transverse contact ratio are limited by the minimum tooth tip thickness. For a case of hardened teeth, it should be sufficient to avoid the hardening through the tooth tip. For gears made out of soft metals and plastics, it should be sufficient to exclude tooth tip bending. Minimum relative tooth

Table 1 Maximum drive contact ratios for reversible asymmetric gears

Number of teeth, $z_{1,2}$	10	15	20	30	40	50
Drive contact ratio, $\varepsilon_{\alpha d}$	1.53	1.931	2.288	2.924	3.49	4.015
Coast contact ratio, $\varepsilon_{\alpha c}$	1.0	1.0	1.0	1.0	1.0	1.0
Drive flank pressure angle, α°_{wd}	25.67	22.02	19.77	17.02	15.34	14.157
Coast flank pressure angle, α°_{wc}	20.58	15.79	13.19	10.47	8.73	7.692
Drive flank tooth tip angle, $\alpha^\circ_{ad1,2}$	43.87	38.97	35.71	31.48	28.75	26.77
Coast flank tooth tip angle, $\alpha^\circ_{ac1,2}$	41.51	36.19	32.85	28.71	26.03	24.147
Drive flank lowest involute angle, $\alpha^\circ_{pd1,2}$	0.0	0.0	0.0	0.0	0.0	0.0
Asymmetry factor, K	1.039	1.038	1.035	1.028	1.025	1.022

tip thickness typically is $m_{a1,2} = 0.25-0.3$. Practical minimal contact ratio for conventional spur gears is about $\varepsilon_{\alpha min} = 1.1-11.15$. For the high contact ratio (HCR) gears, it is about $\varepsilon_{\alpha\, min} = 2.05-2.1$. These minimal contact ratio values are chosen to avoid its reduction below 1.0 for conventional spur gears and below 2.0 for the HCR spur gears, because of manufacturing and assembly tolerances, and tooth tip chamfers or radii. These conditions also identify the practical maximum pressure angle. The practical minimal pressure angle for symmetric gears is defined by the beginning of the tooth involute undercut, when the involute profile angles at the lowest contact points $\alpha_{pd1,2} = 0°$, and where the transverse contact ratio reaches its maximum value $\varepsilon_{\alpha\, max}$.

Practical maximum drive flank pressure angles for conventional and HCR asymmetric gears are shown in Tables 2 and 3 [18].

The maximum drive pressure angle values in Tables 2 and 3 assume some possible small undercut of the coast flank near the root, especially for gears with a low number of teeth (15–30). However, this does not reduce the coast flank contact below $\varepsilon_{\alpha c} = 1.0$.

Table 2 Practical α_{wdmax} for conventional asymmetric gears ($m_{a1,2} = 0.3, \varepsilon_{\alpha d} = 1.1, \alpha_{wc} = 15°$)

z_2	z_1						
	15	20	30	40	50	70	100
15	43.5						
20	44.5	45.5					
30	45.9	46.4	47.3				
40	47	47.3	47.7	48.2			
50	47.6	47.8	48	48.3	48.9		
70	48	48.2	48.6	48.7	49	49.5	
100	48.7	48.9	49.2	49.5	49.1	49.6	50

Table 3 Practical α_{wdmax} for HCR asymmetric gears ($m_{a1,2} = 0.3, \varepsilon_{\alpha d} = 2.1, \alpha_{wc} = 15°$)

z_2	z_1						
	20	25	30	40	50	70	100
20	19.3						
25	20.5	21.5					
30	21.5	22.4	23				
40	23	23.6	24.1	25			
50	24.1	24.6	25	25.6	26.1		
70	25.5	25.8	26.1	26.6	26.9	27.5	
100	26.7	27	27.2	27.5	27.7	28.1	28.5

5 Tooth Geometry Optimization

5.1 Asymmetry Factor K Selection for Reversible Asymmetric Tooth Gears

There are many gear drives where a gear pair transmits the load in both load directions, but with significantly different magnitude and duration (Fig. 9). In this case, the asymmetry factor K for a gear pair is defined by equalizing potential accumulated tooth surface damage defined by operating contact stress and number of the tooth flank load cycles. In other words, the contact stress safety factor S_H should be the same for the drive and coast tooth flanks. This condition can be presented as

$$S_H = \frac{\sigma_{HPd}}{\sigma_{Hd}} = \frac{\sigma_{HPc}}{\sigma_{Hc}}, \tag{37}$$

where:

σ_{Hd} and σ_{Hc} operating contact stresses for the drive and coast tooth flanks,
σ_{HPd} and σ_{HPc} permissible contact stresses for the drive and coast tooth flanks that depend on the number of load cycles

Then, from (37)

$$\frac{\sigma_{Hd}}{\sigma_{Hc}} = \frac{\sigma_{HPd}}{\sigma_{HPc}}. \tag{38}$$

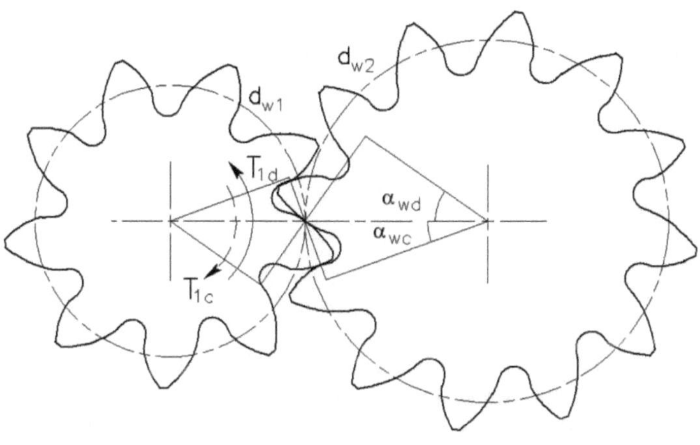

Fig. 9 Asymmetric gear pair; T_{1d} and T_{1c}—pinion torque applied to the drive and coast tooth flanks

The contact stress at the pitch point [19] is

$$\sigma_H = z_H z_E z_\varepsilon z_\beta \sqrt{\frac{F_t}{d_{w1} b_w} \frac{u \pm 1}{u}}, \tag{39}$$

where:

$z_H = \sqrt{\frac{2\cos(\beta_b)\cos(\alpha_{wt})}{\cos(\alpha_t)^2 \sin(\alpha_{wt})}}$ zone factor that, for the directly designed spur gears, is

$$z_H = \frac{2}{\sqrt{\sin(2\alpha_w)}}; \tag{40}$$

- Z_E elasticity factor that takes into account gear material properties (modulus of elasticity and Poisson's ratio);
- Z_ε contact ratio factor, its conservative value for spur gears is $Z_\varepsilon = 1.0$;
- Z_β helix factor, for spur gears $Z_\beta = 1.0$;
- F_t nominal tangent load, that, at the pitch diameter d_{w1}, is $F_t = \frac{2T_1}{d_{w1}}$;
- T_1 pinion torque;
- b_w contact face width;

sign "+" for external gearing, sign "−" for internal gearing.

Then, for the directly designed spur gears, the contact stress at the pitch point can be presented as

$$\sigma_H = z_E \frac{2}{d_{w1}} \sqrt{\frac{2T_1}{b_w \sin(2\alpha_w)} \frac{u \pm 1}{u}}. \tag{41}$$

Some parameters of this equation, Z_E, d_{w1}, b_w, and u, do not depend on the rotation direction and Eq. (38) for the pitch point contact can be presented as

$$\frac{\sin(2\alpha_{wc})}{\sin(2\alpha_{wd})} = A, \tag{42}$$

where a parameter A is

$$A = \frac{T_{1c}}{T_{1d}} \left(\frac{\sigma_{HPd}}{\sigma_{HPc}}\right)^2. \tag{43}$$

According to [19], "The permissible stress at limited service life or the safety factor in the limited life stress range is determined using life factor Z_{NT}". This allows for replacing the permissible contact stresses in Eq. (43) for the life factors

$$A = \frac{T_{1c}}{T_{1d}}\left(\frac{Z_{NTd}}{Z_{NTc}}\right)^2. \quad (44)$$

When the parameter A is defined and the drive pressure angle is selected, the coast pressure angle is calculated by Eq. (43) and the asymmetry coefficient K from the common solution of (42) and (23)

$$K = \frac{\sqrt{1 + \sqrt{1 - A^2(\sin 2\alpha_{wd})^2}}}{\sqrt{2}\cos\alpha_{wd}}. \quad (45)$$

If the gear tooth is equally loaded in both main and reversed load application directions, then both the coefficient A and the asymmetry factor K are equal to 1.0 and the gear teeth are symmetric.

Example #1 The drive pinion torque T_{1d} is two times greater than the coast pinion torque T_{1c}. The drive tooth flank has 10^9 load cycles and the coast tooth flank has 10^6 load cycles during the life of the gear drive. From the S-N curve [19] for steel gears, an approximate ratio of the life factors $Z_{NTd}/Z_{NTc} = 0.85$. Then, the coefficient $A = 0.85^2/2 = 0.36$. Assuming the drive pressure angle is $\alpha_{wd} = 36^o$, the coast pressure angle from Eq. (42) is $\alpha_{wc} = 10^o$ and the asymmetry factor from Eq. (45) is $K = 1.22$.

5.2 *Root Fillet Optimization*

In Direct Gear Design, the tooth fillet is designed after the involute flank parameters are completely defined. One goal is to achieve a minimum of stress concentration on the tooth fillet profile. In other words, the maximum bending stress should be evenly distributed along the large portion of the root fillet. The initial root fillet profile is a trajectory of the mating gear tooth tip in the tight (zero backlash) mesh. This allows for the exclusion of interference with the mating gear tooth.

The fillet optimization method was developed by Shekhtman [20, 21]. It utilizes the following calculation processes:

- Definition of a set of mathematical functions used to describe the optimized fillet profile. Such a set may contain the trigonometric, polynomial, hyperbolic, exponential, and other functions and their combinations. Parameters of these functions are defined during the optimization process.
- FEA with the triangle linear elements is used to calculate stress. This kind of finite elements make possible the achievement of satisfactory optimization results within a reasonable time.
- A random search method [22] is used to define the next step in the multi-parametric iteration process of the fillet profile optimization.

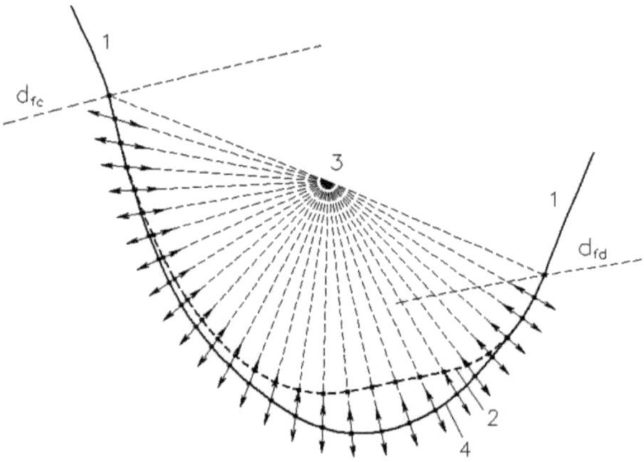

Fig. 10 Tooth fillet profile optimization; *1* involute tooth flanks, *2* initial fillet profile, *3* fillet center, *4* optimized fillet profile, *dfd*, *dfc*—drive and coast flank form circle diameters accordingly

This fillet optimization method establishes the approximate fillet center (Fig. 10). It is defined as the center of the best-fitted circular arc and is connected to the finite element nodes located on the initial fillet profile. The first and last finite element nodes of the initial fillet profile located on the formed diameter circle cannot be moved during the optimization process. The rest of the initial fillet finite element nodes are moved along the straight lines that pass through the fillet center. The bending stresses are calculated for every fillet profile configuration iteration. Variable parameters of the fillet profile functions that describe the fillet profile for the next iteration are defined depending on the stress calculation results of the previous iteration. If it provided stress reduction, the optimization process moves the fillet nodes in the same direction. If stress was increased, the nodes are moved in opposite directions. After the specified number of iterations, the optimization process no longer results in the optimized fillet profile. When more finite element nodes are placed on the fillet profile, stress calculation results are even more accurate, but this requires more iterations, and the fillet profile optimization takes more time. During the optimization process, the fillet nodes cannot be moved inside the initial fillet profile because this may cause interference with the mating gear tooth tip. This is one of the main constraints of fillet optimization.

Figure 11 presents the gear tooth stress distribution comparison before and after root fillet optimization.

Figure 12 presents a comparison of different tooth root fillet profile options. The involute flanks, face widths, tooth load and its application point are the same for all fillet profile options. Results of the FEA stress calculation, along with other root fillet parameters, are shown in Table 4. Calculation results for fillet profile option #1 generated by the standard 20° pressure angle rack profile are considered to be the 100 % benchmark values. Parameters of other fillet profile options are defined relative to option #1.

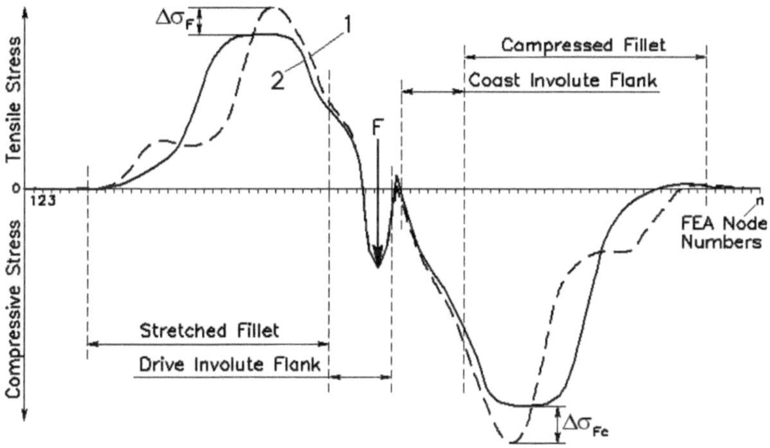

Fig. 11 Tooth profile stress distribution charts before (*1*) and after (*2*) root fillet optimization, $\Delta\sigma_F$ tensile stress reduction, $\Delta\sigma_{Fc}$ compressive stress reduction

The root fillet profile comparison results presented in Table 4 indicate considerable root stress concentration reduction provided by the fillet optimization. At the maximum tensile stress, the point of the optimized fillet has significantly larger fillet radius *Rf*, as well as the smaller distance H and root clearance C. It has the lowest maximum bending stress, which is evenly distributed along the large portion of the fillet profile. Other fillet profiles have significantly greater and more sharply concentrated maximum stress.

Analysis of the fillet optimization results had indicated that the optimized fillet profile practically does not depend on the force value and its application point on the involute flank, except in the case when the application point is located very close to the form diameter. In this case, compression under the applied force may affect the optimized fillet profile. Such load application should not be considered for fillet optimization, because it induces minimal tensile stress in the root fillet in comparison to other load application points along the tooth flank.

The gears with optimized root fillets are shown in Fig. 13.

6 Analytical and Experimental Comparison Symmetric and Asymmetric Tooth Gears

Directly designed gears with an asymmetric involute gear tooth form were analyzed to determine their bending and contact stresses relative to symmetric involute gear tooth form, which is representative of the helicopter main drive gears developed by the Rotorcraft division of the Boeing Company [23]. Asymmetric and baseline (symmetric) toothed gear test specimens were designed, fabricated and tested to

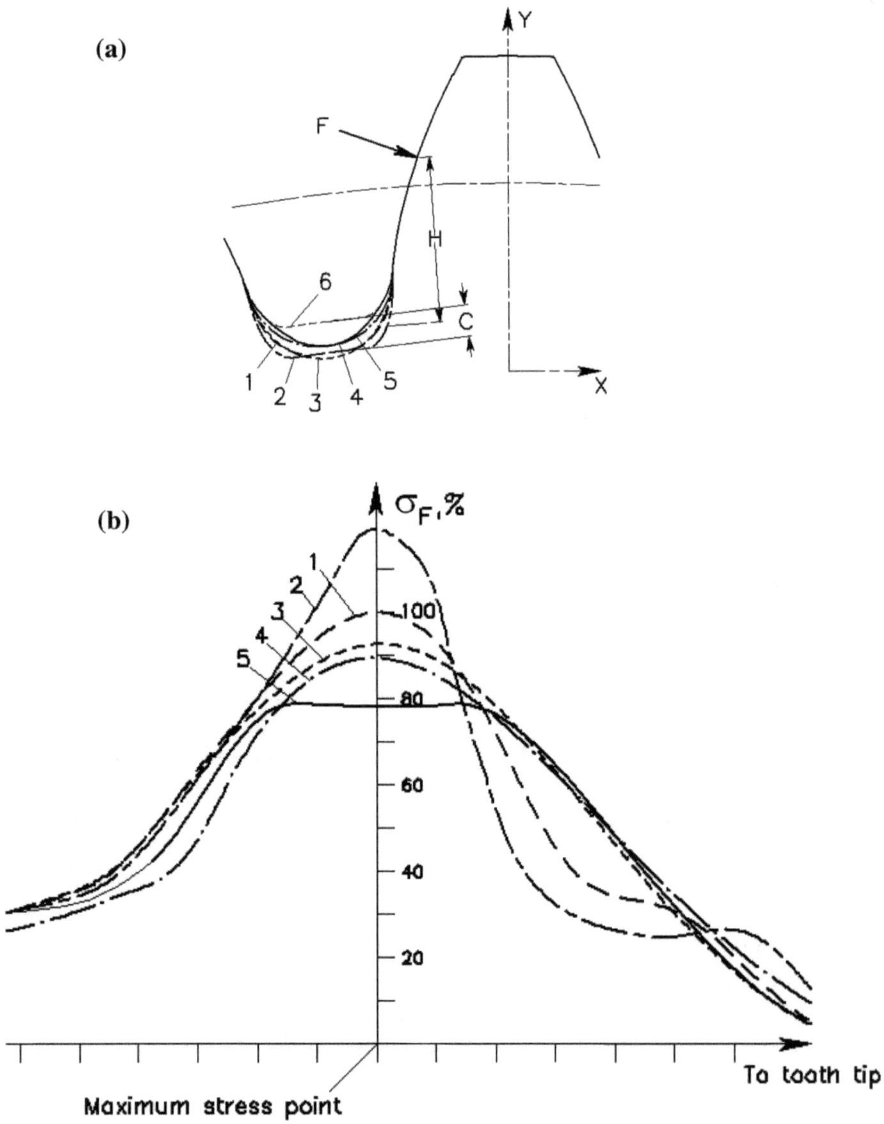

Fig. 12 Root fillet comparison: **a** gear tooth with different fillet profiles; **b** stress chart along the fillet; *1* fillet profile generated by the standard coarse pitch rack with the tip radius 0.3 m; *2* fillet profile generated by the standard fine diametral pitch rack with the tip radius equal to zero; *3* fillet profile generated by the full tip radius rack; *4* circular fillet profile; *5* optimized fillet profile; *6* trajectory of the mating gear tooth tip in tight mesh; *F* applied load; *H* radial distance between load application and max. stress points; *C* radial clearance; R_f fillet curvature radius at the max. stress point; σ_F tensile stress

Table 4 Fillet profile comparison (Fig. 12)

	Rack with tip radius R = 0.3 m	Rack with tip radius R = 0	Rack with full tip radius	Circular fillet profile	Optimized fillet
Fillet profile #	1	2	3	4	5
R_f (%)	100	58	118	121	273
H (%)	100	103	100	88	82
C (%)	100	100	118	79	76
σ_{Fmax} (%)	100	119	90	88	78

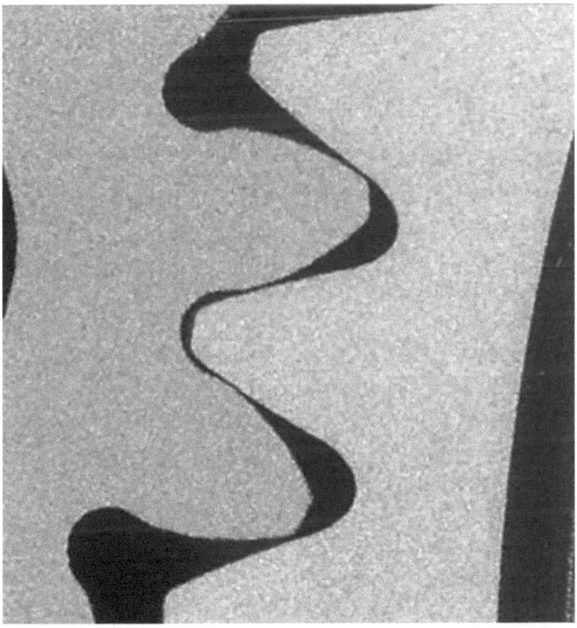

Fig. 13 Gears with optimized root fillets

experimentally determine their single-tooth bending fatigue strength and scuffing resistance. The gear test specimens are presented in Fig. 14.

The gear parameters and FEA-calculated bending stresses for the single tooth bending fatigue (STBF) test gears are presented in Table 5.

Similarly, the scuffing test gears are within the design experience range of typical main transmission helicopter power gears. The gear parameters for the scuffing test gear specimens are presented in Table 6.

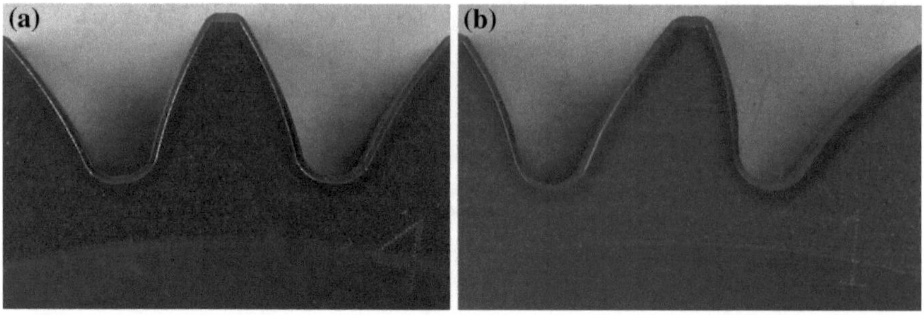

Fig. 14 Test specimen gear tooth profiles; **a** baseline gear teeth, **b** symmetric gear teeth with optimized fillet, **c** asymmetric gear teeth. (Courtesy of Boeing Co, Philadelphia, PA)

Table 5 STBF test gear specimen parameters

	Symmetric gears with circular fillets (baseline)	Asymmetric gears with circular fillets
Number of teeth of both mating gears	32	32
Diametral pitch (1/in)	5.333	5.333
Pressure angle	25°	35°/15°[a]
Pitch diameter (in)	6.000	6.000
Base diameter (in)	5.4378	4.9149/5.7956[a]
Outside diameter (in)	6.3975	6.3864
Root diameter (in)	5.571	5.558
Form diameter (in)	5.6939	5.6581/5.8110[a]
Circular tooth thickness (in)	0.2895	0.2895
Face width (in)	0.375	0.375
Torque (in–lb)	5000	5000
Load application radius (in)	3.06	3.06
Calculated maximum Bending stress (psi)	57,887	54,703 (−5.5 %)

[a]Drive/coast flank parameter

Fatigue results of the single tooth bending fatigue tests of the asymmetric tooth and the baseline specimens are presented in Fig. 15.

Figure 16 shows the scuffing results for baseline and asymmetric gears. The 35° pressure angle asymmetric gears showed an improvement of approximately 25 % in mean scuffing load (torque) compared to the baseline symmetric tooth specimens. The Mean—3 Sigma levels are also shown, based on a population of 8 baseline data points and 6 asymmetric data points.

Test results demonstrated higher bending fatigue strength for both asymmetric tooth forms compared to baseline designs. Scuffing resistance was significantly increased for the asymmetric tooth form compared to a traditional symmetric involute tooth design.

Table 6 Scuffing test gear specimen parameters

	Symmetric gears with circular fillets (baseline)	Asymmetric gears with circular fillets
Number of teeth of both mating gears	30	30
Diametral pitch (1/in)	5.000	5.000
Pressure angle	25°	35°/18°
Pitch diameter (in)	6.000	6.000
Base diameter (in)	5.4378	4.9149/5.7063[a]
Outside diameter (in)	6.400 max	6.403 max
Root diameter (in)	5.459 max	5.510
Form diameter (in)	5.6864	5.6415/5.7607[a]
Circular tooth Thickness (in)	0.3096	0.3096
Face width (in)	0.50	0.50
Drive contact ratio	1.417	1.25
Torque (in-lb)	6000	6000
Calculated maximum Contact stress (psi)	193,180	174,100 (−9.9 %)

[a]Drive/coast flank parameter

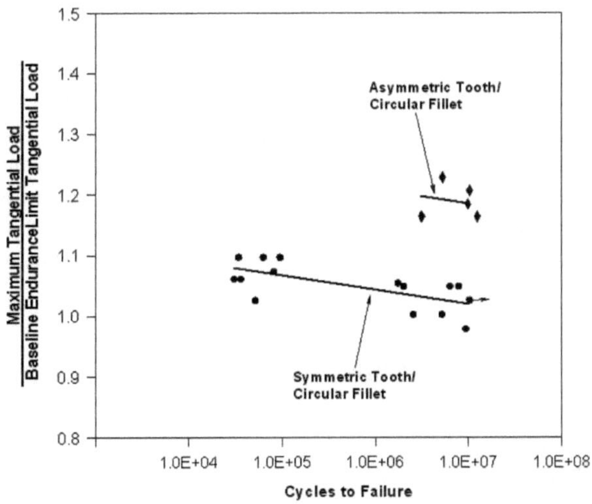

Fig. 15 STBF data for asymmetric gears and optimized root fillet gears along with baseline symmetric tooth/circular fillet test data

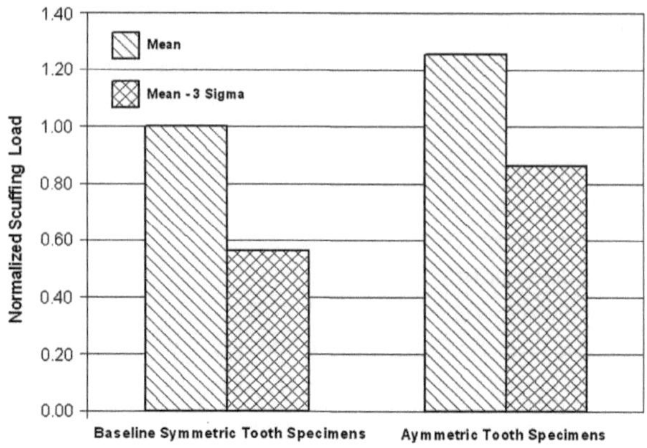

Fig. 16 Results of baseline symmetric and asymmetric gear scuffing tests

7 Implementation of Asymmetric Tooth Gears

The first application of gears with asymmetric teeth in the aerospace industry was for the TV7-117S turboprop engine gearbox [24]. The engine and gearbox were developed by the Klimov Corporation (St. Petersburg, Russia) for a commuter airplane Ilyushin Il-114. The main characteristics of its gearbox are presented in Table 7.

The TV7-117S gearbox arrangement is shown in Fig. 17. The first planetary-differential stage has three planet gears. The second "star" type coaxial stage has five planet (idler) gears and a stationary planet carrier. The first stage sun gear is connected to the engine turbine shaft via spline. Its ring gear is connected with the second stage sun gear and its planet carrier is connected to the second stage ring gear and the output propeller shaft. This arrangement makes it possible to transmit about 33 % of the engine power through the first stage carrier directly to the propeller shaft, bypassing the second stage. This allows for reducing the size and weight of the second stage, because it transmits only 67 % of the engine power

Table 7 TV7–117S turboprop engine data

Input turbine (RPM)	17,500
Output prop (RPM)	1200
Total gear ratio	14.6:1
Overall dimensions (mm):	
• Diameter	520
• Length	645
Gearbox weight (N)	1050
Maximum output power (hp)	2800
Extreme output power (hp)	3500

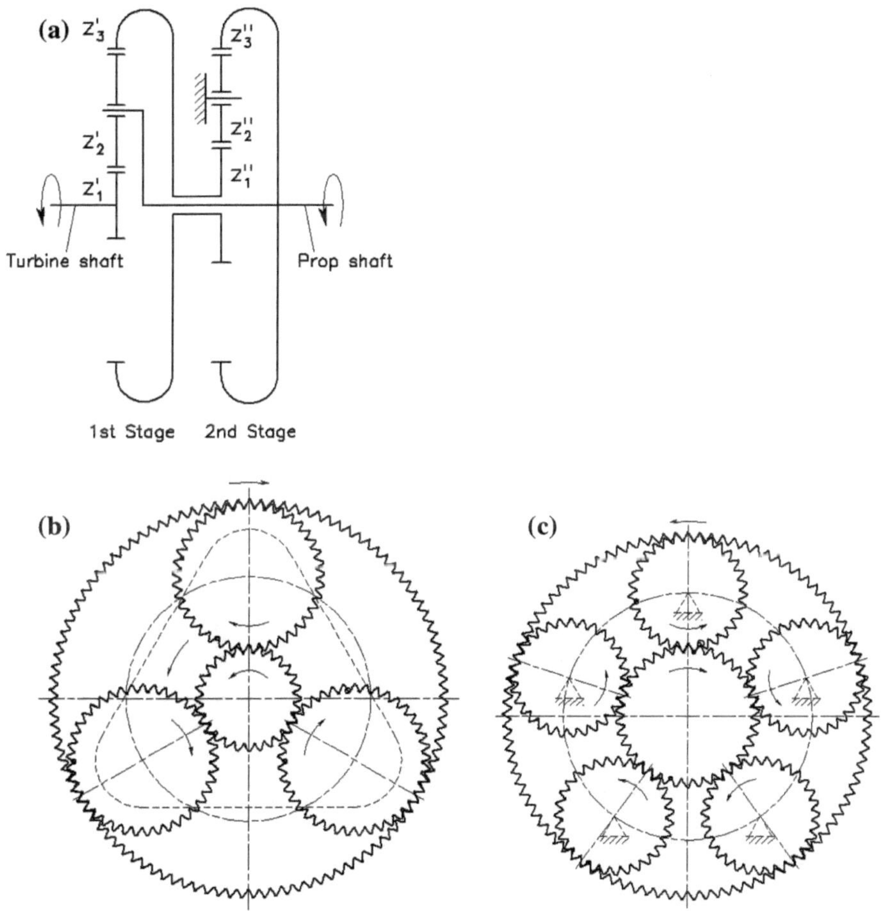

Fig. 17 Gearbox arrangement (**a**), first (**b**) and second (**c**) stages with rotation directions (view from input shaft)

from the first stage ring gear to the second stage sun gear, and then through the planets to the second stage ring attached to the propeller shaft.

All gears have an asymmetric tooth profile. Gear geometry parameters, operating torques and stresses are presented in Table 8.

All gears were made out of the forged blanks of the steel 20KH3MVF (EI-415). Its chemical composition includes: Fe—base material, C—0.15 to 0.20 %, S—<0.025 %, P—<0.030 %, Si—0.17 to 0.37 %, Mn—0.25 to 0.50 %, Cr—2.8 to 3.3 %, Mo—0.35 to 0.55 %, W—0.30 to 0.50 %, Co—0.60 to 0.85 %, Ni—<0.5 %.

Table 8 Gear geometry data

Stage	1st			2nd		
Gear	Sun	Planet	Ring	Sun	Planet	Ring
Number of gears	1	3	1	1	5	1
Nominal number of teeth	28	41	107	38	31	97
Module (mm)	3.000	3.000	3.000	3.362	3.362	3.362
Nominal drive pressure angle (°)	33	33 & 25	25	33	33 & 25	25
Nominal coast pressure angle (°)	25	25 & 33	33	25	25 & 33	33
Nominal pitch diameter (mm)	84.000	123.000	321.000	127.756	104.222	326.114
Drive base diameter (mm)	70.448	103.156 & 111.476	290.925	107.145	87.408 & 94.457	295.56
Coast base diameter (mm)	76.13	111.476 & 103.156	269.213	115.786	94.457 & 87.408	273.502
Tooth tip diameter (mm)	90.02/90.16	128.44/128.6	323.88/324.11	134.07/134.23	110.93/111.07	329.67/329.9
Root diameter (mm)	76.55/77.05	114.55/115.05	337.50/337.7	118.56/119.06	95.45/95.95	345.00/345.2
Tooth thickness at nominal pitch diameter (mm)	4.773/4.814	4.325/4.365	−0.667/ −0.621	4.972/5.018	5.253/5.299	−1.104/ −1.059
Face width (mm)	34.75/ 35	31.75/32	25.48/26	37.75/38	34.75/35.00	27.48/28.00
Center distance (mm)	103.50 ± 0.01			116.00 ± 0.01		

Tolerance analysis results

Operating drive pressure angle (°)	32.98/33.02	29.87/29.93	32.99/33.03	29.88/29.93
Operating coast pressure angle (°)	24.97/25.03	36.64/36.68	24.98/25.04	36.65/36.68
Operating drive contact ratio	1.18/1.26	1.33/1.36	1.20/1.28	1.35/1.38
Operating coast contact ratio	1.33/1/42	1.18/1.21	1.36/1.44	1.21/1.24
Operating normal backlash (mm)	0.196/0.322	0.197/.406	0.189/0.32	0.206/0.414

(continued)

Table 8 (continued)

Stress analysis results									
Maximum power (hp)		2800							
RPM		17500	−11132		−3063	−3063	3755	1200	
Torque per mesh (Nm)		374	548		1430	858	700	2191	
Bending Stress (MPa)	Tension	240	257	280	298	289	318	345	371
	Compression	−384	−403	−306	−355	−470	−500	−385	−430
Contact stress (MPa)		960		604		1043		809	

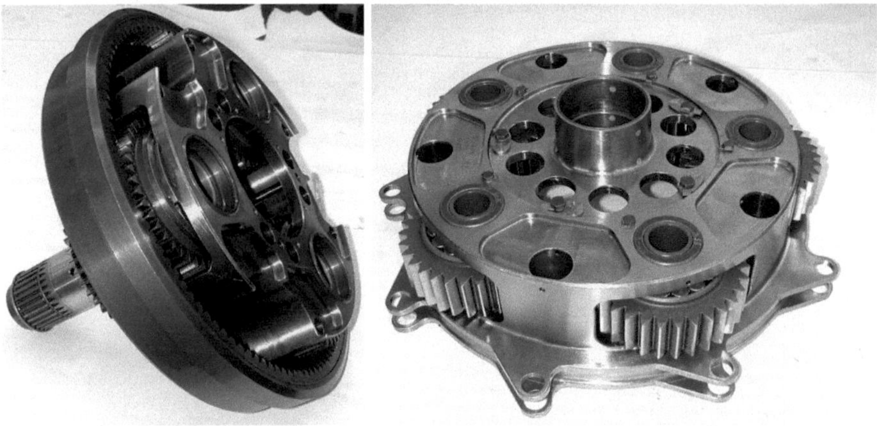

Fig. 18 First and second stage assemblies

Application of the asymmetric teeth helped to provide extremely low weight to output torque ratio, significantly reduced the noise and vibration level, and cut down on the duration and expense of operational development [24].

Photos of the gear assemblies of the TV7-117S gearbox are shown in Fig. 18.

References

1. Woodbury RS (1958) The first epicyclic gear teeth. ISIS 49(4):375–377
2. Willis R (1841) Principles of mechanism. John W. Parker, London
3. Leutwiler OA (1917) Element of machine design. McGraw-Hill Book Company, London
4. Bolotovsky IA, Vasil'eva OF, Kotelnikov VP (1984) Involute gears with asymmetric teeth. Vestnik Mashinostroeniya 4:15–17
5. DiFrancesco G, Marini S (1997) Structural analysis of asymmetrical teeth: reduction of size and weight. Gear Technology, pp 47–51
6. Gang G, Nakanishi T (2001) Enhancement of bending load carrying capacity of gears using an asymmetric involute tooth. In: Paper presented at the JSME international conference on motion and transmissions (MPT2001-Fukuoka), Fukuoka, Japan
7. Karpat F, Cavdar K, Babalik FC (2005) Computer aided analysis of involute spur gears with asymmetric teeth. VDI Berichte 1904(I):145–163
8. Brecher C, Schafer J (2005) Potentials of asymmetric tooth geometries for the optimization of involute cylindrical gears. VDI Berichte 1904(I):705–720
9. Yang SC (2007) Study on an internal gear with asymmetric involute teeth. Mech Mach Theory 42:977–994
10. Pedersen NL (2010) Improving bending stress in spur gears using asymmetric gears and shape optimization. Mech Mach Theory 45:1707–1720
11. Wang S, Liu GR, Zhang GY, Chen L (2011) Design of asymmetric gear and accurate bending stress analysis using the ES-PIM with triangular mesh. Int J Comput Methods 8(4):759–772
12. Vulgakov EB, Vasina LM (1978) Involute gears in generalized parameters. Mashinostroenie, Moscow
13. Vulgakov EB, Rivkin GV (1976) Design of gears with asymmetric tooth profile. Mashinovedenie 5:35–39

14. Kapelevich AL (1984) Research and development of geometry of modernized involute gears. Ph.D. Dissertation, Moscow State Technical University
15. Kapelevich AL (1987) Synthesis of asymmetric involute gearing gears. Mashinovedenie 1:62–67
16. Vulgakov EB, Kapelevich AL (1986) Asymmetric gear transmissions: Possible developments. Vestnik Mashinostroeniya 4:14–16
17. Kapelevich AL (2000) Geometry and design of involute spur gears with asymmetric teeth. Mech Mach Theory 35:117–130
18. Kapelevich AL (2013) Direct gear design. CRC Press
19. Standard ISO 6336 (2006) Calculation of load capacity of spur and helical gears
20. Kapelevich AL, Shekhtman YV (2003) Direct gear design: bending stress minimization. Gear Technology, pp 44–49
21. Kapelevich AL, Shekhtman YV (2009) Tooth fillet profile optimization for gears with symmetric and asymmetric teeth. Gear Technology, pp 73–79
22. Rastrigin LA (ed) (1969) Random search theory and application. Zinatne, Riga
23. Brown FW, Davidson SR, Hanes DB, Weires DJ, Kapelevich AL (2011) Analysis and testing of gears with asymmetric involute tooth form and optimized fillet form for potential application in helicopter main drives. Gear Technology, pp 46–55
24. Novikov AS, Paikin AG, Dorofeyev VL, Ananiev VM, Kapelevich AL (2008) Application of gears with asymmetric teeth in turboprop engine gearbox. Gear Technology, pp 60–65

Analogs of Axes of Meshing in General Type Worm Gearing

S. Lagutin

Abstract The paper describes the meshing space for worm gears with general type worms, including double-enveloping and bevel worms. A pair of conjugate lines is revealed in this space. Position of these lines is defined by position of surface of movement of the cutting edge which generates a worm but does not depend on this edge profile. The revealed lines are pre-assigned to belong to the meshing surface and, in this respect, they are similar to the axes of meshing of gears with a cylindrical worm. Applying the properties of the pointed lines allows, to some extent, for simplification of the analysis of the known and synthesis of new types of worm gears.

Keywords Worm gear · Cylindrical worm · Worm gearwheel · Modified double-enveloping worm gear · Axis of meshing

1 General Type Worm Gears

The concept of a general type worm gear originally appeared in [2]. Later, it was developed in detail in [7, 8].

The basic elements of any worm gear are the driving worm 1 and the driven gearwheel 2, which rotate around their axes O_1–O_1 and O_2–O_2 with angular velocities ω_1 and ω_2. In the case when the axes of elements 1 and 2 of a gear are perpendicular with respect to each other, the gear is called orthogonal; otherwise, it is non-orthogonal. Figure 1 shows the pitch surfaces of elements of a general type worm gear in its orthogonal version.

Active surfaces of worm *1* threads are generated by the motion of tool *3* (cutter or grinding wheel). The teeth of gearwheel *2* are cut by hob *4*, which rotates about the axis O_4–O_4 with angular velocity ω_4. The generating surface of a hob (gener-

S. Lagutin (✉)
Electrostal Plant of Heavy Machines, Electrostal, Russia
e-mail: sergeylagutin@gmail.com

V. Goldfarb and N. Barmina (eds.), *Theory and Practice of Gearing and Transmissions*, Mechanisms and Machine Science 34,
DOI 10.1007/978-3-319-19740-1_8

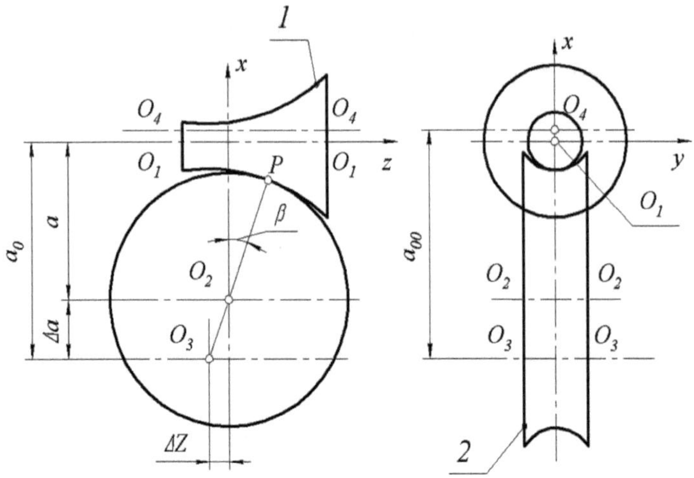

Fig. 1 Scheme of general type worm gear generation

ating worm) is generated by the motion of tool *5*. The generating edges of tools *3*, *5* either rotate with angular velocities ω_3, ω_5 during the cutting process of the double-enveloping worm or have translational motion with velocities V_3, V_5 when cutting cylindrical, bevel, hyperboloid and flat worms.

The type of worm is defined by the relative position of axes O_2–O_2 and O_3–O_3. Especially in case of their coincidence, a classic double-enveloping worm of the Cone type is generated [2]. Different versions of the modified double-enveloping worm-gear pair [11, 18] are obtained in cases when shifts Δa and Δz are small in relation to center distance a of a gear and also when the gear ratio of the machine-tool gearing is changed with respect to the operating one and when the cutting edge of tool *3* is shifted from the mean plane of a gearwheel. A double-enveloping worm-gear pair with a basic cylindrical gearwheel [14] and a worm gear with the screw motion of a helical gearwheel [10] can be considered to be versions of a general type worm gear.

When the axis O_3–O_3 goes to infinity, a double-enveloping worm transforms into a bevel one with the cone angle $\beta = \text{atan}\left[\lim(\Delta z/\Delta a)\right]$, cylindrical worm $(\beta = 0)$ or face worm $(\beta = \pi/2)$.

In worm and double-enveloping worm gears the contact zone for active surfaces of worm threads and gearwheel teeth intersects the plane, passing through the axis O_1–O_1 of a worm and the center perpendicular. In spiroid gears, the meshing zone is placed outside of the pointed plane, their basic distinguishing feature [3, 4]. Also, in the case of gears with two meshing zones, when one of the zones is located close to the axial perpendicular line, the second one is outside of it [6].

Five elements enumerated above constitute one operating gearing *1–2* and three machine-tool meshings *3–1*, *4–2*, *5–3*. For each of them the equation of meshing can be written as:

in operating gearing

$$(\mathbf{n} \cdot \mathbf{V}_{12}) = 0 \tag{1}$$

in worm cutting

$$(\mathbf{n} \cdot \mathbf{V}_{13}) = 0 \tag{2}$$

in gearwheel tooth cutting

$$(\mathbf{n} \cdot \mathbf{V}_{42}) = 0 \tag{3}$$

in shaping of the generating worm

$$(\mathbf{n} \cdot \mathbf{V}_{35}) = 0 \tag{4}$$

Here, **n** is the vector of a common normal to meshing surfaces at the predesigned point of contact, $\mathbf{V}_{12}$, $\mathbf{V}_{13}$, $\mathbf{V}_{24}$, $\mathbf{V}_{35}$ are vectors of relative velocities of elements at the same point, and $\mathbf{V}_{12}=\mathbf{V}_1-\mathbf{V}_2$ etc.

The system of Eqs. (1)–(4) is invariant to the coordinate system, and in this way, it describes in brief and obvious terms the necessary and sufficient conditions that should be satisfied by the surfaces of elements and parameters of their relative motion in order to provide the conjugate gearing.

Worm gears with linear contact are synthesized according to the second Olivier-Gochman principle, according to which the active surface of a worm should be congruent to the generating surface and $\mathbf{V}_{12}=\mathbf{V}_{42}$. In this case, Eqs. (1)–(4) are valid along the whole surface of meshing.

At the synthesis of gears with localized contact, the deviations of geometry and layout of elements should be chosen so that Eqs. (1)–(4) comply at the predesigned point for approximated gearing or along the contact line for gears with a constant gear ratio.

Methods of providing the localized contact in a gear are related to a complex of design solutions at the stage of designing the gear itself, as well as tool support for manufacturing of a worm, gearwheel and hob [9, 15]. Particularly, the profile crowning is achieved by intended deviations of the basic profile of the operating or generating worm thread near the edge of an active surface. In this case, the pitch surfaces of these worms and also the parameters of the machine-tool and operating gearing coincide as in the case of synthesis of meshing with theoretically linear contact.

Longitudinal crowning is provided by modification of the shape and dimensions of pitch surfaces of worms 1 and 3 more often by increasing the generating worm diameter with respect to the operating worm. Worm parameters should be interrelated here by the equation

$$(\mathbf{n}\,\mathbf{V}_3)=(\mathbf{n}\,\mathbf{V}_5), \tag{5}$$

derived from the system (1)–(4) by replacing relative velocity vectors with the difference of linear velocity vectors of corresponding elements and further elimination of linear velocity vectors of worms *1*, *4* and gearwheel *2* [8].

When increasing the diameter of the generating worm relative to the operating one, their axes do not coincide in gearing *1–2* and *4–2*. In this case, four vectors of relative velocities $\mathbf{V}_{12}$, $\mathbf{V}_{31}$, $\mathbf{V}_{42}$, $\mathbf{V}_{53}$ should be coplanar at the predesigned point. This means that two mixed products of these vectors should be equal to zero simultaneously:

$$\mathbf{V}_{12}\cdot\mathbf{V}_{42}\times\mathbf{V}_{53}=0,\quad \mathbf{V}_{13}\cdot\mathbf{V}_{12}\times\mathbf{V}_{42}=0 \tag{6}$$

Equations (6) represent the necessary and sufficient conditions for choosing the mutual position of axes and also ratios of angular velocities $\omega_1\ldots\omega_5$ or helix parameter values for the operating $P_1=V_3/\omega_1$ and generating $p_4=V_5/\omega_4$ worms. These requirements are less strict than the usual requirement of collinearity of vectors $\mathbf{V}_{31}$ and $\mathbf{V}_{54}$, that is, the tangency of thread lines of the operating and generating worms, thus widening the possibilities of synthesis of gears with localized contact.

As for cylindrical worms, substituting the values of vectors $\mathbf{V}_3=p_1\boldsymbol{\omega}_1$ and $\mathbf{V}_5=p_4\boldsymbol{\omega}_4$ into (5) and opening scalar products, we can obtain the direct relation between numbers z of threads, axial modules $m_x=2p/z$, worm lead angles γ and normal profile angles α_n of worms *1* and *4* as follows:

$$m_{x1}\cos\gamma_1\cos\alpha_{n1}=m_{x4}\cos\gamma_4\cos\alpha_{n4} \tag{7}$$

In particular, from Eq. (7) it follows that, by variation of the normal angle α_{n1} of the worm thread profile relative to a profile angle α_{n4} of a hob, we can choose the axial modules of a worm and a hob equal to each other and select them from a standard range. Specific examples of synthesis of gears with cylindrical worms based on conditions (5) and (6) are considered in detail in [9, 15, 16].

2 Axes of Meshing and Their Analogs in General Type Worm Gears

As is known, investigation of worm gears with cylindrical worms is considerably simplified by applying the properties of axes of meshing—a pair of conjugate straight lines, lying on the meshing surface and intersected by all contact normals of a gear [1, 12, 13]. Their position in the fixed space is independent of the thread axial profile that allows estimating the arrangement of contact lines without detailed research of the meshing surface [6]. In every point of axis of meshing, the relative velocity vector is tangent to the worm helix. On the one hand, this property allows for choosing them as predesigned points of tangency of pitch surfaces [13]. On the other hand, it simplifies defining such quality characteristics of gearing as the specific sliding and instant efficiency [17]. The purpose of this paper is to determine lines in some way analogous to axes of meshing on the meshing surface of a general type worm gear.

Let us refer to Fig. 2, where skew lines $O_1{-}O_1, O_2{-}O_2$ and $O_3{-}O_3$ represent axes around which a globoid worm, a double-enveloping gearwheel and a tool are rotating with angular velocities ω_1, ω_2 and ω_3 correspondingly. Nominal and machine tool center distances are indicated as a and a_0. Position of axis $O_3{-}O_3$ is also specified by value Δz, and the distance between the plane of cutting edge rotation and the machine tool axial perpendicular, by value h. The right Cartesian coordinate system XYZ is rigidly bound to axes $O_1 - O_1$ of a worm and $O_2 - O_2$ of a gearwheel. The axis of a tool is pivoted about the worm axis to the angle φ to such a position where the rotation plane of the cutting edge passes through the assigned point K of the fixed space.

The lines to be determined in this paper are the geometric locus of points of the fixed space where the vector of relative velocity of gear elements $\mathbf{V}_{12}=\mathbf{V}_1-\mathbf{V}_2$ is tangent to the worm helix, i.e., it is collinear to the vector $\mathbf{V}_{13}=\mathbf{V}_1-\mathbf{V}_3$ of relative velocity in machine-tool gearing during the worm generation process.

The normal vector $\mathbf{n}$ to the worm surface can be expressed by a vector $\mathbf{t}$ of tangent to the cutting edge of a tool as vector product

$$\mathbf{n} = \mathbf{t}\times\mathbf{V}_{13}, \tag{8}$$

Let us substitute the expression (8) and the condition of collinearity of relative velocities vectors

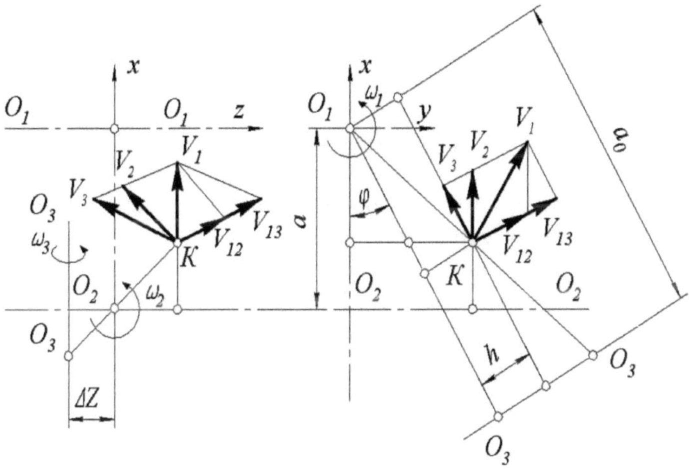

Fig. 2 To derivation of calculation dependences

$$\mathbf{V}_{12} = \boldsymbol{\lambda}\,\mathbf{V}_{13}(\boldsymbol{\lambda} \neq \mathbf{0}) \tag{9}$$

into the equation of meshing $\mathbf{n}\mathbf{V}_{12}$=0. Then, this equation becomes the identity

$$\lambda \mathbf{t}\,\mathbf{V}_{13}\mathbf{V}_{13} = 0.$$

This means that, among the set of points defined by the equation of meshing, there is a subset of points satisfying condition (9), i.e., the lines to be sought are pre-assigned to belong to the meshing surface.

Having projected vectors $\mathbf{V}_{12}$ and $\mathbf{V}_{13}$ on axes of the assigned coordinate system, we present the condition of their collinearity in the form $V_{12x}/V_{13x} = V_{12y}/V_{13y} = V_{12z}/V_{13z}$.

By substituting the expressions for projections of linear velocities $\mathbf{V}_1$ $\mathbf{V}_2$ and $\mathbf{V}_3$ in this system with coordinates of the desired point and angular velocities, we obtain:

$$\frac{\omega_1 y - \omega_2 z}{\omega_1 y - \omega_3 (z + \Delta z)\cos\varphi} = \frac{\omega_1 x}{\omega_1 x - \omega_3 (z + \Delta z)\sin\varphi} = \frac{\omega_2 (a + x)}{\omega_3 (a_0 + x\sec\varphi - htan\varphi)} \tag{10}$$

Having transformed system (10) and designated u_{12}=ω_1/ω_2; u_{13}=ω_1/ω_3, we get expressions to define coordinates x and z of the line to be sought by an arbitrary parameter φ

$$\frac{u_{12}}{a + x} = \frac{u_{13}}{a_0 + x\sec\varphi - h\tan\varphi} - \frac{\Delta z \sin\varphi\cos\varphi}{a_0 x\cos\varphi - a(x - h\sin\varphi)}, \tag{11}$$

$$\mathrm{z} = \Delta \mathrm{z}\frac{(a{+}x)(x{-}h\sin\varphi)}{a_0 x\cos\varphi{-}a(x{-}h\sin\varphi)} \tag{12}$$

The third coordinate y is defined directly from the drawings given in Fig. 2 by the expression:

$$y = h\sec\varphi - x\tan\varphi. \tag{13}$$

In explicit form the abscissa x is defined from the quadric equation obtained from the transformation of expression (11),

$$aX^2 - (a_0 - u_{23}a + u_{21}\Delta z\sin\varphi)X + u_{22}a_0 = 0 \tag{14}$$

where parameter X is defined as

$$X = \frac{a_0 - a\sec\varphi - \mathrm{h}\tan\varphi}{a{+}x} + \sec\varphi$$

Since Eq. (14) gives two values of the parameter X, system (12)–(14) defines two conjugate lines on the meshing surface of a general type worm gear. It is quite obvious that when $\varphi = 0$, one of them passes through the point where relative velocity vectors $\mathbf{V_{12}}$ and $\mathbf{V_{13}}$ coincide both in value and direction. Similarly to designations accepted in [12, 13] for axes of meshing in a gear with a cylindrical worm, we name this line II–II, and the assigned point as P_2, correspondingly. Its coordinates are defined directly from (11)–(13) by the expressions:

$$x_{P2\cdot} = -(a_0 - au_{23})/(1 - u_{23}),\ y_{p2} = h,\ z_{p2} = \Delta z/(1 - u_{23}). \qquad (15)$$

The axis of meshing I–I with $\varphi = 0$ passes through point P_1 located on the plane $x = 0$. Its z-axis, after disclosing indeterminacy in system (10), is defined by the expression:

$$z_{P1} = hu_{12}\frac{a_0 - au_{23}}{a_0 - a} + \frac{\Delta za}{a_0 - a} \qquad (16)$$

Both conjugate lines in the general case are spatial curves of the second order, and both consist of two branches, one of which is obtained by changing φ from $-\pi/2$ to $+\pi/2$, and the second one, in the range $\pi/2 < \varphi < 3\pi/2$. Henceforth, these second branches are indicated as I′–I′ and II′–II′, correspondingly.

It should be noted that the analogy of the determined lines with axis of meshing is not complete, because, in the general case, they are not crossed by all contact normals of a gear. However, since the position of these lines does not depend on the shape of the cutting edge generating a worm, their existence allows for facilitation of the analysis of the surface of meshing and its orientation in the space, as well as simplification of the definition of certain quality characteristics of a gear.

At the same time, condition (9) gives us the ability to solve tasks of synthesis of new types of worm gears. In this case, according to the assigned center distance a, gear ratio u_{12} and coordinates x, y, z of some considered point of contact, having arbitrarily chosen two of the four values h, Δa, Δz, u_{13}, another two values should be defined from system (10). The passage of the meshing surface through the designed point and tangency of gear pitch surfaces at this point will thereby be provided.

In special cases of a general type worm gear when one of the modification parameters equals zero or infinity, expressions (11)–(13) can be simplified considerably. Let us analyze some of these cases.

3 Modified Double-Enveloping Worm Gears

Let us analyze the gear where a worm is modified by method *hua* [18], i.e., by increasing the machine tool interaxial distance a_0 and gear ratio u_{13} relative to parameters a and u_{12} of a gear, and also by correction of the tool cutting edge relative to the mean plane to the value h. In this case, the value $\Delta z = 0$ and system of Eqs. (4) and (5) are separated into two systems. The first solution will be found by equating the denominator of expression (12) to zero, whence

$$x_1 = (a\,h\,\sin\varphi)/(a - a_0\cos\varphi) \tag{17}$$

To define appropriate values of the z-coordinate of line I–I, we replace Δz in (11) by z according to (12), and having substituted x_1 from (17) into this expression, we solve it relative to z:

$$z_1 = h(u_{12}\,a_0 - u_{13}\,a)/(a_0\cos\varphi - a) \tag{18}$$

The y-coordinate of the desired line here and further (unless otherwise specified) is defined by formula (13).

The second solution of system (11), (12) in the case of $\Delta z = 0$ is obvious:

$$z_2 = 0;\;\; x_2(a_0 - au_{23} - h\tan\varphi)/(u_{23} - \sec\varphi), \tag{19}$$

i.e., line II–II belongs to the plane that passes through the gearwheel axis and center perpendicular.

If we add $h = 0$ to the conditions being considered, i.e., proceed to a gear modified by method *au* [11], then according to (17), abscissa x_1 can be different from zero only when $\cos\varphi_c = a/a_0 = \text{const}$. Hence,

$$y_1/x_1 = -\sqrt{a_0^2 - a^2}/a,\; z_1/x_1 = (u_{13}a - u_{12}a_0)$$

i.e., line I–I is a straight line that passes through the origin of coordinates and forms the angle φ_c with the plane *YZ* (Fig. 3a). Line II–II is defined in this case by the expressions

$$x_2 = (a_0 - a\,u_{23})/(u_{23} - \sec\varphi),\; y_2 = -x_2\tan\varphi$$

and corresponds to a hyperbola with asymptotes:

$$x_a = \frac{(a_0 - au_{23})u_{23}}{u_{23} - 1} \pm \frac{y_a}{\sqrt{u_{23}^2 - 1}} \tag{20}$$

It should be noted that the extreme point of a curve line of thread longitudinal modification by the method *au* will be shifted in the direction of worm exit from meshing, as is required by GOST 17696-72 in the case when point P_2 lies inside the pitch globoid of a worm.

For the gear modified by method *ah* [18], i.e., when $u_{13} = u_{12}$, transformation of expressions (17)–(19) presents no difficulties. We pay attention only to the fact that, in this case, point P_2 passes to infinity.

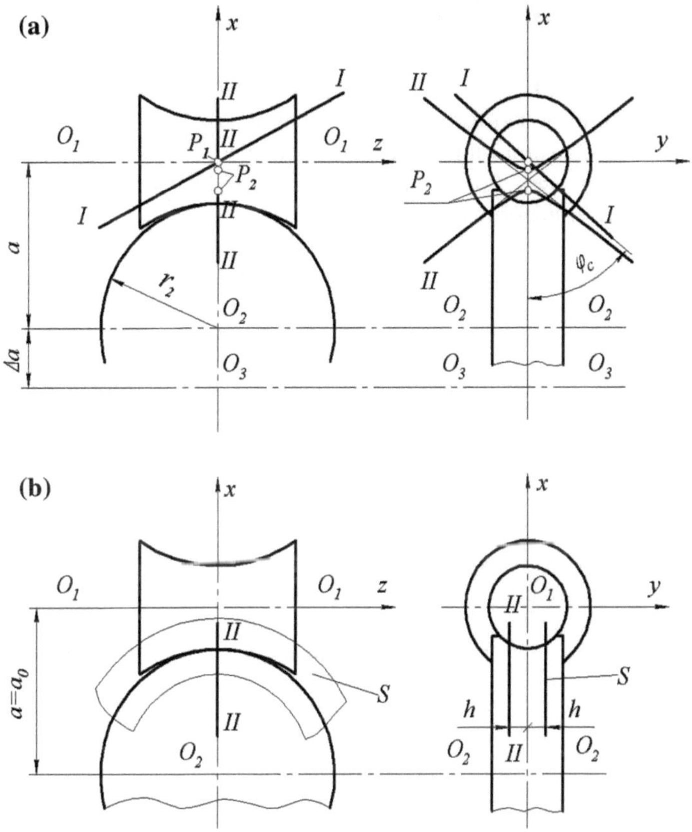

Fig. 3 Conjugate lines of meshing for a double-enveloping worm gear modified by **a**—method *au*; **b**—method *ah*

In a gear modified only by shift h of the cutting edge (Fig. 3b), i.e., when $u_{13} = u_{12}$ and $\Delta a = 0$, any point lying on the surface of its rotation satisfies the condition of the collinearity of vectors $\mathbf{V_{12}}$ and $\mathbf{V_{13}}$, i.e., special lines spread into plane $y = h$ (plane S) dividing the generated and undercut zones of the active surface of the gearwheel tooth. In a classic double-enveloping worm gear the plane S coincides with its mean plane. In addition, Eqs. (13) and (19) in the given case are also satisfied by the straight line II–II ($z_2 = 0$, $y_2 = -h$) lying on the second cavity of the surface of meshing.

4 Gears with Bevel Worms

The bevel worm is resulted from the general type worm in the case when the axis 0_3–0_3 passes to infinity and for all that the relation $\Delta z/\Delta a = \tan\beta$ remains constant and velocity of translation motion of cutting edge $V_3 = \omega_2 r_2$, where r_2 is the pitch radius of the gearwheel. For such a gear, Eqs. (11) and (12), after disclosing indeterminacy, are transformed to:

$$\frac{1}{a+x} = \frac{1}{r_2\cos\beta} - \frac{\tan\beta\sin\varphi}{u_{12}x}, \tag{21}$$

$$z = \tan\varphi\frac{(a+x)(x-h\sin\varphi)}{x\cos\varphi} \tag{22}$$

In Fig. 4, conjugate lines of meshing for the gear with a bevel worm with $h = 0$ are represented. These lines were plotted according to Eqs. (13), (21) and (22). Abscissas x_a of asymptotes of these lines are defined as roots of the equation of the second order (14) where $\sin\varphi = \pm 1$, and the angle of their inclination to the z-axis is defined by the relation

$$z_a/y_a = \pm\tan\beta[(a+x_a)/x_a)]$$

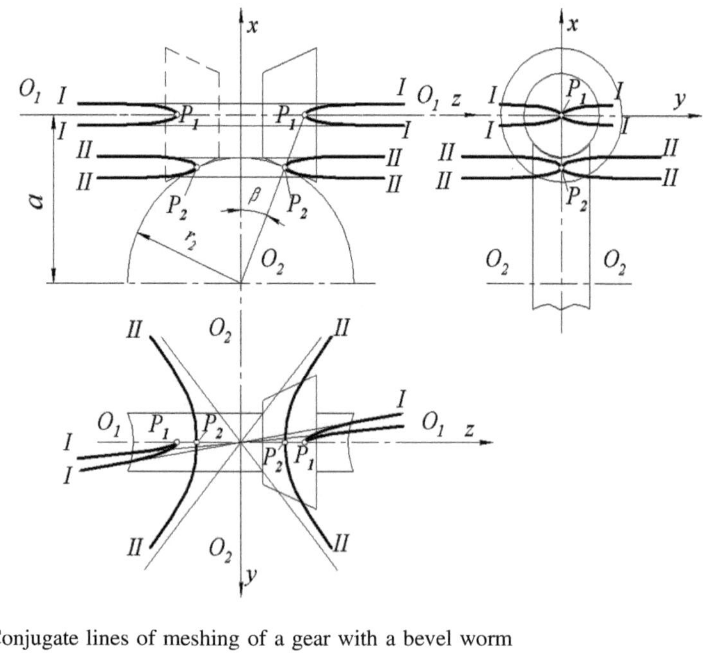

Fig. 4 Conjugate lines of meshing of a gear with a bevel worm

During synthesis of gears with a bevel worm and a double-enveloping worm gearwheel, the contact zone of the active surfaces should be chosen in the neighborhood of point $\boldsymbol{P_2}$. The symmetry of lines II–II and II′–II′ relative to the center perpendicular line allows, in this case, for mounting two bevel worms rigidly on one shaft conjugated with one gearwheel. During synthesis of spiroid gears, the worm parameters should be chosen so that line I–I or I′–I′ could pass through the desirable contact zone.

It should be noted that in all the considered examples the cutting edge that generates a double-enveloping or bevel worm was taken in a plane ($h = const$). However, the obtained dependencies also remain true when substituting into them any other law $h = h\ (r_3)$, where r_3 is the distance from an arbitrary point of the cutting edge to its axis of rotation.

5 Gears with Face Worms

It should be noted that, with increase of the angle β, points P_1 and P_1' move off the center perpendicular line. Specifically, in a gear with a face worm when β= 90° (Fig. 5), the singular line I–I passes to infinity. In this case, line II–II is described by equations

$$x_2 = -\ a; \qquad z_2 = \frac{au_{12}r_2\sec\varphi}{au_{12}+r_2\sin\varphi} \tag{23}$$

and is a planar hyperbola with asymptotes $y_a/z_a = (h \pm a)\ \{au_{12} \pm r_2)/au_2r_2$.

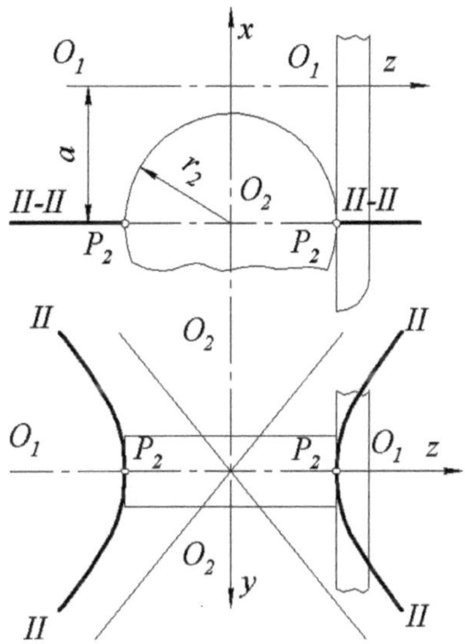

Fig. 5 Conjugate lines of meshing of a gear with a face worm

In machine drives, the worm gears with a face worm are very seldom used. However, a face generating worm or a spiral-disc milling cutter is effectively used for cutting cylindrical gearwheels with arched teeth. In this case, the presence of the line being described by expressions (23) on the contact surface can noticeably relieve the research of machine-tool gearing.

6 Gears with Cylindrical Worms

Equations of the considered lines of meshing for gears with cylindrical worms are obtained from Eqs. (13), (21) and (22) if the angle β is considered to be equal to zero. In this case, Eq. (11) at any φ is satisfied by two values x

$$x_1 = 0, x_2 = r_2 - a \tag{24}$$

After substituting these values into Eqs. (13), (22) and evaluation of indeterminate form in the case of $x_1 = 0$, the relation z/y independently of the value h is represented in the form

$$\frac{z_1}{y_1} = u_{12}\left(\frac{a}{r_2} - 1\right), \frac{z_2}{y_2} = 0 \tag{25}$$

As would be expected, Eqs. (24) and (25) describe the known pair of axes of meshing of an orthogonal worm gear.

In Fig. 6, the gear with a cylindrical worm and a double rim gearwheel is shown [6]. The zone of contact of the worm with a globoid rim passes across axis II–II and, with the face rim, is located in the neighborhood of point P_1 belonging to axis I–I.

Fig. 6 Gear with a cylindrical worm and two zones of meshing

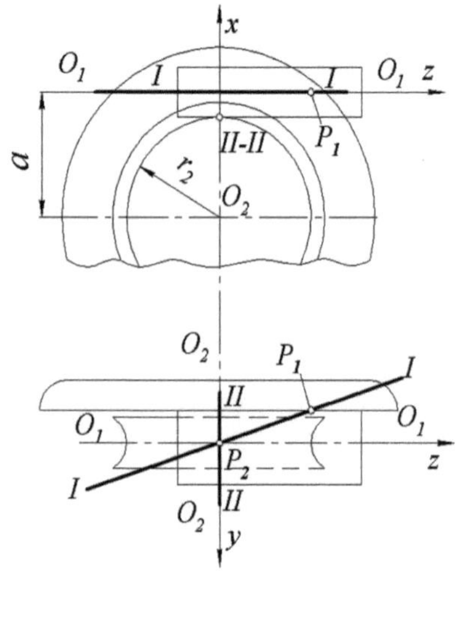

It should be noted that vector Eqs. (1)–(6), (8) and (9) described in this paper are independent of the coordinate system selection and all consequent conclusions are true with any skew angle of worm and gearwheel axes.

However, in order to obtain the design equations which define the position of the determined lines, the vector equations should be projected to the axes of some coordinate system. In the given work this system was linked with reciprocally perpendicular axes of a worm and a gearwheel. Therefore, the design equations are related only to orthogonal gears.

For general type non-orthogonal gears, the design equations defining the position of analogs to axes of meshing should include skew angles, becoming more complex.

The pair of axes of meshing for non-orthogonal gears with a cylindrical worm was found in [12, 13]. The detailed research of such gears was accomplished in [5]. In particular, in this paper, such a range of combinations of the basic parameters of gears was discovered where the quadratic equation, which is equivalent to Eq. (14), has no solution and axes of meshing do not exist. Moreover, it was shown that, paradoxically, the best operating characteristics for a gear are provided exactly within this range.

7 Conclusion

Meshing space for a worm gear with general type worms, including double-enveloping and bevel ones, is considered in this paper. A pair of conjugate lines is determined in this space. Their position is defined by the position of the surface of rotation or translational motion of the cutting edge which generates the worm and is independent of this edge profile.

The determined lines are pre-assigned to belong to the meshing surface, and in this essence, they are analogous to axes of meshing of gears with cylindrical worms. Applying the properties of the pointed lines allows, to a certain extent, for simplifying the analysis of known and synthesis of new general type worm gears.

The research of general type worm gears executed for today can in no way be regarded as complete, and investigations in this area should be continued.

References

1. Argyris J, Litvin FL, Peng A, Stadtfeld HJ (1998) Axes of meshing and their application in theory of gearing. Comput Methods Appl Mech Eng 163(1–4):293–310
2. Diker YaI, Sagin LI (1960) Basics of globoid worm gears production. In: Proceedings of TSNIITMASH, vol 96. MashGIZ, Moscow (in Russian)
3. Georgiev AK, Goldfarb VI (1972) Aspects of the geometric theory and the results of research spiroid gears with cylindrical worms. In: Mechanics of machines, (31–32). Nauka, Moscow, pp70–80 (in Russian)

4. Goldfarb VI (1995) Theory of design and practice of development of spiroid gearing. In: Proceedings of congress gear transmissions'95, vol 2. Sofia, pp 1–5
5. Goldfarb VI, Trubachev ES, Puzanov VU (2009) New possibilities of non-orthogonal worm gears. In: Proceedings of the 3rd international conference "power transmissions 09". Chalkidiki, pp 139–145
6. Korostelev LV, Baltadzhi SA, Lagutin SA (1973) Worm gears with two areas of meshing. In: Theory of transmissions in machines. Nauka, Moscow, pp 15–20 (in Russian)
7. Korostelev LV, Baltadzhi SA, Lagutin SA (1978) Conjugate lines of action in general type worm gears. Mashinovedenie (5):49–56 (in Russian)
8. Lagutin SA (1999) Local synthesis of general type worm gearing and its applications. In: Proceedings of the 4th world congress on gearing and power transmissions, vol 1. Paris, pp 501–506
9. Lagutin SA, Verhovski AV, Dolotov SV (2006) Technological design of worm gears with a localized contact. In: Proceedings of 2nd international conference "power transmissions-2006". Novi Sad, Serbia, pp 177–182
10. Lagutin SA, Gudov EA (2010) Special types of worm gears for rolling mills. In: Proceedings of the international conference on gears. VDI-Berichte 2108.2, Munich, pp 1337–1347
11. Lagutin S, Gudov E, Fedotov B (2011) Manufacturing and load rating of modified globoid gears. Balk J Mech Transm (BJMT) 1(2):45–53
12. Litvin FL (1968) Theory of gearing, 2nd edn. Nauka, Moscow (in Russian)
13. Litvin FL (1998) Development of gear technology and theory of gearing, NASA reference publication 1406, ARL-TR-1500, 114 p
14. Pavlov VA (1968) Globoid gears with initial helical wheel. Gears and worm gears. Mashinostroenie, Leningrad, pp 46–58 (in Russian)
15. Sandler AI, Lagutin SA, Verhovski AV (2008) Manufacturing of worm gears. Mashinostroenie, Moscow, 272 p (in Russian)
16. Sandler AI, Lagutin SA, Gudov EA (2014) Longitudinal contact localization in worm gears. Russ Eng Res 34:480–486
17. Shcherbakov VN, Lagutin SA (1970) Features of contact on the axes of meshing in worm gears. Theory Mech Mach (8):37–42. Vyshcha Shkola, Kharkiv (in Russian)
18. Zak PS (1989) Modification of globoid gears with localized contact by the machine adjustments. Vestnik Mashinostroeniya (1):8–13 (in Russian)

Development of Kinematic Method of Theory of Gearing to Determine Areas of Tooth Flanks Produced by Jogs of Generating Solids

D. Babichev

Abstract The paper describes application of kinematical method of the theory of gearing to determine surfaces and lines produced by generating method related not only to smooth generating surfaces and lines but with the presence of singular points—jogs—in points of adjacent areas. Three types of jogs are considered: jog at the point of intersection of two plane profiles; jog along the line of intersection of two surfaces; jog at the point of intersection of three surfaces. Three types of geometrical images generated by these three types of jogs are proposed: fan, wedge, bunch of normal lines. Efficiency of applying these images when analyzing the process of tooth generation and meshing is shown. Mathematical models are developed and presented for generating processes in the presence of jogs on generating surfaces and lines. Examples of computer modeling with application of these models are given. It is emphasized that the developed method operates much faster than non-differential methods which have been considered to be applicable only when computer modeling the generating processes in the presence of jogs.

Keywords Theory of gearing · Kinematic method · Gearing · Tooth machining · Methods of enveloping and generating

1 Problem Analysis and Investigation Tasks

1.1 Problem Analysis

The scientific foundation of designing gears and tooth-machining tools is the theory of gearing (TG). Its main investigation objects are surfaces produced by generating methods, that is, at complex relative motions of elements of gears and in machine-tool gearing. There are two groups of methods of such a generation

D. Babichev (✉)
Tyumen State Oil and Gas University, Tyumen, Russia
e-mail: babichevdt@rambler.ru

V. Goldfarb and N. Barmina (eds.), *Theory and Practice of Gearing and Transmissions*, Mechanisms and Machine Science 34,
DOI 10.1007/978-3-319-19740-1_9

process analysis in TG: differential and non-differential. The foundation of differential methods [1–6] is the theory of envelopes. Its kinematical interpretation means that points are determined on the generating surface, for which the vector of relative velocity $\mathbf{V}_{12}$ is perpendicular to the vector $\mathbf{n}$ normal to the generating surface, that is, points, where the equation of meshing is $\mathbf{V}_{12} \cdot \mathbf{n} = 0$. The foundation of non-differential methods [6, 7] is the direct time tracking the position of the generating element with respect to the element with the surface to be generated. It also involves memorizing the coordinates of points penetrating more deeply into the workpiece solid than the previously produced surface.

Figure 1 illustrates the above given concepts, presenting results of analysis of the generation of the involute tooth by these two methods. The gearwheel with internal teeth ($z_2 = 20$) was produced by a shaping cutter ($z_0 = 15$, $m = 5$).

Figure 1a presents three areas of the tooth profile obtained by a classical kinematic method. The active part of the involute profile of the shaping cutter was divided here into 10 parts, with its top cutting edge (arc of circumference) divided into 4 parts. According to the classical method of the theory of gearing, 27 points were determined for the gearwheel, the points being conjugated with 27 points on the shaping cutter. It is shown in Fig. 1a that there are two gaps between three determined segments of profiles: fillet surfaces are absent. This happened because singular points met on the generating element (let us call the extreme points at the intersections of the flank and top cutting edges "*jogs of the generating profile*"). According to all laws of differential geometry, it is impossible to determine the tangent and normal to the line or to the surface at such a singular point. Even leading experts on the theory of generation have concluded here (see, for instance, [6, p. 464 and Fig. 14.21]) that the envelope does not exist for singular points of the generating element and it is impossible to determine differentially the part of the real surface on the produced unit which is generated by singular points.

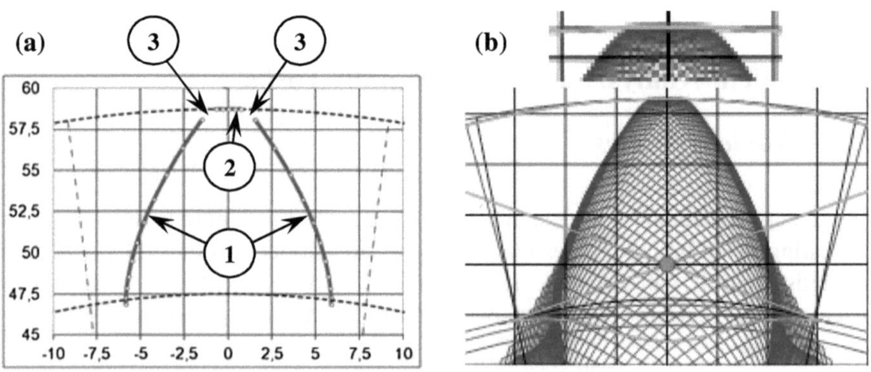

Fig. 1 Two methods of analyzing the generation process for gearwheels with internal teeth: **a** differential (kinematic) method; **b** non-differential method; *1* involute areas of gearwheel tooth profile; *2* arc of addendum circle; *3* transient curves, generated by jogs in Fig. 1b

Note that, for the given example (Fig. 1a), the kinematic method did not reveal any trouble related to the possibility of truncating the top of the gearwheel tooth by a shaping cutter within the zone of secondary cutting.

In Fig. 1b, the gearwheel tooth profile was determined by a non-differential method, by tracking the position of the shaping cutter tooth profile with respect to the gearwheel workpiece. The whole profile of one root of the gearwheel was determined here as it would be in reality:

- arc of the gearwheel root circle (absolutely the same as in Fig. 1a);
- two fillet surfaces generated by jogs of the shaping cutter profile (no fillet surfaces were obtained in Fig. 1a);
- involute areas of tooth profiles (similar to those in Fig. 1a, but truncated on tooth addendums);
- lines of the involute truncation on tooth addendums (caused by the same jogs).

Note that fillet surfaces and truncation lines are not smooth lines: Prof. G.I. Sheveleva proposed calling them "covering" [6]. The irregular character of covering lines can be clearly seen on the truncation lines in Fig. 1b.

Let us compare the two methods under consideration. Differential methods require less calculation (sometimes by one or even two orders). They allow for determining the curvature but cannot reveal truncations of the generated teeth and, therefore, are considered less reliable. Another drawback of differential methods is that they do not provide a real surface generated on the produced unit, but only some abstract "thin film" which can be self-intersected and located inside the solid of the generating element.

The urgent problem of the theory of gearing is the creation of methods of analyzing the generation process combining the reliability of non-differential methods with the advantages of differential ones. An important step in this direction may be development of a kinematic method of determining the surfaces and lines produced according to the generating method by surfaces and lines containing various jogs. This paper presents fundamentals of such a kinematic method developed and tested by the author [8–10].

1.2 Possible Ways of Solving the Problem

The concept of jogs as specific features of generating surfaces most likely comes from P.R. Rodin. In 1977, he wrote [11, p. 52]: «*… the surface of the part comprising a number of adjacent areas can be considered as a single surface. The point of surface profile jog located at the boundary of adjacent areas can be considered as the area of the circumference arc with the radius tending to zero*» . But, as far as we know, there was no further development of this idea and no profound mathematical models were created based on this idea.

In order to outline the ways of solving the problem of the development of mathematical models of generating by jogs, let us consider the formation of the involute gearwheel root by a tooth of the rack shown in Fig. 2. The rack tooth is the broken line consisting of three segments of straight lines (AB–BC–CD) forming two jogs at boundaries of adjacent areas (points B and C). Three segments of the given straight lines form three separate curves at the root: segments of involutes ab' and $c'd$ and the arc of circumference $b''c''$. Rack tooth jogs at points B and C form two transient curves, $b'b''$ and $c''c'$, which are segments of two curves. These five exact segments form a single profile of one gearwheel root. Both jogs—points B and C—are moving with respect to the gearwheel along similar trajectories, which are equidistant lines of the elongated involutes. In the position shown in Fig. 2, point B is tangent to the transient curve $b'b''$, that is, it is the contact point, which point C is not. The classical TG cannot explain its reason mathematically and does not allow for determining the points on transient curves by differential methods.

Let us show how one can determine whether the jog has made contact with the generated surface or not. It was mentioned above that classical geometry implies that there are no tangent and normal lines in singular points of surfaces and lines. Geometry does not forbid supposing that at the intersection point of two lines an infinite set of normal lines forming a circle exists. But when we round the intersection areas of the straight line BC with flank lines AB (at point B) and CD (at point D) by circumference arcs, at any rounding radius, including $R \rightarrow 0$, they will be *circumference arcs* (but not circles!). The angles of these arcs do not depend on the value of the rounding radius R and they are equal to $(90° - \alpha)$ for the example in Fig. 2. Normal lines to rounding arcs also form sectors of circles rather than the whole circles. When we transmit the point along the rounding from one straight line to the other (adjacent one) within the jog, the normal line smoothly rotates by the rounding angle, thus forming a set of normal lines, resembling a fan. That is why such a set of normal lines was called *the fan of normal lines*. Fans of normal lines are present not only at points of plane line intersection, but at all points of surface intersection. This is demonstrated particularly clearly in Fig. 10 in Sect. 6.

So why is point B in Fig. 2 tangent to the fillet surfaces and therefore the generating one while point C is not tangent and therefore not generating? The

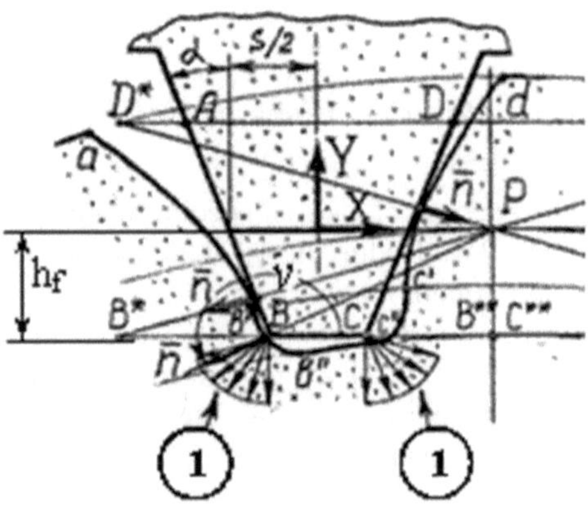

1 – fans of normal lines for two plane jogs

Fig. 2 Explanation of why point *B* is a contact point and point *C* is not

answer is simple: one of the normal lines of the fan passes through pitch point P at point B, that is, the main theorem of planar gearing is observed at point B, and for this, the only normal line in the fan, the equation of meshing is equal to zero: $\mathbf{V}_{12} \cdot \mathbf{n} = 0$. That is why, in this position of elements, point B forms a fillet surface at the tooth root. At point C, the fan itself does not contain the normal line passing through pitch point P, and that is why point C does not make contact with the fillet surfaces in this position.

Development of mathematical models by application of sets of normal lines in jogs of generating surfaces and lines is a possible way of solving the stated problem of development of the kinematic method of determining the surfaces and lines produced by jogs of generating solids.

1.3 Proposed Solution

In order to solve the stated problem, it is necessary to:

- analyze the techniques and calculation equations applied in analyzing the processes of generating by kinematic methods;
- classify the types of jogs and reveal the geometric images produced by them (one image is the fan of normal lines described above);
- reveal and consider the general features of analysis of the processes of generating by jogs;
- develop the theory of generating by jogs, obtain fundamental calculation equations and create the basic algorithms for determining the surfaces produced by all types of jogs;
- implement the techniques and algorithms by computer-aided means, solve the sample tasks by means of developed computer programs and evaluate the obtained results;

Solving the enumerated tasks is the main purpose of the present paper.

2 Equations of Generating According to the Classical Theory of Gearing

Prior to presenting the equations of generating, let us give that explanation which is considered to be essential. Tooth surfaces and profiles are not considered as abstract thin films and lines, but as boundaries of two environments: material of the produced unit and the ambient space. This statement is supported mathematically by <u>directing</u> the normal line to the surface or profile <u>*out of the unit solid*</u>. It allows for performing the calculation of curvature radii according to simple rules: the curvature is positive for the convex solid and negative for the concave solid. If the reduced curvature in the gearing is positive, the tangency of solids is correct; if it is negative, the tangency is impossible, since the conjugated surface is produced inside the solid of the generating element. This phenomenon is well-known to gear-machining experts, since it causes tooth undercutting in machine-tool gearing.

Equations are given below for surfaces and profiles generated by enveloping methods for the three most common types of gearing: spatial single-parametric, plane single-parametric and spatial double-parametric. Equations correspond to a differential kinematic method of investigation. Problems of calculating the curvatures are not considered in this paper, having been addressed, in particular, in [12].

Equations of the surface P_2 *generated by the method of single-parametric enveloping* can be presented as [9]:

$$\mathbf{r}_1 = \mathbf{r}_1(u, v); \tag{1a}$$

$$V_N = F(u, v, \varphi) = \mathbf{V}_{12} \cdot \mathbf{n} = 0; \tag{1b}$$

$$\mathbf{r}_2 = \mathbf{r}_2(u, v, \varphi) = \mathbf{M}_{21}(\varphi) \cdot \mathbf{r}_1(u, v); \tag{1c}$$

$$\mathbf{n}_2 = \mathbf{n}_2(u, v, \varphi) = -\mathbf{M}_{21}(\varphi) \cdot \mathbf{n}_1(u, v). \tag{1d}$$

where component of Eqs. (1a, 1b, 1c, 1d) vectors are determined by analytic operations:

$$\mathbf{n}_1 = \mathbf{n}_1(u, v) = W_N(\mathbf{r}_1(u, v)); \tag{2a}$$

$$\mathbf{V}_{12} = \mathbf{V}_{12}(u, v, \varphi) = \mathbf{V}_{12}(\mathbf{r}_1, \varphi) = W_V(\mathbf{M}_{21}(\varphi), \mathbf{r}_1(u, v)). \tag{2b}$$

Here, (1a) is the equation of generating surface P_1; (1b) is the equation of meshing (V_N is the velocity of feeding-in [2]); (1c) is the equation of coordinate transformation; (1d) is the unit normal vector to surface P_2 (the "–" sign means that normal lines are directed towards conjugated surfaces); (2a) is the unit normal vector to surface P_1; (2b) is the relative velocity; u, v are two curvilinear coordinates on surface P_1; φ is the parameter of enveloping; $\mathbf{M}_{21}$ is the transformation matrix of point coordinates and vector projections from a system of coordinates $X_1Y_1Z_1$ into a system $X_2Y_2Z_2$; W_N is some operand from methods of differential geometry, its participation in Eq. (1a) of surface P_1 resulting in equations calculating the vector projections of the normal line $\mathbf{n_1}$ to surface P_1; W_V is the operand of vector and matrix operations, including differentiating [4], its participation in matrix $\mathbf{M}_{21}$ and Eq. (1a) of surface P_1 resulting in equations calculating the vector projections of the relative velocity $\mathbf{V}_{12}$.

Note that it is acceptable in the theory of gearing [4, 6] to describe conjugated surface P_2 only by Eqs. (1a)–(1c). We represent equation P_2 in complete and operator form (1a)–(2b), which is bulkier but also more correct. Such notation contains *all operations* (including processes W_N and W_V, at least according to names, but not contents) to be executed *with all data and equations* by the generating surface $\{\mathbf{r_1}(u,v)\}$ and movements of elements $\{\mathbf{M_{21}}(\varphi)\}$ in order to obtain the conjugated surface $\{\mathbf{r}_2\}$. That is, this form of notation allows for *reflecting all processes and algorithms* concurrent to the generation of surfaces and contact of solids. The complete operator form in this paper is also suitable since we consider peculiarities of determining normal lines to conjugated surfaces both at points of jogs and to the generated surface. We therefore consider the operand W_N in the Eq. (2a) and apply (1d).

Note also a fact well-known to all experts in the theory of gearing: classical equations of the enveloping surface (1a)–(1c), like Eqs. (1a)–(2b), assign only *the structure of the mathematical model* in complete and operator form. But they do not *assign the contents and sequence of actions* to be executed by the researcher when analyzing a specific gearing. These contents and sequence of action can be explained more simply by means of the operator form of Eqs. (1a)–(2b):

Step 1. Make a calculation scheme to derive equations of the generating surface and obtain Eq. (1a).

Step 2. Applying the methods of differential geometry (operand W_N in (2a)), it is necessary to obtain formulas to calculate projections of the unit normal vector (the normal line in jogs is determined according to radically different techniques described below). One of the tasks of this paper is, therefore, to formalize the contents of operand W_N.

Step 3. Make a calculation scheme for gearing and obtain transformation equations for coordinates of points and vector projections, that is, to make matrix $\mathbf{M}_{21}$.

Step 4. Obtain Eq. (2b) to determine the vector of relative velocity (to apply operand W_V) and derive the equation of meshing (1b).

Step 5. Develop algorithms for solving and calculating the Eqs. (1a)–(2b) and make a computer program.

The equation of profile P_2 in planar gearing generated by a method of single-parametric enveloping will be obtained through Eqs. (1a)–(2b) by deleting one of their curvilinear coordinates:

$$\mathbf{r}_1 = \mathbf{r}_1(v); \tag{3a}$$

$$V_N = F(v, \varphi) = \mathbf{V}_{12} \cdot \mathbf{n} = 0; \tag{3b}$$

$$\mathbf{r}_2 = \mathbf{r}_2(v, \varphi) = \mathbf{M}_{21}(\varphi) \cdot \mathbf{r}_1(v); \tag{3c}$$

$$\mathbf{n}_2 = \mathbf{n}_2(v, \varphi) = -\mathbf{M}_{21}(\varphi) \cdot \mathbf{n}_1(v). \tag{3d}$$

where component of (1) vectors are obtained by analytic operations:

$$\mathbf{n}_1 = \mathbf{n}_1(v) = W_N(\mathbf{r}_1(v)); \tag{4a}$$

$$\mathbf{V}_{12} = \mathbf{V}_{12}(v, \varphi) = \mathbf{V}_{12}(\mathbf{r}_1, \varphi) = W_V(\mathbf{M}_{21}(\varphi), \mathbf{r}_1(v)). \tag{4b}$$

Note that it is acceptable to designate the curvilinear coordinate of the line as u rather than v practically in all works within the theory of gearing. We chose the parameter v since we wanted to emphasize that a jog can appear on the line and the curvilinear coordinate of the jog will be the angle v rather than the distance u along the line. Later in the paper, a rule is followed: the curvilinear coordinate where a jog can appear is designated as v and the "smooth" curvilinear coordinate without jogs is designated as u.

The equation of surface P_2, generated by method of double-parametric enveloping, is similar to Eqs. (1a)–(2b) for a single-parametric enveloping:

$$\mathbf{r}_1 = \mathbf{r}_1(u, v); \tag{5a}$$

$$\left.\begin{aligned} V_N^{\varphi} &= F_1(u, v, \varphi, S) = \mathbf{V}_{12}^{\varphi} \cdot \mathbf{n} = 0 \\ V_N^{S} &= F_2(u, v, \varphi, S) = \mathbf{V}_{12}^{S} \cdot \mathbf{n} = 0 \end{aligned}\right\} \tag{5b}$$

$$\mathbf{r}_2 = \mathbf{r}_2(u, v, \varphi, S) = \mathbf{M}_{21}(\varphi, S) \cdot \mathbf{r}_1(u, v); \tag{5c}$$

$$\mathbf{n}_2 = \mathbf{n}_2(u, v, \varphi, S) = -\mathbf{M}_{21}(\varphi, S) \cdot \mathbf{n}_1(u, v). \tag{5d}$$

$$\mathbf{n}_1 = \mathbf{n}_1(u, v) = W_N(\mathbf{r}_1(u, v)); \tag{6a}$$

$$\left.\begin{aligned} \mathbf{V}_{12}^{\varphi} &= \mathbf{V}_{12}^{\varphi}(\mathbf{r}_1, \varphi, S) = W_V(\mathbf{M}_{21}(\varphi, S), \mathbf{r}_1(u, v)) \\ \mathbf{V}_{12}^{S} &= \mathbf{V}_{12}^{S}(\mathbf{r}_1, \varphi, S) = W_V(\mathbf{M}_{21}(\varphi, S), \mathbf{r}_1(u, v)) \end{aligned}\right\} \tag{6b}$$

In (5a)–(6b), there are two parameters of enveloping (φ and S) and, therefore, two relative velocities and two equations of meshing, as we see when we compare (6b) with (2b) and (5b) with (1b).

3 Types of Jog and Geometrical Images Produced by Them

Figure 3 shows all possible types of jog and geometrical images produced by them. There are three types of jog:

1—*Plane jog* (jog of the plane profile). This is where one will find the plane fan of normal lines.

2—*Edge jog* (jog of the solid surface along the line). This is where one will find the spatial wedge of normal lines. The wedge can be very complicated for non-planar intersecting surfaces. Face surfaces of the wedge are always planes (not always parallel) and flanks are ruled surfaces of the general type. The rib producing the wedge is the spatial line in the general case.

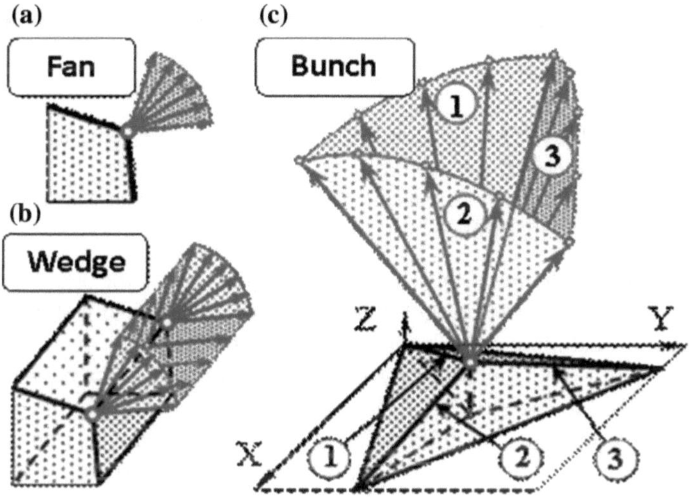

Fig. 3 Types of jog and geometrical images produced by them: **a** plane jog and fan of normal lines; **b** edge jog and wedge of normal lines; **c** vertex jog and bunch of normal lines

3—*Vertex jog* (jog of the solid surface in its vertex). This is where one will find the spatial bunch of normal lines. Let us enumerate the important features of such a jog:

- the vertex is always the intersection point of *three* surfaces; the tetrahedral vertex is the result of coincidence of two vertexes of two trihedral pyramids; the pentahedral vertex is the result of coincidence of three vertexes of trihedral pyramids; and so on.
- the bunch of normal lines is always limited by plane faces. For a trihedral vertex, it is the sector cut off from the sphere by three planes (1, 2, 3) passing through the vertex and perpendicular to "its own" ribs.

In all types of spatial gearing, jogs of solid surfaces are two-dimensional objects with two curvilinear coordinates, for instance, u along the jog line and v across it. In all types of planar gearing, jogs of profiles are one-dimensional objects with one curvilinear coordinate v along the profile. When the curvilinear coordinate v is varied, the Cartesian coordinates x, y, z of the point in the jog remain the same, but projections n_x, n_y, n_z of the normal vector to the surface are altered.

4 Peculiarities of Developing Mathematical Models When Generating by Jogs

According to Eqs. (1a)–(6b), the coordinates of points on the generating surface and the normal line to this surface should be calculated when analyzing the generating processes. In classical theory of gearing, the Eqs. (1a), (3a) and (5a) for calculating $\mathbf{r}_1$ and $\mathbf{n}_1$ are obtained as follows:

- the calculation scheme is drawn to derive equations of the generating line or surface: coordinate systems, and sketch the line or surface with all its parameters pointed out, including curvilinear coordinates u and v;
- the equation of the generating line is derived as radius-vector $\mathbf{r}_1 = \mathbf{r}_1(v)$ or the generating surface as $\mathbf{r}_1 = \mathbf{r}_1(u, v)$;
- according to methods of differential geometry, the formulas for unit normal vectors are obtained as $\mathbf{n}_1 = \mathbf{n}_1(v)$ or $\mathbf{n}_1 = \mathbf{n}_1(u, v)$. as shown in Eqs. (2a), (4a) and (6a) as participation of operand W_N in vector $\mathbf{r}_1 = \mathbf{r}_1(v)$ or $\mathbf{r}_1 = \mathbf{r}_1(u, v)$ of the generating line or surface.

As for the jog of the surface, Eqs. (2a), (4a) and (6a) cannot be obtained from formulas (1a), (3a) and (5a) by methods of differential geometry, since at the jog, $\frac{\partial \mathbf{r}_1}{\partial v} = 0$. But they can be obtained by means of geometrical images: fan, wedge and bunch of normal lines. For the plane jog at point B (see Fig. 2), equations of radius vector $\mathbf{r}_1$ and unit normal vector $\mathbf{n}$ can thus only be written with the help of Fig. 2:

$$\left.\begin{array}{ll} x = -\frac{S}{2} + h_f \cdot \tan \alpha, & n_x = \cos v, \\ y = -h_f; & n_y = \sin v \end{array}\right\}; \quad \text{where} \quad (180^\circ + \alpha) \le v \le 270^\circ \qquad (7)$$

where v is the curvilinear coordinate, and S, h_f and α are other parameters of the jog.

The oddness of Eq. (7) is in their ambivalence:

- since curvilinear coordinate v is not present in formulas for calculating the coordinates, then *it is the point*;
- since curvilinear coordinate v is present in formulas for calculating projections n_x and n_y of the unit normal vector, then *it is the line*. But expressions for calculating n_x and n_y cannot be derived from formulas for calculating x and y by methods of differential geometry, since $\frac{\partial x}{\partial v} = \frac{\partial y}{\partial v} = 0$.

In general, formulas (7) are the equation of the plane jog: the fan of normal lines is located at the point with coordinates $\{x, y\}$, the fan being the one-dimensional set (function of one variable—curvilinear coordinate v) and having *the features of the line* within generating processes. Substituting Eq. (7) of the plane jog into (3b), one can solve the equation of meshing within profiling processes: with respect to either the parameter of enveloping φ, or the curvilinear coordinate v.

Equations of two-dimensional jogs can be similarly written:

- the edge jog will contain the line of jog $\mathbf{r}_1 = \mathbf{r}_1(u)$ and surface of the wedge of normal lines $\mathbf{n}_1 = \mathbf{n}_1(u, v)$;
- the vertex jog with vertex $\mathbf{r}_1 = const$ and the surface of the bunch of normal lines $\mathbf{n}_1 = \mathbf{n}_1(u, v)$.

Having obtained the equation of meshing (1b) or the system of two Eq. (5b) in a case of two-parametric enveloping (Eqs. 1b or 5b), it (they) can also be solved here with respect to the parameter of enveloping φ, S and with respect to curvilinear coordinates u, v. Two different versions for solving the task of determining the surface generated by methods of enveloping are possible here:

Version 1—practically classical approach of the theory of gearing: assign two of three or four parameters determining the position of the point of a generating element in space (u, v, φ, S); solve the equations of meshing (1b) or (5b) to determine the rest missing one or two parameters; calculate $\mathbf{r}_1$, $\mathbf{n}_1$, $\mathbf{r}_2$, $\mathbf{n}_2$ according to formulas (1a, 2a, 1c, 1d) or according to (5a, 6a, 5c, 5d). If the point is at the jog, then the normal line to the generating element does not disappear. It is the distinguishing feature of the classical approach.

Version 2—new and rather different from the classical approach:

- assign displacements of elements and "smooth" curvilinear coordinate u along the jog;
- check if there is at least one contact point at this point of jog; if "no", then change for the next tenant; if "yes", then determine the contact normal line or contact fan of normal lines at this point and calculate $\mathbf{r}_2$, $\mathbf{n}_1$, $\mathbf{n}_2$.

Distinguished features of this version are:

- equations for check-up are necessary to clarify whether or not there is a contact point on the fragment of the generating element (classical approach does not imply this);
- determination of the contact point at jogs is specific: the contact normal line is determined first and then its corresponding curvilinear coordinate (the order of calculations is reversed in the classical approach).

These features will be considered in the subsequent Sects. 5–7 and basic calculation relations and algorithms for two versions of determining the point on the generated surfaces and profiles will be given there.

5 Generation of Lines by Jogs of Profiles in Planar Gearing

5.1 Theory of Generation by Plane Jog

Figure 4 shows the plane jog. There are three points at the jog: A and B are, respectively, the origin and end of the jog; C is the designed point which is assigned, for instance, by the curvilinear coordinate v.

Initial data of the jog (common for two versions of calculation):

Parameters of position: x_C, y_C are coordinates of the jog; Ψ_C is the direction of the normal line at designed point C; p_T is the feature denoting the position of the solid when moving from the origin to the end of the jog ($p_T = +1$—to the right; $p_T = -1$—to the left).

Parameter of dimension: v_{AB} is the angle of the jog (since for $v_{AB} > 0$, the jog is convex, and for $v_{AB} < 0$, it is concave, the sign of v_{AB} is thus also the parameter of shape).

Parameter of boundary: v_C is the "distance" of the current point C from the jog origin.

Figure 5 shows all possible types of plane jog and the features corresponding to these types: position of the solid p_T and sign of the jog angle v_{AB}. Let us consider the formulas and algorithms for two versions of determining the profile generated by the method of enveloping by plane jog.

Version 1 (classical approach of TG). The additional information for analysis is the value of curvilinear coordinate v at the current point C of the jog. Calculations are executed consequently according to formulas (8) and (9) given below:

$$\left.\begin{array}{ll} p_R = if \quad v_{AB} < 0 \quad then \ -1 \quad else \ 1; & p_B = -p_T \cdot p_R; \\ v_{\max} = abs(v_{AB}); \quad \psi_A = \psi_C - p_B \cdot v_C; & k = p_R \cdot \left(10^{10} \ldots 10^{12}\right). \end{array}\right\} \tag{8}$$

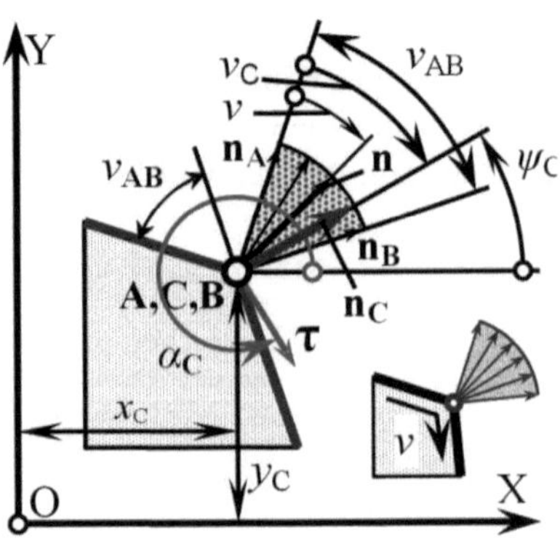

Fig. 4 Parameters of the plane jog

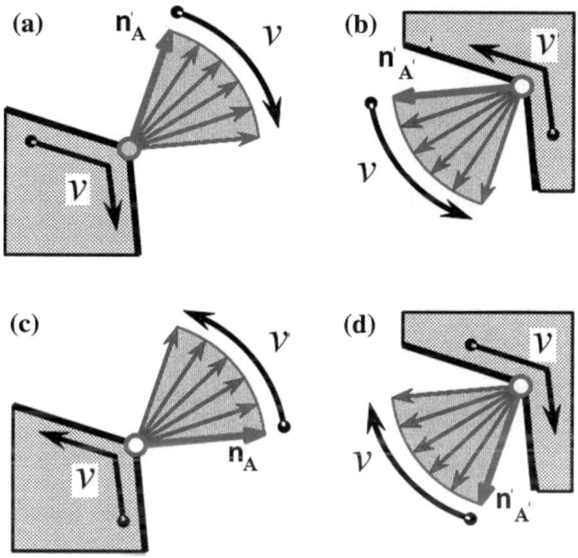

Fig. 5 Types of plane jog and the values of their parameters: **a** $p_T = 1$, $\nu_{AB} > 0$; **b** $p_T = 1$, $\nu_{AB} < 0$; **c** $p_T = -1$, $\nu_{AB} > 0$; **d** $p_T = -1$, $\nu_{AB} < 0$. $\mathbf{n}_A$ is the normal line at the fan origin; arrows along the solid lines are the direction of increasing the coordinate v

where p_B is the auxiliary feature of direction of the normal line revolution in the fan ($p_B = +1$—counter-clockwise, $p_B = -1$—clockwise); k is the curvature at the jog (it is necessary for calculating the radius of curvature of the generated profile, which is not considered in this paper).

$$\left.\begin{array}{r} v = if \quad v<0 \quad then \quad 0 \quad else \quad if \quad v > v_{\max} \quad then \quad v_{\max} \quad else \quad v; \\ \psi = \psi_A + p_B \cdot u; \quad n_x = \cos\psi, \quad n_y = \sin\psi; \quad \tau_x = -p_T \cdot n_y; \quad \tau_y = p_T \cdot n_x. \end{array}\right\} \tag{9}$$

The main result of these calculations is the projections of the unit's normal vector in the system of coordinates of the element with a jog. That is, formulas and algorithm in (8) and (9) are the mathematical description of operand W_N of the Eq. (4a).

Unit vectors of the line tangent to the jog can be required in calculation of curvatures.

Version 2 (author's approach). The additional information is the vector of relative velocity $\mathbf{V}_{12}$ at the assigned point of jog. The mathematical description of the same operand W_N of the Eq. (4a) is given below for another version of the calculations. This algorithm for determining the contact (generating) point at the jog is more complicated, but it involves solving the equation of meshing, unlike version 1. Version 2 is more reliable and gives an unambiguous solution for the plane jog. The unambiguity of the solution is based on two evident facts. First, the angle of the fan is always $\nu_{AB} < 180°$, as seen in Fig. 5. Second, the vector $\mathbf{V}_{12}$ can not be perpendicular to more than one vector within the fan of normal lines with the angle $\nu_{AB} < 180^\circ$ (two vectors can appear only within the fan of normal lines for $\nu_{AB} > 180°$). The following rule therefore arises: if velocities at the origin of the jog

($V_{nA} = \mathbf{V}_{12} \cdot \mathbf{n}_A$) and at its end ($V_{nB} = \mathbf{V}_{12} \cdot \mathbf{n}_B$) have opposite signs, this jog is the generating one and will include the normal line **n**, for which the equation of meshing $\mathbf{V}_{12} \cdot \mathbf{n} = 0$ is true.

The sequence of searching such a contact normal line when calculating according to version 2 is as follows:

1. Execute calculations according to formulas (8) and then according to the formulas below (10):

$$\left.\begin{aligned} \psi_B = \psi_A + p_B \cdot v_{\max}; \qquad n_{Ax} &= \cos\psi_A, \qquad n_{Ay} = \sin\psi_A, \\ n_{Bx} &= \cos\psi_B, \qquad n_{By} = \sin\psi_B. \end{aligned}\right\} \tag{10}$$

2. Determine the normal line **N** in the middle of the jog:

$$N_x = n_{Ax} + n_{Bx}; \quad N_y = n_{Ay} + n_{By}. \tag{11}$$

3. Calculate velocities of feeding-in at both the origin and the end of the jog:

$$\left.\begin{aligned} V_{nA} &= V_{12x} \cdot n_{Ax} + V_{12y} \cdot n_{Ay}; \\ V_{nB} &= V_{12x} \cdot n_{Bx} + V_{12y} \cdot n_{By}. \end{aligned}\right\} \tag{12}$$

4. Clarify if the jog is the generating one, determine the feature p_F of presence of the generating point at the jog:

$$p_F = if\ V_{nA} \cdot V_{nB} > 0\ then\ 0\ else\ 1 \tag{13}$$

5. If $p_F = 1$, the jog is the generating one and the unit vector of the contact normal line is determined:
   If $V_{12} = 0$, *then*
   the model of the generating jog is produced (the reverse fan of normal lines), *otherwise* the unit vector of the contact normal line is determined:

$$\left.\begin{aligned} k &= sign(N_x \cdot V_{12y} - N_y \cdot V_{12x}); \\ n_x &= \frac{k \cdot V_{12y}}{V_{12}}, \qquad n_y = -\frac{k \cdot V_{12x}}{V_{12}}. \end{aligned}\right\} \tag{14}$$

   and the curvilinear coordinate v of the contact point is determined:

$$\left.\begin{aligned} v &= \psi - \arctan\tfrac{n_y}{n_x}; \\ v &= v + if\ v < 0\ then\ 2\pi\ else\ if\ v > \pi\ then\ -2\pi\ else\ 0 \end{aligned}\right\}. \tag{15}$$

One important, though evident, note: none of the formulas and algorithms given in p. 5.1 depend on the shape of lines generating the jog and on the motion of elements. That is, all these formulas and algorithms are suitable for all types of planar gearing with one parameter of enveloping, both for gears and cam mechanisms and machine-tool gearing.

5.2 Examples of Investigating Planar Gearing

According to the algorithms described above, several research programs have been developed to execute the computer-aided simulation of processes of generation by plane jogs. The task was assigned to check the validity of the proposed technique by example of well-studied types of gearing. Let us look at only two examples for solving the test tasks.

5.2.1 Example 1. Generation of the Cylindrical Gearwheel Root

Figure 6 shows the planar gearing with four phases of interaction of generating segment L_1 (horizontal straight line with jogs of 70° at its ends) and its conjugate profile L_2. The motion of elements is executed from the left to the right. At phase 1, the conjugate profiles do not make contact, since there is no contact normal line at the jog; that is why the segment L_1 had not been generated yet. Phase 2 is the beginning of contact between L_1 and L_2: the contacting normal line passing through the pitch point appeared at the jog on L_1; segment L_1 became the generating one. At phase 3, the generating process by the fillet surface on L_2 by the left jog continues. At phase 4 (on segment L_1), there are three generating points at once, each of them generating its segment on L_2, with three contact normal lines being visible.

Note that, having assigned only 6 numbers for the generating segment (coordinates of two ends and two angles of jog), one can obtain all three joint curve lines conjugated with the segment by means of the basic computer program. But version 1 should be applied here and the equation of meshing should be solved with respect to the enveloping parameter—the rack displacement. This does not create any difficulties, since the equation of meshing is linear with respect to S. The same task within the classical approach would involve derivation and solving the equations to determine the points of the fillet surface conjugation with the root circumference.

5.2.2 Example 2. Cutting the Involute Gearwheels with Internal Teeth

Simulation results for this process are shown in Figs. 7, 8 and 9.

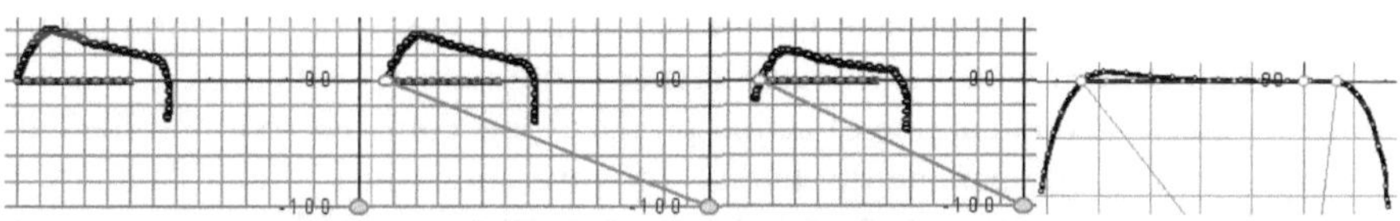

Fig. 6 Profiling by the straight line segment of the rack with jogs at the ends

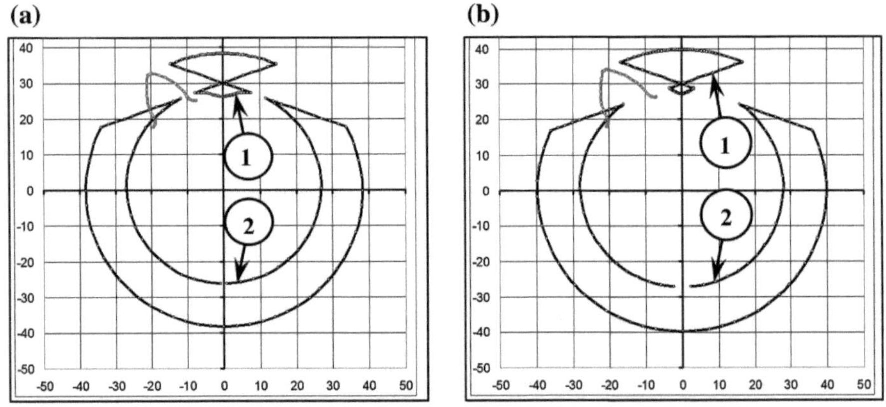

Fig. 7 Lines of action at single-parametric enveloping: z_0 = 15, m = 4, $x_{0a)} < x_{0b)}$

Figure 7 shows the shaping cutter (with z_0 = 15 and m = 4) tooth profiles and theoretical lines of action in two types of machine-tool gearing "shaping cutter-gearwheel". Meshes in Fig. 7a and 7b differ by application of shaping cutters with different shift coefficients: the shift coefficient x_0 of the shaping cutter in Fig. 7b is greater. Lines of action are plotted for *the whole profile* of one tooth of the shaping cutter, that is, for the profile located within one angular pitch of the shaping cutter—between axes of symmetry of two neighboring roots. Nine geometrical elements belong to this tooth profile: half of the root circumference arc, fillet surface, segment of the involute, jog, arc of top circumference, jog, segment of the involute, fillet surface, half of the root circumference arc. There are two lines of action in each Fig., two roots of the equation of meshing (3d) (when solving it with respect to the parameter φ). Both theoretical lines of action are closed curves. In Fig. 7b, there are small breaks in the lines of action. This is because omitting the points generated by the circumference arc in the shaping cutter root when modeling, that is, the angle of the considered shaping cutter tooth profile, was accepted as being less than the angular pitch.

Note that we did not mention the tooth number z_2 for which theoretical lines of action were plotted in Fig. 7. This was done because lines of action remain the same at any tooth number z_2. Even the type of gearwheel teeth (internal or external) is not relevant here. According to the theory of gearing, the shape and position of the theoretical line of action depend only on the generating profile and its element centrode (in this case, on r_{w1}) and nothing else. In practice, the line of action is the set of segments of the theoretical line of action inside the workpiece solid (in this case, beyond the gearwheel top circumferences with radii r_{a2} which are not shown in Fig. 7). Each theoretical line of action produces one complete tooth profile—the envelope of a single-parametric family of generating lines. Such families of envelopes are shown in Fig. 8. In Fig. 8a (on a small scale) and Fig. 8b (on a large scale), two families of tooth profiles are shown for the gearwheel with z_2 = 20 for which the corresponding theoretical lines of action are shown in Fig. 7a. Figure 8c

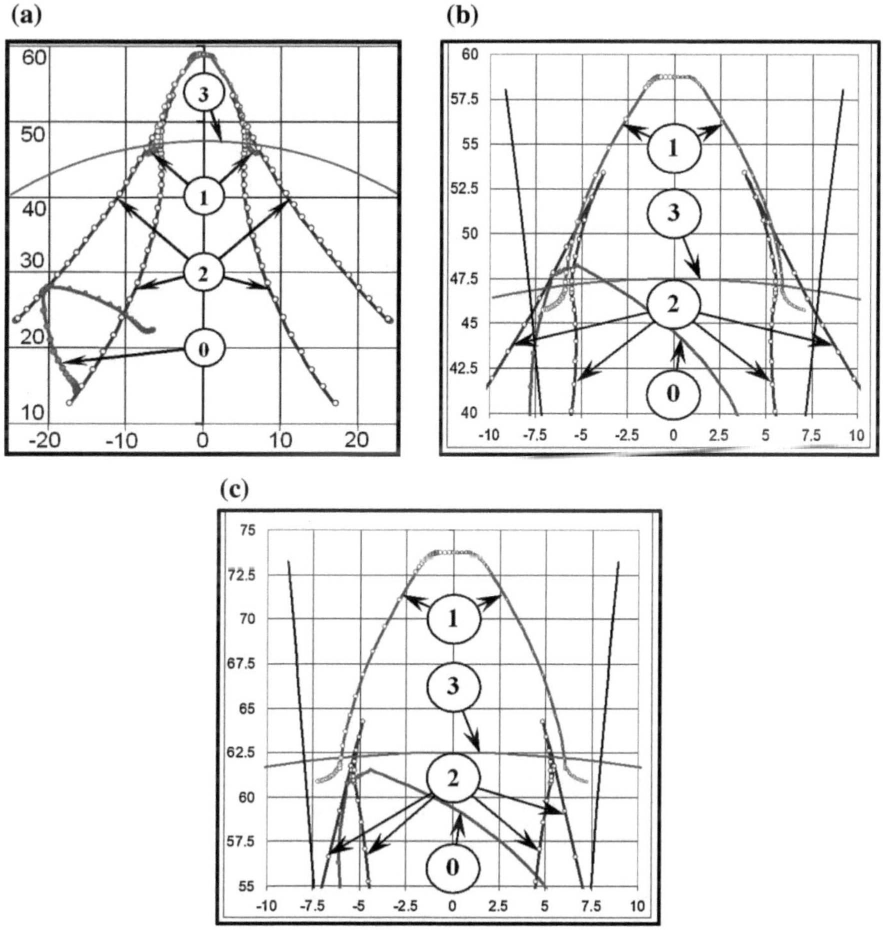

Fig. 8 Envelopes of a single-parametric family of generating lines. **a** $z_0 = 15$, $z_2 = 20$, $m = 5$ мм, **b** $z_0 = 15$, $z_2 = 20$, $m = 5$ мм, **c** $z_0 = 15$, $z_2 = 26$, $m = 5$ мм

also shows two families of gearwheel profiles, but with $z_2 = 26$ for which the corresponding theoretical lines of action are shown in Fig. 7b. The whole theoretical profile of one gearwheel root is generated at lines of action 1 shown in Fig. 7. Lines of action 2 correspond to a zone of secondary cutting and envelopes are generated there which can intersect the theoretical root profile leading to cutoff for the part of the generated tooth addendum. This is demonstrated in Fig. 8c.

Note that no one before us plotted either theoretical lines of action for jogs or enveloping profiles thus generated. The same is true for all areas of theoretical lines of action.

The following elements are designated in Fig. 8 by numbers:

0—profile of the shaping cutter.

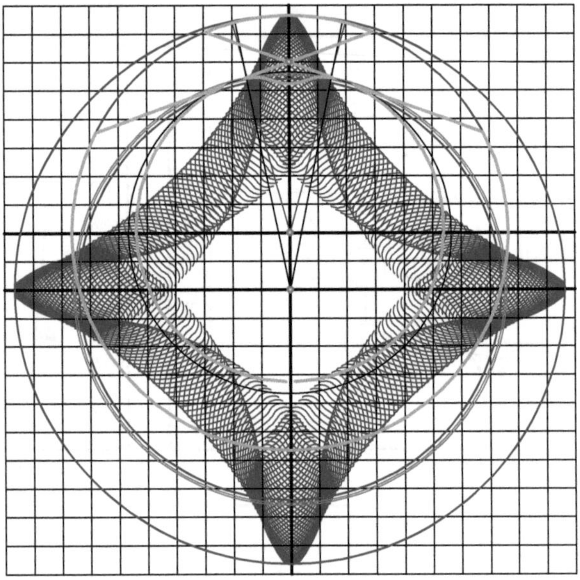

Fig. 9 Closed lines of action produce closed envelopes

1—complete profile of the gearwheel tooth generated at the theoretical line of action, designated as 1 in Fig. 7. Profile 1 also contains "the surplus" segment generated by the shaping cutter root by both the fillet surface and the arc of the root circumference.

2—parts of the gearwheel tooth profile generated at the theoretical line of action designated as 2 in Fig. 7. All parts of profile 2 shown in Fig. 8 contain three segments (demonstrated in Fig. 8b, c):

- **middle**—generated by a part of the involute profile of the shaping cutter and is itself the involute similar to segment 1 but shifted a little; these segments are well demonstrated in Fig. 8b, c;
- **coming to middle**—generated by the jog at the shaping cutter looking at the machined flank of the gearwheel tooth; it contacts the middle segment at its upper part and deviates from the middle segment along the closer tooth flank threatening the addendum cutoff;
- **concurrent to middle**—generated by the fillet surface and root circumference arc at the shaping cutter; it contacts the middle segment at its lower part and does not threaten the addendum cutoff.

3—gearwheel top circumference will "cut" the part of the gearwheel addendum, that is, will provide the radial clearance in machine-tool gearing and remove the part of the theoretical line of action.

Figure 9 shows the lines of action (the same as in Fig. 7a) and the generating process by direct enveloping is shown for the gearwheel with $z_2 = 20$. By the way, Fig. 1b shows the fragment of the Fig. 9. That is, for one and the same gearing, Figs. 8a and 9 demonstrate the gearwheel tooth profile determined by one of two methods of envelope defining: either by kinematic method with accounting for the presence of jogs (Fig. 8a) or by method of direct enveloping (Fig. 9). Envelopes for gearwheel tooth profiles are generated when rotating the shaping cutter by one revolution in the case of Fig. 8a and by 4 revolutions in the case of Fig. 9.

Profiles 2 shown in Fig. 8a are also visible in Fig. 9. It is practically impossible to see the tooth profile 1 shown in Fig. 8a in Fig. 9.

Analyzing theoretical lines of action in Fig. 7, one can define that:

- there are angles of shaping cutter rotation at instants of beginning and end of generating one tooth at the gearwheel (according to instants of contact of the shaping cutter profile with the closed line of action 1).
- angles of shaping cutter rotation corresponding to generation of the envelope at the zone of secondary cutting at the line of action 2 are close to 360°.
- there will be no cutoff at the zone of secondary cutting if all the line of action is outside the workpiece solid (in this case, it will be inside the gearwheel top circumference). To prove it, circumference r_{a2} should be plotted in Fig. 7.
- two involutes (segments of right and left branches) will be present at envelopes corresponding to lines 1 and 2, this number implying the number of segments of straight lines on these lines of action.
- outer curve lines on all lines of action are arcs of the shaping cutter top circumferences (r_{a1}); inner curve lines on lines of action 2 are not the circumference arcs with the radius r_{f1}. This is evident from Fig. 9.
- two closed theoretical lines of action produce two closed profiles at the gear rim—see Fig. 9. The generated profiles become closed if the generating process is kept on until the initial and final points of the generated profile coincide. In Fig. 9, such a coincidence takes place after machining four roots of the gearwheel by one tooth of the shaping cutter. Conditions at which lines of action and generated profiles become closed and at which one line of action is broken on two are considered in [9, 10]. These works touch upon fundamentals of the alternative theory of the generating process taking into account the presence of jogs on generating surfaces and lines.

As is known, two reasons exist in tooth-machining for top cutoff for the gearwheel tooth [13]: at the zone of secondary cutting and tool supply and withdrawal. In order to reveal the tooth top cutoff, intentionally derived equations [13] are applied rather than searching the envelopes through methods of the theory of gearing. The method is shown above for evaluating the risk of tooth top cutoff at the zone of secondary cutting by kinematic method of searching the envelopes (taking into account present jogs at the generating profile). Application of this method also allows for analyzing the risk of tooth top cutoff within the tool supply and withdrawal (see [8, 9]). For this purpose, lines and surfaces should be produced that are generated by methods of multi-parametric enveloping with application of jogs: plane, edge and vertex [8]. In this case, planar gearing should be analyzed for two parameters of enveloping and taking into account plane jogs. This paper does not involve the separate investigative task of studying the cutoff within the tool supply and withdrawal.

6 Generation of Surfaces by Edge Jogs of Solids

Figure 10 shows the edge jog of the solid surface. The curve line is drawn on adjacent surfaces and the jog. This curve line is taken to be the coordinate v-line. Unit normal vectors to the surface are shown along this line. Let us pay attention to a number of circumstances: (a) the curve line outlining the unit normal vector with its end is a continuous line producing *a number of normal lines on the surface and a fan of normal lines at the jog*; (b) this spatial line has jogs at areas "joining" the fan and a number of normal lines; (c) the fan of normal lines to the rib is a plane object and this plane is perpendicular to the rib. The main parameters of the rib are the same as for the plane jog but accounting for the three-dimensional space. They are: (a) vectors: points on the rib (**r**), unit normal vectors at the beginning ($\mathbf{n}_A$) and at the end ($\mathbf{n}_B$) of jog, unit normal vector of the tangent ($\boldsymbol{\tau}$) to the rib; (b) maximum curvilinear coordinate (v_{max}), curvature (k) at the jog; (c) feature p_R of the jog convexity.

Note, that there is an obvious relation between v_{max}, and also between unit normal vectors $\mathbf{n}_A$, $\mathbf{n}_B$ and $\boldsymbol{\tau}$: $k \cdot \boldsymbol{\tau} = (\mathbf{n}_A \times \mathbf{n}_B)$, where $k = \pm sin(v_{max})$. That is, the number of basic parameters is redundant and can be divided into two parts:

(a) main parameters this is the minimum number unambiguously determining the geometry of the considered object;
(b) additional parameters which reduce the number of calculations—it is reasonable to determine them once at the very beginning, and not to calculate them many times through the main parameters. In this case, the main parameters for the edge jog are: **r**, $\mathbf{n}_A$, $\mathbf{n}_B$ and p_R and the additional parameters here are: $\boldsymbol{\tau}$, v_{max} and k.

Calculation of the current point parameters assigned by a curvilinear coordinate v (for version 1 of determining the conjugate profile, it is the classical approach of TG):

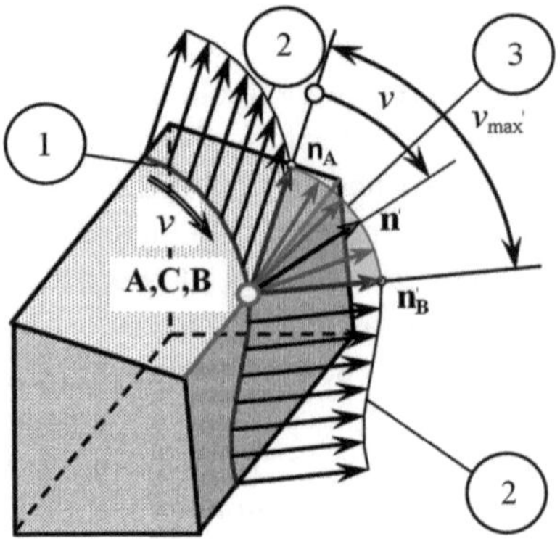

Fig. 10 Parameters of the edge jog: *1* v-line; *2* a number of normal lines; *3* fan of normal lines; $\mathbf{n_A}$, $\mathbf{n_B}$—main unit vectors of the fan of normal lines

1. Check if the three vectors $\mathbf{n}_A$, $\mathbf{n}_B$ and $\boldsymbol{\tau}$ are right-handed (if $\boldsymbol{\tau}$ is not determined by $\boldsymbol{\tau} = \mathbf{n}_A \times \mathbf{n}_B$):

$$p_P = sign((\mathbf{n}_A \times \mathbf{n}_B) \cdot \boldsymbol{\tau}) \tag{16}$$

2. Determine the auxiliary vector $\mathbf{C}$ perpendicular to $\mathbf{n}_A$ and belonging to the plane of the fan of normal lines (it is necessary to project the determined vector $\mathbf{n}$ of the unit normal vector onto two orthogonal bases $\mathbf{n}_A$ and $\mathbf{C}$):

$$\mathbf{C} = p_P \cdot (\boldsymbol{\tau} \times \mathbf{n}_A) \tag{17}$$

3. Calculate the unit normal vector:

$$\mathbf{n} = \mathbf{n}_A \cdot \cos v + \mathbf{C} \cdot \sin v \tag{18}$$

Determining the contact (generating) point at the jog (for version 2 of determining the conjugate profile, use the author's method):

1. Determine the feature p_F—the presence or absence of the generating point on the rib:

$$p_F = if((\mathbf{V}_{12} \cdot \mathbf{n}_A) \cdot (\mathbf{V}_{12} \cdot \mathbf{n}_B)) > 0 \;\; then \;\; 0 \;\; else \;\; 1 \tag{19}$$

2. For $p_F = 1$, the generating point is present and the unit normal vector is calculated for this point:

$$\mathbf{n} = \frac{\boldsymbol{\tau} \times \mathbf{V}_{12}}{V_{12}}; \qquad \mathbf{n} = sign(\mathbf{n} \cdot (\mathbf{n}_A + \mathbf{n}_B)) \cdot \mathbf{n} \tag{20}$$

Note: If $\boldsymbol{\tau} \times \mathbf{V}_{12} = 0$ (vector of relative velocity is directed along the rib), then all points of the jog (the fan of normal lines) are generating.

Let us show the method for developing the mathematical model of a part of the generating surface with edge jogs on its boundaries. Figure 11 shows the design scheme for a part of the cylindrical surface as a stripe with edge jogs along longitudinal sides. Such a fragment of generating surfaces is widely spread in gears (tops of worm threads and cylindrical gearwheel teeth) and tools (hobs and disk mills, etc.), and that is why it should be one of typical spatial elements in systems of computer-aided modeling of generating processes. The fragment of the surface with jogs is the analog of a rectilinear segment with jogs at its end, applied in Example 1 of p. 5.2.

1. Initial parameters of the cylinder fragment:

- r_C, z_{C0}, θ_C,$v_C = S_{AC}$ are four parameters of positions of the designed (for instance, middle) point C on the surface of the cylinder and on the stripe;
- $v_{\max} = S_{AB}$, b_1 are two linear dimensions of the stripe;
- β_C, λ_C is orientation of the fragment within a cylinder (for worms and cylindrical gearwheels, λ_C can be assigned for any β_C; for stripes with lines of edge jogs parallel to the plane XOY, it should be $\beta_C = \lambda_C = 90°$);
- α_{Afan}, α_{Bfan} are angles of jogs (on the plane perpendicular to the rib);

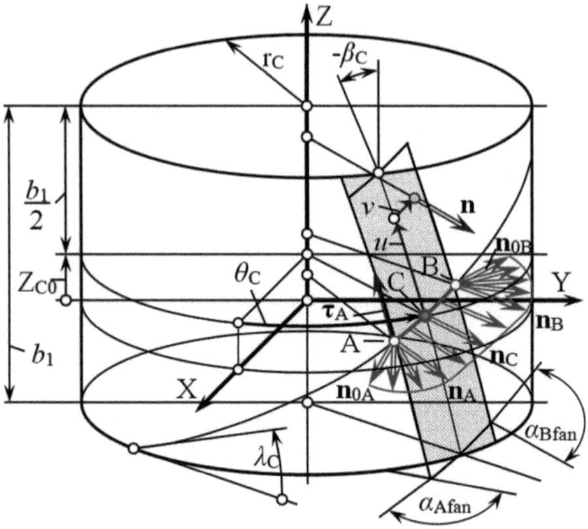

Fig. 11 Generating element —a stripe of the cylinder limited by jogs

- p_K is the feature of direction of the curvilinear coordinate v across the stripe (for $(1-v)$, it increases from point A to point B; for $(-1-v)$, it increases from point B to point A); let us assign in the formulas a derivation that $p_K = 1$ in Fig. 11.

2. *Curvilinear coordinates at the fragment of the cylinder and ribs of the jog*

Both the fragment of the cylindrical surface and the two jogs at its edges are two-dimensional geometric objects. That is why they have two curvilinear coordinates each: u and v. Coordinate u is the linear segment and is directed along the stripe; it is the distance from the designed point C to the current v-line (see Fig. 11). Coordinate v is the linear segment at the part of the surface and is the angular coordinate at jogs A and B. At these jogs, it is the coordinate v_f.

Let us clarify the limits of variation of curvilinear coordinates u, v and v_f at jogs and on the stripe. The preliminary calculations are:

$$If(abs(\beta) \approx 90^\circ \text{ and } abs(\beta) \approx 90^\circ) then (u_{C\min} = -0.5 \cdot b;\ u_{C\max} = 0.5 \cdot b)$$
$$else\left(u_{C\min} = -\frac{0.5 \cdot b + z_{C_0}}{\cos\beta};\ u_{C\max} = \frac{0.5 \cdot b - z_{C0}}{\cos\beta}\right) \tag{21}$$

$$\left.\begin{aligned} \Delta u_A &= sign(\beta) \cdot v_C \cdot \sin\lambda_C, & u_{A\min} &= u_{C\min} + \Delta u_A, & u_{A\max} &= u_{C\max} + \Delta u_A; \\ \Delta u_B &= -sign(\beta) \cdot (v_{\max} - v_C) \cdot \sin\lambda_C; & u_{B\min} &= u_{C\min} + \Delta u_B, & u_{B\max} &= u_{C\max} + \Delta u_B. \end{aligned}\right\} \tag{22}$$

The limits of variation of curvilinear coordinates u, v and v_f are:

(a) At jog A: $u_{A\min} \le u \le u_{A\max}$; $0 \le v_f \le \alpha_{Afan}$ ($v_f = 0$ for the unit vector $\mathbf{n}_A$).
(b) At jog B: $u_{B\min} \le u \le u_{B\max}$; $0 \le v_f \le \alpha_{Bfan}$ ($v_f = 0$ for the unit vector $\mathbf{n}_B$).

(c) On the stripe: $\Delta u = \frac{\Delta u_B - \Delta u_A}{v_{\max}} \cdot v$; $(u_{\min} + \Delta u) \leq u \leq (u_{\max} + \Delta u)$; $-v_C \leq v \leq (v_{\max} - v_C)$ ($v = 0$ for point C (see Fig. 11)).

If curvilinear coordinates u, v and v_f are varied within the pointed limits during calculations, then the determined points and unit normal vectors will cover all the surface of the stripe and jogs—from one edge to the other, along and across.

3. Formulas for preliminary calculation of jog parameters:
Coordinates of points A, B, C and their unit normal vectors to the cylinder surface are:

$$\theta_A = \theta_C - \frac{v_C \cdot \cos\lambda_C}{r_C}, \quad \theta_B = \theta_A + \frac{v_{\max} \cdot \cos\lambda_C}{r_C}. \tag{23}$$

$$\left.\begin{aligned} n_{Cx} &= \cos\theta_C, \\ n_{Cy} &= \sin\theta_C, \\ n_{Cz} &= 0; \end{aligned}\right\}(a) \quad \left.\begin{aligned} n_{Ax} &= \cos\theta_A, \\ n_{Ay} &= \sin\theta_A, \\ n_{Az} &= 0; \end{aligned}\right\}(b) \quad \left.\begin{aligned} n_{Bx} &= \cos\theta_B, \\ n_{By} &= \sin\theta_B, \\ n_{Bz} &= 0. \end{aligned}\right\}(c) \tag{24}$$

$$\left.\begin{aligned} x_C &= r_C \cdot n_{Cx}, \\ y_C &= r_C \cdot n_{Cy}, \\ z_C &= z_{C0}; \end{aligned}\right\}(a) \quad \left.\begin{aligned} x_A &= r_C \cdot n_{Ax}, \\ y_A &= r_C \cdot n_{Ay}, \\ z_A &= z_C - v_C \cdot \sin\lambda_C; \end{aligned}\right\}(b) \quad \left.\begin{aligned} x_B &= r_C \cdot n_{Bx}, \\ y_B &= r_C \cdot n_{By}, \\ z_B &= z_A + v_{\max} \cdot \sin\lambda_C. \end{aligned}\right\}(c) \tag{25}$$

Unit normal vectors of tangents to ribs (along the tooth line):

$$\left.\begin{aligned} \tau_{Ax} &= -n_{Ay} \cdot \sin\beta_C, \\ \tau_{Ay} &= n_{Ax} \cdot \sin\beta_C, \\ \tau_{Az} &= \cos\beta_C; \end{aligned}\right\}(a) \quad \left.\begin{aligned} \tau_{Bx} &= -n_{By} \cdot \sin\beta_C, \\ \tau_{By} &= n_{Bx} \cdot \sin\beta_C, \\ \tau_{Bz} &= \cos\beta_C. \end{aligned}\right\}(b) \tag{26}$$

Unit vectors $\mathbf{n}_{0A}$ and $\mathbf{n}_{0B}$ for perpendicular vectors to unit vectors $\mathbf{n}$ and $\boldsymbol{\tau}$ at points A and B (unit vectors $\mathbf{n}_{0A}$ and $\mathbf{n}_A$ belong to the plane of the fan of normal lines to the rib at point A) are:

$$\mathbf{n}_{0A} = \mathbf{n}_A \times \boldsymbol{\tau}_A; \qquad \mathbf{n}_{0B} = -\mathbf{n}_B \times \boldsymbol{\tau}_B. \tag{27}$$

4. Coordinates of the current point and its parameters of the surface:
Angle α of line AB turning around axis Z due to the shift of this line along the stripe by distance u, that is, because of variation of the curvilinear coordinate u, is:

$$\alpha = \frac{u \cdot \sin\beta_c}{r_c} \tag{28}$$

Coordinates of the point and unit normal vector for jog A:

$$\left.\begin{aligned} x = x_A \cdot \cos\alpha - y_A \cdot \sin\alpha, \quad y = y_A \cdot \cos\alpha + x_A \cdot \sin\alpha, \quad z = z_A + u \cdot \cos\beta; \\ \mathbf{n} = \mathbf{n}_A \cdot \cos(\alpha - v_f) + \mathbf{n}_{0A} \cdot \sin(\alpha - v_f). \end{aligned}\right\} \tag{29}$$

Coordinates of the point and unit normal vector for jog B:

$$\left.\begin{aligned} x = x_B \cdot \cos\alpha - y_B \cdot \sin\alpha, \quad y = y_B \cdot \cos\alpha + x_B \cdot \sin\alpha, \quad z = z_B + u \cdot \cos\beta; \\ \mathbf{n} = \mathbf{n}_B \cdot \cos(\alpha + v_f) + \mathbf{n}_{0B} \cdot \sin(\alpha + v_f). \end{aligned}\right\} \tag{30}$$

Coordinates of the point and unit normal vector for the stripe:

$$\left.\begin{aligned} \theta = \theta_C + \frac{(v \cdot \cos\lambda + u \cdot \sin\beta_c)}{r_c}; \quad n_x = \cos\theta, \quad n_y = \sin\theta, \quad n_z = 0; \\ x = r_C \cdot n_x, \quad y = r_C \cdot n_y, \quad z = z_C + u \cdot \sin\lambda_C + v \cdot \cos\beta_C. \end{aligned}\right\} \tag{31}$$

7 Generation of Surfaces by Vertex Jogs of Solids

Figure 12 shows the jog of the solid surface as a trihedral vertex. The following objects are present at the jog: (a) three intersecting ribs (1, 2, 3); (b) three secant planes, each being one of three faces of the bunch of normal and perpendicular to one of ribs (they are correspondingly denoted as 1, 2, 3); (c) three unit normal vectors ($\mathbf{n}_1$, $\mathbf{n}_2$, $\mathbf{n}_3$) to each of the faces of the jog surface. These three unit vectors are simultaneously the ribs of a trihedral bunch of normal. Note that the direction of the path tracing the number triples of all elements of the considered jog (faces, ribs, secant planes and unit normal) is always the same: in Fig. 12, it is counter-clockwise. Figure 12 shows that the vertex jog of the solid surface generates, as a rule, the edge jog by producing in motion the fan of normals $\{\mathbf{n}_A \ldots \mathbf{n}_B\}$ perpendicular to the vector of relative velocity $\mathbf{V}_{12}$.

The basic parameters of the vertex are: radius-vector of the vertex ($\mathbf{r}$), unit normal vectors ($\mathbf{n}_1$, $\mathbf{n}_2$, $\mathbf{n}_3$) and features of convexity of three ribs (p_{R1}, p_{R2}, p_{R3}). Depending on the combination of convexity features, there are four types of jog functioning as trihedral angles producing bunches of normals: vertex, threshold, curve, pit (see Fig. 13). The basic type of jog in generating tasks is the vertex (all $p_R = +1$), while the jog called the threshold appears less often (in this case, for one of ribs $p_R = -1$).

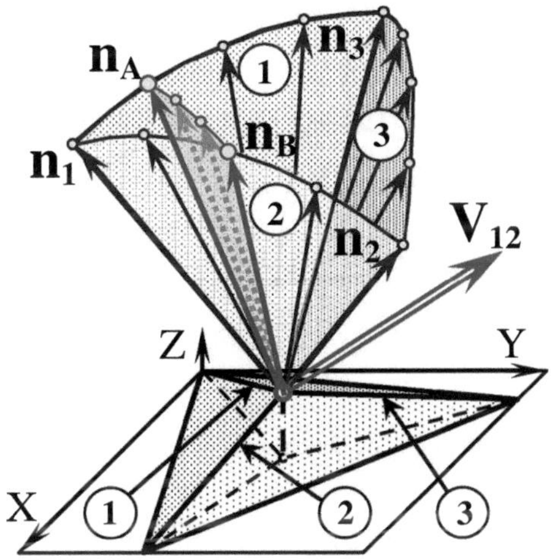

Fig. 12 Generating the rib by vertex jog within a single-parametric enveloping: $\mathbf{n}_1$, $\mathbf{n}_2$, $\mathbf{n}_3$ are normal lines to faces of the vertex (bunch of normals); $\mathbf{n}_A$, $\mathbf{n}_B$ are boundary normals of the generated rib (fan of normals); $\mathbf{V}_{12}$ is the relative velocity

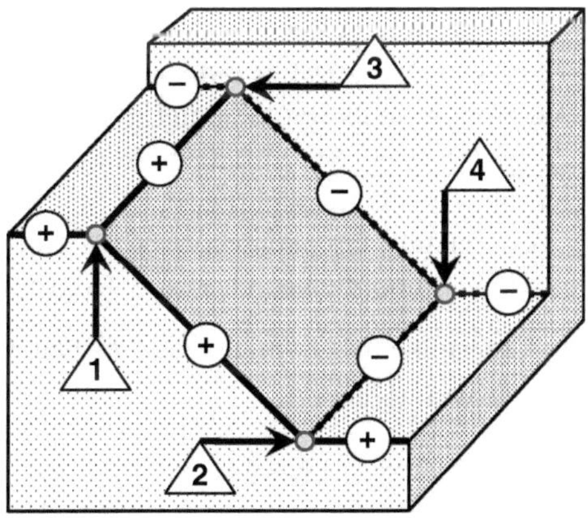

Fig. 13 Four types of angular jogs of solids: 1 *vertex* (+ + +); 2 *threshold* (+ + –); 3 *curve* (+ – –); 4 *pit* (– – –). Convexity of three ribs is shown in brackets: convex rib (+); concave rib (–)

Additional basic parameters are calculated according to basic ones:

1. Unit vectors $\boldsymbol{\tau}_i$ of tangents to ribs:

$$\left.\begin{aligned} &\mathbf{T}_1 = \mathbf{n}_3 \times \mathbf{n}_1, \quad \mathbf{T}_2 = \mathbf{n}_1 \times \mathbf{n}_2, \quad \mathbf{T}_3 = \mathbf{n}_2 \times \mathbf{n}_3; \\ &k_i = \sqrt{T_{ix}^2 + T_{iy}^2 + T_{iz}^2}, \quad \boldsymbol{\tau}_i = k_i \cdot \mathbf{T}_i, \quad (i = 1, 2, 3). \end{aligned}\right\} \tag{32}$$

2. Angles of fans limiting the bunch of normal lines:

$$\left.\begin{aligned} &\cos v_{max\,1} = \mathbf{n}_3 \cdot \mathbf{n}_1, \quad \cos v_{max\,2} = \mathbf{n}_1 \cdot \mathbf{n}_2, \quad \cos v_{max\,3} = \mathbf{n}_2 \cdot \mathbf{n}_3; \\ &\sin v_{max\,i} = k_i, \quad v_{max\,i} = arctan\frac{\cos v_{max\,i}}{\sin v_{max\,i}}, \quad [0 \le v_{max\,i} \le 180^\circ], \quad (i = 1, 2, 3). \end{aligned}\right\} \tag{33}$$

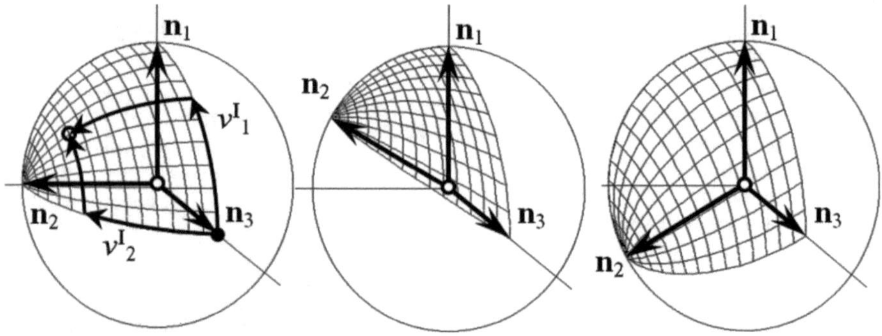

Fig. 14 Ends of bunch unit normals generating the spherical triangles

In order to assign the point on the considered angular jog and determine its unit normal vector (that is, inside the bunch of normals), one should determine the curvilinear system of coordinates for such a jog, that is, at the bunch of normals. Since the set of points of ends of bunch unit normals represents a spherical triangle (see Fig. 14), the task is reduced to assignment of a point inside the plane triangle. We propose to assign the point in the triangle by two coordinates fixed to the unit: ν_1^{I} (analog to ν_1) and ν_2^{I} (analog to ν_2), as shown in Fig. 15 (the point with $\nu_1^{\mathrm{I}} = 0.2$ and $\nu_2^{\mathrm{I}} = 0.4$ is marked there).

Calculating the parameters of the current point of the angular jog, assigned by curvilinear coordinates ν_1^{I} and ν_2^{I} fixed to the unit:

1. Determine unit normal vector $\mathbf{n}_o$ belonging to fan $\mathbf{n}_3$-$\mathbf{n}_1$ perpendicular to rib N1 (see Fig. 15):

$$\nu_1 = \nu_1^{\mathrm{I}} \cdot \nu_{\max 1}; \quad \mathbf{c}_3 = \boldsymbol{\tau} \times \mathbf{n}_2; \quad \mathbf{n}_0 = \mathbf{n}_3 \cdot \cos \nu_1 + \mathbf{c}_3 \cdot \sin \nu_1. \qquad (34)$$

Fig. 15 Curvilinear coordinates within the bunch of normals fixed to the unit

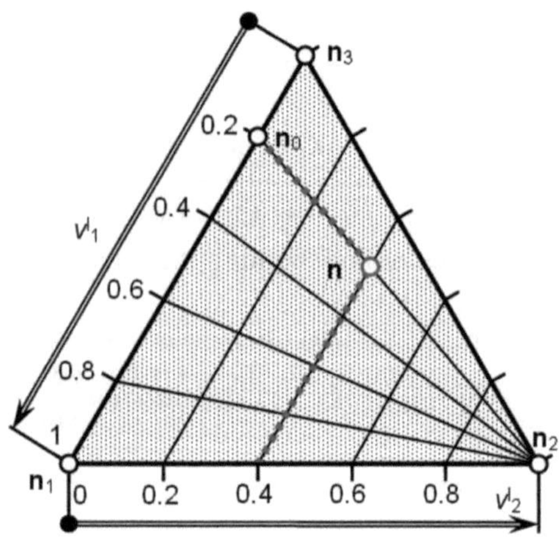

2. Determine the unit normal vector to the fan generated by $\mathbf{n}_0$ and $\mathbf{n}_2$:

$$\mathbf{T}_0 = \mathbf{n}_0 \times \mathbf{n}_2, \quad k_0 = \sqrt{T_{0x}^2 + T_{0y}^2 + T_{0z}^2}, \quad \boldsymbol{\tau} = k_0 \cdot \mathbf{T}_0. \tag{35}$$

3. Determine the angle of this fan:

$$\cos v_{max0} = \mathbf{n}_0 \cdot \mathbf{n}_2; \qquad v_{max0} = \arctan \frac{\cos v_{max0}}{k_0}; \qquad [0 \le v_{max0} \le 180°]. \tag{36}$$

4. Determine the unit normal vector $\mathbf{n}$ at the current point assigned by coordinates $v_1^{\mathbf{I}}$ and $v_2^{\mathbf{I}}$:

$$v_0 = v_2^{\mathrm{I}} \cdot v_{\max 0}; \qquad \mathbf{c} = \boldsymbol{\tau} \times \mathbf{n}_0; \qquad \mathbf{n} = \mathbf{n}_0 \cdot \cos v_0 + \mathbf{c} \cdot \sin v_0 \tag{37}$$

Determining the contact (generating) points within the angular jog at one parameter of enveloping. As before, besides basic parameters of the jog, one should know the vector $\mathbf{V}_{12}$ of relative velocity. According to physical essence and Fig. 12, only those unit normal vectors will be perpendicular to the vector $\mathbf{V}_{12}$ of relative velocity that belong to the plane perpendicular to the vector $\mathbf{V}_{12}$. That is, the set of contact unit vectors is the fan of normals generated by unit vectors $\mathbf{n}_\mathrm{A}$ and $\mathbf{n}_\mathrm{B}$ (Fig. 12) located in two of three planes which are boundaries of the bunch of normals. When moving towards $\mathbf{V}_{12}$, the fan of normals $\mathbf{n}_\mathrm{A}$-$\mathbf{n}_\mathrm{B}$ will be generating the rib, i.e., the wedge of normals. This proves that the enveloping of the family of vertexes of the generating element is the rib (if the vertex is generating). Therefore, when determining the contact points within the angular jog, basic parameters of the fan of normals at the rib are subjected to determination: two unit normal vectors ($\mathbf{n}_\mathrm{A}$, $\mathbf{n}_\mathrm{B}$) and one feature of the rib convexity (p_R). This will be the contact fan of normal lines—the feature, i.e., the line of contact of the generating and generated elements.

Algorithm for determining the contact fan of normal lines:

1. Check the presence of generating points within three fans, limiting the bunch of normals—find the features of ribs $p_{\mathrm{F1}}, p_{\mathrm{F2}}, p_{\mathrm{F3}}$ generating according to Eq. (19).
2. If $\Sigma p_\mathrm{F} = 0$, then the vertex is not generating.
3. If $\Sigma p_\mathrm{F} = 3$, then the vector of relative velocity $\mathbf{V}_{12}$ is directed along one of three ribs of the jog. The fan 1, 2 or 3 in Fig. 12 perpendicular to this rib will be the generating one. This fan is determined according to the condition of parallelism of the rib and vector $\mathbf{V}_{12}$: $\boldsymbol{\tau} \times \mathbf{V}_{12} = 0$.
4. If $\Sigma p_\mathrm{F} = 2$, then two unit vectors $\mathbf{n}_\mathrm{A}$ and $\mathbf{n}_\mathrm{B}$ should be determined for two fans for which $p_\mathrm{F} = 1$.
5. If $\Sigma p_\mathrm{F} = 1$, it proves the error of rounding (or in the program). However, one contact normal can be determined in this case (it should be very close to one of angular unit normal vectors $\{\mathbf{n}_1, \mathbf{n}_2, \mathbf{n}_3\}$).

The rule of determining the feature p_R of convexity of the rib generated by the vertex is as follows: it is opposite to the feature p_R of that rib for which there is no contact point for its fan of normals. For example, in Fig. 12, $\mathbf{n}_\mathrm{A}$ and $\mathbf{n}_\mathrm{B}$ belong to fans of ribs N1 and N2, therefore, for the generated rib, it is $p_\mathrm{R} = -p_{\mathrm{R3}} = -1$.

Another very important rule is: being the envelopes of the family of vertexes, ribs are produced outside the solid of the generating element only when both contact normals belong to fans of convex ribs. That is, for four types of angular jogs shown in Fig. 13:

(a) vertex 1 always generates concave ribs outside its solid;
(b) threshold 2 generates the convex rib outside the solid but only if contact normals $\mathbf{n}_A$ and $\mathbf{n}_B$ belong to fans of two ribs with the positive curvature;
(c) curve 3 and pit 4 always generate ribs inside their solids.

It is reasonable to determine ribs and other elements produced inside the solid of the generating element in order to obtain one continuous enveloping surface which is more easily applied (rather than for separate fragments) to generate the real surface: with jogs, cutoffs, etc. This is related to surfaces produced by all types of jogs and also by smooth generating elements.

8 Summary

1. Two groups of methods are widely applied in analyzing the generating processes: non-differential and differential kinematic. The classical kinematic method operates faster and allows for determining the curvatures but is also less reliable, since it does not allow for determining surfaces generated by singular points of generating surfaces and lines—jogs at intersections of their adjacent areas.
2. It was determined that there are three types of jogs and, therefore, three geometrical images produced by them: fan, wedge and bunch of normal lines. It was proposed by mathematical models to consider the jog of the profile as a line with one curvilinear coordinate rather than as the point; to consider jogs of surfaces as surfaces with two curvilinear coordinates rather than as lines.
3. The methodology was proposed to develop the kinematic method of the theory of gearing to determine surfaces and lines produced by generating not only for smooth generating surfaces and lines but in the presence of singular points—jogs at areas of intersection of adjacent segments. The foundation of the methodology and developed techniques of operating the jogs is to determine such a normal line within a fan, wedge or bunch of normal lines, which would be perpendicular to the vector of relative velocity at this point of the jog. One of the fundamental statements of differential geometry, which states that "tangent lines and normal lines to lines and surfaces are absent at their singular points", is therefore disavowed.
4. The advantage of the proposed technique is its versatility. For example, the shape of the line of action within the zone of secondary cutting depends on the shape of the generating profile in planar gearing. That is why equations derived to reveal cutoffs in involute gearing [13] cannot be applied for other types of gears: Wildhaber-Novikov, cycloidal and other. At the same time, the developed

techniques and computer-aided support allow for analyzing all types of tooth interference and cutoffs for all types of gearing.

5. Based on developed programs which implement the proposed method, the computer-aided modeling of the generating process was executed for plane and spatial gearing. It was determined that:
   - There are two theoretical lines of action in planar gearing with rotating elements. If they are determined for the *whole* generating tooth profile located between axes of symmetry of two neighboring roots of the generating element, they are both *closed* non-smooth lines.
   - A different conjugate profile corresponds to each theoretical line of action. If these two profiles intersect inside the workpiece solid, the cutoff appears for the part of the tooth to be cut.
6. The proposed technique is recommended for application in computer-aided programs of modeling the generating processes, especially in spatial gearing, including worm-type gears [14].
7. Possible and progressive directions of the stated method development:
   - Application of multi-parametric meshes to reveal cutoff in machine-tool gearing within the tool supply and withdrawal, to check the possibility of assembly-disassembly by radial displacement of gear elements in operating gearing. Fundamentals of developing these models are considered in [8, 9]. This paper does not include the stated tasks, since it is the theme of a separate investigation.
   - Investigation of curvatures in meshes with jogs of solids, including the analysis of curvatures by classical methods of the theory of gearing.
8. The author is not familiar with foreign works in which approaches are applied similar to those stated in this paper. According to works [15, 16 and oth.] and proceedings of gear conferences in recent years, the conclusion can be made that foreign gear experts, including Russian-speaking ones, do not know about the technique of *determining the surfaces generated by solid jogs according to the kinematic method.* This despite the fact that more than 10 years have passed since the first publications about fundamentals of this method [8–10].

References

1. Gochman HI (1886) Theory of gearing, generalized and developed by analysis. Master of mechanics thesis, Odessa, 232 pp
2. Shishkov VA (1951) Surface formation by cutting according to generation method. M.: Mashgiz, 150 pp
3. Kolchin NI (1957) Analytical fundamentals of differential method of gearing investigation. In: Proceedings of the seminar on the theory of machines and mechanisms of USSR AS, N64
4. Litvin FL (1968) Theory of gearing. M.: Nauka, 584 pp

5. Zalgaller VA (1975) Theory of envelopes. M.: Nauka, 104 pp
6. Sheveleva GI (1999) Theory of generation and contact of moving bodies. M.: Mosstankin, 494 pp
7. Nesmelov IP, Goldfarb VI (1983) Non-differential approach to solving the enveloping task / Mechanics of machines. N61. M.: Nauka, pp 3–10
8. Babichev DT (2004) To application of multi-parametric envelopes at computer-aided simulation of formation processes in operating and machine-tool gearing, Theory and practice of gearing. In: Proceedings of the scientific technical conference with international participation. Izhevsk, pp 302–315
9. Babichev DT (2005) Development of the theory of gearing and surface formation on basis of new geometric kinematic representations. Dr. Sc. Tech. thesis, Tyumen, 421 pp
10. Babichev DT (2003) Fundamentals of alternative theory of formation based on new geometric concepts. In: International conference "Drive engineering 03", Bulgaria, Sophia, pp 270–275
11. Rodin PR (1977) Fundamentals of surface formation by cutting. Kiev: Visha shkola, 192 pp
12. Babichev DT (2011) Acceleration of introduction is an important factor of the process of surfaces formation by means of bending. In: The 7th international conference and research and development of mechanical elements and systems: IRMES 2011: 8.6. Zlatibor, Serbia, pp 611–618
13. Gavrilenko VA (1969) Fundamentals of theory of involute gearing. M.: Mashinostroyeniye, 432 pp
14. Goldfarb VI, Glavatskikh DV, Trubachev ES, Kuznetsov AS, Lukin EV, Ivanov DE, Puzanov YuV (2011) Spiroid gearboxes for pipeline valves. M.: Veche Publ., 222 pp
15. Litvin FL, Fuentes A (2004) Gear geometry and applied theory, 2nd edn. Cambridge University Press, 800 pp
16. Radzevich SP (2012) Theory of gearing: kinetics, geometry, and synthesis. CRS Press, 695 pp

Damage Types, Load Capacity and Efficiency of Crossed Helical Gears with Wheels from Sintered Steel

A. Miltenović

Abstract Crossed helical gears are used in cars and many household appliances. The trend towards increased comfort in motor vehicles has led to the utilization of more than a hundred servo-drives in luxury class automobiles. Important advantages of crossed helical gears are their easy and inexpensive design, good noise performance and high ratio that can be realized in one step. Sintered steel is a very favorable material for wheels in crossed helical gears. The hardening obtained after the sinter process will affect the microstructure of the sintered steel, so that the wear load capacity can increase to higher values. This report shows results of iron-based sintered material Fe1.5Cr0.2Mo in the case of crossed helical gears concerning wear resistance and other damage types under different speed, torque and lubricants. As material variants, samples with additional treatment, such as pyrohydrolysis, case hardening, shot peening, sinter-hardening, and 2 % copper addition, are used. The calculation method is given for the determination of wear load capacity of the worm with a helical gear made of Fe1.5Cr0.2Mo sintered steel with sinter-hardening treatment. These results provide product developers with the first important clues for indicators for calculation of the worm with a helical gear. The paper also presents an analytical and FEA procedure for determination of coefficient of friction in meshing zone of crossed helical gears. Obtained results from numerical simulation were compared with experimentally obtained results.

Keywords Sintered steel · Crossed helical gears · Wear · Load capacity · Damage types · Efficiency · Coefficient of friction · FEA

A. Miltenović (✉)
Faculty of Mechanical Engineering, University of Nis, Nis, Serbia
e-mail: aleksandar.miltenovic@masfak.ni.ac.rs

V. Goldfarb and N. Barmina (eds.), *Theory and Practice of Gearing and Transmissions*, Mechanisms and Machine Science 34,
DOI 10.1007/978-3-319-19740-1_10

1 Introduction

Crossed helical gears are used, for example, in automotive auxiliary drive units such as window lifters, seat adjustments, windscreen wipers, and also in home appliances. The trend towards increased comfort in motor vehicles has led to the utilization of more than a hundred servo-drives in luxury class automobiles (Fig. 1). The important advantages of crossed helical gears are their easy and inexpensive design, good noise performance and high ratio that can be realized in one step.

Sintering is a method established for making metal objects from powders, and every single step of this process has an influence on a certain material property. The parts made by powder metals are the main group of products in powder metallurgy. They are very profitable products in a highly developed industry, being used in automotive and machinery, as well as in many other areas of the metalworking industry. Compared with other types of production, the marked share of sintered parts is not so big, but it shows steady growth. Over the time, sintered parts have developed from unimportant parts to very important functional parts, such as the one investigated here. These parts must always transfer high static loads, but often dynamic loads as well.

A review of conventional procedures for manufacturing parts using the sintering process can be found, for example, in [1]. The base material for sintered steel is iron powder in its elementary form. To obtain the corresponding features, by using special techniques, different metal powders are mixed with the pure iron powder.

The pressing process requires a precise amount of powder to be measured into the tool (Fig. 2) which will subsequently be compressed until reduction of the volume by a factor between 2 and 2.5 is achieved. The pressure applied by the

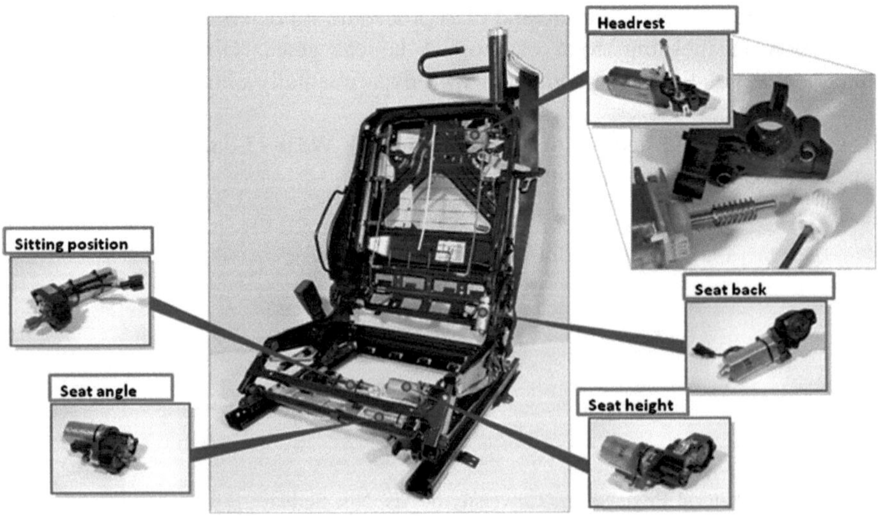

Fig. 1 Crossed helical gears in drive units for seat adjusting in BMW E46 C [17]

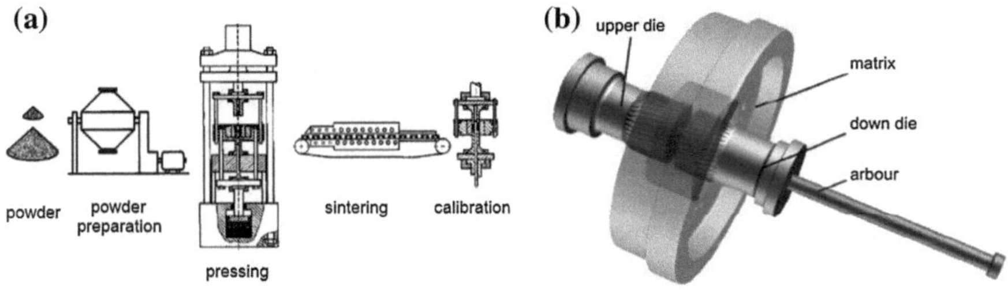

Fig. 2 **a** Technological process of single step sintering, **b** tool for making gears [2]

punch determines the final density of material; for example, a density of 7.5 g/cm^3 can be obtained by applying 800 N/mm^2.

The raw parts obtained by pressing are further subjected to a thermal process in the flow furnace (heating, sintering, cooling). For parts with direct hardening, the sintering process is usually performed at a temperature of 1120 °C in an atmosphere of reducing gases. In this way, new crystals are formed during sintering due to the diffusion processes at the grain boundary, resulting in a heterogeneous mixture. The sintering process ends with a cooling phase.

The properties of the sintered parts primarily depend on the density and composition of the alloys. Another influence which cannot be predicted is the manufacturing process, as it can be different in many cases.

The hardening after the manufacturing process can expand the range of possible applications of the powder metal parts and/or optimize their properties. This can happen for the contours that are made by molding technologies, the final shapes of which have to be obtained by additional machining processing on sintered parts or with various joining methods. Additionally, sintered materials with porosity below 10 % can be produced with wear or corrosion protection, as well as with decorative coatings. Sintered iron and steel can also be hardened directly or after carbonization. For materials with appropriate composition, a precipitation hardening is also possible, and is especially used for bodies that contain impregnated sintered metal.

In these applications, combinations with pinions or worms made of steel and gear wheels made of plastic are widely used. The gear wheels are made by injection moulding at low cost in large quantities. Due to the low carrying capacity of gear wheels made of plastic, they can transfer only small torque. Otherwise, the size of the transmission would be relatively large.

The use of gear wheels made of sintered metal can increase the load capacity of crossed helical gears. As in the case of plastic gear wheels, the large scale production of sintered metal gear wheels requires a special tool and no additional post-production costs.

The work by Wendt [2] is the first to focus on the study of load capacity for crossed helical gears with a steel/sintered metal material combination. Wendt researches the influence of density on the load capacity of crossed helical gears over 120 tests with a test period of 300 h. His work provides approximate equations for the calculation of safety factors for tooth damage such as pitting and wear.

2 Characteristics of Materials

2.1 Chemical Composition of Sintered Steel and Material Variants

The material combination of steel and sintered metal has only been investigated in a few research projects. For example, researchers from the company Höganäs AB in Sweden [3] investigated the sintered metal Astaloy Mo (Fe0.85Mo) and Astaloy CrL (Fe1.5Cr0.2Mo).

The basic raw material for sintered steel is iron powder. Iron powder is mixed with different metal powders by applying special alloy mixing techniques. A homogeneous powder mixture is important for uniform cross-sectional properties within the part. Copper increases the strength and yield strength, but decreases the elongation at break. Nickel improves the strength and relieves the weldability. Carbon (graphite), in small amounts, increases the strength and hardness and improves the subsequent heat treatment. Phosphorus improves the strength and elongation, but causes high sintering shrinkage.

The chemical composition of sintered steel is shown in Table 1. The basis for all tested materials is the iron-based powder Fe1.5Cr0.2Mo. The material variants are shown in Table 2 and Fig. 3. A detailed description of additional treatment methods is given in [4].

2.2 Material Tests

In order to determine the properties of various material versions, extensive material tests have taken place. To determine the exact Hertz surface pressure in the tooth contact, the E-module and Poisson's ratio are required [5]. These values can go as far as 0.2 % of the compression limit determined in a compression test. The tensile test allows for the determination of 0.2 %—yield strength and compression limit. It also determines the density of 10 wheels per batch.

2.2.1 Density of Test Gears

Measurement of the density of helical gears is done according to the buoyancy method. By means of a hydrostatic balance, one can determine the weight of helical gears in air m_L with the density ρ_s, as well as the weight in water m_w with the corresponding density $\rho_{Fl.}$

The difference between the two weighings is delivered according to the principle of Archimedes which states that the weight of the displaced fluid is directly proportional to the volume of the immersed body.

Table 1 Chemical composition of sintered steel Fe1.5Cr0.2Mo (%) [13]

Element	Measuring point 1	Measuring point 2	Measuring point 3	Measuring point 4	Measuring point 5
C	0.254	0.243	0.238	0.253	0.273
Si	0.054	0.047	0.043	0.057	0.053
Mn	0.163	0.161	0.161	0.161	0.161
P	0.009	0.009	0.009	0.009	0.009
S	0.001	0.001	0.001	0.001	0.002
Cr	1.521	1.517	1.526	1.51	1.509
Ni	0.026	0.026	0.026	0.026	0.026
Mo	0.21	0.208	0.209	0.211	0.21
Cu	0.066	0.066	0.067	0.067	0.067
Al	<0.001	0.002	<0.001	<0.001	<0.001
Ti	<0.001	<0.001	<0.001	<0.001	<0.001
V	0.008	0.008	0.008	0.008	0.008
Nb	0.007	0.007	0.007	0.006	0.006
W	0.015	0.015	0.016	0.014	0.013
Co	0.007	0.007	0.007	0.007	0.007
Zr	0.002	0.002	0.002	0.002	0.001
N	0.0248	0.0235	0.0243	0.0244	0.0283
B	0.0003	0.0003	0.0003	0.0003	0.0003
Mg	0.0003	0.0003	0.0003	0.0003	0.0001
Ca	0.0055	0.0042	0.0024	0.0071	0.0063
As	0.002	0.002	0.002	0.002	0.002
Sn	<0.001	<0.001	<0.001	<0.001	<0.001
Pb	0.0021	0.0023	0.0023	0.0018	0.0018
Bi	0.002	0.002	0.003	0.002	0.002
Ce	0.002	0.002	0.002	0.002	0.002
Zn	0.003	0.003	0.002	0.003	0.003

Table 2 Material variants of the sintered steel Fe1.5Cr0.2Mo

	Additional treatment	Density (g/cm^3)	Temperature of sintering (°C)	Dimensional change A (%)
S1	Without	7.50	1120	0.16
S2	Case-hardening	7.49	1120	0.16
S3	Case-hardening + shot peening	7.49	1120	0.16
S4	Pyrohydrolysis	7.50	1120	0.16
S5	Sinter-hardening	7.43	1120	0.16
S6	2 % copper addition	7.43	1120	0.64

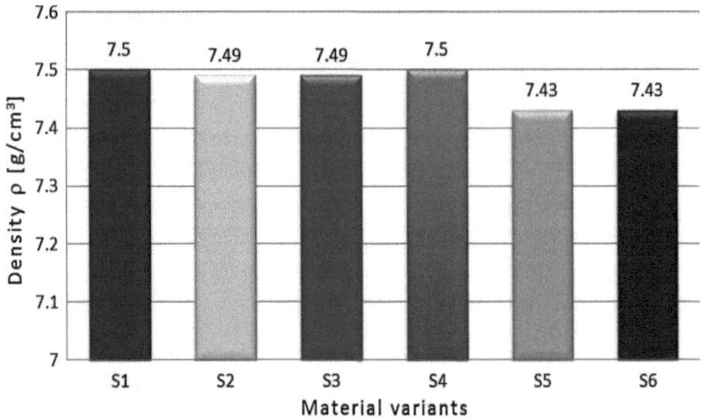

Fig. 3 Comparison of the density of all material variants [13]

$$V_S = \frac{m_L - m_W}{\rho_{Fl}} \quad (1)$$

The density of helical gears is calculated according to the following formula:

$$\rho_S = \frac{m_L}{m_L - m_W} \rho_{Fl} \quad (2)$$

The density specification for sintered helical gears made from the base material without undergoing any further treatment (S1) is 7.5 g/cm^3. Figure 3 shows the comparison of density mean values for various types of helical gears.

The density of sintered helical gears is in accordance with the specified tolerance to the required density. For the hardened materials (S2) and hardened and shot peened materials (S3), as well as for water vapour-treated materials (S4), the densities are approximately equal to those of the materials which have not undergone any additional treatment (S1)—7.5 g/cm^3. A slightly smaller density can be found with the sinter-hardened material and with those with 2 % copper addition. The alloying process by which 2 % of copper is added into the base powder results in a lower density. This is due to the increase in the wheel size with the same amount of powder.

2.2.2 Hardness Test

The micro-hardness of the gears is measured at nine locations which can be recognized in the arc shown in Fig. 4. The Vickers hardness test HV0.1 is the one performed with a test load of 0.9807 N., which is common for such micro-hardness measurements. Figure 4 shows the arrangement of the measuring points. The preparation of cross-section samples is performed by means of the wire EDM and a subsequent grinding, such as a polishing treatment.

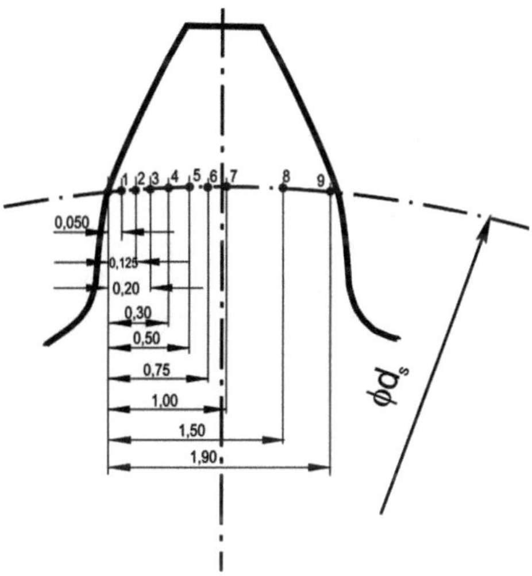

Fig. 4 The positions of measuring points for micro-hardness. where d_s is a pitch diameter

The micro-hardness profile of the gears is variable with respect to teeth thickness. The largest micro-hardness is normally present in the regions at the edge of the layer hardened with martensite (Points 1 and 2).

The comparison of micro-hardness measurements (mean values) of helical gears with different material variants is shown in Fig. 5. As might have been expected, the material variants with hardening or case-hardening and shot peening have the largest micro-hardness mean values. The smallest micro-hardness averages are found in the material variants without treatment or with steam treatment. The micro-hardness profile of gears is variable with respect to teeth thickness. The largest micro-hardness is normally present in the regions at the edge of the layer hardened with martensite (Points 1 and 2).

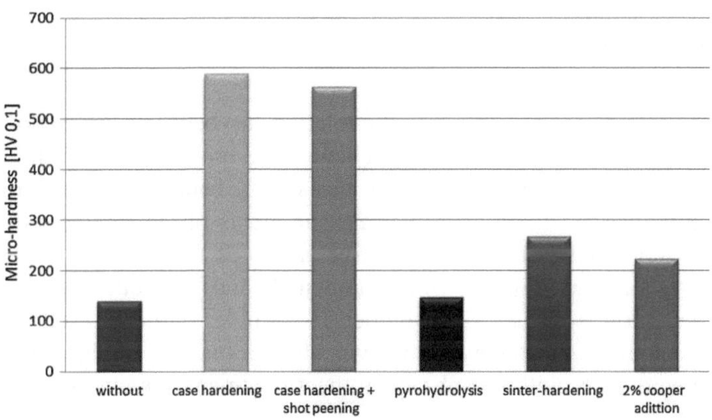

Fig. 5 Comparison of the mean values of micro-hardness [13]

2.2.3 E–Module

Usually, the E-modules of metals are determined in either tensile or compression tests. The compression test for this measurement is selected on the basis of a cuboid geometry sample with a volume of 10 mm × 10 mm × 15 mm. The samples were taken from the gears by means of the wire EDM process. Subsequently, the experiment showed only a tendency, but did not provide precise results.

A sample rod is compressed by a hydraulic testing machine at a continuous speed. From the predetermined cross-sectional area and the length of the sample, the measured compressive force and the measured elongation of the sample, the E-module E can be calculated with the aid of Hooke's law according to Formula 3.

$$E = \frac{\sigma}{\varepsilon} \tag{3}$$

σ Stress in N/mm^2
ε Elongation

The compressive force is calculated according to the following equation:

$$\sigma = \frac{F}{A} \tag{4}$$

F Force in N
A Surface in mm^2

The relative elongation ε is calculated according to Eq. 5:

$$\varepsilon = \frac{\Delta l}{l_0} \tag{5}$$

Δl Change in length in mm
l_0 Original length in mm

Since, up to a defined yield point, the relationship between force and elongation is linear, one can use Hook's Law to calculate the E-module. The starting values for each measurement are characterised by the setting behavior of the material. For every material variant, there were two measurements of the sample size.

In Table 3, one can see the comparative values of the modules of elasticity for the following material variants: "sinter-hardened" and "without treatment". Due to the hardened surface in the "sinter-hardened" samples and the resulting structural changes, its module of elasticity is somewhat higher than the module of elasticity of normal sintered gears without subsequent heat treatment. On average, this increase is 10.1 %.

Table 3 The comparative values of the E-modules

Material	E-module (N/mm^2)
Without additional treatment	184,950
Sinter-hardening	203,760

The E-module of the gears with material variant "without treatment" reaches approximately 88 % of the E-module of steel, while the material variant "sinter-hardened" reaches 97 %.

2.2.4 Poisson'S Ratio

Poisson's ratio ν is a dimensionless material constant that indicates how this ratio behaves in relation to the longitudinal extension of a material. It is only valid for linear elastic load cases. Typical values for steel are 0.2–0.33. Equation 6 describes the calculation of Poisson's ratio.

$$\nu = -\frac{\Delta d/d_0}{\Delta l/l_0} \tag{6}$$

Δd Thickness change in mm
d_0 Original thickness in mm
Δl Change in length in mm
l_0 Original length in mm

In order to determine Poisson's ratio, one can use the modules of elasticity of the sample measurement. Mean values of Poisson's ratio are shown in Fig. 6.

2.2.5 Tensile Test

With the help of a uniaxial tensile test, it is possible to testify to the yield strength $Rp_{0.2}$, the tensile strength R_m, and the uniform elongation A_g, as well as to the

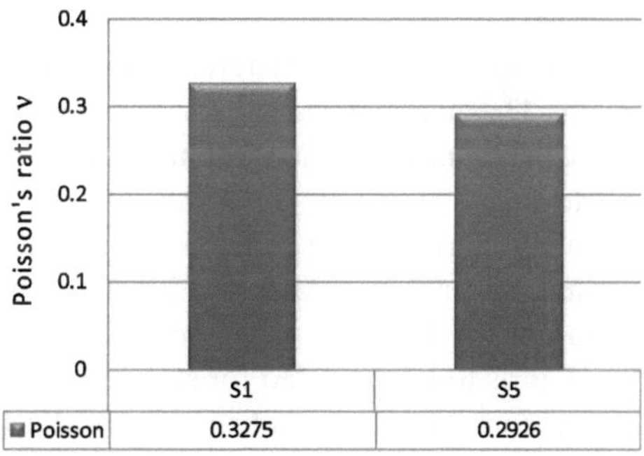

Fig. 6 Mean values of Poisson's ratio [13]

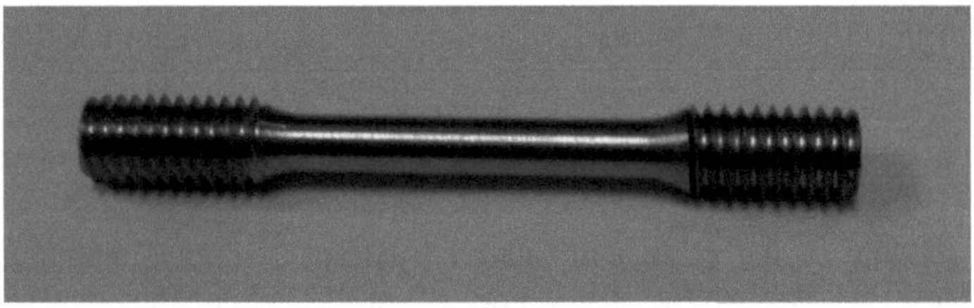

Fig. 7 Tensile specimen before the test

elongation A_5 of any material. For this purpose, a cylindrical material sample [6] with thread at both ends was used. In the course of these experiments, the material variants S1 and S5 were used. Since the samples were taken from the centre of the gear, an examination of the surface-treated materials was not possible. The diameter d_0 in the centre of the sample was 3 mm. The original length l_0 of the relevant sample portion equals ten times the diameter. Figure 7 shows a tensile specimen before the test.

The elongation at the break and the uniform elongation can be calculated from the force and the axial strain by using the following formulas. The elongation at break A_5 of the initial length l_0 is related to the elongation at the break of the tensile specimen. The uniform elongation A_g of the initial length l_0 is related to the tensile specimen plastic elongation under stress with the maximum force.

$$A_5 = \frac{l_5 - l_0}{l_0} \cdot 100\ \% \tag{7}$$

l_5 The elongation at the break of a tensile specimen in mm
l_0 The original length in mm

$$A_g = \frac{l_{mp} - l_0}{l_0} \cdot 100\ \% \tag{8}$$

l_{mp} The elongation at the break of the tensile specimen at maximum force in mm

The yield strength and the tensile strength can be read from the stress-strain diagram.

The six samples of material density come from three different gears of the same batch. The average values for the tensile strength R_m, the 0.2 %—yield strength $R_{p0.2}$. The uniform elongation A_g and the elongation at break A_5 are listed in Table 4.

In comparison to the wheels of the material variant S1, the wheels of the material variant S5 have a greater tensile strength by a factor of 1.61 and a smaller elongation at the break by a factor of 2.38. The measurement of the elongation at the break is as to be expected for sintered material, but only slightly larger than uniform elongation. This fact confirms the brittle fracture behavior of the material used.

Table 4 Measured values of the tensile test

Designation	R_m (N/mm^2)	$R_{p0.2}$ (N/mm^2)	A_g (%)	A_5 (%)
S1: without additional treatment 1	395	216	11.5	13.8
S1: without additional treatment 2	389	214	11.4	12.6
S5: sinter-hardened 1	643	462	5.3	6.2
S5: sinter-hardened 2	620	440	4.8	5.1

3 Test Conditions

The practical tests were carried out by using five test benches with a center-to-center distance of 30 mm. The transmission of the asynchronous motor was mounted on the test bench and the output torque was applied via a magnetic particle brake. Measure points of test benches are shown in Fig. 8.

The test transmitter is made of aluminum. The axial section of the worm shaft is shown on the left side of Fig. 9 and the axial section of the gear shaft is shown on the right.

The bearing of the worm shaft (1) is achieved with two angular contact ball bearings (3) in X-arrangement. The worm (1) is below the wheel (2). The temperature in the oil sump is measured with a nickel-chromium-nickel thermal element (4). The thermal element (4) measures the sump temperature directly under the worm in the area of gearing contact. The sealing of the test housing is provided by a shaft seal (5).

The wheel shaft (9) is supported with two groove ball bearings (6) (Fig. 9, right). The lubrication of the seals closed by the bearing allows for sufficient lubrication and absorbs the dynamic load [7, 8]. The entire worm is in the oil.

The time duration of a test run is set to 40 h. The output torque of the first test run is 12 Nm, and for each subsequent test is T_2 increased by 4 Nm. After every 20

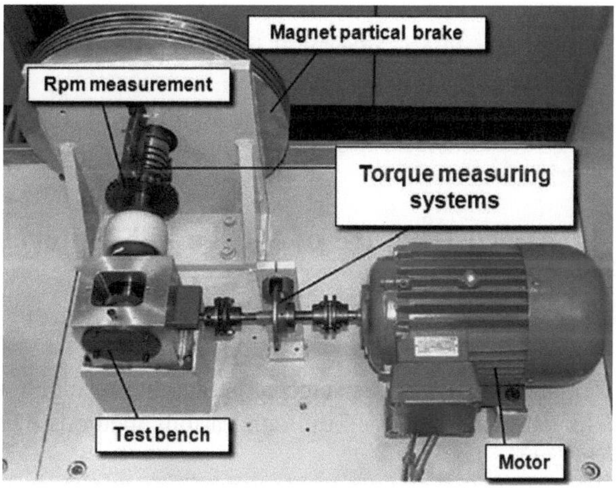

Fig. 8 Measure points of test benches

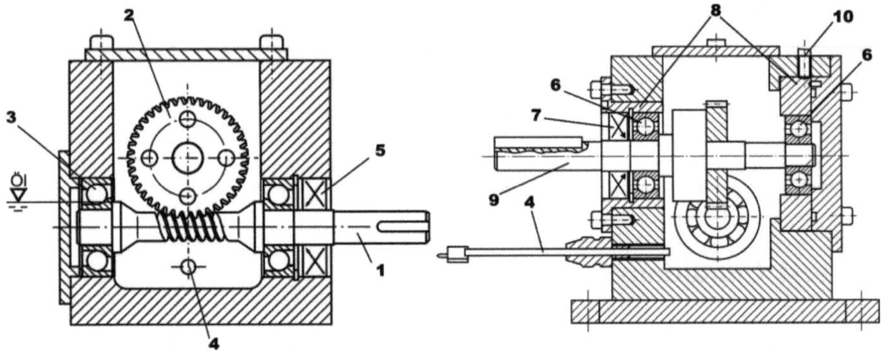

Fig. 9 Sectional drawings of test transmission

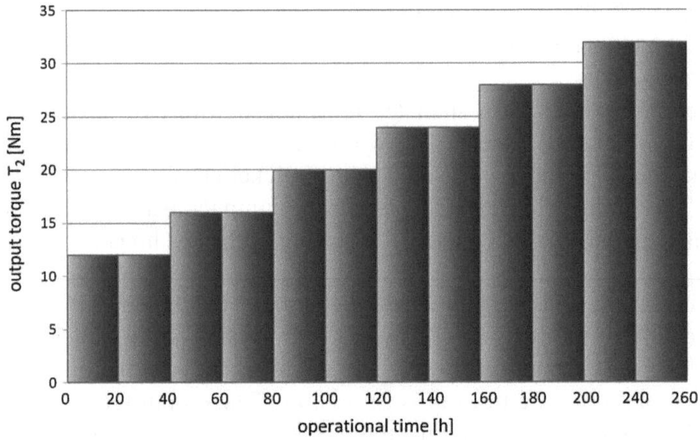

Fig. 10 Experimental procedure

h, the helix line is measured in order to determine the width of the wear surface (Fig. 10).

All data for the test gear pair are given in Table 5.

4 Microstructure of the Tested Materials

4.1 Additional Hardening

To increase wear resistance, the hardness, along with the tensile strength and elongation, is achieved through various methods of hardening of the sintered parts (Fig. 11).

Table 5 Data for the test gear pair

Parameter		Unit	Worm	Wheel
Flank form. Reference profile		[−]	ZI (DIN 3975)	DIN 867
Number of teeth	$z_{1.2}$	[−]	1	40
Centre distance	a	[mm]	30	
Normal module	m_n	[mm]	1.252	
Transverse module	m_{t2}	[mm]	1.2608	
Transmission ratio	i	[−]	40	
Helix angle	$\beta_{1.2}$	[°]	82.493	7.507
Pressure angle	α	[°]	20	
Gear width	$b_{1.2}$	[mm]	25	10
Pitch diameter	$d_{s1.2}$	[mm]	9.57	50.43
Tip diameter	$d_{a1.2}$	[mm]	12.09	52.95
Addendum coefficient	$h_{ap1.2}$	[−]	1.2615	1.26
Tooth thickness at pitch cylinder	$s_{n1.2}$	[mm]	1.963	1.963
Tooth thickness at tip cylinder	$s_{an1.2}$	[mm]	1.036	0.944
Material		[−]	16MnCr5	Fe1.5Cr0.2Mo
Speed	n_1	[min^{-1}]	1500–10,000	
Torque	T_2	[Nm]		12–36
Lubrication		[−]	oil: Klüber GH6 1500	

Fig. 11 Worm and wheels of different material variants

Basically, sintered steels can also be case hardened. A carburizing of a surface layer is possible when the relative density is at least 85 % of the theoretical density and the porosity is then less than 15 %. While the case hardness depth for steels made by melting metallurgy is mainly influenced by the carburization temperature and time, the porosity is an additional influence for PM parts. The carburization depth for sintered components is therefore highly dependent on the density. Only when the density is higher than 7.3 g/cm^3 and the porosity is closed can a unique case depth be achieved with traditional gas carburizing. Open porosity provides a large surface area because the carburizing agent penetrates into the pores and a diffusion of carbon into the surface of the porous material occurs.

In order to improve the mechanical properties, surface compaction methods such as shot peening, deep rolling, and laser shock compression, as well as different heat treatment processes, are used. The purpose of this procedure is to increase the strength and wear resistance, as well as to achieve a more favorable structure for improved strength carburizing [9].

The pyrohydrolysis treatment of sintered parts based on iron in water vapor is a proven technique for increasing the corrosion resistance, hardness and wear resistance. At a reaction temperature of 450–570 °C, a thick oxide layer of approximately 10 μm will form on the surface of the workpiece. Since the iron oxide is also formed in the pores, it will result in an increase in hardness and tensile strength compared with the untreated condition. However, by increasing the density, the process is not as positive as expected. After the pyrohydrolysis treatment, the elongation can be reduced by 50 %.

The solubility of copper in iron is temperature dependent. The hereby achievable precipitation hardening (solution annealing. quenching and tempering) causes a significant increase in strength and a higher ductility. An increase of the copper content leads to an increase of the tensile strength and to a reduction of the fracture strain. On the other hand, the tensile strength and the fracture strain increase through greater material density.

The sinter-hardening treatment is a single-stage process in which the thermal treatment is performed directly after the sintering. In order to achieve the desired mechanical properties, it is possible to change the microstructure and the amount of the formed martensite structure by controlling the cooling rate after the sintering process. It is possible to form more martensite structure by using the right alloyed materials. This significantly increases the hardness and improves mechanical properties.

The advantages of the sinter-hardening treatment: cost savings obtained when the additional cost of alloying is less than that of other treatment costs; high dimensional accuracy; good shape retention (compared with other hardening treatments, shape of parts would have only minimal deformation during the sintering process); no defects (cracks do not occur during sinter-hardening treatment because parts are not subjected to the drastic rate of temperature change); time saving; readily impregnated (since there is no entrapped quench oil, the gears can be impregnated with lubrication oil); good appearance (gears produced with this hardening process have cleaner and more uniformly colored surfaces).

4.2 Microstructure

In order to study the material properties, the gears were divided into sections. These section pieces were subsequently embedded, grounded and polished. Pictures of this prepared mettalographic specimen were taken with the scanning electron microscope after they were etched in alcoholic nitric acid (2 vol.% HNO3) [4].

The micrograph of the variant without additional treatment (Fig. 12) shows a basic structure with 70 % ferrite and 30 % pearlite. Cementite in the pearlite grains loses the lamellar shape and gets a granular shape instead, i.e., coagulate. Material should be softer and more durable in regard to the ferritic core structure.

The micrograph of the variant with a 2 % copper addition (Fig. 13) shows a uniform ferrite and pearlite microstructure. In grains with pearlite, cementite lamellas are changed into the granular shape. The increase of the pearlite phase of the granular form is a consequence of the presence of copper. Good mechanical properties are expected due to the uniform distribution of the unconnected pores.

The micrograph picture of the model with the pyrohydrolysis treatment (Fig. 14) shows a uniform, homogeneous fine-grained ferrite and pearlite microstructure. This is a homogeneous microstructure with enough pearlite phase and with good strength and toughness [10].

The sinter-hardening process structure leads to a homogeneous microstructure—tempered state with grains in the ferritic core microstructure (Fig. 15).

Figure 16a shows the core structure of the tempered martensite, respectively bainite. Figure 16b shows the edge area—the hardened layer that has the martensitic structure. Altogether, the picture shows a homogeneous microstructure with finely distributed needles.

The microstructure on Fig. 17 shows the edge area of the hardened layer with the martensitic structure having smaller martensite-needles than on the variant, which has only hardening treatment. The core has a martensite structure respectively bainite structure. Overall, it shows a homogeneous microstructure that has fine

Fig. 12 The scanning of electron micrographs for a sample without any treatment

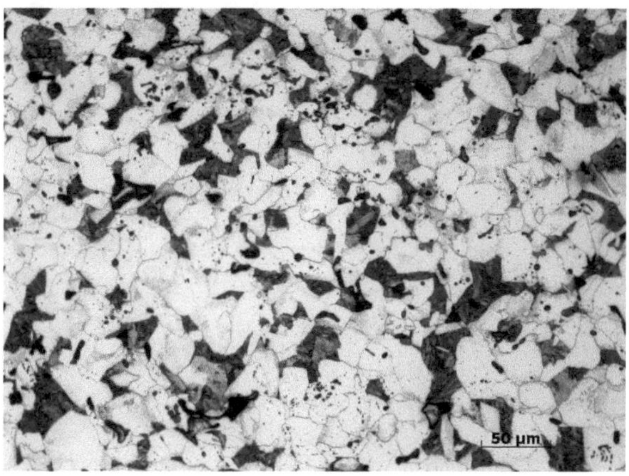

Fig. 13 The scanning of electron micrographs for a sample with 2 % copper addition

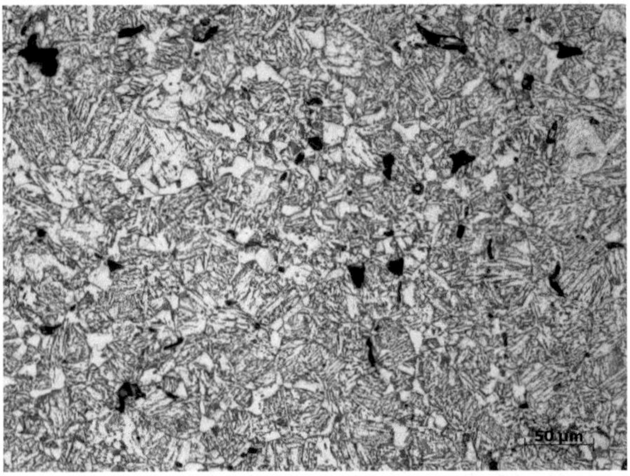

Fig. 14 The scanning electron micrographs for a sample with pyrohydrolysis treatment

Fig. 15 The scanning electron micrographs for a sample with sinter-hardening treatment

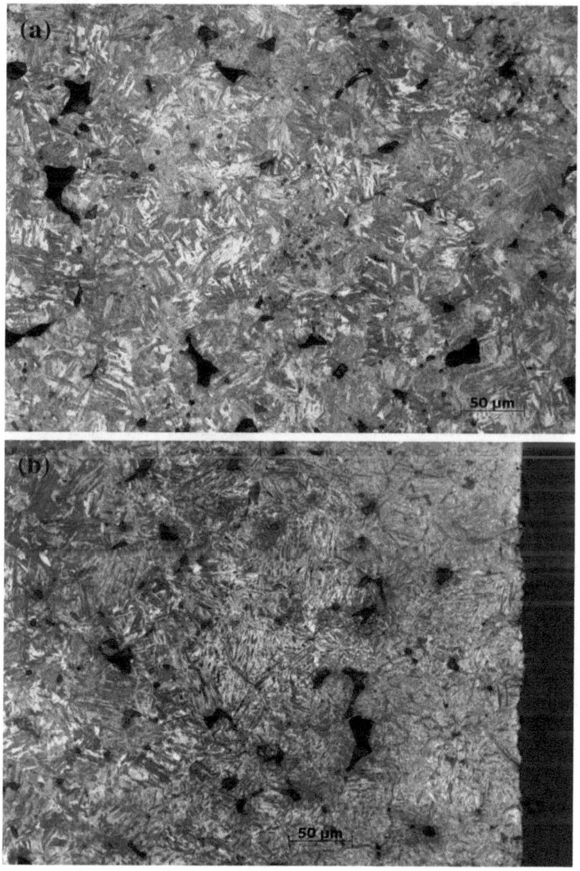

Fig. 16 The scanning electron micrographs for a sample with case hardening treatment, **a** core, **b** edge area

needle distribution. Pores with different sizes are clearly visible in the structure, their distribution being relatively homogenous.

5 Lubrication

For the sake of comparison, lubrication tests were done under same parameters and with different lubricants: synthetic oil, mineral oil and grease.

5.1 Synthetic Oil

Synthetic oil made by the Klüber company (Klübersynth GH6 15000) was used in the tests. This oil can withstand a high scuffing load capacity and has good wear

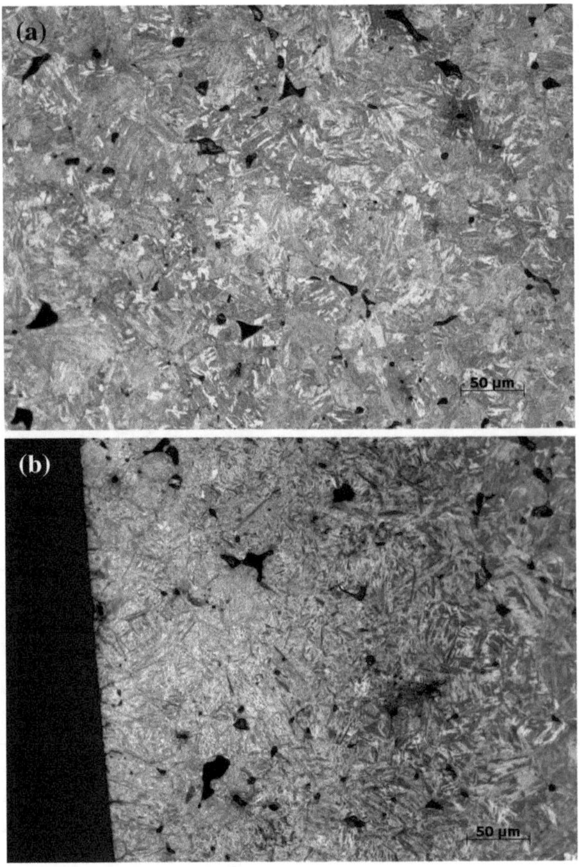

Fig. 17 The scanning electron micrographs for a sample with case hardening and shot peening treatments. **a** Core, **b** edge area

protection. Furthermore, the oil reduces friction and displays flat viscosity-temperature behaviour. This oil is based on polyglycol and is mainly used in transmissions with the material combination of steel/steel. The additive GH6 reduces the friction coefficient and wear, especially for worm gears with the material combination steel/bronze. Very low wear intensities for worm gears can be achieved with this oil. Specifically in this case, Klübersynth GH6 oil with a viscosity of ν_{40} = 1500 mm^2/s was used. Synthetic oil with additives has good anti-scoring properties [11]. Table 6 shows the characteristics of the lubricants used.

5.2 Mineral Oil

Castrol Optigear BM 1500 was used in the experiment as the mineral oil with the same viscosity as Klübersynth GH6 1500. High performance gear oil Castrol Optigear BM contains the mineral oil-based additive package MICROFLUX Trans

Table 6 Data for lubricants

Parameter	Unit	Klübersynth GH6 1500	Optigear BM 1500	Klübersynth G34-130
Temperature range	[°C]	−20 to 160	−10 to 90	−30 to 130
Density (20 °C)	[g/cm^3]	1.08	0.93	0.87
Kin. base oil viscosity DIN 51562 ν_{40}	[mm^2/s]	1500	1507	150
Kin. base oil viscosity DIN 51562 ν_{100}	[mm^2/s]	231	75.6	16
Consistency class DIN 51 818. NLGI (grease)				0

(MFT). This combination of load-active agents and additives adjusts to varying loads and actively prevents wear. Difficulties in the run-in phase can be reduced and problems with wear, material fatigue on surfaces (pitting) and micropitting can be avoided.

5.3 Grease

Klübersynth G34-130 grease for small gears was used in the grease tests. Klübersynth G34-130 is a grease based on synthetic hydrocarbon oil. Its other ingredients are mineral oil and lithium soap, special polyureas which serve as consistency factors. The grease has good anti-wear properties and can be used with the combination of steel and plastic.

Figure 18 presents the dependence of kinematic viscosity on temperature for the selected lubricants. The oils Klüber GH6 1500 and Castrol Optigear BM 1500 have

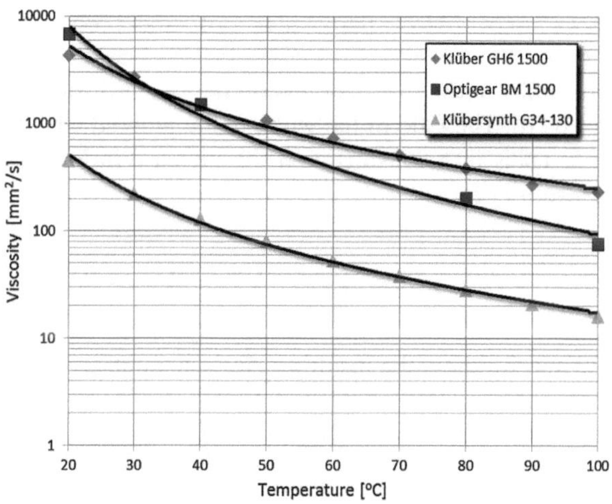

Fig. 18 Viscosity-temperature behavior for the lubricants used

the same viscosity at a temperature of 40 °C. At higher temperatures, the viscosity of the mineral oil Castrol Optigear BM 1500 is smaller than the synthetic oil Klüber GH6 1500. The lubricants' dependence on viscosity is very important at low temperatures. They should still be able to flow at high temperatures and be very viscous at extremely high temperatures. The best oil in this sense is the Klüber GH6 1500 oil.

6 Hertzian Contact Stress

Hertzian contact stress has a significant influence on the wear rate and width of the wear surface. Hertzian contact stress in crossed helical gears can be determined as the Hertzian contact stress of the globoid wheel. It should be taken into consideration that the Hertzian contact stress on the crossed helical gear depends on the width of the wear surface. Therefore, a correlation between the width of the wear surface b_V and the dimensionless ratio p_m* [4] is introduced. Hertzian contact stress by the dimensionless parameter of the average Hertzian contact stress $p_{m.V}$* is taken into account. The new Hertzian contact stress σ_{Hm} can be calculated for the average Hertzian contact stress $p_{m.V}$* according to Eq. 9, depending on the output torque T_2, the E-Module E_{red} and centre distance a_s.

$$\sigma_{Hm} = \frac{4}{\pi} \cdot \sqrt{\frac{p^*_{m,V} \cdot T_2 \cdot 1000 \cdot E_{red}}{a_s^3}} \tag{9}$$

According to the tests (Table 3) for sintered steel Fe1.5Cr0.2Mo with sintered-hardening (S5), the value of E-Module is $E_2 = 203{,}760\ \text{N/mm}^2$. Therefore, for the material combination of a worm made of 16MnCr5 and a wheel made of sintered steel Fe1.5Cr0.2Mo, the value of the reduced E-Module is $E_{red} = 227{,}288\ \text{N/mm}^2$.

The Hertzian contact stress that depends on the width of the teeth $p_{m.V}$* can be calculated using Eq. 10. The ratio of the width of the globoid wheel b_{2H} to the width of wear surface b_v takes into consideration the increase in Hertzian contact stress with decreasing width [12].

$$p^*_{m,V} = \left(\frac{b_{2H}}{b_v}\right)^{0{,}8614} \cdot p^*_m \tag{10}$$

Figures 19, 20 and 21 show the resulting Hertzian contact stress of all tests with different lubricants and n_1 = 1500; 5000 and 10,000 min^{-1}. According to Fig. 10, in the first load step, output torque is 12 Nm (8 Nm for 10,000 min^{-1}), while for each subsequent step, output torque T_2 is increased by 4 Nm. and the time duration of load level is set at 40 h.

Higher torque leads to an increase in pressure.

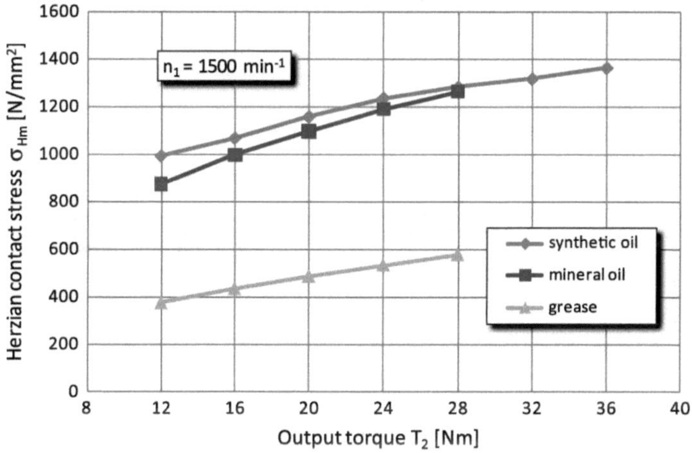

Fig. 19 Hertzian contact stress σ_{Hm} for duration of the experiment with different lubrication and v_{gs} = 0.76 m/s

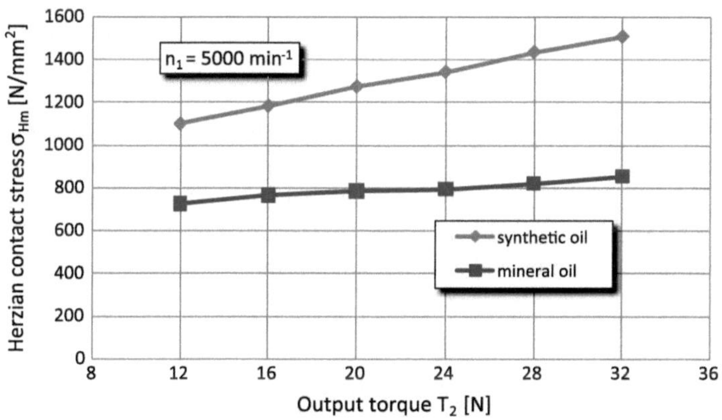

Fig. 20 Hertzian contact stress σ_{Hm} for duration of the experiment with different lubrication and v_{gs} = 2.53 m/s

There is no great difference in the pressure values of Hertzian contact stress for lubrication with mineral and synthetic oil for the rotation speed n_1 = 1500 min^{-1} or sliding velocity v_{gs} = 0.76 m/s (Fig. 19). For grease lubrication, Hertzian contact stress is, on average, half the size in relation to mineral and synthetic lubricating oil.

There is a great difference in pressure values of Hertzian contact stress for lubrication with mineral and synthetic oil for the rotation speed n_1 = 5000 min^{-1} or sliding velocity v_{gs} = 2.53 m/s (Fig. 20). The values of Hertzian contact stress with synthetic oil are about 65 % higher compared to lubrication with mineral oil. Hertzian contact stresses for lubrication with synthetic oil goes up to 1400 N/mm^2.

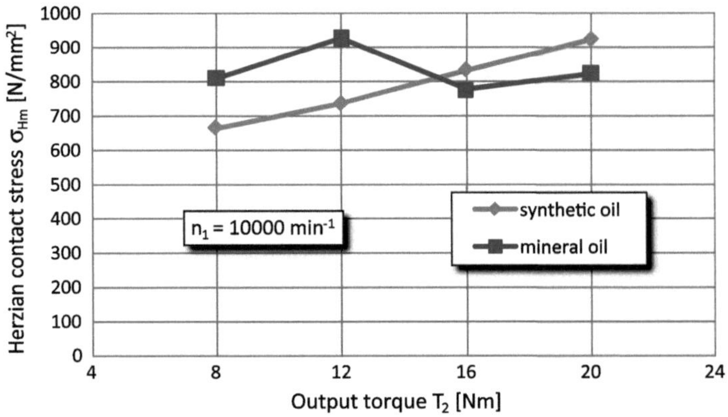

Fig. 21 Hertzian contact stress σ_{Hm} for duration of the experiment with different lubrication and $v_{gs} = 5.05$ m/s

For the rotation speed $n_1 = 10{,}000\ \text{min}^{-1}$ or sliding velocity $v_{gs} = 5.05$ m/s up to the number of load changes $N_L = 0.9 \times 10^6$, values of Hertzian contact stress are greater for mineral oil lubrication as compared to synthetic lubricating oil (Fig. 21). Thereafter, we see a sharp increase in wear rate and reduction of pressure. Lubrication with synthetic oil has a smaller increase in wear rate and less reduction of pressure.

Based on the foregoing analysis, it can be concluded that Hertzian contact stress depends on the choice of lubricant. Synthetic lubricating oil is the most favourable from the aspect of wear, and in terms of achieving hydrodynamic lubrication. Optimal lubrication conditions were obtained for the rotation speed of $n_1 = 5000\ \text{min}^{-1}$ or sliding velocity $v_{gs} = 2.53$ m/s.

7 Damage Types and Load Capacity

7.1 Wear

Wear describes the continuous loss of material from the surface of a basic body which has a relative movement with respect to a solid, liquid or gaseous mating with which it is in contact.

Contrary to hardness or tensile strength, wear is not a specific material property but a system property which depends on the particular tribological system. In our case, the elements of the tribological system are: the gear wheel (basic body), the worm (opposed body) and the lubricant (intermediate component).

Worm and wheel are in contact at a point. During operation, a change in the tooth flank of the wheel appears due to wear. The worm forms on the tooth flank of the wheel, a wear surface that has a shape which is identical to the worm gear flank.

Wear progress widens the wear surface, which leads to a lower Hertzian contact stress in the tooth contact. After a certain period of operation under intensive wear progress, the steady state occurs, where a necessary oil layer exists, so that the wear progress is minimal. Therefore, the basic criterion of work capacity of crossed helical gears is satisfactory wear resistance in the exploitation conditions.

Figures 22, 23 and 24 show a comparison of the wear surfaces of wheels for n_1 = 1500, 5000 and 10,000 min^{-1}. The output moment of the first test run was 12 Nm for n = 1500 and 5000 min^{-1} and 8 Nm for n = 10,000 min^{-1}, with each subsequent test (40 h) being T_2 increased by 4 Nm.

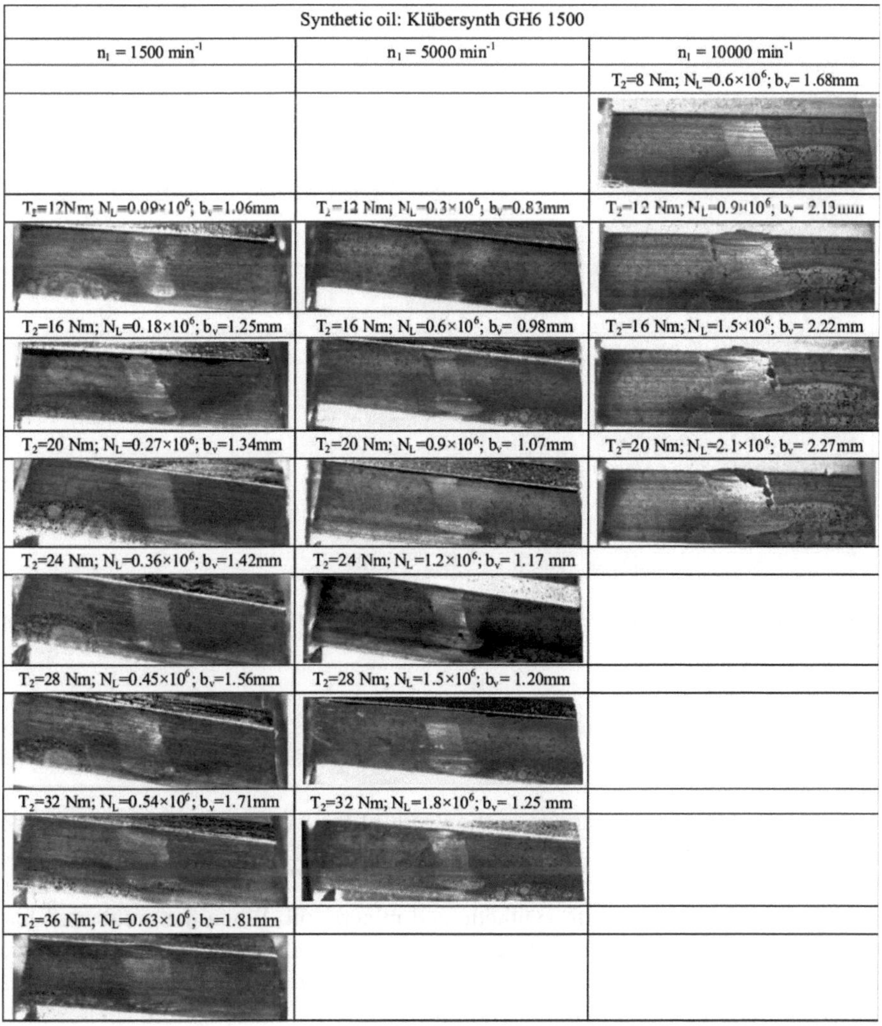

Fig. 22 Comparison of the wear surfaces of wheels for n_1 = 1500, 5000 and 10,000 min^{-1} and synthetic oil GH6 1500 [12]

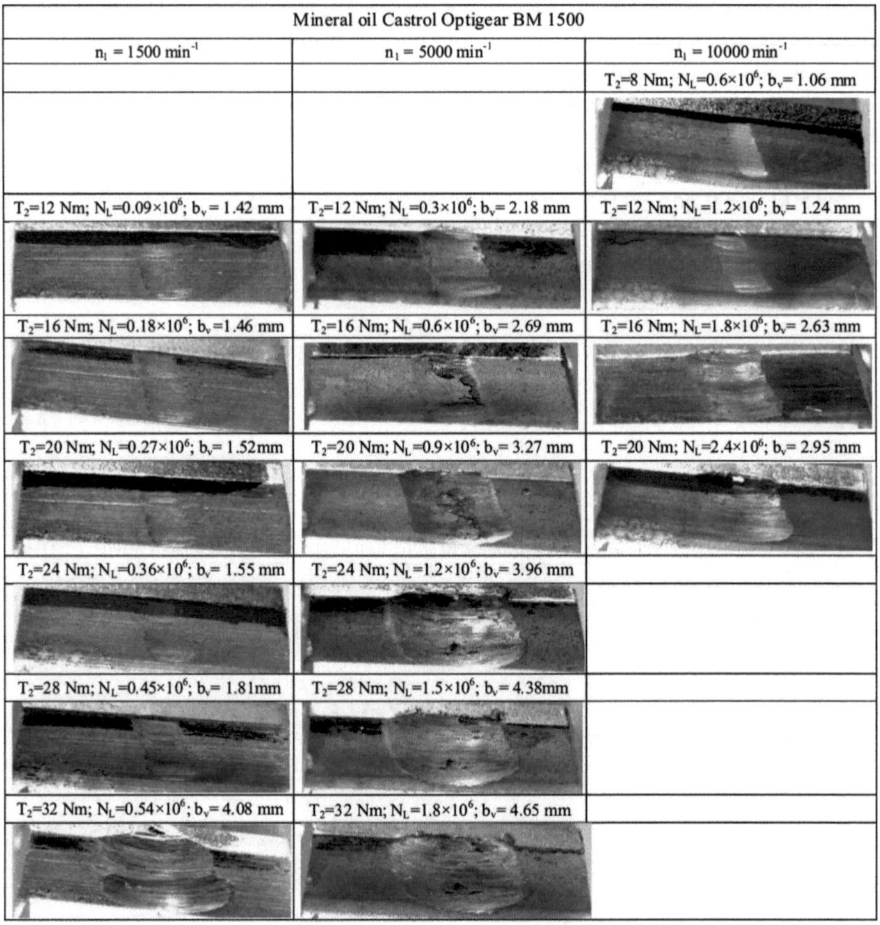

Fig. 23 Comparison of the wear surfaces of wheels for n_1 = 1500, 5000 and 10,000 min^{-1} and mineral oil [12]

The greatest wear resistance was observed in wheels lubricated with synthetic oil GH6 1500, while the lowest one was found in wheels lubricated with Klübersynth G34-130 grease.

Figure 25 gives a comparison of the width of the wear surfaces for wheels with different lubricants and T_2 = 20 Nm; t = 120 h; n_1 = 1500 min^{-1} (v_{gs} = 0.76 m/s).

There was no significant difference in the width of the wear surfaces up to t = 200 h by lubrication with synthetic and mineral oil. When the number of load cycles reached $N_L = 0.5 \times 10^6$ in tests with mineral oil, a sudden increase in wear and destruction of the flank due to scuffing occurred. The highest width of wear surface occurred in tests with mineral oil.

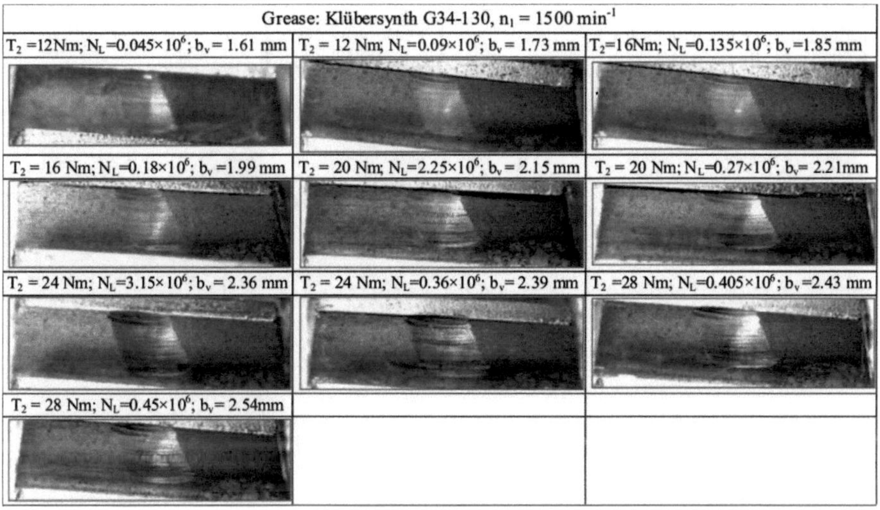

Fig. 24 Comparison of the wear surfaces of wheels for n_1 = 1500 min^{-1} and grease [12]

Fig. 25 Wear width of the wheel lubricated with mineral oil, synthetic oil and grease for n_1 = 1500 min^{-1}

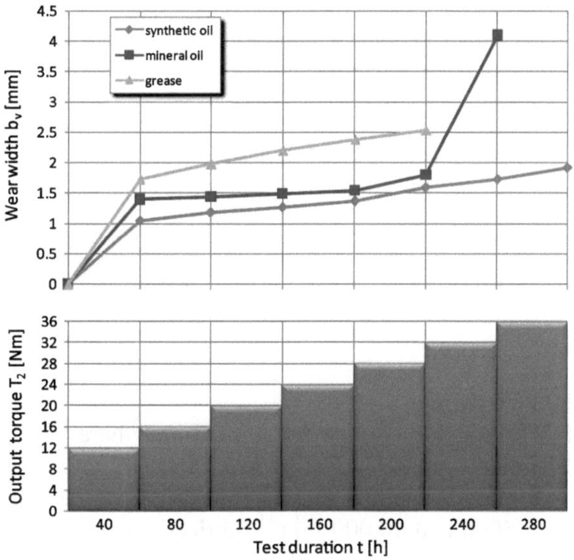

The comparison of wear surface for T_2 = 20 Nm after t = 120 h (Fig. 26) shows that wear is higher by 14 % in tests with mineral oil and by 65 % in tests with grease, in regard to the tests with synthetic oil.

At high speeds, grease cannot maintain a solid consistency and cannot remain between teeth during operation due to the high centrifugal forces. A sufficient

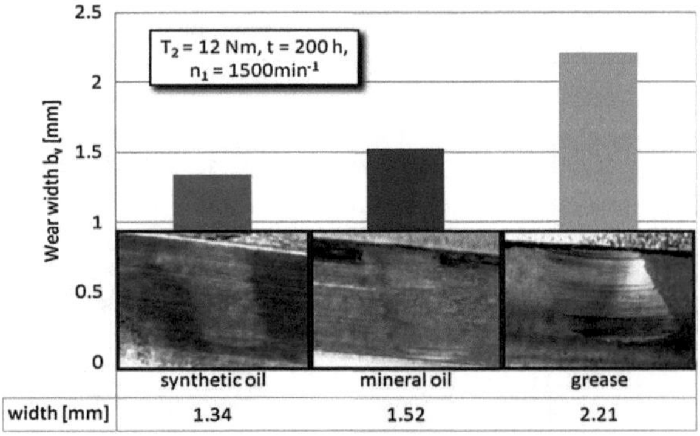

Fig. 26 Comparison of the width of the wear surfaces for wheels with different lubricants and $T_2 = 20$ Nm; $t = 120$ h; $n_1 = 1500\ \text{min}^{-1}$ [5]

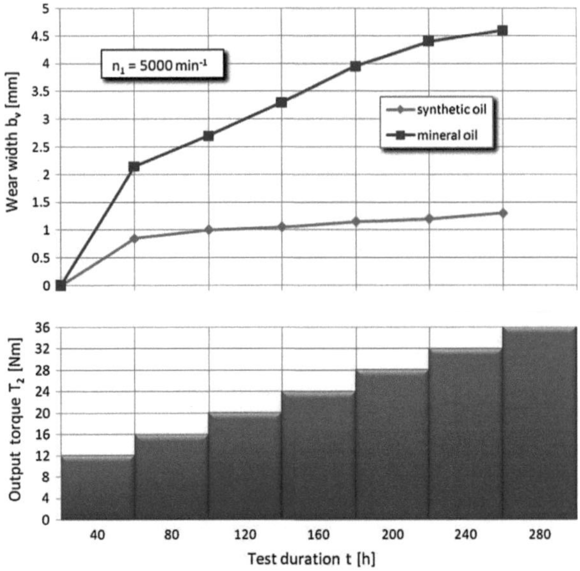

Fig. 27 Width of wear surface of wheels lubricated with mineral and synthetic oil for $n_1 = 5000\ \text{min}^{-1}$

lubrication film cannot be formed and there can be no positive effect on wear. Therefore, grease cannot be used for high speeds.

Figures 27 and 28 show a comparison of the wear width of the wear surface for $n_1 = 5000$ and 10,000 min^{-1} lubricated with mineral and synthetic oil.

There is a considerable difference in the width of the wear surface for $n_1 = 5000\ \text{min}^{-1}$, which increased steadily during the examination. Compared to the synthetic oil, the wear for lubrication with mineral oil increased for the number of load changes $N_L = 0.3 \times 10^6$ for 2.6 times, and for the $N_L = 1.8 \times 10^6$ for 3.7 times. Wear was higher by 3.3 times for mineral oil than for synthetic oil. The trend of increasing wear was, therefore, much higher for lubrication with mineral oil compared to synthetic oil.

Fig. 28 Width of wear surface of wheels lubricated with mineral and synthetic oil for n_1 = 10,000 min^{-1}

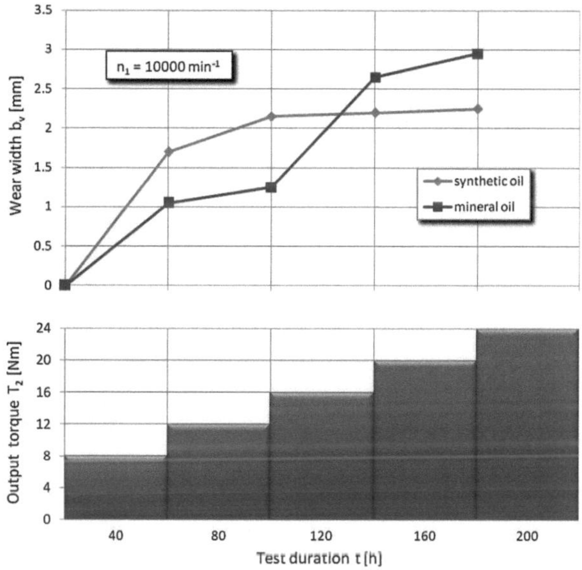

In the beginning, there was less wear for lubrication with mineral oil and n_1 = 10,000 min^{-1} than for lubrication with synthetic oil. Wear grew rapidly for $N_L = 1.5 \times 10^6$ and lubrication with mineral oil became higher for $N_L = 1.9 \times 10^6$ after 120 h than with lubrication with synthetic oil.

Figure 29 shows the wear width of the wear surface at a wheel from material S5 with different lubricants for input speed n_1 = 1500 min^{-1} [13].

7.2 *Pitting*

Significant pressure on a surface does not lead to a sudden failure of drive, but over time, small holes (pits) emerge in the shape of a shell on the tooth flank. Pit peak always points in the sliding direction. This damage occurs through a cyclic fatigue due to repeated elastic and plastic deformations of the surface. The holes occur only after a sufficiently large number of overrollings (from ca. 5×10^4 load cycles). If only initial pitting is present, the situation is not dangerous. Destructive pitting destroys the flank and causes failure due to noise and fatigue. The pitting occurred on wheels made from materials S4. S5 and S6 by an output torque of 16 and 32 Nm after the trial time period of 120–240 h.

Figure 30 shows initial pitting on the tooth flank of a wheel with different material variants. In trials with material S1—without additional treatment, initial pitting occurred under input speed n_1 = 5000 min^{-1} and output torque of 36 Nm. In trials with material S5—sinter-hardening, initial pitting occurred on the wheel tooth

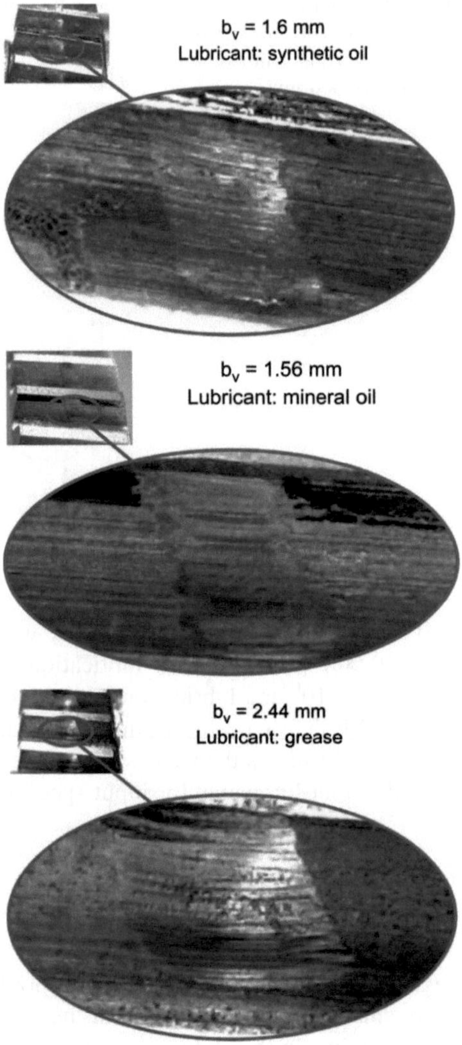

Fig. 29 Wear surfaces on a wheel of material S5—sinter-hardening with different lubricants and $T_2 = 28$ Nm; $t = 200$ h; $n_1 = 1500$ min^{-1}

flank under input speed $n_1 = 5000$ min^{-1} and output torque of 16 Nm, and only with mineral oil.

Figure 31 shows destructive pitting on a tooth flank with different material variants. In trials with material S5—sinter-hardening, destructive pitting occurred under input speed $n_1 = 5000$ min^{-1} and output torque of 20 Nm and mineral oil. In trials with material S4—"pyrohydrolysis", destructive pitting occurred after at least 180 h under input speed $n_1 = 5000$ min^{-1} and output torque of 28 Nm.

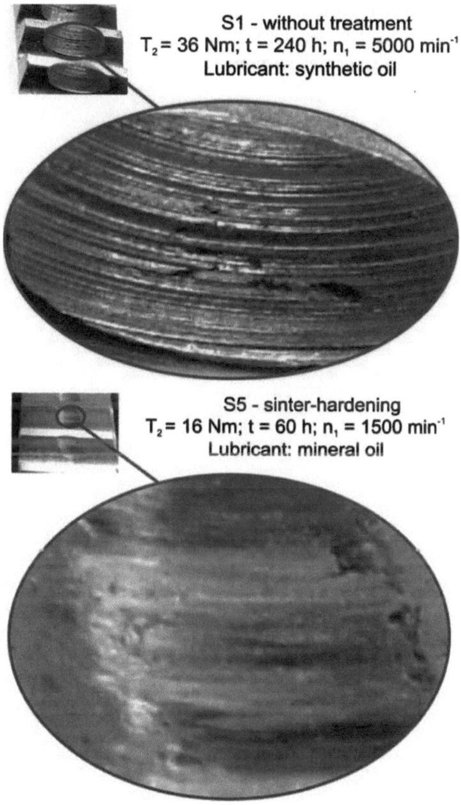

Fig. 30 Initial pitting on wheel tooth surface from different material variants [18]

7.3 *Scuffing*

There is a difference between cold and warm scuffing. Both damage types are caused by a lack of lubricant in contact between the teeth. Cold scuffing is relatively rarely seen. It occurs mainly at low speed (<4 m/s) and between teeth that are having relatively high hardness and rough quality of contact surfaces. Warm scuffing occurs due to great pressure and high sliding velocity between tooth flanks. Under such a load, combined effects occur which lead to the increase in temperature that disrupts the lubricant film between tooth flanks, making the contact direct and dry. This can cause a short local welding of the flanks which damages both of them. Warm scuffing is characterized by strip-shaped bands in the direction of the tooth height, and with the strongest expression in the tooth addendum and tooth root. Scuffing on high-speed gears increases the temperature and tooth forces, eventually leading to shaft fracture due to high damage on the tooth flanks.

Figure 32 shows the scuffing on wheel tooth flanks for different material variants. With the exception of S1, scuffing occurred on all material variants. On wheels made from materials S2 and S3, under load of T_2 = 16 Nm and T_2 = 20 Nm, the phenomenon of scuffing and significant increase of wear surface and gear forces

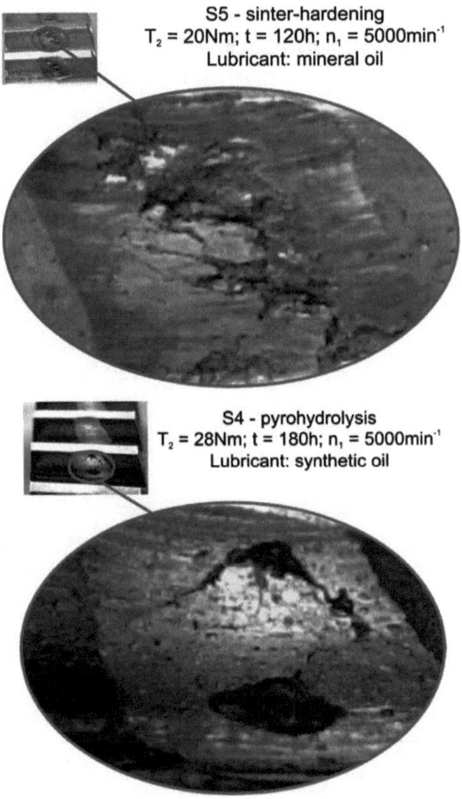

Fig. 31 Destructive pitting on a wheel tooth flank with different material variants

were observed. Under load of output torque T_2 = 28 Nm, and input speed n_1 = 5000 min^{-1}, scuffing occurred for S4 and S6. Gear loads rose and, suddenly, the failure of the worm shaft occurred. On material variant S4 "pyrohydrolysis", scuffing occurred on the tooth root in trials with input speed n_1 = 5000 min^{-1} and output torque of T_2 = 16 Nm, and a very short running time of ca. 10 min. Scuffing also occurred on wheel tooth flanks with speed n_1 = 5000 min^{-1} and output torque of T_2 = 36 Nm for material variant S5 sinter-hardening. On trials with input speed n_1 = 10,000 min^{-1}, scuffing also occurred under smaller output torque [14].

7.4 Scoring

Compared with worm materials, the wheel sintered material has lower hardness. In trials, damage also occurred on the harder opposite body. Scoring was observed in the circumferential direction of the wheel tooth flanks and on the wear wheel surface. This damage on the worm caused increased wear on the wheel, which could lead to early failure of transmission.

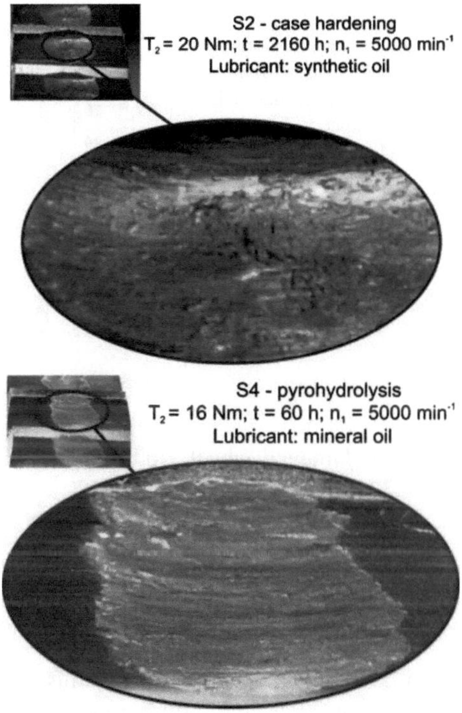

Fig. 32 Scuffing on the tooth flanks of a wheel for different material variants [18]

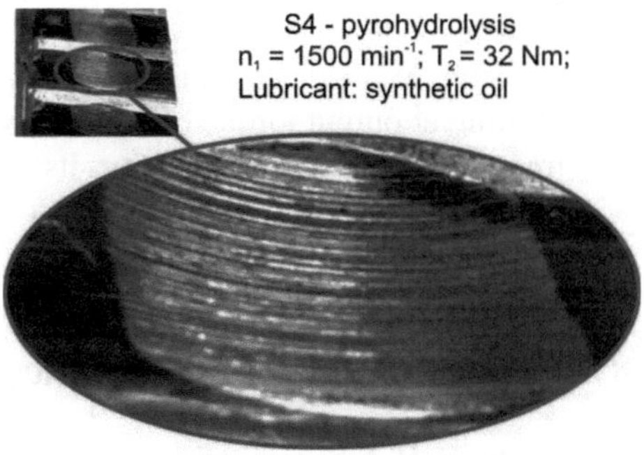

Fig. 33 Scoring on wheel [18]

The grooves are caused by abrasive wear particles of the worm, which are released due to the high Hertzian pressure in tooth contact. The wear particles in the continued operation on the wheel surface cause greater scoring damage. The contamination of the oil by wear particles from the wheel also speeds up the progression of scoring.

The scoring occurred on all sliding velocities and generally more at higher torques. For material S4 "pyrohydrolysis", high scoring on the tooth flank was observed at input speed $n_1 = 1500\ \text{min}^{-1}$ and output torque of 24 Nm. Figure 33

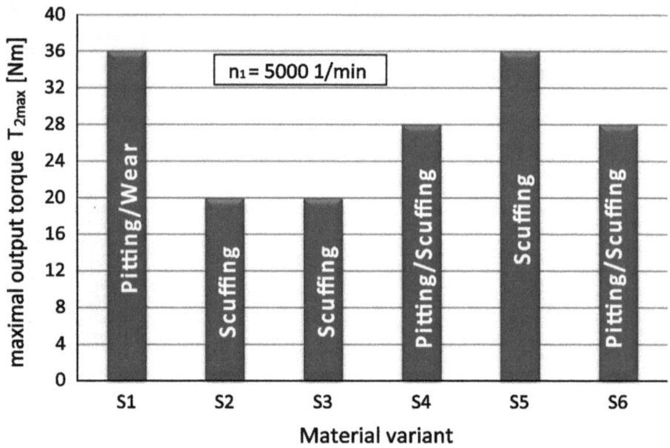

Fig. 34 Comparison of maximum transmissible torque with critical damage type for different materials and input speed $n_1 = 5000\ min^{-1}$

shows a damaged wheel tooth flank of material variant S4 with $n_1 = 1500\ min^{-1}$, $T_2 = 32$ Nm and synthetic oil [10].

7.5 Load Capacity

Figure 34 shows maximum transmissible torque, as well as the type of critical damage, in a bar chart. With an output torque of $T_2 = 20$ Nm, materials S2 (case hardening) and S3 (case hardening and shot peening) were damaged due to scuffing. Materials S4 (pyrohydrolysis) and S6 (2 % copper addition) were damaged by pitting when output torque was $T_2 = 24$ Nm, and by scuffing at the value of $T_2 = 28$ Nm. Without an additional treatment, S1 had the most critical wear and some pitting at output torque of $T_2 = 28$ Nm. The wheels S5 (sinter-hardening) had the greatest load carrying capacity, its maximum transmissible torque being 32–36 Nm, and the scuffing being the most dangerous damage form in this case.

8 Experimental and Theoretical Research of Efficiency

8.1 Overall Efficiency

The overall efficiency quantifies the efficiency of power transmission. The calculation of the overall efficiency for transmissions comes from torque. For crossed helical gears, this is applied according to Eq. (11), which shows the relationship between the measured input and output torque as a function of the overall efficiency η_{ges}.

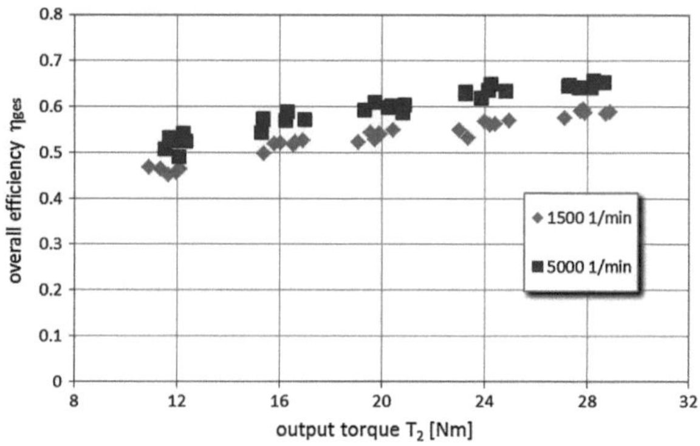

Fig. 35 Overall efficiency η_{ges} of crossed helical gears with wheels made of Fe1.5Cr0.2Mo sintered steel with sinter-hardening treatment [19]

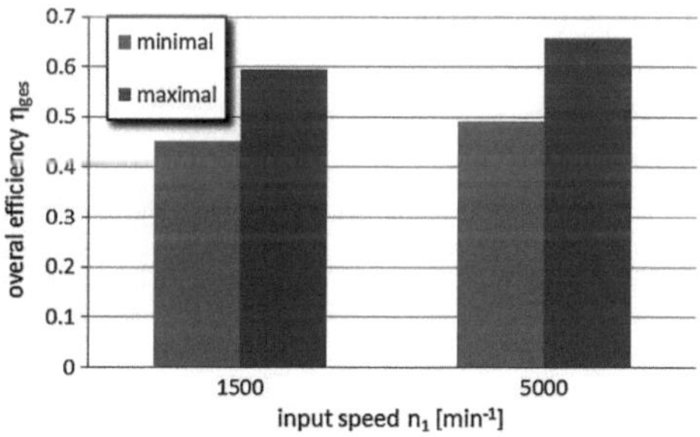

Fig. 36 Comparison of minimal/maximal overall efficiency of crossed helical gears with wheels made of Fe1.5Cr0.2Mo sintered steel with sinter-hardening treatment

$$\eta_{ges} = \frac{P_2}{P_1} = \frac{T_2}{T_1 \cdot i} = \frac{P_1 - P_V}{P_1} = 1 - \frac{P_V}{P_1} \tag{11}$$

According to Eq. (12), the total power loss P_V is the sum of the power loss of gear pair P_{VZ}, the bearing power losses of the worm shaft P_{VD1} and the wheel shaft power loss P_{VL2} and power losses of seals P_{VD1} and P_{VD2} (Figs. 35 and 36).

$$P_V = P_{VZ} + P_{VL1} + P_{VL2} + P_{VD1} + P_{VD2} = P_{V_Z} + P_{VL} + P_{VD} \tag{12}$$

By measuring the output T_2 and input torque T_1 (Fig. 8), the overall efficiency of the crossed helical gear η_{ges} can be determined. Figures 35 and 36 show the overall efficiency η_{ges} over a test period regarding the speed n_1 for sintered steel wheels. The efficiency for all experiments is in the range of 45.1–65.7 %.

The tests with input speeds $n_1 = 5000\ \text{min}^{-1}$ in the lower torque range provide the highest levels of efficiency, compared to the tests with input speeds of $1500\ \text{min}^{-1}$. By tests with input speed of $1500\ \text{min}^{-1}$ and output torque $T_2 = 32$ Nm at load cycles $N_L = 0.54 \times 10^6$, strong scoring and pitting occur on the tooth flank (Fig. 37). The resulting high surface roughness leads to increased friction in the contact area, and higher temperatures and lower efficiency.

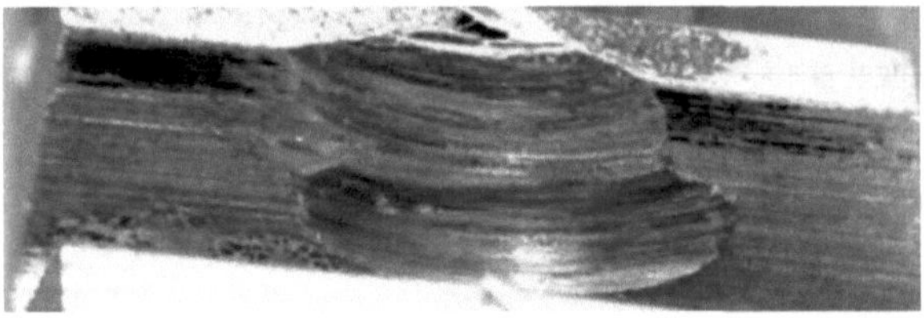

Fig. 37 Wear surface of wheel made of Fe1.5Cr0.2Mo sintered steel with hardening treatment with $n_1 = 1500\ \text{min}^{-1}$, output torque $T_2 = 32$ Nm, number of cycles $N_L = 0.54 \times 10^6$ and mineral oil lubrication [19]

Fig. 38 Distribution of power losses for transmission with a speed of 5000 min^{-1} (average values) [19]

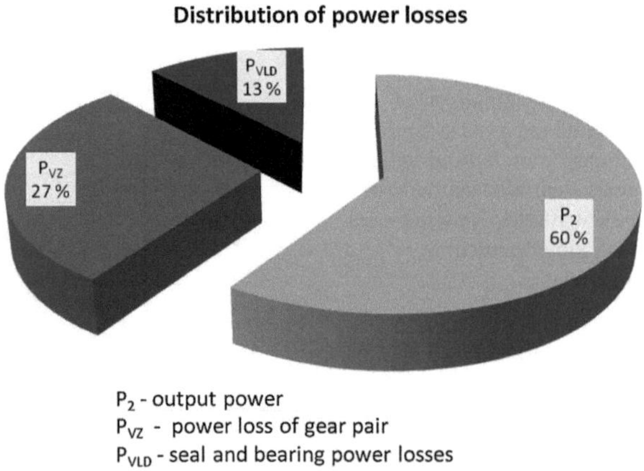

Figure 38 shows the distribution of power losses (average values) for sintered metal with hardening treatment for input speed of 5000 min^{-1}. Compared with bearing and seal power losses, power loss of a gear pair for input speed of 5000 min^{-1} is approximately 2.1 times higher.

Figure 39 shows the share of total power loss (average values) for an input speed of 1500 and 5000 min^{-1}. Power loss of gear pair P_{VZ} for $n_1 = 5000\ \text{min}^{-1}$ in comparison with $n_1 = 1500\ \text{min}^{-1}$ is approximately 2.1 times higher. Power losses of bearings and seals P_{VLD} for $n_1 = 5000\ \text{min}^{-1}$ compared to $n_1 = 1500\ \text{min}^{-1}$ are approximately 4.4 times higher [14].

Meshing efficiency η_z can be determined with the power loss of gear pair P_{VZ} and output power P_2.

$$\eta_z = \frac{P_2}{P_2 + P_{Vz}} \tag{13}$$

Meshing efficiency η_z can also be calculated with the averaged tooth friction coefficient μ_{zm} and the crossed-axes angle $\Sigma = 90°$:

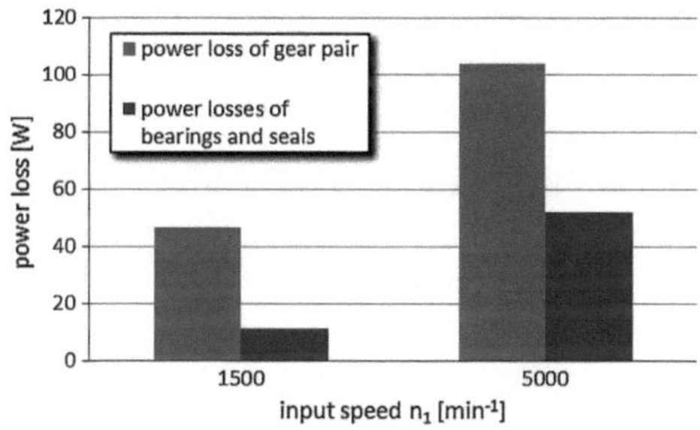

Fig. 39 Share of the total power loss at different operating conditions (average values) [20]

$$\eta_z = \frac{\tan \beta_{s2}}{\tan\left(\beta_{s2} + \arctan \mu_{zm}\right)} \tag{14}$$

8.2 *Influence of Temperature*

The oil is an important element in a machine transmission, and it can provide optimal lubrication only up to a maximum limit temperature. The sump temperature significantly influences the progress of wear of the crossed helical gears, since the viscosity of the lubricant depends on the oil sump temperature. A high viscosity reduces the friction in the area of tooth contact. The oil sump over temperature $\Delta\vartheta_S$ is the result of the difference of the oil sump temperature ϑ_S and room temperature ϑ_0.

$$\Delta\vartheta_s = \vartheta_S - \vartheta_0 \tag{15}$$

The power losses at constant speed and constant torque remain the same. Therefore, there can be a direct correlation between oil sump temperature and gear mesh power loss or the friction in the tooth contact. Further, the viscosity of the lubricant has a great impact on the oil sump temperature. Lubricants with higher viscosity lead to a thicker oil film, but also lead to higher temperatures and a decrease in viscosity.

According to DIN 3996 [15], wheel mass temperature ϑ_M can be calculated according to Eq. 16. Wheel mass temperature ϑ_M depends on oil sump ϑ_S and the oil sump over temperature $\Delta\vartheta$:

$$\vartheta_M = \vartheta_S + \Delta\vartheta \tag{16}$$

The oil sump over temperature $\Delta\vartheta$ can be approximately calculated according to Eq. (17):

$$\Delta\vartheta = \frac{P_{Vz}}{\alpha_L \cdot A_R} \tag{17}$$

According to (13) and (14), meshing power loss P_{VZ} can be calculated in the function of the coefficient of friction μ_{zm}, and helix angle β_{s2} can be determined on the crossed helical gear and the output power P_2:

$$P_{VZ} = \left(\frac{\tan(\beta_{s2} + \arctan \mu_{zm})}{\tan \beta_{s2}} - 1\right) \cdot P_2 \tag{18}$$

The coefficient of friction μ_{zm} is calculated as the product of the basic coefficient of friction μ_{OT} and factor $Y_{T\rho}$.

$$\mu_{zm} = \mu_{0T} \cdot Y_{T\rho} \tag{19}$$

The basic coefficient of friction μ_{OT} can be determined according to [15]. The factor $Y_{T\rho}$ depends on the material density of the worm with a helical gear and is, for the density $\rho \geq 7.2$ g/cm^3, according to [2], $Y_{T\rho} = 1$.

The relevant cooling surface of the gear pair A_R is given by Eq. (20):

$$A_R = b_{2R} \cdot d_{s2} \cdot 10^{-6} \tag{20}$$

The heat transfer coefficient α_L can be determined by Eq. (21):

$$\alpha_L = c_k \cdot (1940 + 15 \cdot n_1) \tag{21}$$

Here, it is $c_k = 0.8$ for a non-diving wheel and $c_k = 1$ for a diving wheel.

8.3 *Prediction of Wheel Tooth Mass Temperature by FEA*

Figure 40 shows the procedure for obtaining numerical results. Geometry data with wear rate was given in program SGGEN 2, which generated geometry and mesh in MSC.Patran. This model is exported into Ansys in which the setup of boundary conditions with all experimental and numerical data are given.

Based on the analytically determined coefficient of tooth friction, the heat generation in the meshing zone of crossed helical gears was predicted by FEA procedure. The software used for this analysis was the ANSYS Workbench, version 14. Analysis was defined as direct coupled transient structural-thermal analysis. As an accurate distribution of the temperature was requested, the high order finite element was used in analysis, namely SOLID 226.

The rate of frictional dissipation in the contact elements in ANSYS is evaluated using the frictional heating factor and is given by:

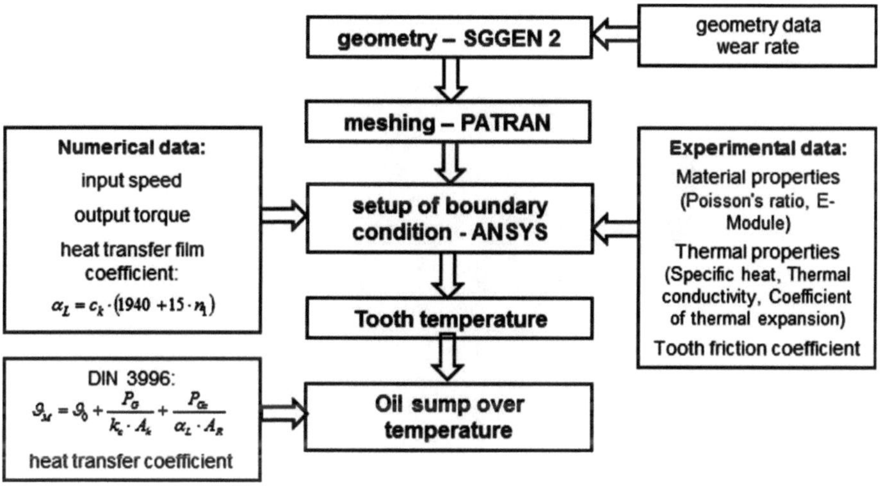

Fig. 40 Procedure for obtaining numerical results [16]

$$q = FGHT \cdot \tau \cdot v \tag{22}$$

where τ is the equivalent frictional stress, v the sliding rate and *FGTH* the fraction of frictional dissipated energy converted into heat (the default value of 1 was used). The properties of the materials used in the analysis are listed in Table 7.

As a higher order element was used due to computational limitations, only a fraction of the wheel model was considered in analysis. The rotation of the worm and worm wheel was defined in time interval via joint condition in ANSYS Workbench in order to simulate the rotational speed of 1500 min^{-1} (Figs. 41 and 42).

The analysis was performed for two different output torques: 12 and 24 Nm. The coefficient of tooth friction was defined based on Eq. (12). Table 8 shows the comparison of wheel mass temperature obtained analytically and by simulation [16]. As one can see, there is good agreement of noted temperatures. As the load increases, the difference between the simulation and calculated value of temperature decreases due to better prediction of contact surface by simulation (Fig. 43). To obtain better prediction of contact surface, as well as the surface pressure, it is necessary to further improve the mesh in the contact area and to use the nonlinear material model (bilinear hardening) which was not possible due to computational limitations.

Table 7 Thermal properties of materials used in analysis

	Specific heat (J/kgK)	Thermal conductivity (W/mK)	Coefficient of thermal expansion (C^{-1})
16MnCr5	434	60.5	1.2×10^{-05}
Fe1.5Cr0.2Mo		40	0.9×10^{-05}

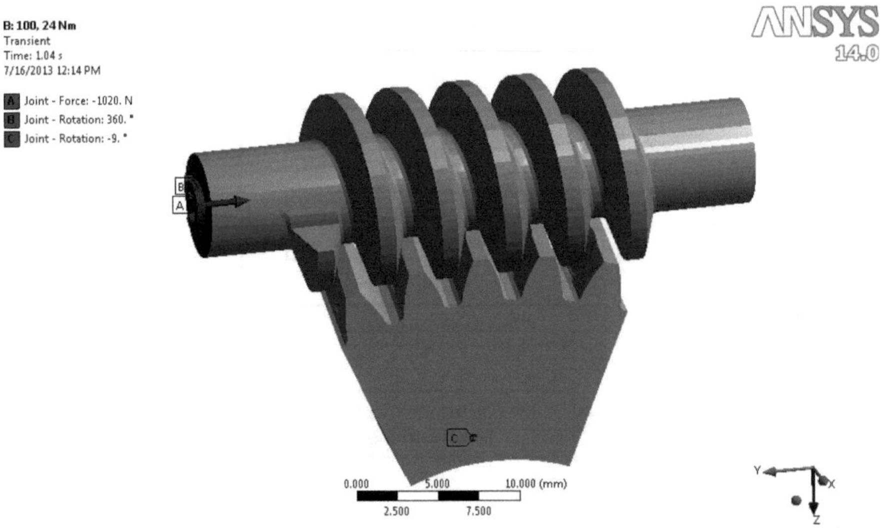

Fig. 41 Analysis setup and boundary conditions for prediction of heat generation in contact

Figure 43 shows the comparison of average oil over sump temperature $\Delta\vartheta_S$ obtained experimentally and by simulation. Oil sump temperature was determined over the experimental period of 280 h for speed of 1500 min^{-1} and the output torque T_2 for wheels made of sintered metal with sinter-hardening.

Figure 43 shows that there is a linear dependence between the oil sump temperature $\Delta\vartheta_S$ and the output torque T_2. Oil sump temperatures increased with the increase in torque, resulting in higher overall power losses. The lowest oil sump temperature was $\Delta\vartheta_S = 25$ K at the lowest speed of 1500 min^{-1} and minimum torque $T_2 = 12$ Nm. From the figure, it is clear that there is relatively good agreement between the experiment and simulation. Again, the simulation results differ more from those experimentally obtained with the decrease of load.

9 Summary

The modern technology of sintering parts has a number of advantages over the traditional method of making parts in mechanical engineering. The advantages of this technology are reflected primarily in the lower cost of production, while at the same time, the quality of parts meets all the requirements of operating conditions. After producing sintered parts, it is possible to carry out the final thermal treatment, which can affect the structure of the sintered steel, and thus develop a high resistance in relation to wear and pitting.

The paper presents the performed experimental research relating to the characteristics of sintered steel depending on various after treatments: case-hardening,

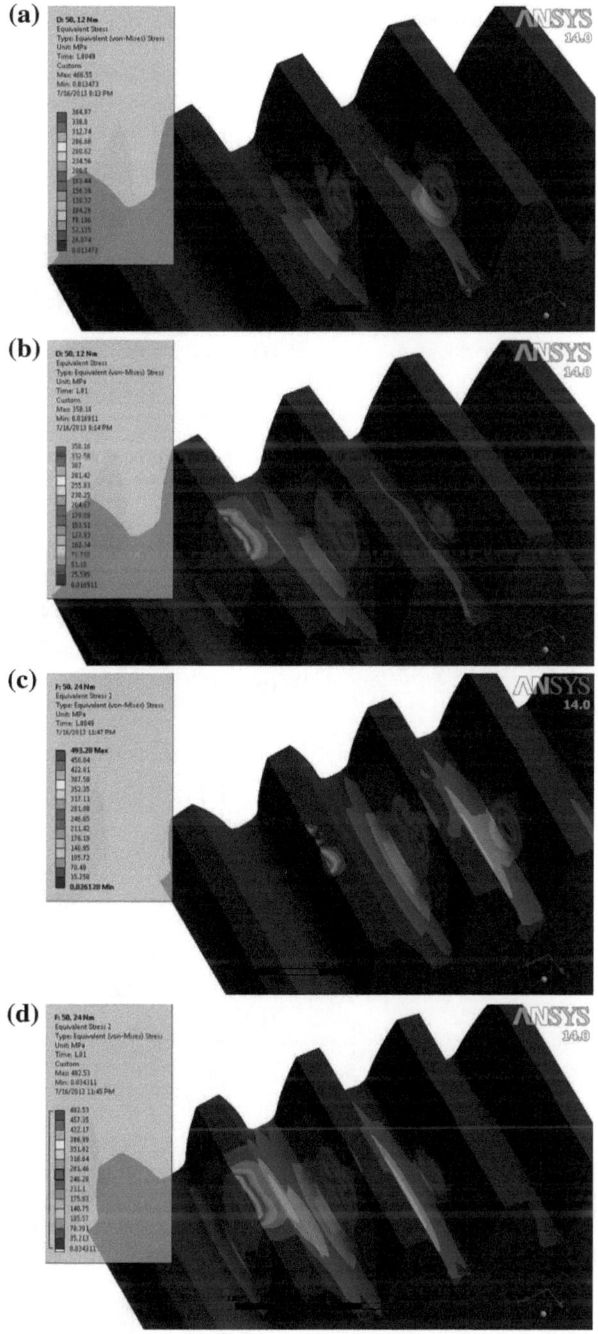

◀ **Fig. 42** Equivalent stress (**a**), (**b**), (**c**), (**d**) and contact thermal flux (**e**), (**f**), (**g**), (**h**) over time in meshing of crossed helical gears for $n_1 = 1500\ min^{-1}$ [16]. **a** Equivalent stress $T_2 = 12$ Nm, time 0.0049 (1/8 of full revolution), **b** Equivalent stress $T_2 = 12$ Nm, time 0.01 (1/4 of full revolution), **c** Equivalent stress $T_2 = 24$ Nm, time 0.0049(1/8), **d** Equivalent stress $T_2 = 24$ Nm, time 0.01 (1/4), **e** Thermal contact flux $T_2 = 12$ Nm, time 0.0049 (1/8), **f** Thermal contact flux $T_2 = 12$ Nm, time 0.01 (1/4), **g** Thermal contact flux $T_2 = 242$ Nm, time 0.0049 (1/8), **h** Thermal contact flux $T_2 = 24$ Nm, time 0.01 (1/4)

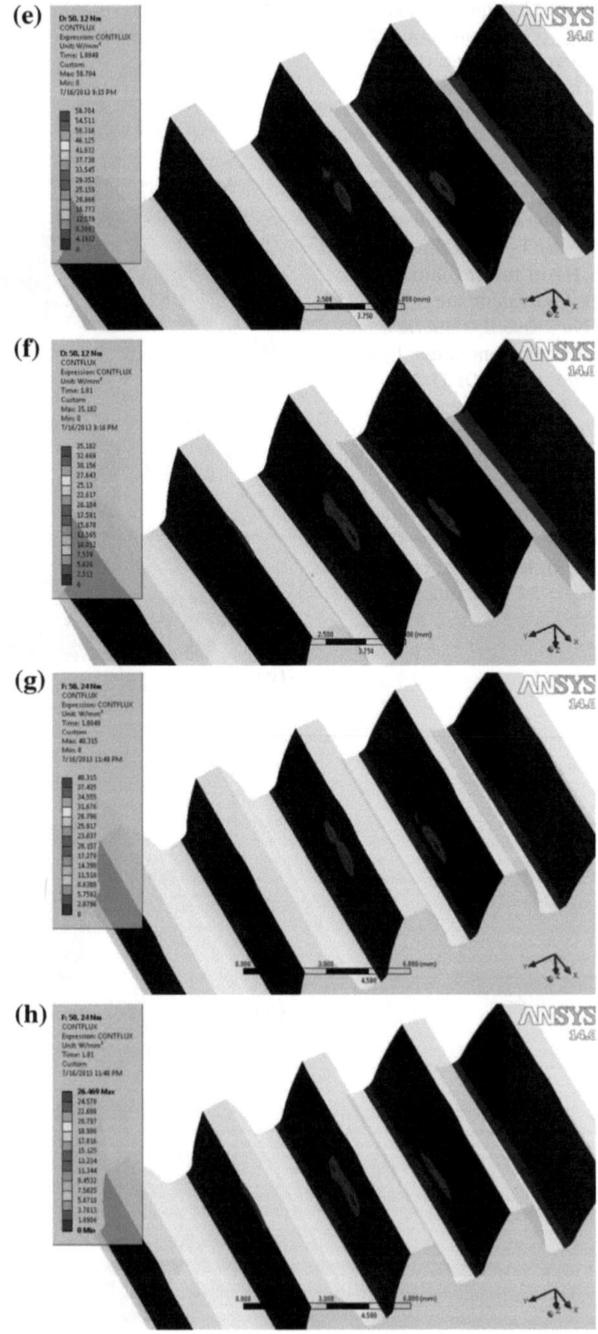

Fig. 42 (continued)

Table 8 Comparison of wheel mass temperature obtained analytically and by simulation

Load (Nm)	Analytically determined wheel tooth temperature (°C)	Wheel mass temperature predicted by simulation (°C)
12	98	108
24	112	119

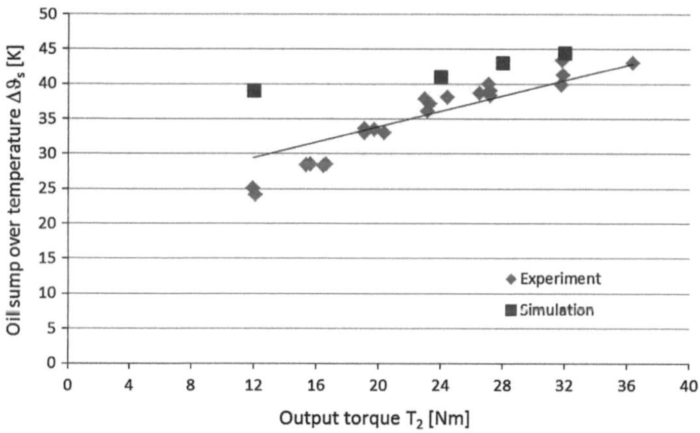

Fig. 43 Comparison of the oil sump over temperatures of crossed helical gears lubricated for $n_1 = 1500\ \text{min}^{-1}$ obtained experimentally and with simulation [16]

case-hardening plus shot peening, pyrohydrolysis, sinter-hardening and 2 % copper addition. The research has included the following material properties: chemical composition, density, hardness, E-module, and Poisson's ration, as well as static strength properties.

The micro-hardness measurements have shown that the greatest hardness is found, as expected, in the material variants S2—case-hardening and S3—case-hardening plus shot peening. The lowest hardness is found in the material variants S1—without additional treatment and S4—pyrohydrolysis.

The E-module experiment has shown that, in the material variant S5—sinter-hardening, the E-module reaches as far as 97 % of the values of the steel E-module, which is satisfactory when considering the porosity of sintered steel.

The microstructure was improved by different treatment methods of sintered steels. In the case hardening, a thick edge layer with martensitic structure is created, going 0.3 mm deep into the material. The core of the material has a mixed structure of tempered bainite and martensite. It is evident that the microstructure is relatively homogeneous without the sharpness of the edge area transitions to the core. The structure has pores of different shapes which are evenly distributed. The samples with case hardening and shot peening show similar properties.

The micrographic pictures of the material sample with pyrohydrolysis treatment shows a uniform, fine-grained ferrite and pearlite structure. The sinter-hardening treatment leads to a homogeneous microstructure—improved with martensite grains in a basic ferritic structure.

The micrographic picture of the variant with 2 % copper addition shows a uniform ferrite and pearlite structure. Pearlite and cementite have grain structure form. The pores are homogeneously distributed and they have no connections. The variant with the addition of copper has much larger deformations than the other variants. The increase in diameter of the crossed helical gears is, for example, 0.64–0.16 % when compared with other variants.

For the normal work of a crossed helical gear, it is necessary that a gear pair has sufficient wear resistance in exploitation conditions. The research in this paper shows that gears made of sintered metal Fe1.5Cr0.2Mo with sinter-hardening treatment fulfil technical requests from the aspect of wear. Lubrication has great influence on working characteristics in exploitation conditions.

The growing trend of wear is always greater for lubrication with mineral compared to synthetic oil. For example, there is a considerable difference at the speed of $n_1 = 5000\ min^{-1}$ for the wear surface width of a gear tooth, which increased steadily during the test. The wear for lubrication with mineral oil was higher than that with synthetic oil by a factor of 3.3 times on average. A steady-state appearance with minimal wear was clearly observed for lubrication with synthetic oil, which was not the case for lubrication with mineral oil.

The pitting was observed in wheels of material variants S4, S5 and S6 by output torque from 16 to 32 Nm. The initial pitting on the tooth flank occurred on material variant S1—without additional treatment (trials with $T_2 = 36$ Nm; $t = 250$ h; $n_1 = 5000\ min^{-1}$; lubricant: synthetic oil) and S5—sinter-hardening ($T_2 = 16$ Nm; $t = 60$ h; $n_1 = 5000\ min^{-1}$; lubricant: mineral oil). Progressive pitting on the tooth flank occurred in material variant S4—"pyrohydrolysis" ($T_2 = 28$ Nm; $t = 180$ h; $n_1 = 5000\ min^{-1}$; lubricant: synthetic oil) and S5—sinter-hardening ($T_2 = 20$Nm; $t = 120$ h; $n_1 = 5000\ min^{-1}$; lubricant: mineral oil).

With the exception of S1, scuffing occurred on all material variants (S2, S3, S4, S5, and S6). The scuffing was the critical type of damage for wheels from material variants S2 and S3 under the output torque of $T_2 = 20$ Nm and by S5 under the output torque of $T_2 = 36$ Nm. For wheels of material variant S5, scuffing occurred for trials with $n_1 = 5000\ min^{-1}$ and higher output torque and for trials with input speed $n_1 = 10{,}000\ min^{-1}$ at smaller output torque. For wheels of material variants S4 and S6, the critical type of damage was the combination of pitting and scuffing.

The scoring occurred at all sliding velocities and mostly at higher torques. For material variant S4 "pyrohydrolysis" and at $n_1 = 5000\ min^{-1}$, by torque greater than 24 Nm, a very strong damage on the tooth flanks due to scoring was observed.

Relative sliding of the meshing teeth under high-speed conditions results in significant frictional heat generation, which is determined by the load sustained, rotational speed and coefficient of friction. Heat generated is conducted into the gear teeth and is also taken away by the cooling lubricant. The frictional heat flux generated on the tooth flank and heat transfer on the gear surface and tooth flanks have a significant influence on the temperature variation and distributions.

Consequently, they affect the surface failure, such as hot scoring, wear, surface pitting, and thermal cracking.

The paper presents equations for determination of oil sump temperature and the coefficient of friction, as well as overall efficiency. Equations are based on experimental results for material combination of crossed helical gears 16MnCr5 and Fe1.5Mo0.2Cr sintered steel with sinter hardening.

The paper also presents a new procedure for prediction of heat generation in the meshing of crossed helical gears. The new procedure is based on direct coupling of the thermal and structural fields in transient FEA and usage of the analytically determined coefficient of friction. The obtained simulation results are in good agreement with analytically determined values, as well as the experimental ones. The proposed procedure is applicable in the early phases of design, as well as study of the effect of wear and other geometrical and assembly deviances on thermal stability of the power transmitter. Furthermore, it can be used for the same purposes with other gear types.

References

1. Kotthoff G (2003) Neu Verfahren zur Tragfähigkeitssteigerung von gesinterten Zahnrädern. Dissertation RWTH Aachen
2. Wendt T (2008) Tragfähigkeit von Schraubradgetrieben mit Schraubrädern aus Sintermetall. Dissertation Ruhr-Universität Bochum
3. Dizdar S, Johansson P (2007) Höganäs AB. 263 83 Höganäs. Sweden: P/M Materials for Gear Application; EURO PM2007. Tolouse, 16 Oct 2007
4. Predki W, Miltenović A (2010) Influence of hardening on the microstructure and the wear capacity of gears made of Fe1.5Cr0.2Mo sintered steel. Int J Sci Sinter 42:183–191
5. A. Miltenović, V. Nikolić, M. Milovančević, M. Banić: Experimental and FEM analysis of sintered steel worm gear wear. TRANSACTIONS OF FAMENA XXXVI-4 (2012). Faculty of mechanical engineering and naval architecture. ISSN 1333-1124. pp. 85-96 Zagreb 2012
6. DINEN10002: Metallische Werkstoffe—Zugversuch, 12/2001
7. Banić M, Stamenković D, Milošević M, Miltenović A (2013) Tribology aspect of rubber shock absorbers development. Tribol Ind 35(3):242–248
8. Stamenković D, Nikolić M, Milošević M, Banić M, Miltenović A, Mijajlović M (2014) Tribological aspect of rubber based parts used in engineering. Tribol Ind 36(1):9–16
9. Engstrom U, Klekovkin A, Berg S, Edwards B, Frayman L, Hinzmann G, Whitehouse D (2005) Advances in powder metallurgy and particulate materials. MPIF, Princeton
10. Miltenović A, Mitrović R, Banić M (2013) Crossed helical gears with wheels from sintered steel with pyrohydrolysis. Adv Mater Res 633:197–208. doi:10.4028
11. Rackov M, Milovančević M, Kanović Ž, Vereš M, Rafa K, Banić M, Miltenović A (2014) Optimization of HCR gearing geometry using generalized particle swarm optimization algorithm, Tehnički vjesnik 21(4)
12. Miltenović A (2011) Verschleißtragfähigkeitsberechnung von Schraubradgetrieben mit Schraubrädern aus Sintermetall. Dissertation Ruhr-Universität Bochum
13. Miltenović Đ, Tica M, Miltenović A, Banić M, Živković S, Mišković Ž (2014) Pitting of teeth flanks of crossed helical gears made from sintered steel. Transactions of Famena XXXVIII-4. Faculty of mechanical engineering and naval architecture. Zagreb 2014, pp 77–88
14. Miltenović A, Kuzmanović S, Miltenović V, Tica M, Rackov M (2012) Thermal stability of crossed helical gears with wheel made from sintered steel. Int Sci J Therm Sci 7–12

15. DIN3996: Tragfähigkeitsberechnung von Zylinder-Schneckengetrieben mit sich rechtwinklig kreuzenden Achsen. 8/2005
16. Miltenović A, Banić M, Miltenović V (2014) FEM and experimental determination coefficient of friction of crossed helical gears with wheels from sintered steel. In: International symposium theory and practice of gearing—2014, 21–23 Jan 2014, Izhevsk, Russia
17. Wassermann J (2005) Einflussgrößen auf die Tragfähigkeit von Schraubradgetrieben der Werkstoffpaarung Stahl/Kunststoff. Dissertation Ruhr-Universität Bochum
18. Miltenović A, Predki W (2011) Damage types of crossed helical gears with wheels from sintered steel. Int J Sci Sinter 43:205–214
19. Miltenović A, Nikolić V, Mitrović R (2012) Efficiency of crossed helical gears with wheels made of sintered steel Fe1.5Cr0.2Mo with sinter-hardening treatment. Transactions OF Famena XXXVI-2 (2012), pp 31–40. Faculty of mechanical engineering and naval architecture. ISSN 1333–1124. Zagreb 2012
20. Tomović R, Miltenović V, Banić M, Miltenović A (2010) Vibraton response of rigid rotor in unloaded rolling element bearing. Int J Mech Sci (ISSN 0020–7403)

Modern Methods of Calculation and Increasing the Load-Carrying Capacity of Surface-Hardened Gears of Transmissions and Drives

E. Tesker and S. Tesker

Abstract Creation of high-performance machines and equipment implies special requirements for the most loaded parts, units and aggregates, their engineering level and quality influencing the operating characteristics of machines in general. Surface-hardened gearwheels, the feature of which is non-uniform distribution of physical and mechanical properties in a surface-hardened layer sustaining high contact and bending loads, become more widely applied in machine drives and transmissions and in the gearbox industry. That is why ensuring the reliability and functional characteristics of the most loaded power gears of drive mechanisms and transmissions is an urgent problem.

Keywords Load-carrying capacity · Surface-hardened gears · Drives · Transmissions

1 Contact Strength of Parts with Hardened Surface Layer or Wear-Resistant Coating

1.1 Model of Depth Contact Failure of Loaded Parts with Structurally Non-uniform Hardened Surface Layer

In modern machines, the loaded elements of drives and transmissions are subjected to surface hardening or have special coatings. The feature of the contact of such parts is non-uniform distribution of material properties within the active volume. In addition, distribution of stress fields along the vertical axis and equivalent stress

E. Tesker (✉) · S. Tesker
Volgograd State Technical University, Volgograd, Russia
e-mail: agromash-vlg@rambler.ru

V. Goldfarb and N. Barmina (eds.), *Theory and Practice of Gearing and Transmissions*, Mechanisms and Machine Science 34,
DOI 10.1007/978-3-319-19740-1_11

along the material depth should correspond to the character of the coating non-uniformity.

These features stipulate different resistance to failure of the material of a part depending on the distance from the surface.

It follows from solutions of the contact problem and the principles of stress distribution described above that contact loading of parts can cause primary contact failure both on the surface and at some depth, even for structurally homogeneous material.

In this regard, the classification of contact fatigue is widely accepted according to the principle of location determination of the source of primary failure, classified as surface or depth contact fatigue.

The power characteristics of contact, the state of the surfaces of parts in contact, lubrication conditions and material properties in the contact area influence the conditions of surface and subsurface failures in different ways. That is why the evaluation of contact strength and durability of loaded elements of drives and transmissions of mechanical vehicles require separate analysis of factors that influence surface and depth contact strength. The greatest attention is paid to investigation of depth contact strength, as depth failures are connected with the peeling-off of significant portions of the surface and abrupt increases in specific contact load at non-failed areas that, in turn, lead to further intensive failure development. It is this type of failure that is taken into account during the supervision of defects of parts (gearwheels, bearings, etc.) in the operation process. Many researchers observed the appearance of subsurface fatigue failures during testing. Certain conditions are necessary for the appearance of such failures. It has been established through experimental investigations that location of maximum shearing stresses at definite depth under the surface is not the only reason for the appearance of depth failures. Such failures can appear because of additional stress concentrators, such as microstructure defects or a structurally non-uniform surface layer. In this case, if the cyclic contact stresses are high, the influence of material properties and the stressed state character of an extreme surface layer on the contact endurance is decreased, and the influence of subsurface stress concentrators becomes dominant. This aspect is confirmed, for example, by the dependence of the location of initial fatigue failures on the method of steel manufacture and lubrication conditions of gears. Such dependence was obtained during contact fatigue testing of parts produced by electroslag and vacuum re-melting of steels having different non-metallic inclusions, most prominently, Titanium carbides. Small cracks located at an angle of 45° appeared on the surface after 106 loading cycles in unfavorable lubrication conditions, when the ratio of oil film thickness h0 to the reduced height of micro-roughness *RZ* was $h_0/RZ \le 1.25$. Fatigue cracks were generated close to the surface in the area of micro-roughness.

The material in the area of crack appearance was subjected to structural transformations that were confirmed by the existence of a poorly etched layer located near the area of action of maximum shearing stresses. Under these conditions, difference in durability of the investigated steel grade types was insignificant. The number of surface cracks decreased with improvement of the lubrication conditions characterized by the ratio h_0/RZ and, in addition, subsurface cracks located parallel to the contact surface appeared more often. The dominant appearance of such cracks was at $h_0/RZ = 3$. Upon that, the durability of parts made of vacuum re-melting steel increased significantly, which is explained by lower content of non-metallic inclusions, the source of subsurface crack generation. Consequently, when the oil film thickness does not exceed the reduced height of micro-roughness, stress concentration in areas of actual contact of micro-roughnesses plays a leading role. Steel quality has almost no influence in this case, and part durability is minimal. An increase in oil film thickness led to elimination of contact of micro-roughnesses and a considerable increase in durability of parts made of vacuum re-melting steel. Failure mechanisms were changed simultaneously, namely, the appearance of depth cracks in the area of inclusions instead of surface cracks. The same effect was obtained during testing of gearwheels lubricated with high-viscosity oils and heat-resistant liquids. Subsurface cracks located parallel to the surface also appeared in these tests. Non-metallic inclusions in the area of maximum contact stress action were the source of failure. Lubrication by low-viscosity oils led to the appearance of surface cracks. Complete hydrodynamic lubrication was ensured only at $h_0/R = 2.5$.

The present data show that depth contact failures appear not only due to high contact stresses, but also due to worsening of the physical and mechanical properties of the material in the subsurface area, where maximum equivalent stresses act.

A generalized mathematical model is developed to investigate the conditions under which the depth contact failures appear. The model is based on the fact that contact fatigue is stipulated by features of the stress-strain state of surface-hardened tooth material and also by non-uniform distribution of its physical and mechanical properties within the active volume. According to this model, the design conditions include equations describing the change of contact stresses along the depth at the most loaded contact area at static or cyclic loads, and also analytical relationships characterizing the principles of change in tooth material properties along the thickness of the hardened layer, which characterize the ability to resist the contact loads. Taking all this into account, the criteria defining depth contact strength are as follows:

at static loads

$$\tau_E(z) < \tau_S(z) \text{ or } \sigma_E(z) < \sigma_S(z); \qquad (1)$$

at cyclic loads

$$\tau'_E(z) < \tau_k(z) \text{ or } \sigma'_E(z) < \sigma_{limb}(z), \qquad (2)$$

where τ_E, τ_E' and σ_E, σ_E' are equivalent stresses found according to corresponding strength theories; τ_S, σ_S and τ_k, $\sigma_{lim\ b}$ are yield and contact endurance limits.

Thus, at any point of the tooth material located at a depth of $z = z_i$ from the contact surface, the following strength conditions should be satisfied:

$$\tau_E(z) < \tau_S(z) \text{ or } \sigma_E(z) < \sigma_S(z); \tag{3}$$

$$\tau'_E(z) < \tau_k(z) \text{ or } \sigma'_E(z) < \sigma_{limb}(z). \tag{4}$$

The model is based on the following theoretical assumptions and experimentally proven facts:

(1) A stressed state of tooth material at any contact point can be characterized by the main relationships for stress components obtained from the solutions of the contact problem of elasticity theory for the case of contact between two cylinders with parallel axes. Here, influence of the shearing component and lubrication on the stressed state of depth material layers at teeth contact may not be taken into account for the most typical conditions of operation of well-lubricated (μ = 0.05 to 0.08) highly-stressed gearwheels.
(2) Depth failures of surface-hardened gearwheels result from peculiarities of the stressed state in the contact area and non-uniform distribution along the active volume of physical and mechanical material properties which determine its resistance to contact loads. Also, there are quite definite principles of change in stresses and physical and mechanical material properties along the thickness of surface-hardened material located in the contact area. Such principles can be described by corresponding equations depending on the coordinate z.
(3) Depth contact fatigue of hardened high-strength steels is connected with processes of local micro-non-uniform plastic-elastic deformation, which is why it does not differ from the typical fatigue. Surface-hardened (case-hardened, nitro-carburized, etc.) steels applied for gearwheel manufacturing can be treated as quasi-brittle materials on the basis of numerical experimental data showing the change of microstructure and micro-hardness of depth layers of material with high hardness at cyclic contact loading. Consequently, the main concepts of dislocation theory can be used to explain the mechanisms of depth failures.
(4) Resistance of tooth material to plastic-elastic deformations and contact fatigue is best correlated with its hardness, measured at small loads applied to the indenter. Such measurements allow for detecting physical and mechanical properties of separate structural components located at dangerous contact areas. It can be assumed that there is an almost linear relationship between the yield limit at shear, contact fatigue limit and hardness of the material micro-volumes.
(5) The value of plastic shears at static loads depends on the level of shearing stresses. That is why maximum shearing stresses $\tau_{max} = \tau_{45}{}^{\circ}$ acting along the central symmetry axis z on the areas located at an angle of 45° to this axis are assumed in criterial conditions (1) and (2) to be stresses causing the development of plastic deformations in surface-hardened material during overloads.

(6) Depth contact fatigue below rolling cylinders is caused by alternating-sign contact stresses τ_{yz} acting near the contact area outline ($y \approx \pm b$) in the planes parallel and perpendicular to coordinate axes. Since failure due only to normal or shearing stresses is impossible, criterial Eqs. (2) and (4) contain functions of equivalent shearing stresses calculated according to Mohr's theory. Strength criteria (1) and (2) have physical significance and satisfactory coincidence with the results of experimental investigations of contact failures. When considering problems connected with depth contact strength, the possibility of application of generalized strength criteria for a complex stressed state is investigated for the first time in the design model on the basis of modern achievements of strength science for a complex stressed state. These criteria are invariant, with respect to the stressed state, functions of shearing stresses, maximum (taking into account the sign) normal stress and material constants.

In accordance with the proposed generalized model, a relative material stressed state at any point is characterized independently through the method of equivalent stress calculation by a safety factor value of n, which, like the stresses and material properties, is changed along the thickness of the hardened layer, i.e., may be represented as a function $n(z)$. Thus, the model takes into account the feature of contact loading of surface-hardened material, which consists of the fact that equivalent stresses and resistance of material to contact loads change simultaneously with a change in the distance from the contact surface.

The first indications of plastic deformation at overloads or fatigue cracks appear, initially, in the zones with minimum factors n_{Smin} or n_{Kmin}.

The equations describing the functional relationships of factors $n(z)$, accounting for the accepted strength criteria, are as follows:

at static loads

$$n_s(z) = \frac{\tau_s(z)}{\tau_E(z)} \text{ or } n'_s(z) = \frac{\sigma_s(z)}{\sigma_E(z)}; \tag{5}$$

at cyclic loads

$$n_k(z) = \frac{\tau_k(z)}{\tau'_E(z)} \text{ or } n'_k(z) = \frac{\sigma_{limb}(z)}{\sigma'_E(z)}. \tag{6}$$

Thus, depth contact strength analysis consists of the determination of the minimum value of factor n_{min} and location of the dangerous zone specified by coordinate $z = z_{nmin}$, where $n = n_{min}$. The problem can be solved if the functional relationships, which determine the principles of change for equivalent stresses and characteristics of contact strength of material located in the contact area, are known.

1.2 Principles of Hardness Change in Surface Layer After Surface Hardening

The most widely used types of surface hardening during manufacture of loaded elements of drives and transmissions are case hardening, carbonitriding, nitriding, induction hardening, electromechanical treatment, surface plastic strain hardening and laser quenching.

Hardness change in the hardened layer obtained by chemical heat treatment depends on distribution of Carbon for case hardening or Carbon and Nitrogen at carbonitriding. It is established that, under the correct chemical heat treatment, the content of Carbon (or Carbon and Nitrogen) is decreased gradually from the surface to the material core. This principle is explained by the kinetic features and physical and chemical mechanisms of diffusion processes for the generation of hardened layers. Gradual transition from a hypereutectoid structure on the diffusion layer surface to a ferrite-pearlite structure and further to the structure of initial material leads to smooth micro-hardness decreasing along the thickness of the hardened layer. Experimental curves characterizing principles of change in hardness along the thickness of different hardened layers are shown in Figs. 1 and 2. The relationships specified are obtained during the process of investigation of various steel grades applied for the manufacturing of case-hardened and nitro-carburized gearwheels of tractors, cars, machine-tools, coal-mining machines and other types of machines. It can be seen from the graphs that, in the majority of cases, experimental hardness curves $H(z)$ describe a monotonically decreasing function, which is maximum at the point $z = 0$, i.e., on the surface of the diffusion layer. Some decreasing in micro-hardness appears only when defects are present in the external zone of the diffusion layer at a depth of $0 \le z \le 0.1$ mm.

The following equation satisfying the above-described conditions is accepted as an expression approximating the functional relationship $H(z)$:

$$H = \frac{1}{Az^2 + B}. \tag{7}$$

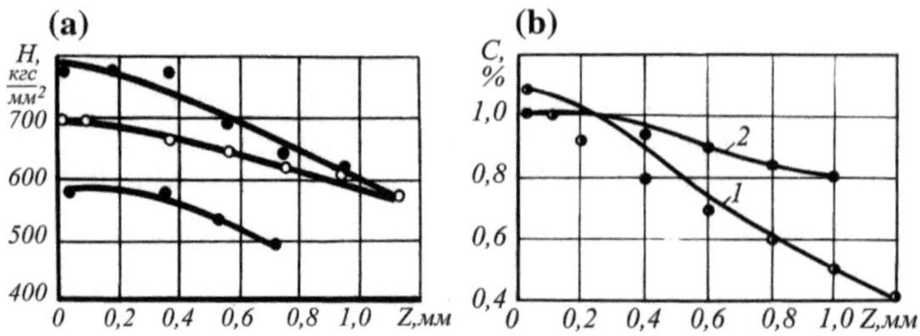

Fig. 1 Distribution of micro-hardness **a** and Carbon percentage **b** along thickness of hardened layers of teeth: *1*, *2*—technological variants of chemical heat treatment (curves are drawn according to formula (8); dots are marked according to experimental data)

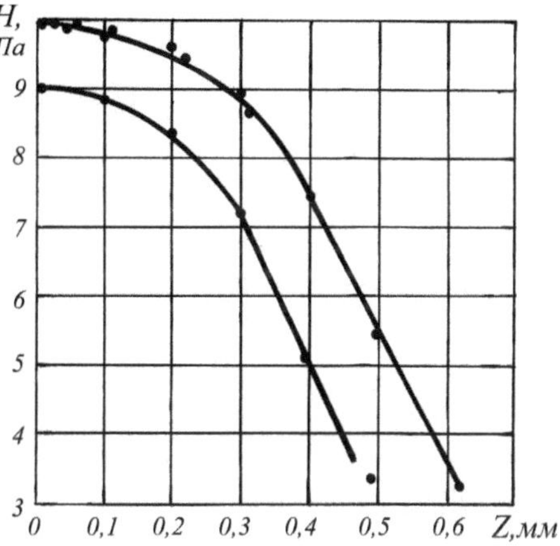

Fig. 2 Distribution of micro-hardness along thickness of the hardened layer treated by laser (dots correspond to values of hardness found according to (9))

Constants *A* and *B* are found from the boundary values of hardness on the surface and in the material core.

Changing the constants *A* and *B* by their values in (4.7), after the transformation, we obtain

$$H = \frac{H_s}{\left(\frac{H_s}{H_c} - 1\right)\left(\frac{z}{\delta_l}\right)^2 + 1} \tag{8}$$

Expression (8) describes principles of change in hardness of case-hardened and nitro-carbirized layers as a function of coordinate z, surface H_s and core H_c hardness of teeth material, and also layer thickness δ_l determined as a value of coordinate z, for which the layer hardness becomes equal to the core hardness.

Calculations according to formula (8) give satisfactory agreement of calculated and experimental hardness values obtained by different investigations.

Approximation of experimental relationships $H(z)$ allowed for obtaining the expression describing principles of change in hardness along the thickness of the laser layer:

$$H = \frac{H_{sl}}{\left(\frac{H_{sl}}{H_{cl}} - 1\right)\left(\frac{z}{\delta_{ll}}\right)^4 + 1}. \tag{9}$$

Here, H_{sl} and H_{cl} are hardness values on the surface and in the core of the material; and δ_{ll} is the thickness of the layer hardened by laser measured from the surface to the area having constant hardness equal to the material core hardness. For quenched and low-tempered steels, the value δ_{ll} is assumed equal to the distance from the surface to the area with minimum hardness equal to the hardness of the tempering area located under the hardened layer.

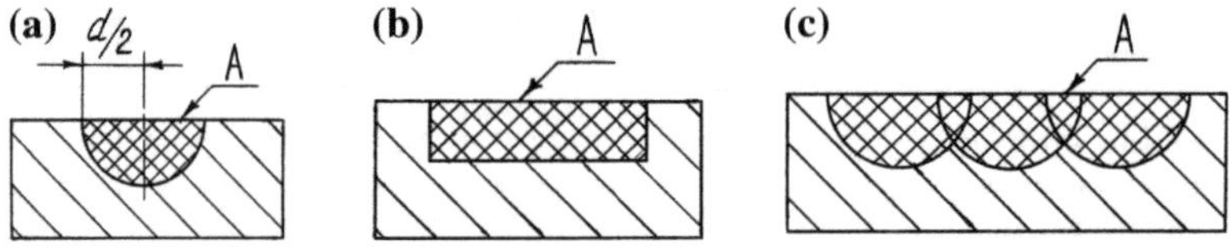

Fig. 3 Location of quenched areas after laser treatment of metal. **a** Profile of quenched area for single run of focused beam. **b** The same, but with application of special devices. **c** location of quenched areas for continuous treatment

The peculiarity of teeth laser quenching is metal surface treatment by a focused beam from a high-concentrated power source. That is why a continuous quenched zone along the width (tooth profile height) during the treatment of considerable surface areas can be obtained only as a result of laser track overlap of each beam running along the treated surface (Fig. 3). The width of the quenched area increasing within one run and, consequently, decreasing the number of runs is achieved by means of special devices for scanning with the beam.

Specific features of laser thermal hardening stipulate structural non-uniformity not only along the thickness, but also along the width of the hardened layer. Distribution of micro-hardness along the width of the laser track in the external area of the hardened layer (measuring was made at a distance of 30 mcm from the surface) is shown in Fig. 4. It can be seen that the hardness abruptly decreases near the zone outline of the metal treated by laser. Such hardness behavior along the laser track width is stipulated by distribution of power density in the pattern of laser emission.

Figure 5 shows hardness distribution for continuous laser quenching, when treatment is carried out with laser track overlap (see Fig. 3c). It can be seen that the hardness abruptly decreases in the overlap zone and its value depends on the degree of overlap. Special experiments allowed for establishing that optimal hardness (see Fig. 5a) and structure at the overlap zone are achieved during treatment with step $S = (0.3{-}0.4)d_b$, where d_b is the diameter of the beam trace measured on the surface of the part.

Fig. 4 Distribution of micro-hardness along the width of the laser track for a single run of the laser beam

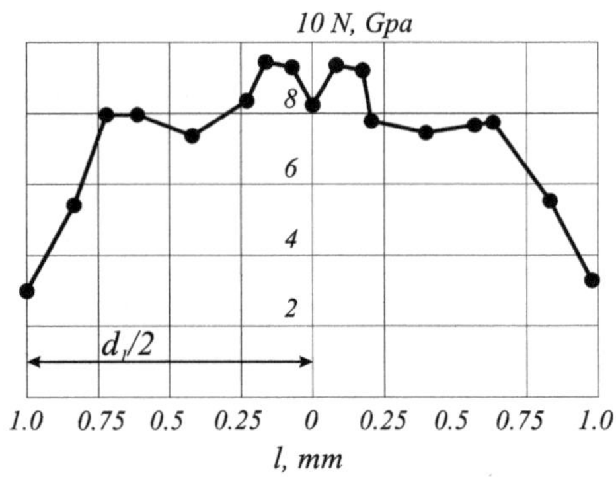

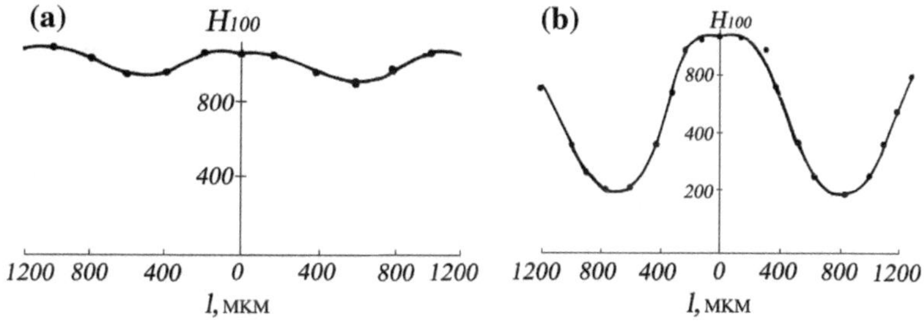

Fig. 5 Hardness change along the width of the laser quenched area for treatment with optimal **a** and not optimal **b** track overlap

1.3 Conditions of Plastic Deformation Appearance and Calculation of Load-Carrying Capacity at Maximum Load

In some cases, short-time overloads at gearwheel operation can cause failure of active teeth surfaces because of residual plastic deformation (punching of the hardened layer) and subsequent distortion of initial profile geometry in the contact area. To prevent this failure, the standard GOST 21354-87 recommends analyzing the teeth for contact strength at maximum load according to the condition

$$\sigma_{Hmax} \leq \sigma_{Hpmax}\,, \tag{10}$$

where σ_{Hmax} is the design stress and σ_{Hpmax} is the allowable stress.

Condition (10) limits maximum contact Hertz stresses, which can cause plastic deformation on active teeth surfaces. However, plastic deformation of metal occurs by shear and the reason for plastic deformation in this case is the shearing stresses τ_{max}. For surface-hardened gearwheels, the first indications of plastic deformation on a layer boundary or in the material core appear for small-thickness hardening only. Indeed, it follows from the elasticity theory that, at the loaded contact of two cylinders, the maximum shearing stresses act in the symmetry plane of the contact area at a depth of $z_{max} = 0.786b$. Then, the risk of hardened layer punching appears at its thickness

$$\delta_l = z_{max}. \tag{11}$$

Analysis of contact loading conditions of surface-hardened gearwheels having module m = (3 to 12) mm, applied in loaded gears of various types of machine and equipment, for example, cars, tractors, etc., shows that, in the majority of cases, even under extreme operational conditions (σ_{Hmax} = 3500 to 4500 MPa), the semi-width of the contact pattern is $b \approx 0.3$–1.0 mm. It follows from this relationship (11) that if the condition (10) is not satisfied, then indication of plastic deformation on the layer boundary or in the material core can appear only for a thickness of the hardened layer $\delta_l \leq 0.8$ mm. Experience shows that the thickness of the hardened layer exceeds 1.5 mm for carbonitriding and case hardening of high-stressed gearwheels, considerably greater than the value obtained according to the relationship (11). That is why, in a more general case, the first indications of plastic deformation can appear both in the hardened layer itself and in the core, even when shearing stresses there are not at a maximum.

Let us apply the model of depth failure described above for solution of a general problem of searching the location of a dangerous zone and development of a method of analysis for depth contact strength at maximum load. According to conditions (5), plastic deformations at peak loads will first appear in the area with minimum value of n_{Smin}. When Mohr's strength theory is applied, the coordinate $z = z_{nsmin}$, defining the location of the dangerous zone can be found from the equation

$$\frac{d}{dz} n_s(z) = \frac{d}{dz}\left[\frac{\tau_s(z)}{\tau_{E\max}}\right] = 0. \tag{12}$$

Thus, location of the dangerous zone is determined by simultaneous accounting of plots $\tau_S(z)$ and $\tau_E(z)$ within the whole contact area placed along the z axis. Functional relationship $\tau_S(z)$ is determined by principle of change in hardness $H(z)$ along the thickness of the hardened layer. Taking into account that shear yield limit values at each layer point at a depth $z = z_i$ may be accepted to be proportional to hardness, we obtain

$$\tau_S(z) = kH(z) \tag{13}$$

Factor k is usually determined on the basis of experimental data. Investigations of a large number of alloy steel samples (more than 20 grades) having hardness $HB > 3000$ (MPa) allowed us to establish that yield limit δ_S has an almost linear relationship (the error doesn't exceed 5 %) with respect to the plastic hardness of metal H, defined in correspondence with the standard GOST 18661-73. According to [1], we have

$$\delta_S = 0.185H \text{ (MPa)}.$$

Converting the Vickers hardness for small loads applied to the indenter and taking into account that $H \approx 1.24\ HV$, we obtain

$$\delta_S \approx 0.23HV(\text{MPa}) \tag{14}$$

As shear yield limit τ_S always exceeds the elongation yield limit expressed in terms of shearing stresses or is equal to it, then it can be accepted that

$$\tau_s \geq \frac{\sigma_s}{2} \text{ or } \tau_s \geq 0.115HV. \tag{15}$$

Thus, for a hardened layer having hardness $HV > 3500$, it can be accepted with accuracy sufficient for practical analysis that $k = 0.12$. Then, taking into account (8) and (9), formulas for shear yield limit variation along the thickness of the hardened layer $0 \leq z \leq \delta_S$ are as follows:
for case hardening and carbonitriding

$$\tau_s(z) = k\frac{H_s}{\left(\frac{H_s}{H_c} - 1\right)\left(\frac{z}{\delta_l}\right)^2 + 1}; \tag{16}$$

for laser treatment

$$\tau_s(z) = k\frac{H_{sl}}{\left(\frac{H_{sl}}{H_{cl}} - 1\right)\left(\frac{z}{\delta_{ll}}\right)^4 + 1}. \tag{17}$$

Strength condition conforms well with Mohr's theory for the analyzed case of contacting, when the first indications of plastic shears leading to strength failure appear in depth material layers at overloads. Equivalent shearing stress in the dangerous zone is

$$\tau_E = \tau_{max} + f\sigma_Z, \tag{18}$$

where τ_{max} and σ_Z are stresses in the dangerous zone and f is the factor determined according to recommendations [2]. It is accepted that f = 0.25 to 0.35, as for quasi-brittle material. The expression for maximum shearing stresses acting in the symmetry plane (y = 0) is found from solutions of the contact problem for initial contact along the line (5):

$$\tau_{\max} = \frac{1}{2}(\sigma_1 - \sigma_3) = -\sigma_H\frac{z}{b}\left(1 - \frac{z}{\sqrt{b^2 + z^2}}\right). \tag{19}$$

Normal stresses σ_y in areas of maximum shearing stress action are found according to (2). Figure 6 shows a diagram illustrating the variation in equivalent shearing stresses $\tau_E(z)$.

Accounting for expressions (16)–(19), condition (12) can be transformed into the following form:
for case-hardened and nitro-carburized layers

$$\frac{d}{dz}\cdot\frac{kH_s}{\left[\left(\frac{H_s}{H_c} - 1\right)\left(\frac{z}{\delta_l}\right)^2 + 1\right]\left[\tau_{\max} - 0.3\sigma_y\right]} = 0; \tag{20}$$

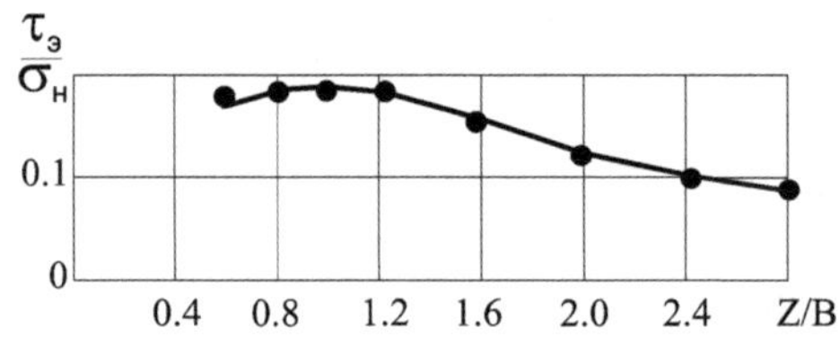

Fig. 6 Dependence $\frac{\tau_E}{\sigma_H}$ on relative coordinate $\frac{z}{b}$ (dots correspond to values $\frac{\tau_E}{\sigma_H}$)

for laser treatment

$$\frac{d}{dz}\cdot\frac{kH_{sl}}{\left[\left(\frac{H_{sl}}{H_{cl}}-1\right)\left(\frac{z}{\delta_{ll}}\right)^{4}+1\right]\left[\tau_{\max}-0.3\sigma_y\right]}=0, \tag{21}$$

where τ_{max} and σ_y are found according to formulas (2) and (19).

Thus, when parameters of hardened layers, such as surface and core hardness and layer thickness, are known from engineering documentation, location of the most loaded material area (coordinate z_{nsmin}) and, consequently, value n_{smin} are determined by functional relationship $\tau_E(z)$, including geometrical parameters of gearwheels and the value of maximum contact stress (according to Hertz) at points of the symmetry axis X of the contact area.

Determination of coordinate $z = z_{nsmin}$ from conditions (20) and (21), which are given in a generalized form, is a complex problem. It follows from simultaneous accounting of plots $\tau_E(z)$ and $\tau_S(z)$ that the dangerous zone, where the first indications of plastic shears can appear, should be located at a depth of $0.8b \le z \le \delta_l$.

For these boundary conditions of coordinate z, the function $\tau_E(z)$ may be approximated by the relationship

$$\tau_E=\frac{\sigma_{H\max}}{0.42\left(z/b\right)^{2}+4.84}. \tag{22}$$

Parameters of function τ_E (z/b) are found according to the least square method. As can be seen from Table 1, values τ_E calculated by formula (22) for $z \ge 0.8b$ are almost the same as values τ_E found by applying the elasticity theory.

It should be noted that the design value of shearing stresses is influenced by factor f, which depends on material properties and, consequently, f can change within a certain range for different parts of the hardened layer. It is impossible to carry out experiments for direct determination of the factor f value at each point of the surface-hardened layer. It is not only connected with difficulties for ordinary tests when implementing the stress variation cycle, which corresponds to formula (18), but, primarily, with the local character of material deformation and failure at contact loads. For such a type of complex stressed state, the factor f depends not on macro-properties of the specimen tested on the whole, but on properties of material macro-volumes.

Table 1 Results of stress calculation

$\frac{z}{b}$	$\frac{\tau_E}{\sigma_{\mathcal{H}\max}}$ according to (22)	$\frac{\tau_E}{\sigma_{\mathcal{H}\max}}$ according to (18)
0.8	0.195	0.181
0.9	0.193	0.187
1.0	0.191	0.191
1.2	0.184	0.187
1.3	0.181	0.184
1.4	0.176	0.179
1.5	0.172	0.176
1.6	0.169	0.171
1.7	0.165	0.166
1.8	0.161	0.162
1.9	0.153	0.152
2.0	0.149	0.148

Numerous experimental data [2] confirm that, for the majority of structural steels, $0.15 \leq z \leq 0.5$. Here, smaller values are related to ductile and larger ones to brittle materials. On this basis, taking into account the established principles of micro-hardness variation along the thickness of the hardened layer and recommendations [2], we accepted a constant value of factor $f = 0.35$ during determination of equivalent stresses (see Table 1) for a part of layer $0 \leq z \leq 0.3\delta_l$ with maximum hardness $HV \geq 6000$ MPa ($HRC \geq 54$). Material hardness within this part of the hardened layer varies insignificantly and differs from the surface hardness by no more than 5–8 %. For the rest of the main part of the hardened layer, $0.3\delta_l < z \leq \delta_l$, where hardness smoothly reduces to the core (35–45 *HRC*), it is accepted that $f = 0.25$ [2]. The specified assumptions in a number of cases can lead to some increase in design equivalent stress level, and it may be accounted for in engineering analysis by choice of a factor of safety. Calculation of equivalent stresses according to the formulas specified above showed that maximum value of stresses $(\tau_E)_{max} = 0.16\sigma_H$ (for $f = 0.35$) are at a depth of $z_{max} \approx (0.8\text{–}1)\,b$, act along the areas normal to a plane *yz*, and are inclined to an angle of 45° with respect to the *z* axis.

On the basis of the variation principle of physical and mechanical properties of material and equivalent shearing stresses, it is supposed that the dangerous zone, where indications of plastic deformation can appear, should be located at a depth of $z_{max} \leq z \leq \delta_l$. For this area of material active volume, expression (5), accounting for (22), has the following form for case hardening (carbonitriding):

$$n_s = k\frac{H_s\left[0.42\left(z/b\right)^2+4.84\right]}{\sigma_{H\max}\left[\left(H_s/H_c-1\right)\left(z/b\right)^2+1\right]}; \qquad (23)$$

for laser treatment:

$$n_s = k\frac{H_s\left[0.42(z/b)^2+4.84\right]}{\sigma_{H\max}\left[(H_{sl}/H_{cl}-1)(z/b)^4+1\right]},\tag{24}$$

where $b = 1.733 \cdot 10^{-5}\ \sigma_{H\ max}\ \rho_W$; ρ_W is reduced curvature radius of contacting surfaces; and σ_{Hmax} is the maximum contact stress caused by action of maximum (peak) load for the whole operational period.

For each particular case of contact loading, it is possible to apply expressions (23) and (24) for specifying conditions, for which indications of plastic deformation can appear and lead to punching (failure) of the surface-hardened layer at some material point located under the surface at a depth of $0.8b \le z \le \delta_l$. It follows from expressions (23) and (24) that the zone of material where factor n_S has minimum value n_{Smin}, will be the dangerous one. Location of the zone is defined by coordinate $z_0 = z_{ZSmin}$. It is possible also to solve an inverse problem by application of (23) and (24), i.e., to specify limiting stresses σ_{Hmax} and parameters of the hardened layer, for which depth stresses do not appear, for a given value of n_{Smin} ensuring necessary depth static contact strength.

It is obvious that, according to condition (5), the sites of plastic shears will appear for $n_{Smin} = 1$, i.e., when equivalent shearing stresses at some point of the earlier active volume get a critical value equal to the shear yield limit of the material. In accordance with the design model, the location of the point at which the first indications of plastic deformation appear is found from the condition of tangency of plots $\tau_E(z)$ and $\tau_S(z)$:

$$\left.\begin{aligned}\tau_S(z_0) &= \tau_E(z_0)\\ \frac{d\tau_S(z_0)}{dz} &= \frac{d\tau_E(z_0)}{dz}\end{aligned}\right\},\tag{25}$$

where $z_0 = z_{ZSmin}$ is the coordinate of the tangency point of curves $\tau_E(z)$ and $\tau_S(z)$.

Calculations according to formulas (23)–(25) are carried out by means of a specially developed program. Variation intervals for values H_S, H_C, δ_l, ρ_W and σ_{Hmax} (Table 2) are chosen to account for the main types and combination of plots $\tau_E(z)$ and $\tau_S(z)$, which may occur at loaded contact of involute surface-hardened gearwheels.

Calculation results allow us to establish the most probable location of dangerous zones $n_S = n_{Smin}$, defined by coordinate z_0, where the first indications of plastic shears might appear under unfavourable conditions. Practically possible locations of plots $\tau_E(z)$ and $\tau_S(z)$, corresponding to the condition

$$n_{S\min} = \frac{\tau_s(z_0)}{\tau_E(z_0)} = 1,$$

are shown in Fig. 7 as a design example.

Table 2 Values of parameters included in equations of plots τ_E *(z)* and τ_S *(z)*

Name of parameter	Designation	Variation interval	
		From	To
Thickness of the hardened layer, mm	δ_l	1.0	4.0
Hardness on the surface, MPa	H_S	6200	10000
Hardness of the core, MPa	H_C	3000	4800
Coordinate *z*, mm	z	0	4.0
Hertz stress, MPa	σ_H	1000	3500
Reduced curvature radius of contacting surfaces, mm	ρ_W	5	100

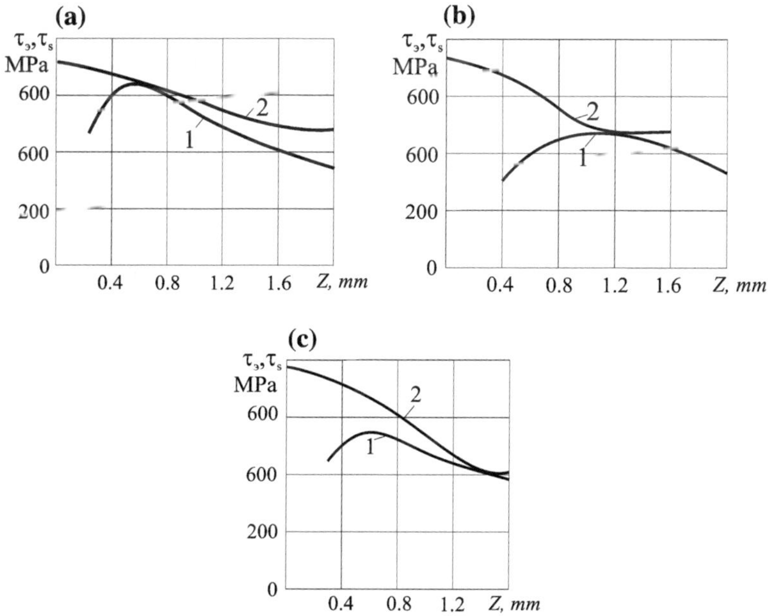

Fig. 7 Relation of plots τ_E *(z)* and τ_S *(z)* for $n(z) = n_{min} = 1$. **a** Dangerous zone is located within the hardened layer for $z_{ZSmin} = z_{max} < \delta_l$ and τ_E $(z = z_{ZS1min}) = \tau_{max.}$ **b** On the boundary of the hardened layer for $z_{ZSmin} = z_{max} = \delta_l$ and τ_E $(z = z_{ZS1min}) = \tau_{max.}$ **c** On the boundary of the hardened layer for $z_{ZSmin} = z_{max} > \delta_l$ and τ_E $(z = z_{ZS1min}) < \tau_{max}$; 1, 2 – plots τ_E *(z)* and τ_S *(z)* correspondingly

Let us describe expressions for n_S in each potentially dangerous zone of the surface-hardened material.

1. The dangerous zone is located in the hardened layer at a depth of $z_O = z_{max} = (0.8–1)b$, where maximum equivalent shearing stresses $(\tau_E)_{max}$ act (see Fig. 7a). Here, the following cases are possible.

(a) $z_O = z_{max} \leq 0.3\delta_l$, i.e., the dangerous zone is located in the area of the hardened layer having maximum hardness. In this zone, $(\tau_E)_{max} = 0.16\sigma_H$ at a depth of $z_{max} \approx 0.8b$. Then, we have for case hardening (carbonitriding)

$$n_{s1} = k\frac{H_S}{0.16 \cdot \sigma_{H\max}\left[\left({}^{H_S}/_{H_C} - 1\right)\left({}^{z_{\max}}/_{b_l}\right)^2 + 1\right]}; \tag{26}$$

for laser treatment

$$n_{s1} = k\frac{H_{Sl}}{0.16 \cdot \sigma_{H\max}\left[\left({}^{H_{Sl}}/_{H_{Cl}} - 1\right)\left({}^{z_{\max}}/_{b_{ll}}\right)^4 + 1\right]}. \tag{27}$$

(b) $0.3\delta_l \leq z_{max} < \delta_l$, i.e., the dangerous zone is located in the deep layers of the hardened material. Then, for case hardening (carbonitriding)

$$n_{s2} = k\frac{H_S}{0.19 \cdot \sigma_{H\max}\left[\left({}^{H_S}/_{H_C} - 1\right)\left({}^{z_{\max}}/_{b_l}\right)^2 + 1\right]}; \tag{28}$$

for laser treatment

$$n_{s2} = k\frac{H_{Sl}}{0.19 \cdot \sigma_{H\max}\left[\left({}^{H_{Sl}}/_{H_{Cl}} - 1\right)\left({}^{z_{\max}}/_{b_{ll}}\right)^4 + 1\right]}. \tag{29}$$

For $z_{ZSmin} = (0.8\text{–}1)b = \delta_l$ (see Fig. 7b), accounting for $k = 1.2$, formulas (28) and (29) have the following form:

$$n_{s2} = 7.0\frac{H_C}{\sigma_{H\max}}; \tag{30}$$

for laser treatment

$$n_{s2} = 7.0\frac{H_{Cl}}{\sigma_{H\max}}. \tag{31}$$

2. The dangerous zone is located on the layer boundary $z_O = \delta_l$, and equivalent stresses are maximum within its limits, i.e., $z_{max} < \delta_l = z_O$ (Fig. 7c). In this case, we have for case hardening (carbonitriding)

$$n_{s3} = k\frac{H_C}{\sigma_{H\max}}\left[0.42\left(\frac{\delta_l}{b}\right)^2 + 4.84\right]; \tag{32}$$

for laser treatment

$$n_{s3} = k\frac{H_{Cl}}{\sigma_{H\max}}\left[0.42\left(\frac{\delta_{ll}}{b}\right)^4 + 4.84\right]. \tag{33}$$

Thus, Eqs. (26)–(33), obtained by means of the model, entirely describe the conditions of plastic deformation formation in the surface-hardened layer for maximum contact loads. Location of a zone, where the first indications of plastic deformation can appear for $n_S = n_{Smin}$, is defined by coordinate $z_{0.}$ Values n_{Smin} and coordinate z_0, determining location of this zone, are found from Eqs. (23) and (24). However, the same problem can be solved in a simpler way, by comparing values of factors n_{Smin}, determined according to formulas (26)–(33). The zone with a minimum of the found factors n_{Smin} will be the dangerous one.

It is necessary to note that in a number of works, occurrence of a yield point for other types of stressed state is analyzed by means of the energy theory of Mises-Hencky, in which resistance of the material to deformation is connected only with octahedral shearing stresses. According to this theory, ductility condition has the following form:

$$\sigma_E\frac{\sqrt{2}}{2}\sqrt{(\sigma_1-\sigma_2)^2+(\sigma_2-\sigma_3)^2+(\sigma_3-\sigma_1)^2}\le\sigma_s. \tag{34}$$

The Mises-Hencky theory does not take into account a varied resistance of real material to tension and compression and the possibility of its hardening during deformation. Besides, it is supposed that the material is ductile and anisotropic. It is obvious that these circumstances must not be applied to the considered case of contact loading of non-uniform surface-hardened tooth material.

Theoretical and experimental investigations carried out lately prove the necessity of accounting for the influence of normal stresses on resistance of the material to deformation. It is confirmed by the analysis given below of depth contact strength in terms of a generalized strength criterion developed by the Institute for Problems of Strength at the Academy of Sciences of the Ukrainian Soviet Socialist Republic and indicating modern achievements in the field of structural material strength for a complex loaded state. We developed a principally new approach to evaluation of depth contact strength, which allows for obtaining, by means of this criterion, the solution to the problem of dangerous zone location, where indications of plastic deformation can appear. As the question of the criterial equation type characterizing the limiting state occurrence has not yet been decided definitely for hardness-uniform material, let us confine ourselves to an approximate solution. This solution is based on the fact that occurrence of a yield point in subsurface layers of the material is defined by equivalent stresses σ_E, depending, as well as on Mohr's theory, on shearing and normal stresses. Here, equivalent stress σ_E means normal stress, producing the stressed state in the material, corresponding to a complex state in terms of contact strength characteristics.

According to the developed model, the conditions causing the first indications of plastic deformation, as in the case discussed above, are defined by relations of plots of equivalent stresses $\sigma_E\ (z)$ and yield limits $\sigma_S\ (z)$, i.e.,

$$n_s(z) = \frac{\sigma_s(z)}{\sigma_E(z)}. \quad (35)$$

In formula (35), the expression for equivalent stresses has the following form:

$$\sigma_E = \chi A \sigma_{H\max} + (1 - \chi)\sigma_{H\max} B, \quad (36)$$

where $A = \frac{\sigma_i}{\sigma_{H\max}}$; $B = \frac{\sigma_1}{\sigma_{H\max}}$.

Taking into account (8), (9) and (36), we obtain for case-hardened (carbonitrated) layer

$$n_s(z) = C \frac{H_S}{\sigma_{H\max}\left[\left({}^{H_S}/_{H_C} - 1\right)\left({}^{z}/_{b_l}\right)^2\right][\chi A + (1 - \chi)B]}; \quad (37)$$

for laser treatment

$$n_s(z) = C \frac{H_{Sl}}{\sigma_{H\max}\left[\left({}^{H_{Sl}}/_{H_{Cl}} - 1\right)\left({}^{z}/_{b_{ll}}\right)^2\right] \cdot [\chi A + (1 - \chi)B]}. \quad (38)$$

In formulas (37) and (38),

$$\sigma_i = \frac{1}{\sqrt{2}}\sqrt{(\sigma_1 - \sigma_2)^2 + (\sigma_2 - \sigma_3)^2 + (\sigma_3 - \sigma_1)^2},$$

where σ_i is the maximum (accounting for the sign) principal normal stress.

Calculation of stresses σ_1, σ_2, σ_3 by formulas (3) showed that, for the most dangerous zone of the material for $z \geq 0.8 \cdot b$, the principal components of stress tensor are equal to $\sigma_1 = -\sigma_y$; $\sigma_2 = -\sigma_X$; $\sigma_3 = -\sigma_z$. It is obvious that the analytical solution of he problem of contact strength for overloads ($\sigma_H = \sigma_{Hmax}$) with application of conditions (36) and (37) should not differ from the solution described above, obtained by means of Mohr's theory. It is stipulated, first of all, by the fact that, in both cases, occurrence of the limiting state is connected with the ability of material to resist both shearing and normal stresses.

Determination of parameter $\chi = \sigma_{St}/\sigma_{Sc}$ (here, σ_{St}, σ_{Sc} are yield limits for tension and compression, correspondingly) is the main difficulty during calculation of equivalent stresses according to formula (36). Values of parameter χ depend on the hardness and can change within the certain range $0.5 \leq \chi \leq 1$ for local volumes of surface-hardened material, located in dangerous zones at different distances from the contact surface. It is almost impossible to carry out experiments for direct determination of contact loading for each definite case. Factor χ is chosen according to the following considerations. It is generally accepted that there is an approximate proportion for quenched steels between yield limit and hardness. This aspect is absolutely true for the specified case as well, when conditions of plastic deformation occurrence are determined by the hardness of local volumes of the material. On the basis of this fact, it can be accepted with accuracy sufficient for practical analysis that $\chi = 0.7$ for the main active volume of quasi-brittle tooth material, located in the area $0.8b \leq z \leq \delta_l$, where appearance of the first indications of plastic deformation is the most probable.

2 Methods of Analysis of Surface-Hardened Involute Gearwheels for Depth Contact Strength

2.1 Methods of Analysis of Case-Hardened, Nitro-Carburized and Laser-Hardened Teeth for Depth Contact Strength at Static Loads (Overloads)

Modern investigations into plastic deformation appearance conditions at static contact loads allowed us to develop a new universal science-based method for analysis of highly-stressed surface-hardened gearwheels of machines of various types. When developing the method of active teeth surface analysis, depth contact strength of the material is assumed from the condition of having prevented residual deformations, punching of the hardened layer and subsequent distortion and failure of teeth profiles.

Depth contact strength is ensured if the condition is fulfilled:

$$n_{S_{\min}} \geq [n_s] \text{ or } \sigma_{H_{\max}} \leq \sigma_{H_{P\max}} \tag{39}$$

where n_{Smin} is the design factor of safety of the depth strength at static loads (overloads); $[n_S]$ is the factor of safety necessary to prevent residual deformation of the tooth material; and σ_{Hpmax} is the stress allowable for single impact without causing residual deformation of the surface layer.

It is possible to carry out the analysis two ways:

(a) the function $n(z)$ is calculated for the given conditions of contact loading through application of general design equations and then strength condition (39) is checked in the zone, where $n(z) = n_{min}$;
(b) simplified calculation with subsequent check of strength condition (39) in each potentially dangerous zone.

Calculation according to the first method is carried out in the following sequence.

1. On the basis of experimental data and according to the formulas, maximum stress σ_{Hmax} is determined, which is produced by the load T_{max}, maximum

among all loads applied to the gear. When determining σ_{Hmax}, all factors influencing the value of load factor k_H are taken into account.

2. Semi-width of contact area b_{Hmax} is determined, which complies with σ_{Hmax} at the pitch point

$$b_{H\max} = 1.733 \cdot 10^{-5} \sigma_{H\max} \cdot \rho_{vhw} \tag{40}$$

where ρ_{vhw} is the reduced radius of curvature of teeth contacting surfaces.

3. Function $n(z)$ for $0.7b_{Hmax} \leq z \leq \delta_l$ is calculated for case-hardened (nitro-carburized) teeth or laser treatment. When allowable factor of safety $[n_s] > 1$, it is accepted as $\sigma_{Hmax}{\cdot}[n_s]$ in formula (39). In this case, variation of plot $\tau_E(z)$ location is taken into account due to a possible (specified by the factor of safety) increase of stress σ_{Hmax} for extreme operation conditions in comparison with the design value.
   Calculation of function $n(z)$, determination of its minimum value and coordinate z_{nsmin}, (where $n = n_{Smin}$), is carried out on a personal computer by a specially developed program.
4. Condition (39) is checked. If the condition is not fulfilled to prevent residual deformations caused by maximum load, it is necessary, first of all, to determine the location of the dangerous zone, i.e., to find coordinate $z = z_{nsmin}$, where $n(z) = n_{Smin}$. Then, the necessary value of material hardness, for which condition (39) is fulfilled, should be calculated in the dangerous zone for a given value of z_{nsmin} for case hardening (nitro-carburizing) and laser treatment. It is this value of hardness $H = H_{ef}$, and also corresponding values H_S, H_C and δ_l, defining location of plot $H(z)$ for the given conditions of contact loading, that are the optimal parameters of the hardened layer obtained by a particular type of hardening.
   Simplified calculation is carried out in the following sequence.

   1. According to paragraphs 1 and 2 (see above), σ_{Hmax} and b_{Hmax} are calculated.
   2. 2.Location of action zone of maximum equivalent stresses is determined.

$$z_{\max} = 0.8b_{H\max} \tag{41}$$

   3. Factors n_{S1min} are determined in each potentially dangerous zone.
      Calculation of function $n(z)$, determination of its minimum value and coordinate z_{nsmin}, (where $n = n_{Smin}$), is carried out on a personal computer by a specially developed program.
   4. For $z_{max} < \delta_l$, the dangerous zone can be located in the hardened layer or on its boundary with the core. That is why n_{Smin} is determined in the specified zones and then (39) is checked. By applying these formulas, it is possible to get expressions for limiting normal stresses $(\sigma_{Hmax})_{lim}$, which correspond to the case, when $n_{Smin} = 1$, for the given conditions of contact loading.

In this case, the limiting state in the hardened layer is determined from the condition

$$(\sigma_{H\max})_{\lim} \approx 7 \cdot H_{ef} \tag{42}$$

For the boundary between the layer and the core, we have

$$(\sigma_{H\max})^3_{\lim} - D(\sigma_{H\max})^2 - k = 0 \tag{43}$$

where $k = 1.6 \cdot 10^6 \left(\frac{H_C \delta_l^2}{\rho_{red}^2}\right)$; $D = 0.42H_C$; and H_{ef} is hardness of the hardened layer at a depth of $z_{max} = 0.8\ b_{Hmax}$.
Equation (43) can be represented as a nomogram (Fig. 8) that considerably simplifies determination of the limiting stresses. The limiting stress is the minimum one $(\sigma_{Hmax})_{lim}$ among values found from formulas (42) and (43).

5. When $z_{max} \geq \delta_l$, the dangerous zone is located on the boundary of the layer or in the material core. In this case, the value n_{Smin} is compared with allowable factor $[n_s]$. At the specified conditions, the load-carrying capacity of the teeth hardened layer is determined by its thickness δ_l and hardness H_C of the material core. Here, the limiting contact stresses for $[n] = 1$ are equal

$$(\sigma_{H\max})_{\lim} = 7 \cdot H_C. \tag{44}$$

Taking into account the relation between Vickers and Rockwell hardness, the strength condition for the boundary of the material layer and the core has the following form:

$$(\sigma_{H\max})_{\lim} = 70 \cdot H_{HRC}. \tag{45}$$

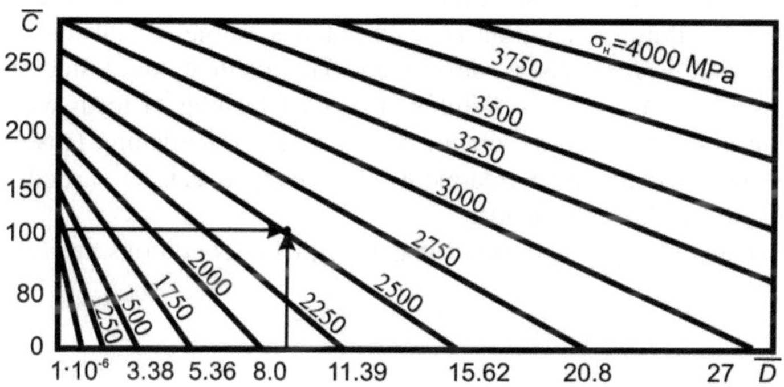

Fig. 8 Nomogram for determination of limiting normal stresses σ_{Hpmax}

Increasing the operating stresses, which exceeds values determined by formulas (42)–(45), leads to the appearance of merely the first indications of plastic deformation in the subsurface layers of the tooth material.

However, as there are constraints for deformation, which takes place in local internal material micro-volumes, the processes of plastic shears do not have essential development for a given level of stress $(\sigma_{Hmax})_{\lim}$. That is why a criterion of an allowable stress level, appearing at single load application, is the limiting stress that causes residual deformation and failure generation (profile distortion), having definite value, on tooth surfaces. Besides, the value of allowable residual deformation (dimple) should be assumed, based on the conditions of normal functioning of the gear. It is obvious that allowable distortion of tooth profiles, caused by the residual deformation, is defined, first of all, by an accepted smoothness rating, i.e., by errors of profile and base pitch. In connection with the specified features of appearance conditions of contact deformations and failures at static loads, it is accepted in formulas (39)–(42) through a simplified calculation method that $[n_s] = 1$, i.e., the allowable contact stresses σ_{Hpmax} for maximum load are equal to the limiting ones $(\sigma_{Hmax})_{lim}$. If it is accepted that $[n_s] > 1$ according to real operation conditions, then the allowable stresses are calculated by the formula

$$\sigma_{Hp\max} = \frac{(\sigma_{H\max})_{\lim}}{[n_s]} \tag{47}$$

It should be noted that sometimes there might appear a necessity of checking the depth strength by condition (39), not only at the pitch point, but at the terminal point of the active part of the line of action. Then, stresses σ_{Hmax} should be calculated for the specified contact point as well. This aspect is, to a greater extent, related to helical gearwheels ($\beta \neq 0$).

2.2 Method of Depth Contact Strength Analysis at Cyclic Loads

On the basis of modern investigations of gear contact strength and conditions of depth contact failure generation at cyclic loads, a new analysis method for surface-hardened gearwheels is developed. The method takes into account all possible cases of depth fatigue failures at contact loading of teeth with a surface layer of non-uniform hardness, obtained as a result of hardening (case hardening, carbonitriding, nitriding, laser treatment). The analysis can be carried out in two ways, in regard to the static loads.

Depth contact strength is ensured if the following conditions are fulfilled:

$$n_{\min} \geq [n] \tag{48}$$

or

$$\sigma_H \leq \sigma_{Hp} \tag{49}$$

where: n_{min} is the design factor of safety of cyclic depth contact strength; $[n]$ is the allowable factor of safety; σ_{Hp} is allowable contact stress, which does not cause dangerous contact fatigue of the tooth material, determined according to the formula

$$\sigma_{Hp} = \frac{\sigma_{H\lim}}{[n]} \cdot Z_N, \tag{50}$$

where σ_{Hlim} is the depth contact endurance limit, corresponding to the basic number of stress cycles; and Z_N is the durability factor.

Calculation according to the first option, including application of general design equations, describing generation conditions of depth fatigue contact failures, is carried out in the following sequence.

1. On the basis of experimental data and according to the design formulas specified above, the cyclic contact stresses σ_H are determined at the pitch point. While determining stresses σ_H, all factors influencing the load factor k_H are taken into account. If it is necessary, the value of stresses at terminal point a_2 of the line of action a_1a_2 is determined.
2. Semi-width of contact area b_H is determined, corresponding to the stress σ_H:

$$b_H = 1.733 \cdot 10^{-5} \cdot \sigma_H \cdot \rho_{red}. \tag{51}$$

   For $[n] > 1$

$$b_H = 1.733 \cdot 10^{-5} \cdot \sigma_H \cdot [n] \cdot \rho_{red}. \tag{52}$$

3. The function $n(z)$ is calculated at $0.7b_H \leq z \leq \delta_l$ for case-hardened (carbonitrided) teeth and laser treatment. The value b_H for $[n] > 1$ is defined in accordance with (52). This condition takes into account a possible increase of stresses σ_H (provided by factor of safety) and, consequently, location variation of plot $\tau_E(z)$.
   Calculation of the function $n(z)$, determination of its minimum value and coordinate $z_{n,min}$, where $n = n_{min}$, is carried out by computer.
4. Condition (48) is checked. If it is not fulfilled, then it is necessary, first of all, to determine the optimal parameters of the hardened layer, ensuring the required cyclic strength factor of safety for preventing the depth contact fatigue failures. Such parameters are: δ_{ef} – effective thickness of the hardened layer, determined as a distance from the surface to the zone, where function $n(z)$ takes minimum value ($\delta_{ef} = z_{n,min}$); H_{ef} – effective hardness of the hardened layer, i.e., hardness of material at a depth of $z_{n,min} = \delta_{ef}$. These parameters, and also values H_S, H_C и δ_l, determine the location of plot $H(z)$, which corresponds to minimum

resistance of case-hardened (carbonitrided) or laser-hardened material to depth fatigue failures.

The simplified calculation method consists of determination of factors n_{min} in each potentially dangerous zone, the possible location of which is found as a result of comparing the total thickness of the hardened layer δ_l and distance z_{max} from the contact surface to the zone of action of maximum equivalent stresses. Calculation is made in the following sequence.

1. Values σ_H and b_H are calculated.
2. Coordinate of the point with maximum equivalent stresses $z_{max} = 0.7b_H$ is determined.
3. Factors n_{min} are determined in each potentially dangerous zone.

3.1. For $z_{max} < \delta_l$, the dangerous zone can be located in the hardened layer or on its boundary with the core. Value n_{min} is calculated in the specified zones and then condition (48) is checked.

The zone with a minimum of the found factors n_{min} will be the dangerous one. It is possible to find limiting (for $n_{min} = 1$) and allowable normal stresses. The limiting stresses for the hardened layer are as follows:

$$(\sigma_H)_{\lim} = 4.5 \frac{H_S}{\left[\left(\frac{H_S}{H_C} - 1\right)\left(\frac{0.7b_H}{\delta_l}\right)^2 + 1\right]}. \quad (53)$$

For the boundary of the hardened layer, the limiting stresses are determined from the equation

$$(\sigma_H)^3_{\lim} = A(\sigma_H)^2_{\lim} - B = 0, \quad (54)$$

where:

$$A = 4.1 \cdot H_C;\ B = \frac{5 \cdot 10^9 \cdot H_C \cdot \delta_l^2}{\rho^2}. \quad (55)$$

Equation (54) can also be represented as a nomogram.

3.2. For $Z_{max} \geq \delta_l$, the dangerous zone is only located under the hardened layer. In this case, n_{min} is determined, and then condition (48) is checked. Here, the limiting stresses are as follows:

$$(\sigma_H)_{lim} = 4.5 \cdot H_C. \quad (56)$$

When depth contact strength is analyzed, allowable stresses σ_{Hp}, meeting the requirements of (49) for all cases of dangerous zone location, are determined by the formula

$$\sigma_{Hp} = \frac{(\sigma_{Hp})_{\lim}}{[n]}. \tag{57}$$

Compared to calculations at static loads, the allowable factor of safety at cyclic loads should satisfy the condition $[n] > 1$. This is connected to the fact that generation of depth fatigue cracks is always accompanied by progressive pitting of working surfaces of parts in contact.

Depth contact fatigue does not differ from ordinary fatigue, and resistance to plastic-elastic deformation and fatigue failure is well-correlated with the micro-hardness of local volumes (H^{ef}). Due to this fact, normal laws of distribution of values σ_H and H^{ef}, found as a result of experimental investigations, in general, allow for the application of the probability methods for depth contact strength analysis, recommended for endurance analysis by factors of safety or for calculation of functions of durability distribution at irregular loading.

Thus, the developed calculation method, including application of function $n(z)$, allows us to obtain a general solution to the depth contact strength problem and thereby to take into account all practically possible cases of location of the dangerous zone, its coordinates and the value n_{min} being found depending on conditions of contact loading and parameters of the hardened layer. Application of the specified analysis method resulted in the possibility of realizing for the first time the complex evaluation of load-carrying capacity of the teeth hardened layer and determining its optimal characteristics in the calculation of surface-hardened gearwheels.

2.3 Application Area of Depth Contact Strength Analysis

The results of testing the case-hardened (nitro-carburized) gearwheels for surface and depth contact fatigue are applied to analysis of the conditions, for which it is necessary to carry out calculations of surface-hardened gearwheels for depth contact strength.

Coordinates of point *A* in Fig. 9 are the value of surface contact endurance limit $\sigma_{Hlimb} = 23$ HRC = 1290–1450 MPa (for HRC 56–63) and the basic number of cycles $N_{HQ} = 1.2 \cdot 10^8$, determined according to the standard GOST 21354 – 87. Location of point *B* corresponds to limiting stresses, for which fatigue failures appear in the material core. The load-carrying capacity of the hardened layer is defined by the depth contact strength in an area limited by lines 1 and 2. Points *B* and *C* correspond to conditions when surface failures may be accompanied by depth cracks. It is obvious that generation of depth fatigue failures in the hardened layer itself will occur for a higher level of stresses and a lower number of stress cycles. For example, fatigue failures inside the case-hardened layer at a depth of $z = z_{max}$ for hardness HV 600 appear when limiting stresses are $(\sigma_H)_{np} = 2700$ MPa (see. Figure 9, point *C*). In this case, the number of cycles *N* up to the appearance of failure is considerably

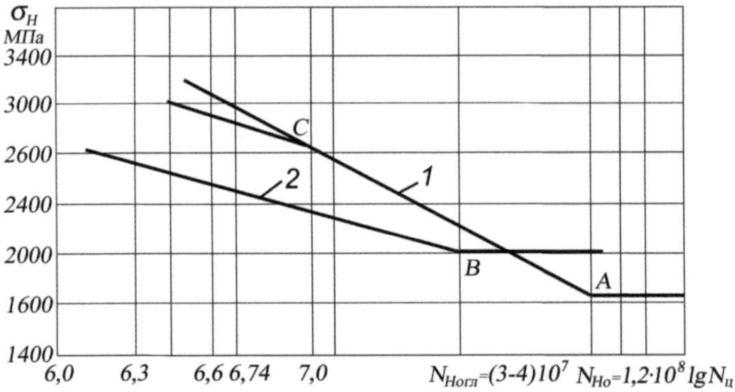

Fig. 9 Location of graphs of fatigue and dangerous stresses areas. *1* for surface failures *2* for depth failures

decreased. The conditions under which surface or depth fatigue failures prevail are defined by relations of corresponding allowable stresses. It follows from Fig. 9. that development of depth failures for sufficient surface contact strength of teeth can be caused by the stress σ_H, which is determined from the condition

$$\frac{(\sigma_H)_{\lim}}{[n]} < \sigma_H < \frac{\sigma_{H\lim_b} \cdot k_{HL}}{S_H} \tag{58}$$

where σ_{Hlimb} is the contact endurance limit of tooth surfaces, corresponding to an equivalent number of stress cycles; k_{HL} is the durability factor (see above); and $S_H = 1.2$ is the factor of safety according to recommendations of the standard GOST 21354-87.

Analysis of condition (58) shows that the value σ_H considerably depends on the properties of the external zone material of the hardened layer and the factor k_{HL}. Let us mention analysis results for the case in which depth failures appear in the hardened layer, i.e., limiting stresses $(\sigma_H)_{lim}$ are determined by formula (53). It is accepted that $[n]$ = 1.5 because of significant risk of depth failures for gear operability. As resistance to contact failures is better correlated with Vickers hardness and ratio $\frac{HV}{HRC}$ = 12 at HRC ≥ 50, then we will obtain for case-hardened (nitro-carburized) teeth

$$\sigma_{H\lim_b} = 23 \cdot \text{HRC} = 1.916 \cdot H_S, \tag{59}$$

where H_S = 670–760 kgf/mm^2 (58–63 HRC) is the hardness of the tooth surface.

After substitution of corresponding values into the formula (58), we have

$$3H^{ef} < \sigma_H < 1.6H_S \cdot k_{HL}, \tag{60}$$

where H^{ef} is the hardness of the layer in the dangerous zone at a thickness of $z = z_{max}$.

$$k_{HL\max} = \sqrt[6]{\frac{N_{HO}}{N_{HE}}} \leq 1.8 \tag{61}$$

When the material has equal resistance to surface and depth failures,

$$\sigma_H = 3H^{\mathrm{ef}} = 1.6 \cdot H_S \cdot k_{HL\max}, \tag{62}$$

then

$$k_{HL\max} = 1.875 \frac{H^{\mathrm{ef}}}{\mathrm{H_S}}. \tag{63}$$

In formula (63), the ratio $\frac{H_{\mathrm{ef}}}{\mathrm{H_S}} < 1$ is true for almost all cases of contact loading of surface-hardened gearwheels. The value k_{HLmax} conforms with restrictions, accepted in the standard GOST 21354-87. But it should not be asserted that restriction $k_{HLmax} \leq 1.8$ must always be true. It is obvious that the factor k_{HLmax} will exceed the specified value for decreasing the factor m of the fatigue curve ($m < 6$) and increasing the basic number of cycles $N_{HO} > 120 \cdot 10^6$. It can be observed in cases of variable stress amplitude, hardness increasing and quality improvement of the surface layer. For example, according to [1], $N_{HO} = 340 \cdot 10^6$ for case-hardened teeth. It was obtained during experimental investigations of the depth strength of case-hardened gearwheels that m = 4–6, depending on the microstructure and hardness of the case-hardened layer. Let us consider one example of applying the relations (60)–(63). Let us assume that the material hardness in the dangerous zone at a depth of $z = z_{max}$, where maximum equivalent stresses $\tau'_{eq} = (\tau'_{eq})_{max}$ act, is equal to $H^{ef} = 630$ kgf/mm^2. Then, according to (61) and (62) for $N_{HO} = 120 \cdot 10^6$, we obtain $\sigma_H = 1890$ MPa; $k_{HLmax} = 1.64$; $N_{HE} \approx 6.2 \cdot 10^6$. Resistance of the hardened layer to surface and depth contact failures should be the same for the specified characteristics of contact loading.

However, due to a difference in mechanisms of generation and the kinetics of surface and depth failure development, even insignificant excess in the level of design stresses $\sigma_H > 1890$ MPa can cause rapid gear breakdown, as depth failures are always progressive. At the same time, simultaneous development of surface pitting might not influence gear operability. Analyzing formulas (60)–(63), one can conclude that, without depth contact strength analysis of highly-stressed surface-hardened gearwheels, it is impossible to estimate the full extent of the load-carrying capacity of the hardened layer, or to find the location of the dangerous zone and the characteristics of the material properties, defining its resistance to contact loads. That is why it is necessary to carry out depth contact strength analysis while designing or checking calculations of surface-hardened gearwheels for all cases when operational loads approach what is allowable by condition of the surface contact strength.

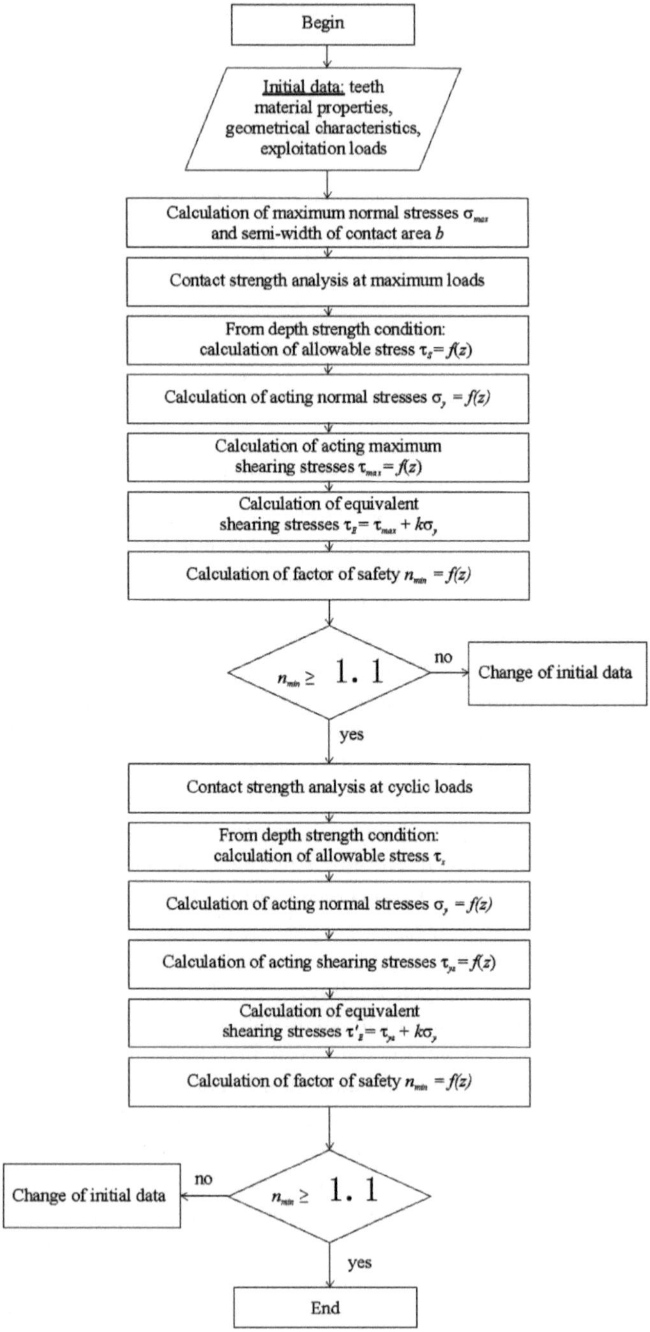

Fig. 10 Computer-aided engineering method of gearwheel analysis for depth contact strength

These calculations also allow us to define the optimal parameters of the hardened layer, which is especially important for the purposeful choice of steel grade and the method of its hardening.

A scheme of the computer-aided method of gearwheel analysis for depth contact strength is shown in Fig. 10.

References

1. System of production state testing. Testing of mechanical engineering products. Classification of mechanical impacts
2. Shirman A Ch, Solovyev AB (1985) Practical vibration-based diagnostics and monitoring of mechanical equipment state, 400 p. Mashinostroeniye, Moscow

Kinematical and Efficiency Analysis of Planetary Gear Trains by Means of Various Graph-Based Approaches

J. Drewniak, J. Kopec and S. Zawislak

Abstract Mathematical, especially graph-theoretical, models of gears can be very useful within the conceptual phase of a gear design activity. In the present paper, three graph methods of modeling of gears are discussed. The applied graphs are as follows: mixed, contour and bond graphs. Some general rules of modeling are roughly explained. Exemplary automotive gear box in case of particular drive (gear) is considered. Some engineering tasks are performed (e.g. kinematical analysis) by means of three discussed graph approaches to show usefulness of the proposed methods. The compatible results were obtained in every case. The bond graph method due to its generality and availability of professional software seem to be most useful.

Keywords Ratio · Rotational velocities · Accelerations · Forces · Automation · Algorithmization

1 Introduction

The task of gear design has been performed successfully for many years. However, it is still a demanding goal, especially in the case of planetary gears, which are elementary units of automotive automatic gear boxes. Design problems include achieving high compactness and efficiency, as well as choice of the final design solution of the gear boxes. Several methods of modeling of gears by means of schematic structures are known, e.g., A. Wolf's schemes (used in Germany) and

J. Drewniak · J. Kopec · S. Zawislak (✉)
University of Bielsko-Biala, Bielsko-Biala, Poland
e-mail: szawislak@ath.bielsko.pl

J. Drewniak
e-mail: jdrewniak@ath.bielsko.pl

J. Kopec
e-mail: jkopec@ath.bielsko.pl

V. Goldfarb and N. Barmina (eds.), *Theory and Practice of Gearing and Transmissions*, Mechanisms and Machine Science 34,
DOI 10.1007/978-3-319-19740-1_12

graphs. Professor Faydor Litvin used versatile mathematical models for modeling of curvature of tooth flanks and surfaces, as well as entire 3-D geometries of gear wheels [1, 2].

Till now, tasks performed upon graph models have included kinematical analysis, dynamical analysis, synthesis, enumeration, efficiency analysis, etc. In the present paper, mixed, contour and bond graphs are discussed and applied. A general idea of modeling of gears by means of graphs is presented in Fig. 1. The upper part of the schema describes the replacement of a conventional approach with a graph approach. The lower part describes the reverse problem, which is not discussed in

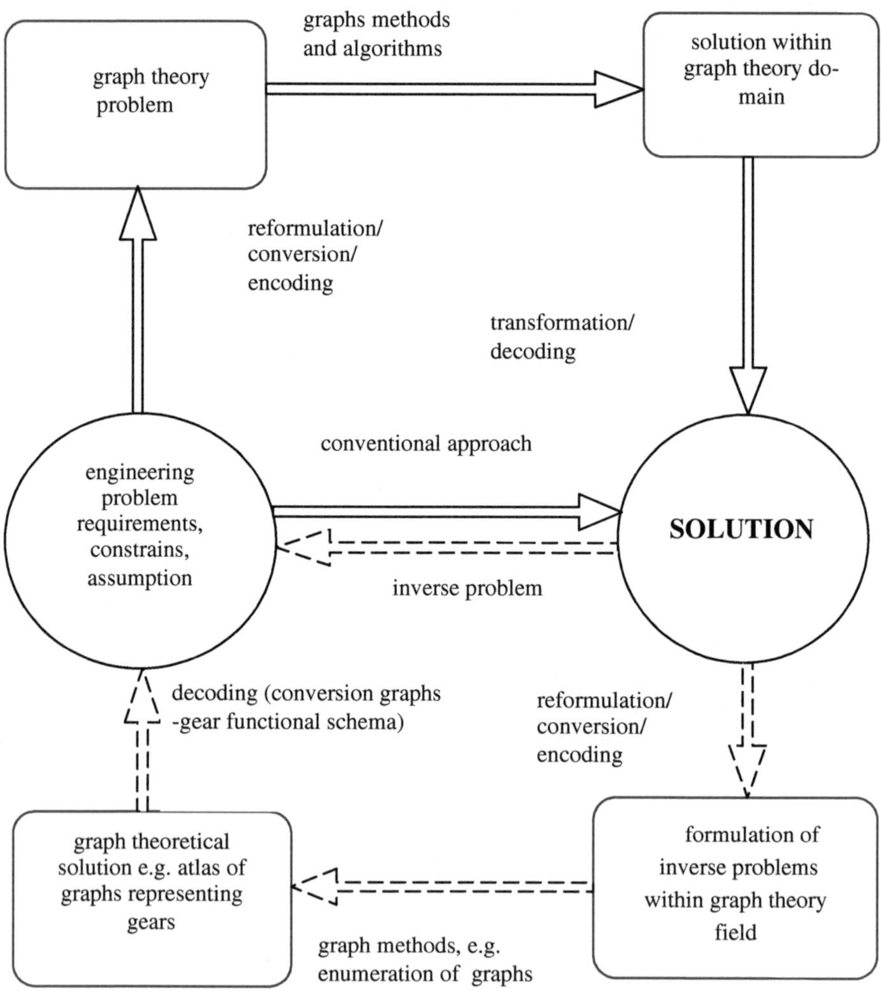

Fig. 1 Block diagram representing idea of graph modeling of mechanical systems

the paper. However, the possibility of solving reverse engineering problems confirms the usefulness of the graph approach.

In all of the above-mentioned graph-based approaches, the modeling process consists of the following steps: (a) abstraction; (b) simplification; (c) graph assignment upon the adequate rules; (d) distinguishing of particular subgraphs and other relevant graph structures; (e) generation of system of equations; (f) solution and comparison of results. These activities—in general, in Fig. 1—were presented as: problem transformation (e.g., abstraction, simplification and graph assignment) and usage of graph methods and algorithms. The meaning of these activities can be described as follows: abstraction means focusing attention on selected aspects and simultaneously neglecting or omitting other ones, e.g., we neglect lubrication, heat emission and vibrations, etc. But it is valid only within the conceptual design phase. Simplification means that we partly neglect even the selected dimensions, but diameters of gear wheels are taken into account.

Graph assignment is an essence of the whole approach. It is connected with precise rules which allow for encoding and decoding of an artifact into its graph model, respectively. The rules of assignment differ for versatile methods in almost every aspect. In a case of mixed and contour graphs, the vertices represent the discrete gear elements, but for bond graphs, the vertices (also called nodes) are chosen from the available set of nodes characteristic for this approach. Assignment of a graph to a gear is not sufficient, as a cross-domain knowledge transfer procedure should be performed which allows for reformulation of the problem, e.g., finding or distinguishing the elementary gear units is equivalent to searching for f-cycles.

In all three discussed methods, generating of system of equations is the benefit of modeling because it is done in an algorithmic way, which allows for avoidance of mistakes. Other features are: repeatability, elegance and generalization. The generation of equations consists of several steps, for example, detection of special subgraphs which have their special codes, i.e., codes of f-cycles for mixed graph and codes of contours for contour graphs. In the case of bond graphs, generation of equations is immanently connected with the type of nodes and bonds for which the equations are assigned. Moreover, the graph models of gears yield their algebraic representation, which allows for storage in computer memory. This, in turn, enables utilization of certain methods of artificial intelligence, e.g., evolutionary algorithms for generating alternative design solutions or just making calculation via standard procedures. The derived algorithms can be and are converted into computer programs [3]. As a consequence, such an approach can accelerate and automatize the design activities of an engineer. Moreover, it gives new, deeper insight into design activity.

2 Related Work

Graphs could be considered as mathematical models of gears due to analysis of relations between their parts. In light of discrete mathematics, every relation can be represented as a graph. Graph theory applications in this field have been intensively developed—and just recently, in fact—as powerful and effective tools. The relevant papers are [3–6] and books [7–15]. In general, there are only a few books in English on the world-wide market connected to the discussed topic. Book [7] is a habilitation thesis of one of the co-authors. The book [8] gives a wide description of the contour graph method, which is, in fact, used for different mechanisms rather than for gears. Book [9] describes widely graph-based methods which are used in the creation of atlases of design solutions, utilizing either the Freudenstein or Tsai graphs. It is connected with several tasks of enumeration, as highlighted in its title. Books [10, 11], written by Murota, are dedicated to the general system analysis for which matroids are also widely used. Books [12–14], written by Kaveh, are related to an analysis of mechanical and civil engineering structures (e.g., frames and trusses). The matroid approach is also discussed in Recski's book [15], and is only related to mechanics in very narrow way. Finally, books [13, 14] are dedicated to bond graph modeling. Book [15] is just a work dedicated to gears in which all engineering aspects of gear design are considered. The topic of the graph-based modeling of mechanical systems has been discussed in a number of works, for example, in book [3], edited in 2010, which included a wide reference review list of more than 500 adequate publications.

In the present paper, we use mixed graphs and contour graphs [8], as well as bond graphs [16, 17]. It is worth underlining that a similar approach is described in paper [4] in which pure Hsu graphs are considered.

In papers [18–20], Hsu and co-workers described some basics of Hsu graph modeling aspects, such as the notation itself and synthesis, as well as searching for so-called embedded structures of mechanisms. The latter problem is connected with the task of generating of the atlases of design solutions via the graph approach. In some cases, infeasible solutions could be generated in an introductory phase of the procedure, so they should be recognized and removed. Another approach to gear modeling by means of graphs is presented by the author in paper [21]. In papers [22, 23], so-called graph grammars are used for the creation of gear functional schemes upon their graphs, i.e., the reverse problem of gear modeling is effectively solved. The wider analysis of references can be found in book [7], as well as in the popular paper [24]. Recent consideration of drive bus systems can be found in [25].

3 Graph-Based Models of Planetary Gears

Based on graph-based models of gears, versatile engineering tasks could be performed, such as analysis, synthesis, optimization and enumeration of an engineering design solution. The last task consists in generation of a family of graphs which are then converted into the functional schemes of gears (inverse problem). In the paper, exemplary gears will be analyzed via three graph methods which give the same results. This allows for comparison of outcome, deeper insight into the design and the possibility of checking the correctness of the obtained results.

A graph *G*(*V*, *E*) is a pair of two finite sets: *V*—vertices and *E*—edges. Vertices are graphically presented as dots, circles or other signs; edges are drawn as straight or curved lines. Usually, elements of set E are called edges and are considered as two element subsets of *V*; $e = \{v1, v2\}$. In some cases, the E elements are treated as pairs. This means that there is a first and a second element of the pair. So, we can establish an order of vertices which causes that an element of set E is called an arc. Its end is marked by means of an arrow. A graph *G* (*V*, *E*, *W*) is called a weighted graph, where W is a weight function, i.e., $W: E \rightarrow (R, N, C)$; (*R*—real numbers, *N*—natural numbers, *C*—codes) which means that weights are assigned to the graph edges.

Underneath, the rules of assignment of the considered graphs are roughly presented. Full description and further details can be found in dissertation [7]; a slightly different approach is also discussed in [4].

3.1 Mixed Graphs

The mixed graphs are an original proposal of the authors of the present paper. In fact, it is a slight generalization of the Hsu and Tsai approaches. The mixed graph *G* (*V*, *E*) has the edge set *E* which consists simultaneously of: (i) edges $e = \{v1, v2\}$ where the sequence of vertices is not considered and (ii) arcs $e = (v1, v2)$ where the first and second element of the pair is distinguished. The vertices are depicted here as small circles or black dots, while the vertices' notions are placed nearby the vertex pictogram. The edges or arcs can be drawn as straight line segments, arcs or curves, using different line types: continuous, dashed or double, which have special encoding purposes.

The phase of abstraction means here that we temporarily focus our attention on kinematics alone. The assignment rules are as follows: the graph vertices represent the chosen gear parts which are connected with rotational movement: sun wheels, pinions, carriers, planetary wheels, drums with internal toothing as well as the housing (support or reference system). The graph edges represent the relationships between these gear elements. The following three relations are distinguished. First relation: "to turn around the main geometrical axis of a gear" is represented via a shaded polygon (which, in fact, is a clique of edges). All the edges representing this relation theoretically could not be drawn as a clique. However, it could cause that the graph is difficult to analyze. Despite the fact that the polygon is drawn, needed edges are explicitly considered in particular detailed considerations, and even drawn in some cases. The relation "to be in mesh" for two neighbor geared co-operating elements is represented by means of the dashed line, and the third relation, "a pair—carrier and planetary wheel", is drawn as a continuous line. The arcs were introduced converting a plain Hsu graph into a mixed graph. The arcs are marked from an input element (connected to the motor) to the output element passing a rotational speed onto a machine or a clutch, and then onto a working machine or car wheels. The path created by the arcs allows for analysis of passing rotational movement and power throughout a gear, as well as for checking the correctness of a graph model. The edges adjacent to the braked element are marked as double lines (continuous or dashed). This allows for spotting redundant elements or temporary redundant elements in automatic gear boxes built upon elementary planetary gears. In light of algebra, the graph is directly connected to a relationship and a matrix or a list. This allows, in turn, for storage of the graph model in computer memory and the ability to make some analyses in an algorithmic manner.

Some detailed comparisons of mixed and contour graphs are given below in Table 2 which is general, but refers to the particular graphs drawn in Fig. 5.

3.2 Contour Graphs

Contour graphs [8] are a relatively narrowly known method of modeling of mechanical systems. The idea of these graphs consists in gathering all closed circles of consecutive sets of gear parts in one subgraph (circle, cycle). Here, the cycles are called contours according to the nomenclature introduced, e.g., in [5], which is consequently applied in the papers of the authors. In the light of graph theory nomenclature, it is the same notion as a cycle, but for the reason of distinguishing these two methods and avoiding misleading interpretations, two names are consequently used. The vertices are drawn as circles, and descriptions of the vertices are placed inside these circles. This allows for distinguish between the two discussed approaches.

The set of contours has to be completed to generate the complete (solvable) system of equations. The codes of cycles are created. These codes are then used for algorithmic writing of the system of equations describing the kinematics and dynamics of the mechanical system or mechanism. The originality of this method consists in consideration of relative rotational velocities. Some of them should be converted into absolute ones. The advantages of this method of modeling are similar to that mentioned above, characteristic for mixed graphs, but here, the equations for acceleration of gear parts and forces acting on particular elements are possible to generate. So, the method is more general.

The following system of equations holds for every contour:

$$\begin{cases} \sum\limits_{(i)} \boldsymbol{\omega}_{i,i-1} = \mathbf{0} \\ \sum\limits_{(i)} \mathbf{r}_{A_i} \times \boldsymbol{\omega}_{i,i-1} + \sum\limits_{(i)} \mathbf{v}^r_{A_{i,i-1}} = \mathbf{0} \\ \sum\limits_{(i)} \boldsymbol{\varepsilon}_{i,i-1} = \mathbf{0} \\ \sum\limits_{(i)} \mathbf{r}_{A_i} \times \boldsymbol{\varepsilon}_{i,i-1} + \sum\limits_{(i)} \mathbf{a}^r_{A_{i,i-1}} + \sum\limits_{(i)} \mathbf{a}^c_{A_{i,i-1}} - \omega_i^2 \mathbf{r}_{A_i A_{i+1}} = \mathbf{0} \end{cases}, \tag{1}$$

where:

ω_{ij} is the the vector of the relative angular velocity of the link i in relation to the link j;

$v^r_{A_{ij}}$ is the the relative linear velocity of the point A_i on the link i in relation to the A_j on the link j;

ε_{ij} is the the vector of the relative angular acceleration of the link i in relation to the link j;

$a^r_{A_{ij}}$ is the the vector of the relative linear acceleration of A_i on the link i in relation to the A_j on the link j;

$a^c_{A_{ij}}$ is the the vector of the Corriolis acceleration of A_i in relation to A_j;

r_{AB} is the the vector from the joint A to the joint B;

r_{A_i} is the the vector from the origin of the co-ordinate system to the joint A_i, and moreover, the following notation is used:

The equality $\omega_{i,0} = \omega_i$ means that, if we omitted the indicator 0, then absolute velocity is considered. The following equation describes the connection between relative and absolute angular velocity:

$$\boldsymbol{\omega}_i = \boldsymbol{\omega}_{i-1} + \boldsymbol{\omega}_{i,i-1}. \quad (2)$$

The first and third equations of the above system are formulated for plain vectors, whereas the second and fourth equations are formulated for vector cross products: radiuses multiplied by adequate variables. Even more equations related to contour graph method can be found in book [5].

In the present paper, the planetary gear is considered, therefore not all summands of the above equations are needed to describe the mechanical essence (meaning) of the analyzed cases, e.g., the angle between radius and rotational velocity is equal to 90°. The general form of equations is suitable for all types of mechanisms in which elements slide one along another, and axes of different elements are not parallel, and the angles between vectors are arbitrary.

3.3 Bond Graphs

Bond graphs [16, 17] are well known graphs used for modeling of versatile mechanical and engineering systems. Their advantage consists in the following factors: (i) they are now a standard approach; (ii) professional software for their usage is available; (iii) they can be applied for simultaneous modeling of mechanical, mechatronic, electrical and even hydraulic systems by means of the same schemas and the same similar, adequate rules for generation of equations. The versatility of the method results from the fact that it describes the power flow in an arbitrary system of any physical domain and uses the Maxwell analogy (force-voltage, velocity-current). The nodes of bond graphs, called further elements, describe inertial, compliant and dissipative components of the engineering system. Another pair of complementary elements (0-nodes, 1-nodes) describes the summation of power in a system (Kirchhoff's circuit laws extended to other systems, i.e., mechanical, thermal, hydraulic systems). We assume that power flows to the nodes representing the physical system through bonds. Each bond has an assigned pair of variables (the effort variable e and the flow one f), the product of which is power.

$$P = e \cdot f \tag{3}$$

The next pair of elements (transformers and gyrators) represents general ideal machines that can convert power within a particular domain (i.e., electric transformer or mechanical gear) or that can convert power between domains (i.e., electric engine, centrifugal pumps, turbine).

The bond graphs are drawn according to general rules using the standard node elements, as well as the standard bonds [16, 17]. The equations are generated automatically, frequently by means of professional software.

However, bond graphs have not been frequently used for the modeling of gears, especially planetary gears. Some details will be given below on those bond graph elements which have been used in the modeling of planetary gears.

The bond graph method is a promising tool for the mechanical engineer who wants to perform a detailed analysis of complicated gears, especially planetary ones, due to the fact that, besides kinematics, dynamics can also be analyzed.

The selected rules for building a kinematical model were applied and presented in Figs. 2 and 3.

The bond graphs in Fig. 2 are the same, whereas the type of meshing internal or external will be distinguished in the adequate derived equations. The bond graph method is useful tool for mechanical engineer, who wants perform detailed analysis of complicated gears, especially planetary ones.

The selected examples for building kinematical model are presented in Fig. 3: the inertial element and the gyrator.

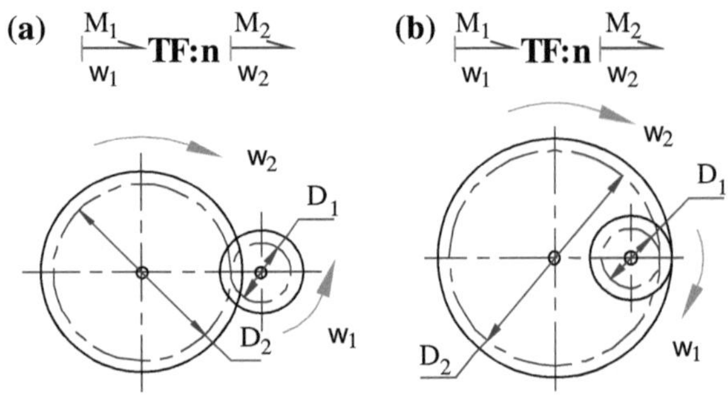

Fig. 2 Bond graph describing cylindrical gears: **a** bond graph for external meshing pair and **b** bond graph for internal meshing pair

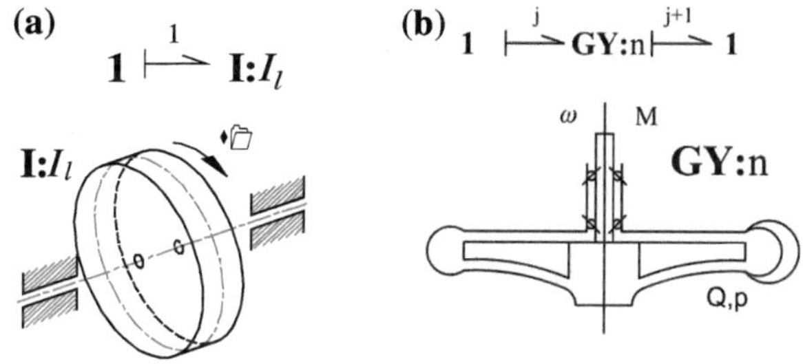

Fig. 3 Bond graph description of: **a** rotating inertial mass and, **b** ideal pump/turbine/hydrodynamic clutch

The bond graph model of the considered system consists of a kinematical schema (set of **1**-nodes and **0**-nodes), the local ratios (**TF** elements, i.e. transformers) and a hydrodynamic coupling or clutch (this coupling is represented by the **GY** element). Detailed standard rules of modeling can be found in the cited books [16, 17].

4 Exemplary Gear and Assigned Graph-Based Models

The discussed graph methods are applied for modeling of the gear presented in Fig. 4a. The duty routines or modes of operation of this gear (shown in Fig. 4) are given in Table 1. The elements of the gear are, in this example, planetary wheels, marked as 2, 5 and 8, and the arms (carriers), denoted by h_1, h_2, and h_3. The sun wheels are marked as 1, 4 and 7, whereas the ring wheels (drums) are marked as 3,6, 9. Some of the elements are mounted together, e.g., $9 = h_2$.

Here, only the chosen drives, i.e., DE2, and the reverse are discussed, but the remaining can be analyzed in the same way. Reverse drive is analyzed in the next subchapter. Further analyses are described in papers [2, 4]. Drives are activated via adequate arrangements of control elements. Activated elements are described by 'ON' in the appropriate column of Table 2. In the considered case, activated control elements are: Fw1, cl3 and b2, respectively, the fly wheel, clutch and brake. Other elements are temporarily redundant. They are drawn (Fig. 4b) by means of thin lines creating a background, whereas active elements are drawn by means of bold lines. Due to the design features and control system states, different parts can work as input and output elements, and can be either temporarily united or they are permanently united, e.g.: carrier h_2 is mounted together with wheel *9* of the internal toothing (aiming for compactness and functionality). Several elements were marked separately to express their function, e.g., the row of teeth on the sun wheel *4* and *7*.

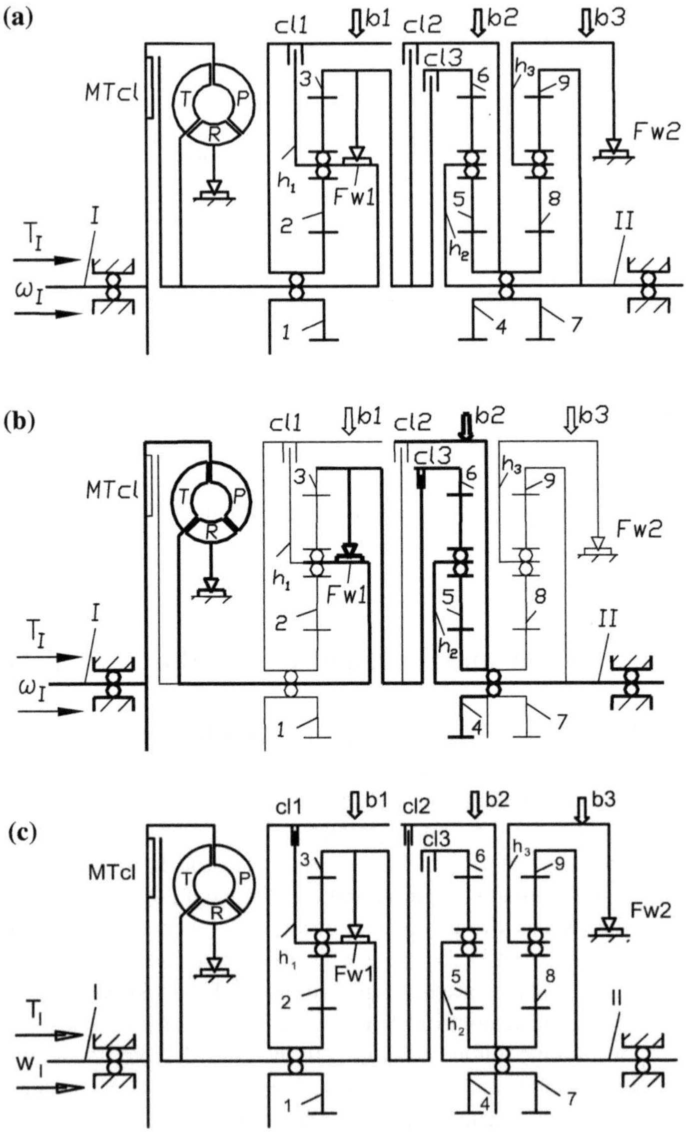

Fig. 4 Gearbox—general schema **a** DE2—operation mode **b** and **c**—reverse drive, respectively (based upon [9]

Table 1 Modes of operation of the discussed gear; based upon [9]

Drive	Activated element								
	MTcl	cl1	b1	Fw1	cl2	cl3	b2	b3	Fw2
DE 1	ON		ON	ON	ON	ON	ON		ON
2	ON			ON	ON	ON			
3				ON		ON			
4						ON			
2. Drive		ON		ON		ON	ON		
1. Drive		ON		ON		ON		ON	
Reverse		ON		ON	ON			ON	

DE—economic, ON—activated

The data according to numbers of teeth are as follows:

$$\begin{array}{l} z_1 = z_4 = z_7 = 18 \\ z_2 = z_5 = z_8 = 27 \\ z_3 = z_6 = z_9 = 72 \end{array}, \tag{4}$$

and the modulus was $m = 2$.

This automotive gear was also analyzed by the authors in papers [3–5]. Some additional data and more extended analysis are included in this paper. Analogical schemes can be drawn for all remaining forward and reverse gears (drives) listed in Table 1.

4.1 Graph-Based Models of the Gear (Drive DE2)

Three graph-based models are considered for the mode of operation DE2 (for the gear shown in Fig. 4b). Other cases can be found in [3, 4].

In the mixed graph (Fig. 5a), we can distinguish so-called f-cycles: *(5, 6)h2* and *(4,5)h2*. The name is connected with the term 'fundamental cycle'. The detailed considerations are omitted here, but can be found in Tsai's book [9]. The code of a particular f-cycle is, in general, as follows: *(i, j)k*—where *i* and *j* are two meshing gear elements and *k* denotes a carrier. The elements in brackets are placed in an increasing order. Upon these codes, the equation system (1) can be written for rotational velocities of the gear elements. The indices are arranged in an algorithmic way based upon the appropriate codes, and either the + or − sign is entered, depending on the internal or external meshing of the geared wheels. The third equation confirms that the element '*4*' is stopped.

Table 2 Comparison of mixed and contour graphs

Notions	Mixed graph	Contour
Subgraph	Cycle	Contour
Drawing/schema of the subgraph	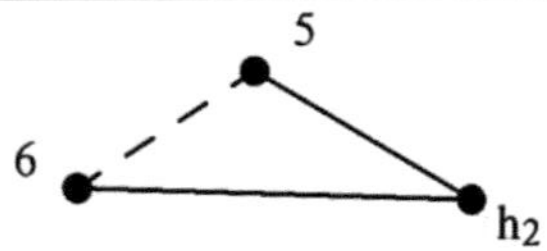	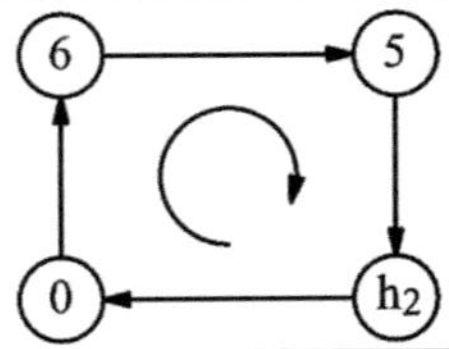
Set of edges/sequence of arcs	$\{(5, 6); (5, h_2); (h_2, 6)\}$	$((0, 6); (6, 5); (5, h_2); (h_2, 0))$
Code of cycle and contour	$(5, 6)\ h_2$ where: $(5, 6)$ dashed line edge h_2 carrier for element no 5	$(0, 6, 5, h_2\, 0)$ sequence of elements' numbers
Generated equation (s)	$\omega_5 - \omega_{h2} = +N_{65}(\omega_6 - \omega_{h2})$	$\boldsymbol{\omega}_{60} + \boldsymbol{\omega}_{56} + \boldsymbol{\omega}_{h2,5} + \boldsymbol{\omega}_{0,h2} = \mathbf{0}$ $\mathbf{0} \times \boldsymbol{\omega}_{60} + \mathbf{r}_6 \times \boldsymbol{\omega}_{56} + (\mathbf{r}_4 + \mathbf{r}_5) \times \boldsymbol{\omega}_{h2,5} + \mathbf{0} \times \boldsymbol{\omega}_{0,h2} = \mathbf{0}$
Set of variables	Direct rotational velocities	Relative rotational velocities
Benefit	Arrangement of indices upon the codes in algorithmic manner, i.e.: reverse order of indices for local ratio N	Arrangement of indices upon the codes in algorithmic manner, i.e.: consecutive pairs of indices in reverse order
Additional explanation/ decomposition	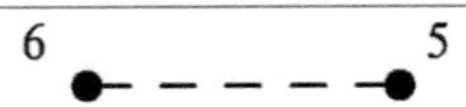 Meshing/ gearing of element no 5 and 6	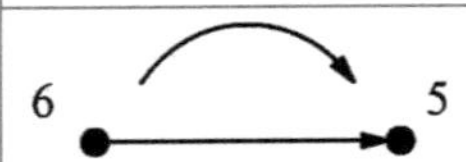
	 Planet-carrier	Compatibility of arc direction and circulation
	 Rotation around main axis	

(continued)

Table 2 (continued)

Notions	Mixed graph	Contour
Functional scheme	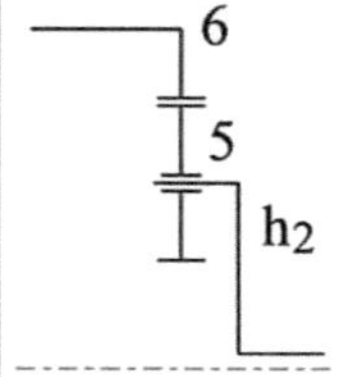	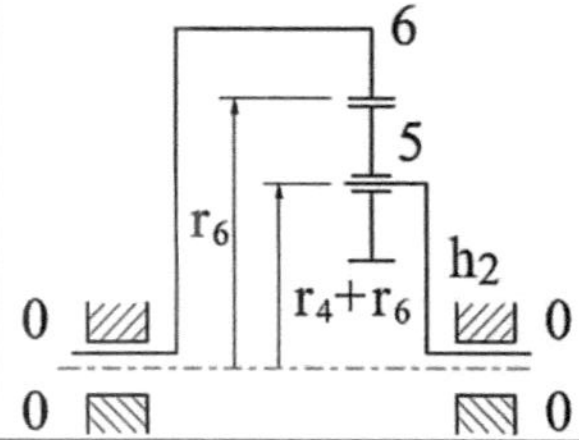
2nd subgraph and its code	(4, 5); h_2 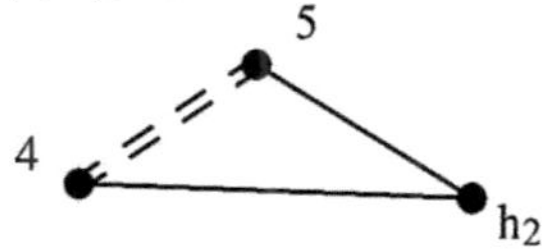	(0, 6, 5, 4, 0) 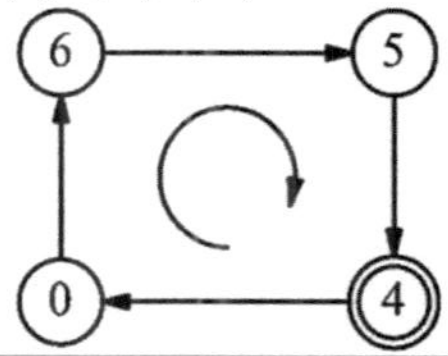
Additional equation	$\omega_4 = 0$	$\omega_{40} = 0$ $\omega_4 = 0$ vector equation is converted into scalar one
Special description of fixed element	Double dashed line indicates that element no 4 is fixed	Double circle indicates fixed element

(continued)

Table 2 (continued)

Notions	Mixed graph	Contour
Path from input to output	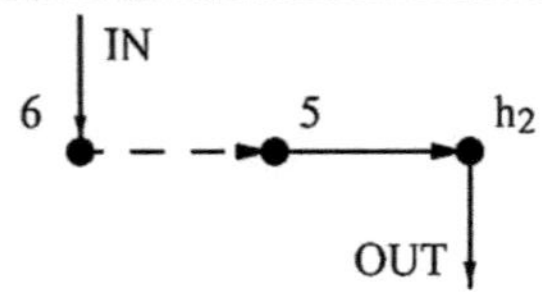	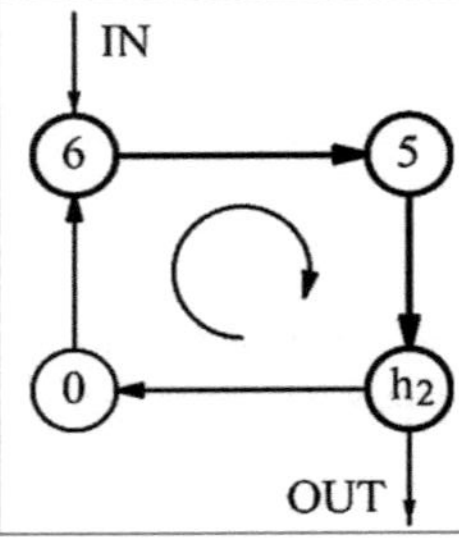

$$\begin{cases} \omega_5 - \omega_{h2} = +N_{65}(\omega_6 - \omega_{h2}) \\ \omega_4 - \omega_{h2} = -N_{54}(\omega_5 - \omega_{h2}). \\ \omega_4 = 0 \end{cases} \tag{5}$$

The solution is as follows:

$$\frac{\omega_6}{\omega_{h2}} = \frac{1}{N_{65}}\left(\frac{1}{N_{54}} + N_{65}\right) = \frac{z_5}{z_6}\left(\frac{z_4}{z_5} + \frac{z_6}{z_5}\right) = \frac{z_4 + z_6}{z_6} = \frac{18 + 72}{72} = 1.25. \tag{6}$$

For the contour graph (Fig. 5b), codes of contours (i.e., directed cycles) can be created as follows: *(0, 6, 5, 4, 0)* and *(0, 6, 5, h2, 0)*. In this method, we consider relative velocities ω_{ij} of an element i in relation to the element j. Every contour is considered as starting and terminating in the *0* vertex, i.e., the support system. This helps in solution of the system of equations. The solution consists in elimination of all relative velocities which are unavoidable in equations but they are not interesting in terms of the general outcomes. Nevertheless, it gives broader insight into the analyzed object.

$$\begin{cases} \boldsymbol{\omega}_{60} + \boldsymbol{\omega}_{56} + \boldsymbol{\omega}_{45} + \boldsymbol{\omega}_{04} = \mathbf{0} \\ \mathbf{r}_6 \times \boldsymbol{\omega}_{56} + \mathbf{r}_4 \times \boldsymbol{\omega}_{45} = \mathbf{0} \\ \boldsymbol{\omega}_{60} + \boldsymbol{\omega}_{56} + \boldsymbol{\omega}_{h2,5} + \boldsymbol{\omega}_{0,h2} = \mathbf{0}. \\ \mathbf{r}_6 \times \boldsymbol{\omega}_{56} + (\mathbf{r}_4 + \mathbf{r}_5) \times \boldsymbol{\omega}_{h2,5} = \mathbf{0} \\ \boldsymbol{\omega}_{40} = \mathbf{0} \end{cases} \tag{7}$$

The system of vector Eq. (4) can be solved taking into account the gear wheel dimensions, i.e., pitch diameters (5), where the pitch diameter of an element i is equal to d_i. In fact, the vector system is converted into a scalar one, taking into account a geometrical property that the vector of rotational speed is perpendicular to the plane of rotational movement, as well as the fact that all vectors in the first and third equations act along the main gear axis of symmetry. All pitch radiuses should be given in accordance with the functional scheme of the considered gear.

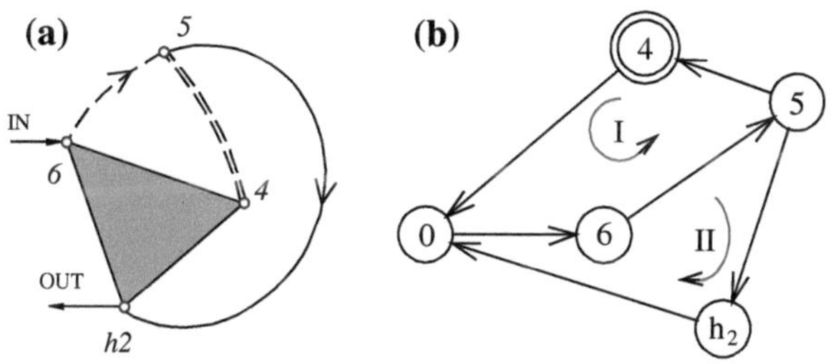

Fig. 5 Graphs assigned to the gear: mixed **a** contour **b** where 0 represents the support

$$\begin{cases} r_4 = 0,5d_4 \\ r_5 = 0,5d_5. \\ r_6 = 0,5d_4 + d_5 \end{cases} \quad (8)$$

In Table 2, the comparison of the mixed and contour graphs is shown—considering some detailed aspects of the discussed methods. In general, they are very similar, and the bond graph is essentially different because it is based upon classical, general bond graph theory. The differences among contour and mixed graph methods consist, for example, in considering slightly different cycles, different kinematic equations are generated for different set of variables, cycle codes are different, etc.

In the last line of Table 2, the paths connecting an input and an output are shown. The input, i.e., the driven element (connected to the motor), is denoted by an additional arrow 'in'. The output element connected to car wheels (or, in the general case, to the working machine) is denoted by the description 'out'. The path of arcs allows for an analysis of passing a rotation and power through the gear.

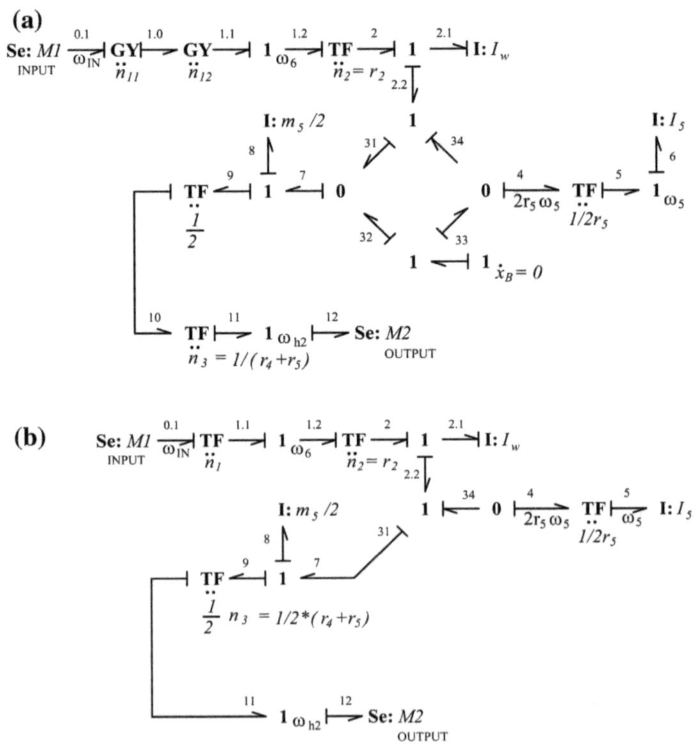

Fig. 6 Two Bond graphs assigned to the DE2 drive of the gear: according to Brown's idea (**a**) [16]; an simplified one (**b**)

The same operating mode of the gear is shown in Fig. 6 by means of bond graphs which are essentially different from those considered previously.

These bond graphs take into account the hydrodynamic clutch. Based upon the bond graph presented in Fig. 6b, the same ratio was calculated. As was stated, further, complete analyses have been published for this gear in [3, 4] and for other gears in [5, 6].

4.2 Willis Method for Drive DE2

For a comparison of results, the Willis method was applied. The local ratio in reference to carrier h_2 can be expressed by the following formula:

$$i_{64}^{h_2} = \frac{\omega_6 - \omega_{h_2}}{\omega_4 - \omega_{h_2}}. \tag{9}$$

The detailed calculations give the value of the same ratio, taking into account the tooth numbers:

$$i_{64}^{h_2} = \left(-\frac{z_5}{z_6}\right)\left(-\frac{z_4}{z_5}\right) = \frac{z_4}{z_6}. \tag{10}$$

Since $\omega_4 = 0$, the aggregated equation has the following form:

$$\frac{z_4}{z_6} = \frac{\omega_6 - \omega_{h_2}}{0 - \omega_{h_2}}. \tag{11}$$

After simple recalculations, the desired ratio is obtained:

$$i_{6,h_2}^{4} = \frac{\omega_6}{\omega_{h_2}} = 1 - \frac{z_4}{z_6} = 1 - \left(\frac{18}{-72}\right) = 1.25. \tag{12}$$

In the Willis method, the number of internal teeth is considered to be a negative value, therefore we have the negative number in the denominator in the last formula. The same value of the ratio was obtained, as before, using graph-based methods.

4.3 Graph-Based Models of the Gear (Reversed Drive)

Additionally, along with the above-analyzed forward drive (DE2), we consider a reverse drive. In this case, we start consideration from the conventional Willis method. We can observe that, due to the special arrangement of control elements (several activated clutches), we have a case in which $\omega_3 = \omega_7 = h_1$. Therefore, as with the input element, we consider sun wheel *7* and the output consists of element *9*. The local ratio can be expressed here by the following formula:

$$i_{7,9}^{h_3} = \frac{\omega_7 - \omega_{h_3}}{\omega_9 - \omega_{h_3}}. \tag{13}$$

Following the detailed considerations, the same notion can be expressed using tooth numbers:

$$i_{7,9}^{h_3} = \left(-\frac{z_8}{z_7}\right)\left(-\frac{z_9}{z_8}\right) = \frac{z_9}{z_7}. \tag{14}$$

Taking into account the condition $\omega_{h_3} = 0$, and after some recalculation, we obtain the ratio for the reverse drive:

$$i_{h_1,9}^{h_3} = i_{h_1,9} \cdot i_{7,9}^{h_3} = = 1 \cdot \frac{z_9}{z_7} = \left(\frac{-72}{18}\right) = -4. \tag{15}$$

The value is negative, as opposed to all forward (ahead) drives, which have positive values of adequate ratios.Below, the reverse drive is analyzed by means of two first graph-based methods. The mixed and contour graphs are presented in Fig. 7. In the case of a mixed graph, f-cycles can be distinguished: *(7,8)*h_3 and *(8,9)*h_3. Stoppage of the gear arm is considered graphically via the double edge of the graph. Based upon the graph and the codes, the system of equations can be written as:

$$\begin{cases} \omega_7 - \omega_{h3} = - N_{87}(\omega_8 - \omega_{h3}) \\ \omega_8 - \omega_{h3} = + N_{98}(\omega_9 - \omega_{h3}) \\ \omega_{h3} = 0 \end{cases}. \tag{16}$$

Taking into account the last equation describing fixing of the arm/carrier, the resultant formula is obtained:

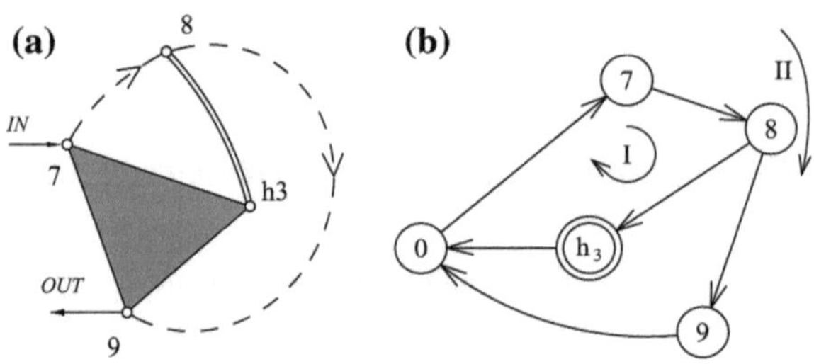

Fig. 7 Graphs assigned to the gear for the reverse drive: mixed (**a**); contour (**b**)

$$\omega_7 = -N_{87}N_{98}\omega_9. \tag{17}$$

Finally, the ratio is derived:

$$\frac{\omega_7}{\omega_9} = -\frac{z_8}{z_7}\cdot\frac{z_9}{z_8} = -\frac{z_9}{z_7} = -\frac{72}{18} = -4. \tag{18}$$

Taking into account the contour graph, the following contours can be distinguished, having codes: *(0, 7, 8, h_3, 0)* and *(0, 7, 8, 9, 0)*. Fixing of the gear arm is graphically represented via a double circle around h_3. Therefore, the following system of equations describes the reverse drive case:

$$\begin{cases} \boldsymbol{\omega}_{70} + \boldsymbol{\omega}_{87} + \boldsymbol{\omega}_{h3,8} + \boldsymbol{\omega}_{0,h3} = \mathbf{0} \\ \mathbf{r}_7 \times \boldsymbol{\omega}_{87} + (\mathbf{r}_7 + \mathbf{r}_8) \times \boldsymbol{\omega}_{h3,8} = \mathbf{0} \\ \boldsymbol{\omega}_{70} + \boldsymbol{\omega}_{87} + \boldsymbol{\omega}_{98} + \boldsymbol{\omega}_{09} = \mathbf{0} \\ \mathbf{r}_7 \times \boldsymbol{\omega}_{87} + (\mathbf{r}_7 + 2\mathbf{r}_8) \times \boldsymbol{\omega}_{98} = \mathbf{0} \\ \boldsymbol{\omega}_{h3,0} = 0 \end{cases}. \tag{19}$$

The consecutive calculations are omitted. The solution of the system is as follows:

$$\frac{\omega_7}{\omega_9} = -\frac{r_7 + 2r_8}{r_7} = -\frac{18 + 2.27}{18} = -\frac{72}{18} = -4. \tag{20}$$

In all analyzed cases, the same results according to the ratios were obtained. The advantage of graph methods lies in their algorithmic nature. This allows for preparation of computer programs for the entirety of activities [7].

5 Efficiency of Meshing for the Selected Drives

5.1 *Efficiency of Meshing for the Second Economic Drive DE2*

We assume that the efficiency of the external meshing is equal to $\eta_{0e} = -0.990$ and the efficiency of internal meshing is equal to $\eta_{0i} = -0.995$, as it has been given in several monographs. Equivalence of the friction power losses P_F of the real gear (driver DE 2) and the power losses P_{F0} of the base gear (considered according to the carrier h_2) is used for further recalculation:

$$P_F = P_{F0}. \tag{21}$$

This equivalence in power loss arises from the fact that these are the same gear, however, the real one is analyzed from the position of unmoving observer while the base gear analysis is from the position of the observer placed on the carrier (where the velocities of links are determined in accordance with the carrier).The loss of friction power P_F of the real gear (for the drive DE 2) is equal:

$$P_F = P_6\left(1 - \eta^4_{6,h2}\right), \tag{22}$$

where power on the input of real gear is as follows:

$$P_6 = T_6\,\omega_6, \tag{23}$$

where T_6 is the given input moment, ω_6 is the given angular velocity of wheel *6* (input), and $\eta^4_{6,\ h2}$ is the determined efficiency of the real gear in the case of the direction of drive from wheel *6* to carrier h_2 in the case of the braking of wheel *4*. The loss of friction power P_{F0} of the base gear for the DE 2 drive is equal to:

$$P_{F0} = P_{60}(1 - \eta_0), \tag{24}$$

where the power on output of the base gear is:

$$P_{60} = T_6\,\omega_{60}, \tag{25}$$

and the efficiency of the base gear can be expressed as:

$$\eta_0 = \eta_{0e} \cdot \eta_{0i}. \tag{26}$$

The angular velocity ω_{60} (according to the convention applied in this subchapter) is the relative velocity of wheel *6* in relation to carrier h_2:

$$\omega_{60} = \omega_6 - \omega_{h2} = \omega_6 \frac{-i_0}{1 - i_o} > 0, \tag{27}$$

where the base ratio is as follows: $i_0 = i^{h2}_{6,4} = -\frac{z_4}{z_6} = -0.25$.Combining the above Eqs. (22–27), after adequate recalculation,—the efficiency for the DE2 drive was obtained:

$$\eta^4_{6,h2} = \frac{1 - i_0\eta_0}{1 - i_0}. \tag{28}$$

For the assumed data and the assumed simple efficiencies, we can write the formula:

$$\eta^4_{6,h2} = \frac{1 + 0.25 \cdot 0.955 \cdot 0.990}{1 + 0.25} = 0.98909. \tag{29}$$

Additionally, the output moment can be calculated:

$$T_{h2} = T_6 \left| i^4_{6,h2} \right| \cdot \eta^4_{6,h2} = 800 \cdot 1.25 \cdot 0.989 = 989\,\mathrm{Nm}. \tag{30}$$

Wider analyses will be performed in further papers by the authors.

5.2 *Efficiency of Meshing for the Reverse Drive*

In the considered gear, in case of the of the reverse drive, the carrier h_3 is fixed, transforming it into a gear having fixed axes. Therefore, its efficiency can be expressed by the formula:

$$\eta^{h3}_{7,9} = \eta_0 = \eta_{0e} \cdot \eta_{0i} = 0.98505, \tag{31}$$

and the output moment can be calculated:

$$T_9 = T_7 \left| i^{h3}_{7,9} \right| \cdot \eta^{h3}_{6,9} = 800 \cdot 4 \cdot 0.985 = 3152\,\mathrm{Nm}. \tag{32}$$

In the present paper, due to the defined length, certain chosen analyses of the considered gear are performed. Other results can be found in other papers of the authors [3, 5, 6].

6 Conclusion

In the present paper, three graph methods were described and utilized for the modeling of an exemplary planetary gear. Consecutive drives of the gear are reached via adequate activation of the system of control elements. The kinematical and efficiency analyses for chosen drives were performed. The traditional Willis method was applied for comparison. Compatibility of results has been observed. The methods can be used within the conceptual phase of a gear design. Usage of several methods simultaneously allows for avoidance of mistakes, as well as allowing for easier comparison of alternative design solutions.

In the paper, the basics of graph approaches were roughly described. Further possibilities could be found in the cited references, e.g., synthesis and generation of atlases of design solutions. In some cases, graphs are uniquely effective tools for the solution of reverse problems.

References

1. Litvin FL, Rahman P, Goldrich RN (1992) Mathematical models for the synthesis and optimization of spiral bevel gear tooth surfaces, National Aeronautics and Space Administration, Scientific and Technical Information Office
2. Litvin FL et al (2005) Topology of modified surfaces of involute helical gears with line contact development for improvement of bearing contact, reduction of transmission errors and stress analysis. J Math Comput Model 42(9–10):1063–1078
3. Drewniak J, Zawislak S (2010) Linear-graph and contour-graph models of planetary gears. JTAM 48(2):415–433
4. Ardte D (2012) Kin-Tak Lam: analysis of kinematics of a gear steering mechanism for the tracked vehicles. J Sci Innov 2(3):183–188
5. Drewniak J, Zawislak S (2013) Graph-based models of compound planetary gear boxes. Solid State Phenom 199:143–148
6. Drewniak J, Zawislak S, Wieczorek A (2013) Analysis method of automatic planetary transmission kinematics (in Polish). Zeszyty Naukowe Politechniki Śląskiej 1825:97–102
7. Zawislak S (2010) The graph-based methodology as an artificial intelligence aid for mechanical engineering design. University of Bielsko-Biala, Poland
8. Marghitu DB (2005) Kinematic chains and machine components design. Elsevier, Amsterdam, San Diego
9. Tsai LW (2001) Mechanism design: enumeration of kinematic structures according to function. CRC, Boca Raton
10. Murota K (1987) Systems analysis by graphs and matroids—structural solvability and controllability. series: algorithms and combinatorics, vol 3, Springer, Berlin
11. Murota K (2010) Matrices and matroids for systems analysis. series: algorithms and combinatorics, vol 20, Springer, Berlin
12. Kaveh A (2004) Structural mechanics: graph and matrix methods, research studies press. Wiley, Exeter, U.K
13. Kaveh A (2006) Optimal Structural analysis. Wiley, Chichester, U.K
14. Kaveh A (2014) Advances in metaheuristic algorithms for optimal design of structures. Springer
15. Recski A (1989) Matroid theory and its applications in electric networks and in statics, series: algorithms and combinatorics, vol 6, Springer, Berlin
16. Brown FT (2001) Engineering systems dynamics: a unified graph-centered approach. Marcel Dekker, New York
17. Merzouki R, Samataray AE et al (2013) Intelligent mechatronic systems. Springer, London
18. Hsu CH, Chang CC, Hsu JJ (1999) Structural synthesis of bevel-gear robotic wrist mechanisms. Proc Natl Sci Council ROC(A) 23(4):518–525
19. Hsu CH (1992) A graph notation for the kinematic analysis of differential gear trains. J Franklin Inst 329(5):859–867
20. Hsu CH, Wu YC (1997) Automatic detection of embedded structure in planetary gear trains. ASME J Mech Des 119:315–318
21. Lang SYT (2005) Graph-theoretic modeling of epicyclic gear system. Mech Mach Theory 40:511–529
22. Li X, Schmidt L et al (2004) Transformation of an EGT grammar: new grammar, new designs. ASME J Mech Des 126(4):753–756
23. Schmidt L, Li X (2004) Grammar-based designer assistance tool for epicyclic gear trains. ASME J Mech Des 126:895–902
24. Wojnarowski J, Kopec J, Zawislak S (2006) Gears and graphs. J Theor Appl Mech 44:139–162
25. Xunming Li et al. (2014) Modelling and simulation of power coupling system in hydraulic hybrid city bus. Energy Power Eng 6(6):119–132

Rational Designs of Planetary Transmissions, Geometry of Gearing and Strength Parameters

F. Plekhanov and V. Goldfarb

Abstract The paper deals with designs and geometry of planetary transmissions with internal gearing with a small difference of gearwheel tooth numbers. Internal planetary gears with non-standard tooth profiles and their strength parameters are considered. Results of the experimental research of these mechanisms are presented.

Keywords Planetary transmissions · Design · Geometry · Strength

1 Introduction

The essential requirements specific to gears include fair mass and dimension parameters, high load carrying capacity, durability and low friction losses [1–3]. These requirements are vital in aerospace engineering, robotics and mechatronic systems. Most often, such high-duty devices involve harmonic gear drives possessing high load carrying capacity at low mass and overall dimensions [4]. However, the pointed gears provide low efficiency (70–85 %) as compared to several types of planetary mechanisms, and in this case, special materials are to be applied in order to ensure the necessary durability of the flexible element (a gearwheel with a thin-walled rim) and of the gear drive as a whole.

F. Plekhanov (✉)
Glazov Institute of Engineering and Economies, Glazov, Russia
e-mail: gfi@gfi.edu.ru

V. Goldfarb
Institute of Mechanics, Kalashnikov Izhevsk State Technical University, Izhevsk, Russia
e-mail: veniamingoldfarb@yahoo.com

V. Goldfarb and N. Barmina (eds.), *Theory and Practice of Gearing and Transmissions*, Mechanisms and Machine Science 34,
DOI 10.1007/978-3-319-19740-1_13

In this connection, it becomes a relevant task to develop and research planetary mechanisms which would have fair mass and dimension parameters at high load-carrying capacity and lifetime along with a higher efficiency.

A very efficient gear with regard to mass and dimension parameters, load-carrying capacity, lifetime, gear ratio ($20 \leq i \leq 200$ in one stage) and efficiency is a planetary gear with an eccentric carrier, two to three satellites, a fixed central gearwheel with internal teeth and the mechanism of torque absorption from the satellite [5, 6]. The difference between tooth numbers of engaged gearwheels is taken to be small (Δz = 1-3), providing high load-carrying capacity due to small clearances in the engagement and several tooth pairs engaged simultaneously. The aspect limiting a wide application of this type of gear is the necessity of using the mechanism of torque absorption with a mangle gear, reducing the efficiency or parallel cranks complicating the design layout. The authors will introduce gears with a roller mechanism of torque absorption without the pointed drawbacks (design layouts have the inventor's certificates).

2 Rational Design of Planetary Transmissions with Internal Engagement

Layouts of planetary gears with the motion from satellites to a low-speed shaft transmitted by rollers do not require application of a complex mechanism of parallel cranks, and are very productive and efficient [7, 8]. Figure 1 shows a roller planetary gear, in which rollers loosely mounted in satellites and rigidly connected to each other and with a low speed shaft are rolled over their conjugated surfaces, thus providing high efficiency of the gear without complicating its design. The diameter

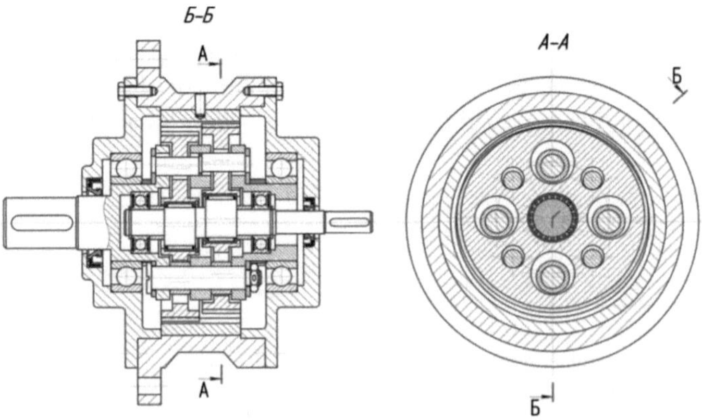

Fig. 1 Planetary gear with internal engagement of gearwheels and mechanism of torque absorption from the satellite (Inventor's certificate N2402709 Russia Federation: МПК F16H1/32. Planetary gear)

of the rollers is less than the diameter of the holes for their mounting by the value of the interaxial distance of the gear. Figure 2 shows the same design but with three satellites and a symmetrical scheme of elements loading where centrifugal forces acting on the satellites are balanced, thus improving the gear's dynamics and reducing its vibration activity [9, 10]. When rollers are hollow and contain rods connecting the disks with the low speed shaft (Fig. 3), a greater numbers of rollers can be placed in satellites and disks, and load-carrying capacity can therefore be increased. Here, rollers are conjugated with only the surfaces of satellites and disks rolling along them but not contacting the tie rods and bushing mounted on them.

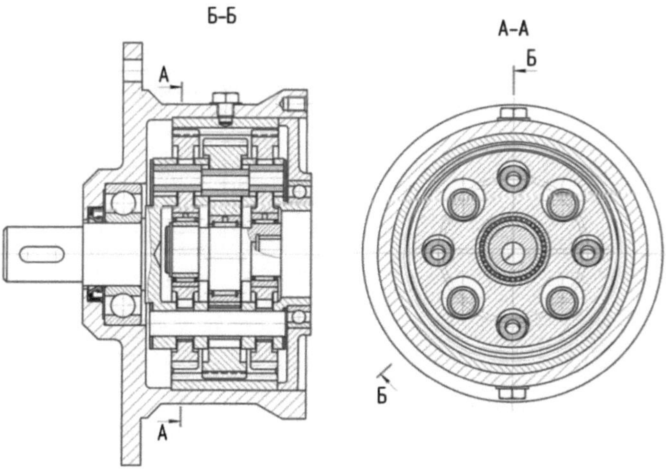

Fig. 2 Planetary gear with roller mechanism of torque absorption from the satellite and symmetrical scheme of elements loading

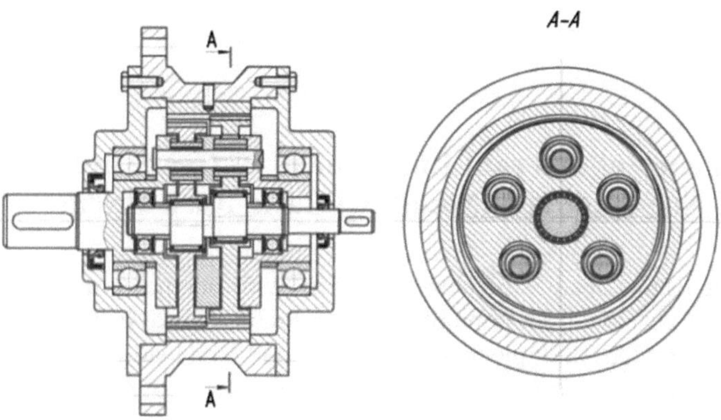

Fig. 3 Planetary gear with hollow rollers (Inventor's certificate N2492376 Russia Federation: MPK F16H1/32. Planetary gear)

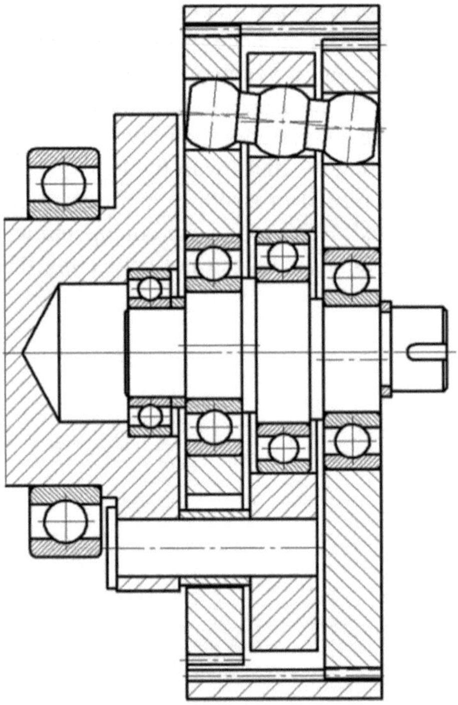

Fig. 4 Planetary gear with swing rollers (Inventor's certificate N2475662 Russia Federation: MPK F16H1/32. Planetary gear)

Figure 4 shows the design of a planetary gear with internal engagement of gearwheels and spherical rollers providing the swing motion. The gear comprises fewer disks of the mechanism of torque absorption and less axial dimension, but friction losses are a little higher in this gear than in the above-mentioned layouts. To implement greater gear ratio ($i = 200 \div 1000$), a combined gear can be used with a high speed stage comprising a sun pinion, satellites, fixed central gearwheel and the driven carrier, and a low speed stage representing one of the above layouts of the roller planetary gear (Fig. 5).

3 Geometry of Internal Engagement of Gearwheels with Non-standard (Modified) Basic Rack Profile

The distinguishing feature of planetary gears with internal gearing with a small difference of gearwheel tooth numbers is the low clearance between teeth. On the one hand, it provides multiple pair engagement and high load-carrying capacity, but on the other hand, it causes a risk of gear scuffing because of the imposition of tooth profiles on each other (interference of the longitudinal edge of the external tooth with the basic surface of the internal tooth) (Fig. 6). To eliminate this negative effect, it is necessary to reduce the depth of tooth entry and increase the pressure

Fig. 5 Combined planetary gear

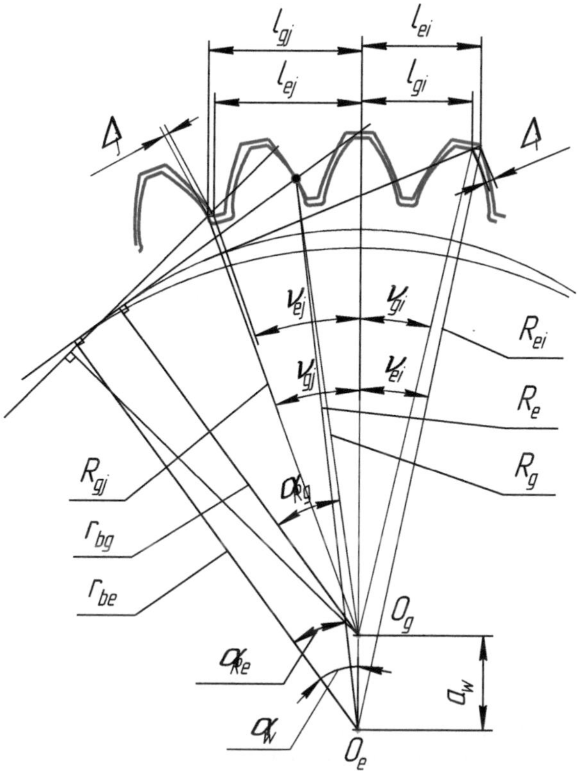

Fig. 6 Internal involute gearing

angle, but this decreases the overlap ratio and smoothness of gear operation, thus reducing the positive effect of multi-pair contact. In this connection, it is essential to select the most reasonable basic rack profile and carry out the geometry synthesis of gearing providing high strength at a guaranteed absence of interference and the value of overlap ratio $\varepsilon \geq 1$. It should be taken into account here that, with a small difference in number of gearwheel teeth and, therefore, small eccentricity of the carrier, the reduced radii of the engaged gearwheel tooth profiles, rollers and surfaces conjugated with them have greater values than necessary to provide their load-carrying capacity. That is why, analysis shows, the load-carrying capacity of gears under investigation is limited in most cases by the bending strength of the teeth. The influence of roller strains on load distribution in their conjugations is also small at the corresponding degree of manufacturing accuracy and reasonably chosen parameters of the mechanism [8].

To reduce the bending strength at the tooth root, it is reasonable to decrease the reference dedendum and increase the curvature radius of the transient curve. Figure 7 shows the modified basic rack profile representing the corrected version of the basic rack profile according to the standard GOST 13755-81 which provides the necessary quality characteristics of the engagement and the gear as a whole. Unlike the standard one, the limiting height of its tooth h_l is taken to be decreased by 0.25 m and the transient curve does not have the rectilinear segment, thus representing the circular arc with the radius ρ_f. Parameters of the pointed basic rack profile (designation of parameters correspond to the standard GOST 13755-81):

$$\alpha = 20^\circ; h_\alpha = m; h_l = 1.75 \text{ m};$$

$$\rho_f = \frac{0.5}{\cos\alpha}[0.5\pi m - 2(h_l - h_a)\tan\alpha] = 0.545 \text{ m};$$

$$c = \rho_f(1 - \sin\alpha) = 0.359 \text{ m}; h_f = h_l - h_a + \rho_f(1 - \sin\alpha) = 1.109 \text{ m};$$

$$h = h_l + c = 2.109 \text{ m}$$

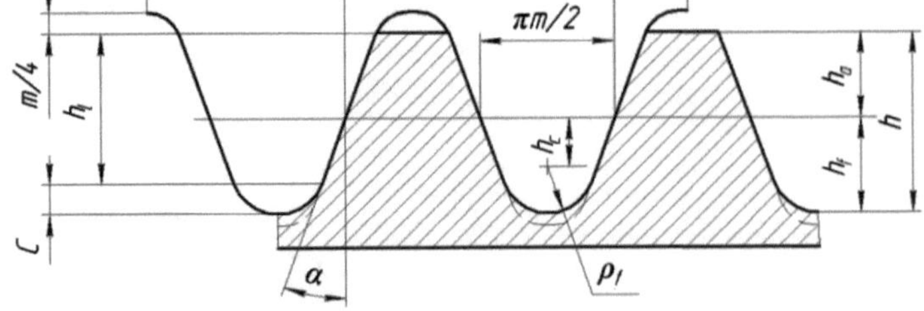

Fig. 7 Modified (hatched) and original basic rack profiles of the satellite (the *dashed line* stands for the basic rack profile according to the standard GOST 13755-81)

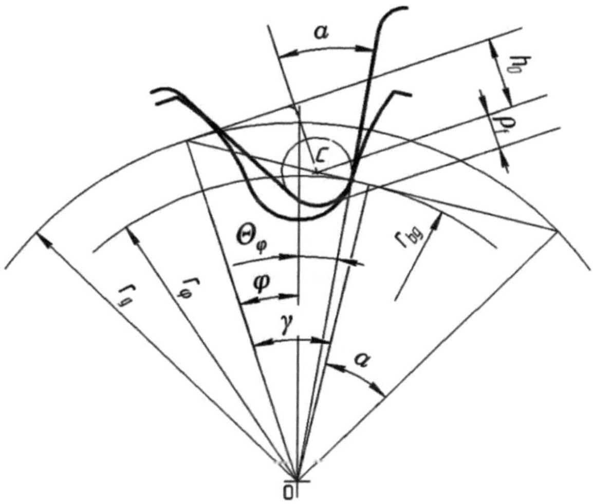

Fig. 8 Generating scheme of the satellite tooth profile with the modified basic rack

$$h_c = h - h_a - \rho_f = 0.564 \text{ m}.$$

In this case, the transient curve of the satellite tooth profile g represents the equidistant line of the rack trochoid, its equations being represented in the following parametric way (Fig. 8):

$$r\left(-\sqrt{r_g^z + \left(\frac{h_o}{\sin(+\rho_f)}\right)^z - zr_g\left(\frac{h_o}{\sin\gamma} + \rho_f\right)\sin\gamma, \alpha}\right)\alpha\theta\varphi - \arcsin\left|\frac{\cos\gamma}{r_f}\frac{h_o}{\sin\gamma} + \rho_f\right| - \frac{h_o}{r_g \tan\gamma}\Bigg) \quad (1)$$

where θ_ϕ and r_ϕ are current values of the polar angle and vector radius of the point of the transient curve; h_0 is the distance from the initial straight line of the tool rack in machine-tool gearing to the centre of its tooth profile curvature at the apex, $h_0 = h_c - x_g m$; ρ_f is the curvature radius of the transient curve of the satellite basic rack profile equal to the curvature radius of the tool rack tooth profile at its apex; γ is the angle between the perpendicular line to the profile normal line at the non-involute segment of the generated tooth and the vector radius of the contact point of the initial straight line of the tool rack with the reference circumference of the satellite,$\gamma = arctan\left(\frac{h_0}{r_g\phi}\right)$; φ is the current value of the angle between the symmetry axis of the satellite dedendum and the vector radius of the contact point of the initial straight line of the tool rack with the reference circumference of the satellite; r_g is the radius of the satellite reference circumference, $r_g = 0.5\, mz_g$; and x_g and z_g are coefficients of the basic rack shift and the satellite tooth number correspondingly.

At the boundary point of the satellite tooth profile l, the angle $\gamma = \alpha$ and

$$r_{\phi} = r_{1g} = \sqrt{\left(0.5mz_g\right)^2 + \left(\frac{h_0}{\sin\alpha} + \rho_f\right)^2 - mz_g\left(h_0 + \rho_f \sin\alpha\right)} \quad (2)$$

At the value $\left(\frac{h_0}{\sin\alpha} + \rho_f\right) \triangleright \left(r_g \sin\alpha\right)$, the tooth undercut takes place (for zero gearwheel with the above-mentioned basic rack, the condition of the undercut absence is the ratio $z_g \geq 13$).

The transient curve of the gearwheel tooth profile cut by a non-standard toll represents the equidistant line of the elongated hypocycloid. Analysis shows that its bending strength is much higher than the satellite tooth strength, which is why, at the geometrical synthesis of gearing, the satellite tooth height must be taken as the minimum possible.

Therefore, the pressure angle α_w and radii r_{ab} of the tooth apex circumference of the fixed gearwheel b and r_{ag} of the satellite g are determined according to the presence of backlash v between the transient surface of the satellite tooth and the gearwheel tooth apex at the area of maximum depth of tooth entry

$$v = 2\left[r_{\varphi b}\theta_{\varphi b} - r_{ab}\left(inv\alpha_{ab} - inv\alpha + \frac{0.5\pi - 2x_b \tan\alpha}{z_b}\right)\right] \geq 0, \quad (3)$$

and according to the required value of the overlap ratio ε, the elimination of the gearwheel tooth edge interference with the transient surface of the satellite tooth (presence of clearance δ), and the interference of the longitudinal edge of the external tooth with the basic surface of the internal tooth (presence of clearance Δ):

$$\varepsilon = \frac{1}{\pi \cos\alpha}\left[\sqrt{r_{ag}^2 - r_{bg}^2} + a_w \sin\alpha_w - \sqrt{r_{ab}^2 - r_{bb}^2}\right] \geq 1, \quad (4)$$

$$\delta = \sqrt{\left(\sqrt{r_{ab}^2 - r_{bb}^2} - a_w \sin\alpha_w\right)^2 + r_{bg}^2} - r_{1g} \geq 0, \quad (5)$$

$$\Delta = r_{ab}\left[\frac{inv\alpha_w}{z_b} - inv\alpha_{ab} + \frac{z_g}{z_b}\left(\Phi_1 + inv\alpha_{ag}\right) - \Phi_2\right] \geq 0. \quad (6)$$

In Eqs. (3)–(6), $\alpha_{ab} = \arccos\left(\frac{mz_b \cos\alpha}{2r_{ab}}\right)$; $r_{\phi b} = r_{ab} - a_w$; $a_w = \frac{m(z_b - z_g)\cos\alpha}{2\cos\alpha_w}$; $\alpha_{ag} = \arccos\left(\frac{mz_g \cos\alpha}{2r_{ag}}\right)$;

$\Phi_1 = \arccos\left(\frac{r_{ab}^2 - r_{ag}^2 - a_w^2}{2a_w r_{ag}}\right)$; $\Phi_2 = \arccos\left(\frac{r_{ab}^2 - r_{ag}^2 + a_w^2}{2a_w r_{ab}}\right)$; $r_{bg} = 0.5mz_g \cos\alpha$; $r_{bb} = 0.5mz_b \cos\alpha$; z_b is the tooth number of the fixed gearwheel b; x_b is the shift factor of the gearwheel b basic rack, $x_b = x_g + (inv\alpha_w - inv\alpha)\frac{z_b - z_g}{2\tan\alpha}$; $\theta_{\varphi b}$ is the polar angle corresponding to the vector radius of the point of the satellite;and $r_{\varphi b}$ is the transient curve and is determined according to Eq. (1).

Moreover, the circumference radius of the fixed gearwheel tooth apexes cannot be less than its limiting value determined according to the presence of the radial clearance between the root of the shaping cutter and the tooth apex of the gearwheel to be cut c_{0-b} = 0.25 m. The absence of tooth edge interference with the transient surface of the shaping cutter tooth is guaranteed and according to the well-known formula of the machine-tool gearing

$$r_{ab} \geq m\left[0.5z_0 - h_a^* + \frac{(z_b - z_0)\cos\alpha}{2\cos\alpha_{w0}}\right] \quad (7)$$

where the pressure angle of the machine-tool gearing α_{w0} is determined according to the involute function:

$inv\alpha_{w0} = \tan\alpha_{w0} - \alpha_{w0} = inv\alpha + 2(x_b - x_0)\frac{\tan\alpha}{z_b - z_0}$, h_a^* is the factor of the gearwheel basic rack addendum ($h_a^* = 1$), z_0 and x_0 are the tooth number and the shift factor of the shaping cutter basic rack.

Therefore, having assigned the gearwheel tooth numbers and values x_g, ε, Δ, z_0, x_0, Eqs. (4), (6) and (7) are applied to determine the radii r_{ag}, r_{ab}, and pressure angle α_w, after which Eqs. (3) and (5) are applied to check the presence of clearances v and δ (at the absence of clearances, the parameters r_{ag}, r_{ab}, and α_w are increasingly corrected).

The performed analysis shows that at $x_0 = x_g = 0$, ε = 1.05; Δ = 0.5 m; $z_0 = 25$; $60 \leq z_g \leq 140$ and the limiting value r_{ab} according to condition (7), the pressure angle does not essentially vary with variation of the tooth number, but it varies a lot with variation of the value $\Delta Z = z_b - z_g$ (Table 1). Necessary clearances v and δ are provided here, guaranteeing the absence of interference and gear wedging.

Table 1 Rational values of the gearwheel pressure angle of a planetary gear

Δz	ε			
	0.8	1	1.2	1.4
1	57.3°	59.0°	61.0°	63.3°
2	47.2°	47.9°	48.8°	50.1°
3	42.1°	42.3°	42.6°	43.4°
4	39.2°	39.2°	39.2°	39.2°
5	36.5°	36.5°	36.5°	36.5°

4 Evaluation of Gearwheel Tooth Strength and Gear Load-Carrying Capacity

Comparative evaluation of tooth strength for gearwheels with modified basic rack and parameters corresponding to the equations of geometrical synthesis given above was carried out by finite-element analysis (Fig. 9). Investigations performed by means of the SolidWorks program showed that the bending strength of non-standard teeth is essentially higher than that of standard ones at similar geometrical parameters of engagement (the bending strength of the internal teeth of the gearwheel is higher by 50 %, and of the satellite, by 10 %).

Table 2 shows the average values of the satellite tooth geometry factor $y_F = \frac{mb\sigma_F}{F_t}$ depending on ΔZ at $60 \leq z_g \leq 140$ and parameters of engagement determined according to the relations given above (here, σ_F is the numerically-determined maximum value of stress at the root of the shortened tooth when the load is applied to its apex, F_t is the circumferential force, and b is the tooth length).

Geometrical parameters of the transmission and values of the satellite tooth geometry factor determined according to the relations given above are reasonable to be applied when composing equations of displacement compatibility of gearing elements and statics equations, their solution allowing for determination of the load-carrying capacity of the planetary mechanism.

Components of the load in the gearing are determined through accounting of the initial value of clearances Δi, Δj between teeth (see Fig. 6) and their deformation

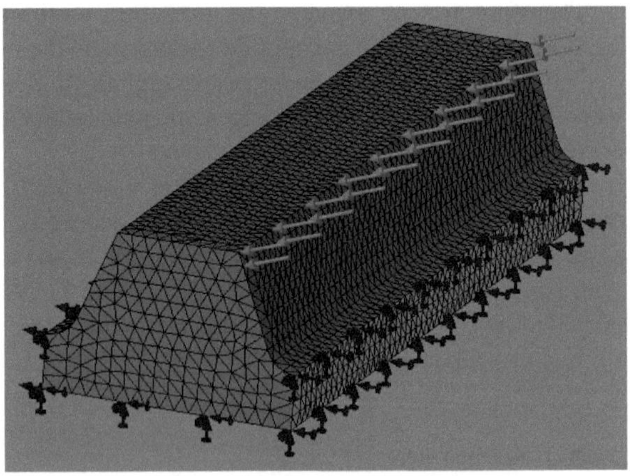

Fig. 9 Model of a satellite tooth by SolidWorks for determining stresses

Table 2 Tooth geometry factor for the satellite with modified basic rack

ΔZ		
1	2	3
2.65	2.34	2.21

when solving equations of displacement compatibility and static equations, allowing us to find out the number of tooth pairs carrying the load, load distribution and to perform the design analysis of the gearing:

$$\left.\begin{aligned} F_0 &= \delta bc = f(\sigma_P), \\ F_{i=1} &= (\delta - \Delta_{i=1})bc, \\ F_{j=1} &= (\delta - \Delta_{j=1})bc, \\ &\ldots\ldots\ldots\ldots\ldots\ldots, \\ F_{i=n} &= (\delta - \Delta_{i=n})bc, \\ F_{j=m} &= (\delta - \Delta_{j=m})bc, \end{aligned}\right\} \quad (8)$$

where F_0 is the normal force of the most loaded tooth pair; F_i is the force in the *i*th tooth pair to the right of the most loaded pair; F_j is the force in the *j*th tooth pair to the left of the most loaded pair; δ is the deformation of the most loaded tooth pair; *b* is the facewidth of the satellite; *c* is the stiffness of the gearing ($c = 0.075E$); *E* is the Young modulus); and σ_P is the allowable stress determined by known strength analysis formulas (σ_{HP}—at contact strength analysis, σ_{FP}—at bending strength analysis).

When the gearing is analyzed according to Hertz contact strength for steel gearwheels

$$f(\sigma_p) = 5.72\frac{\rho b\sigma_{HP}^2}{EK_{H\Sigma}},$$

when the satellite tooth is analyzed according to bending strength

$$f(\sigma_P) = \frac{\sigma_{FP}bm}{Y_F K_{F\Sigma}\cos\alpha}.$$

In these equations, ρ is the reduced radius of curvature of tooth profile; α is the basic rack profile angle; $K_{F\Sigma}$ is the factor taking into account the non-uniform load distribution along the tooth length, geometrical parameters and dynamics of the gear [11, 12]; and Y_F is the tooth geometry factor.

The first equation of system (8) permits for finding the allowable relative unit load according to strength conditions $\Phi_0 = F_0/bm$ at the most loaded tooth of the satellite and its corresponding deformation δ/m. After that relative unit, loads are determined for other teeth engaged with the fixed gearwheel (the number of tooth pairs in contact is determined by the number of positive forces obtained by solving the system of Eq. (8) and taken into account in further analysis). Clearances between tooth pairs involved in equations of displacement compatibility are expressed by the gearing module according to Fig. 6 and accounting for the gearing phase of the most loaded tooth (in terms of bending strength, the engagement phase is unfavorable when the satellite tooth apex is conjugated with the tooth of the fixed gearwheel; in terms of contact strength, it is the engagement phase corresponding to the tooth contact in the middle part of the active segment of engagement line:

$$\Delta_i = \left(R_{ei} \sin \nu_{ei} - l_{gi}\right) \cos\left(\alpha_{ag} - \nu_{gi}\right), \tag{9}$$

$$\Delta_j = \left(R_{gj} \sin \nu_{gj} - l_{ej}\right) \cos\left(\alpha_{ae} - \nu_{ej}\right), \tag{10}$$

where α_{ag}, α_{ae} are the satellite g and fixed gearwheel e tooth profile angles at addendum circumferences).

According to this load distribution factor, representing the portion of the total load per most loaded tooth in engagement:

$$\Psi = \Phi_0 / \left(\sum_{i=1}^{n} \Phi_i + \sum_{j=1}^{m} \Phi_j + \Phi_0 \right). \tag{11}$$

Figure 10 shows the dependence of Ψ on satellite tooth number z_g at different values of Δz, the dependencies being drawn according to (11) for the reversible transmission comprising gearwheels of QT steel with the load-carrying capacity limited by the bending strength of the satellite teeth. The pressure angle and satellite tooth height have been selected to provide the initial overlap factor $\varepsilon \geq 1.05$ and to eliminate interference.

The obtained values of relative unit loads in contact pairs of gearwheel teeth allow for determining the total load and the required module of gearing m at the assigned torque at the output shaft T. Therefore, for a two-satellite layout of the transmission

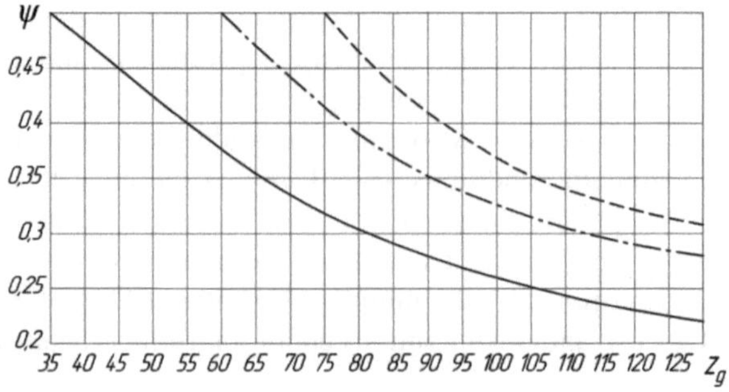

Fig. 10 Load distribution factor Ψ versus tooth number of the satellite z_g and difference of gearwheel tooth numbers Δz:—$\Delta z = 1$; – • – • – • – $\Delta z = 2$; – – – – $\Delta z = 3$

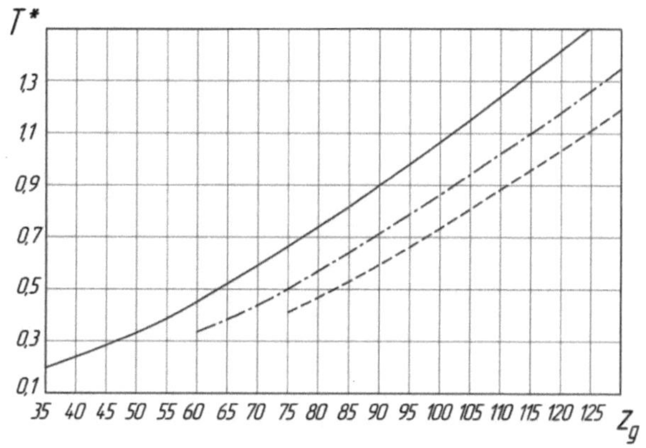

Fig. 11 Relative torque at the output shaft T^* versus z_g and Δz:—$\Delta z = 1$; – • – • – • – $\Delta z = 2$; – – – – $\Delta z = 3$

$$T/(mbr_{bg}) = 2\left(\sum_{i=1}^{n}\Phi_i + \sum_{j=1}^{m}\Phi_j + \Phi_0\right), \tag{12}$$

or as a dimensionless expression

$$T^* = T/\left(Em^3\right) = z_g\overline{b}\cos\alpha\left(\sum_{i=1}^{n}\Phi_i + \sum_{j=1}^{m}\Phi_j + \Phi_0\right)/E, \tag{13}$$

where $\overline{b} = b/m$, r_{bg} is the radius of the satellite base circumference.

Diagrams of values T^* versus z_g and Δz at $\overline{b} = 15$ are shown in Fig. 11. It demonstrates that the load-carrying capacity of the gear is increased with an increase of the satellite tooth number z_g and decrease of the value Δz. It is caused by lowering the clearances in the engagement and increasing the number of tooth pairs carrying the load. It is the reason for a higher load-carrying capacity of a transmission with internal gearing instead of the most wide-spread, conventional three-satellite planetary mechanism. Figure 12 shows diagrams of load distribution between tooth pairs for gearwheels with different tooth numbers.

5 Experimental Investigation of Planetary Transmissions with Internal Gearing

Several layouts of planetary gearwheels with roller mechanisms of mechanism of torque absorption from satellites (one of them is shown in Figs. 13 and 14) have been designed and manufactured based on the performed investigations. They have

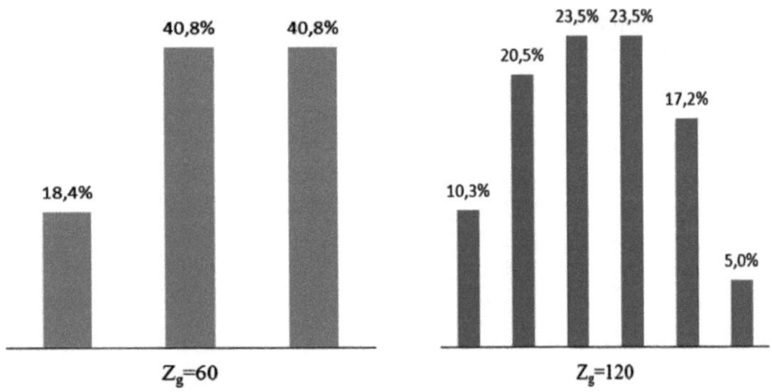

Fig. 12 Diagrams of load distribution between tooth pairs for internal gearing at $\Delta z = 1$

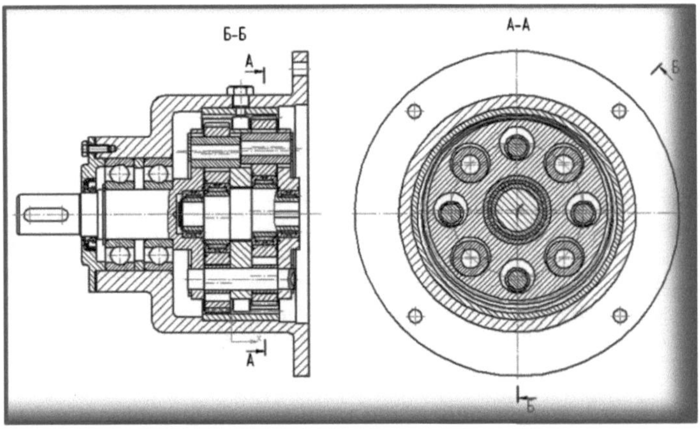

Fig. 13 Tested planetary transmission with roller mechanism of torque absorption from satellites

been tested at a special test-stand (Fig. 15) and the efficiency of these gear mechanisms has been proved, their load-carrying capacity being close to its calculated value.

It is complicated to determine theoretically the friction losses of planetary transmissions under consideration, which was why such an important factor of their effective operation as efficiency has been estimated experimentally. Two transmissions made according to the schemes in Figs. 1 and 4 have been subjected to testing. At nominal torque of the output shaft, the efficiency of the first one proved to be 90–95 %, of the second one—about 10 % less.

Fig. 14 Satellite unit according to the scheme in Fig. 13 ($m = 1.5$; $z_g = 100$; $\Delta z = 2$)

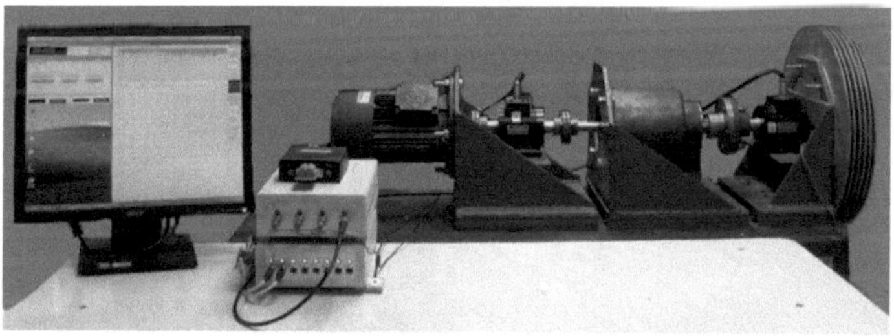

Fig. 15 Test-stand for experimental investigation of coaxial gears

6 Conclusion

The obtained theoretical and experimental relations allow for determining the rational geometrical parameters of the gearing, estimating its strength characteristics, load-carrying capacity of the drive, friction losses and performing the design of efficient planetary transmissions with internal gearing and roller mechanism of torque absorption from satellites.

References

1. Kraynev A, Salamandra K, Raghavan M (2012) Synthesis of the two-stream transmissions. In: Power transmissions: proceedings of the 4th international conference. Sinaia, 20–23 Jun 2012, pp 335–345 (in Russian)
2. Kahraman A, Ligata H, Singh A (2010) Influence of ring gear rim thickness on planetary gear set behavior. J Mech Des 132:021002

3. Parker GR, Agashe V, Vijayakar MS (2000) Dynamic response of a planetary gear system using a finite element contact mechanics model. Trans ASME 122:304–310. doi:10.1115/1.1286189
4. Timofeyev GA, Samoylova MV (2012) Study of geometry and kinematics of a combined epicyclic-and-harmonic mechanism. Vestnik MGTU. Mashinostroyeniye N1:70–80 (in Russian)
5. Lopatin BA, Poluektov EA, Khaustov SA (2011) Generation of approximate internal meshing of spur-bevel gears. Vestnik YuUrGU. Mashinostroyeniye N17:39–48 (in Russian)
6. Ivanov AS, Ermolaev MM, Krikunov DE, Miroshnik AA, Rudnev SK, Chirkin AV (2013) Design of epicyclic-and-pin gearboxes for precision tracking drives. Vestnik mashinostroyeniya N3:9–11 (in Russian)
7. Plekhanov FI (2013) Effect of K-H-V epicyclic gear geometry on its strength factors. Vestnik mashinostroyeniya N3:16–20 (in Russian)
8. Plekhanov FI, Ovsyannikov AV (2011) Load distribution over the roller length in a gear-roller planetary transmission. Russ Eng Res 31(9):825–827 (in Russian)
9. Lin J, Parker GR (1999) Analytical characterization of the unique properties of planetary gear free vibration. Trans ASME 121:316–321
10. Lin J, Parker GR (2001) Natural frequency veering in planetary gears. Mech Struct Mach 29 (4):411–429, doi:10.1081/SME100107620
11. Lin J, Parker GR (2002) Planetary gear parametric instability caused by mesh stiffness variation. J Sound Vib 249(1):38–48. doi:10.1006/jsvi.2001.3848
12. Singh A (2011) A Epicyclic load sharing map—development and validation Mech Mach Theory 46(5):632–646

Compensation of Errors of Alignment Caused by Shaft Deflections in Spiral Bevel Gear Drives

A. Fuentes, R. Ruiz-Orzaez and I. Gonzalez-Perez

Abstract The effect of errors of alignment on the bearing contact of spiral bevel gear drives has been the subject of research for many years. Generally speaking, gear misalignment causes transmission errors and edge contacts, leading to incremental levels of noise and vibration, and a reduction of the gear drive service life. Apart from assembly and/or manufacturing errors, supporting shafts deflections caused by torque transmission constitute an important but predictable source of misalignments in gear drives. In the paper, a procedure of determination of the relative spatial position of spiral bevel gear supporting shafts during torque transmission will be proposed, in order to predict the relative errors of alignment between spiral bevel gears. The obtained errors of alignment will be employed as initial data in the local synthesis method, for the purpose of compensating them through modification of the pinion surface microgeometry. Finally, a numerical example will illustrate the proposed procedure, as well as the advantages of its consideration in the design of advanced spiral bevel gear drives in order to achieve the best contact pattern and function of transmission errors for nominal torque transmission.

Keywords Gear geometry · Spiral bevel gears · Tooth contact analysis · Finite element analysis · Local synthesis · Errors of alignment · Shaft deflections

1 Introduction

Spiral bevel gears are applied in many applications wherein transformation of rotation and torque between intersected axes are required. This type of gear drive provides uniform and quiet performance under high loads, and shows relatively low sliding velocities and high contact ratios due to the additional overlapping tooth action. Consequently, it is well suited to applications in which tooth loads and

A. Fuentes (✉) · R. Ruiz-Orzaez · I. Gonzalez-Perez
Department of Mechanical Engineering, Polytechnic University of Cartagena (UPCT), Cartagena, Spain
e-mail: alfonso.fuentes@upct.es

V. Goldfarb and N. Barmina (eds.), *Theory and Practice of Gearing and Transmissions*, Mechanisms and Machine Science 34,
DOI 10.1007/978-3-319-19740-1_14

operating speeds are high, as in helicopter and truck transmissions, reducers, and many other applications [1, 10, 11, 14].

Gear misalignment causes a shift in the bearing contact towards the edges of gear tooth surfaces, leading to a defective contact pattern. Accordingly, several negative consequences arise, such as an increase in both contact and bending stresses, increments of both wear and heat generation rates, incremental levels of noise and vibration, and reduction of service life of the gear drive. In order to investigate the consequences of gear misalignment, different types of analytical or numerical models have been developed, the generation of finite elements models being one of the adopted approaches [13].

Integration of such computational tools as stress analysis in the design procedures of gears drives is not a novelty. An extended approach was introduced by Faydor L. Litvin, who proposed the application of the Finite Element Method (FEM) as part of an integrated procedure of design and stress analysis of spiral bevel gear drives [2]. Basically, the above-mentioned approach is based on the automatic computational generation of a finite element model of the spiral bevel gear drive that takes into account the finite element mesh of both members of the gear drive, the definition of contacting surfaces, and the establishment of boundary conditions for loading the gear drive [2, 6, 10]. Later, Qi Fan [5] and Shuting Li [9] proposed Loaded Tooth Contact Analysis (LTCA) algorithms based on Finite Element Analysis (FEA), while Mohsen Kolivand proposed a similar approach employing Rayleigh-Ritz-based shell models [8]. Additionally, K. Mao employed a non-linear finite element model in order to investigate mitigation of gear surface contact fatigue wear through microgeometry optimization [12], and Yi-Cheng Chen applied FEA in contact stress analysis of beveloid gear drives [3].

Additionally, during the last 4 years, the automatic generation procedure of finite element models of spiral bevel gear drives proposed initially by Faydor L. Litvin [2, 6, 10] has been improved with the development of finite element models of gear supporting shafts and their consideration in a general gear drive finite element model [7, 13] that also takes into account the bearing positions along the shafts. The aforementioned improvement constitutes a unique tool for determination of predictable misalignment of gears due to power transmission. However, a procedure of determination of relative spatial orientation of gear shafts is needed in order to determine errors of alignment to be compensated by the design of the microgeometry of the pinion tooth surfaces of the gear set.

The main purpose of this work is the proposal of a mathematical procedure based on analytic geometry for the determination of the relative errors of alignment between spiral bevel gears. The relative spatial position of gear supporting shafts provided by the finite element model of the whole spiral bevel gear drive will be employed in the proposed mathematical procedure. In turn, the numerical results provided by the proposed methodology will be introduced as initial data in the local synthesis method proposed by Faydor L. Litvin [2, 6, 10], with the final purpose of modifying gear mating surfaces and, consequently, improving the conditions of meshing and contact of the considered gear drive. The developed research has been illustrated with a numerical example.

2 Development of the Finite Element Model of a Complete Spiral Bevel Gear Drive

In this section, a computational procedure of automatic generation of finite element models of spiral bevel gear drives, including their corresponding supporting shafts, will be described, in order to estimate the effect of shaft deflections caused by power transmission on the shift of the bearing contact. Basically, this is based on the ideas presented at [7, 13], which are applicable to any type of gear drive.

Finite element analysis (FEA) is a widely employed tool in the design procedure of gear drives, and it allows for reaching the following goals:

1. Determination of contact and bending stresses throughout the whole cycle of meshing.
2. Investigation of the formation of bearing contact and transfer of load between contiguous pairs of teeth.
3. Detection of areas of severe contact stress throughout the whole cycle of meshing.

Figure 1 shows the physical model of a spiral bevel gear drive taking into account the gear supporting shafts and bearing positions. Basically, it is comprised of one pinion, one wheel, their corresponding supporting shafts, and bearing

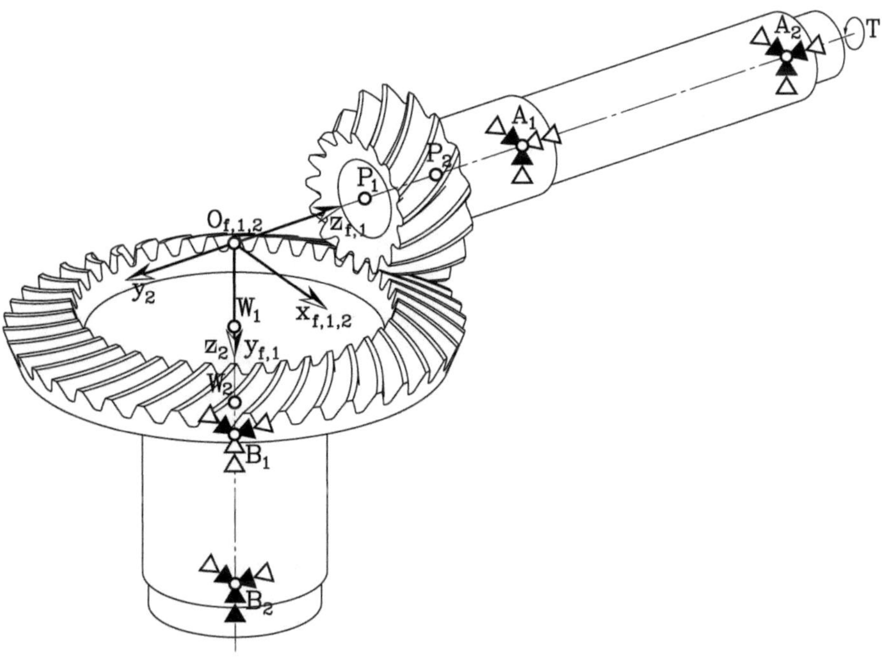

Fig. 1 Physical model of a spiral bevel gear drive

locations. Both pinion and wheel are, in this case, overhung mounted, although the procedure is totally general, and they are considered initially to be perfectly aligned and supported by bearings in which bending is allowed. Additionally, coordinate systems $S_f(x_f, y_f, z_f)$, $S_1(x_1, y_1, z_1)$ and $S_2(x_2, y_2, z_2)$ have been considered. Coordinate system S_f constitutes the fixed coordinate system rigidly connected to the housing and coordinate systems S_1 and S_2 represent fixed coordinate systems the origins of which are located on the pitch cones apices and axes z_1 and z_2 oriented along the pinion and wheel axes of rotation, respectively.

Application of FEM requires the development of a finite element model of the considered spiral bevel gear drive. Based on the physical model illustrated in Fig. 1, the finite element model is generated according to a procedure based on several steps, which are described below:

1. Gear tooth surfaces are computationally generated considering the procedures described in [2, 6, 10]. These surfaces, along with topland and rim surfaces, are considered together in order to obtain the volume of each spiral bevel gear tooth. Finally, the aforementioned volume is discretized into finite elements as illustrated in Fig. 2.

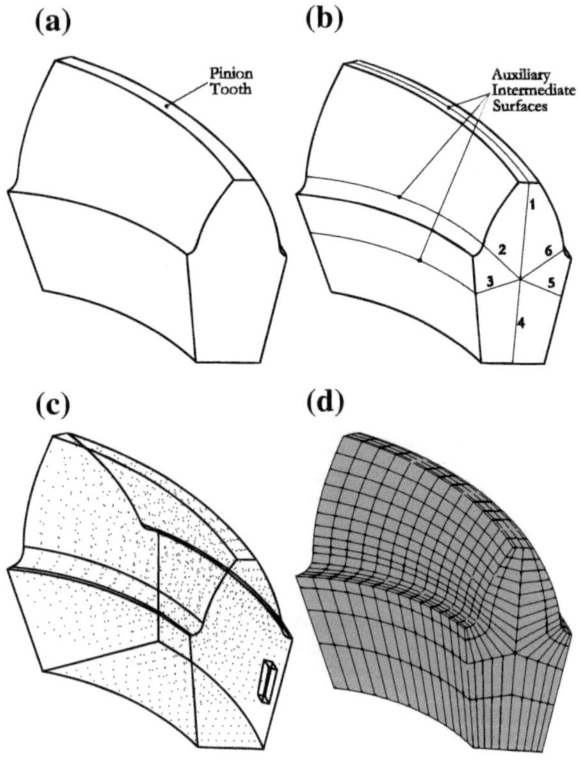

Fig. 2 Illustrations of **a** volume of the designed body, **b** auxiliary intermediate surfaces, **c** determination of nodes for the whole volume, and **d** discretization of the volume by finite elements

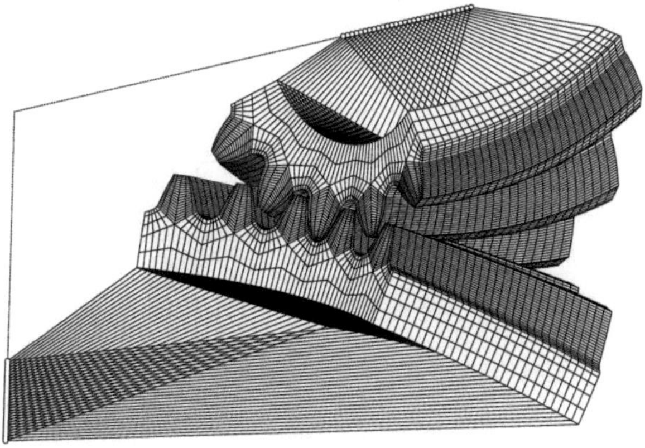

Fig. 3 Finite element mesh of a spiral bevel gear drive employed in FEA with definition of boundary conditions to take into account possible torsional deformation of the gear tooth surfaces

2. The model of one tooth is cloned on both sides in order to obtain the required number of contacting teeth of the gear drive. Figure 3 shows the mesh of a spiral bevel gear drive with five pair of contacting teeth. Finite element models of five pairs of contacting teeth represent a trade-off decision, since associated computational cost is considerably lower than that of analyzing a whole gear drive model and, at the same time, boundary conditions set on the sides and bottom part of the rim are kept far enough from tooth loaded areas in order to increase the accuracy of contact analysis.
3. Spiral bevel gear supporting shafts are modeled with linear three-dimensional beam elements. Employment of linear three-dimensional beam elements instead of three-dimensional solid elements in gear supporting shaft modeling implies lower computational cost with a similar degree of accuracy [13].
4. Boundary conditions at the supporting shaft nodes A_1, A_2, B_1 and B_2, where bearings are located, are schematized in the physical model depicted in Fig. 1. In this figure, black triangles represent restricted degrees of freedom, whereas white triangles represent unrestricted degrees of freedom; additionally, primary triangles denote lineal displacements and secondary triangles denote angular rotations. Regarding lineal displacement, displacements along axes x_1 and y_1 are fixed at node A_1, whereas displacements along axes x_2 and y_2 are fixed at node B_1. At the same time, all displacements are blocked at nodes A_2 and B_2. As far as angular rotation is concerned, rotations about axes x_1, y_1, x_2 and y_2 are allowed at all supporting shaft bearing locations in order to take into account shaft bending, whereas rotations about axes z_1 and z_2 are allowed at A_1, A_2 and B_1, respectively. Finally, a torque T is applied about pinion axis z_1 at node A_2, while rotation about z_2 is blocked at node B_2.
5. Boundary conditions between spiral bevel gear rims and the corresponding portions of their supporting shafts have been defined (see Fig. 3). Specifically, a group of independent rigid edges around the rims have been connected to independent reference nodes on the pinion and gear supporting shafts. The number of rigid edges and reference nodes coincides with the number of nodes

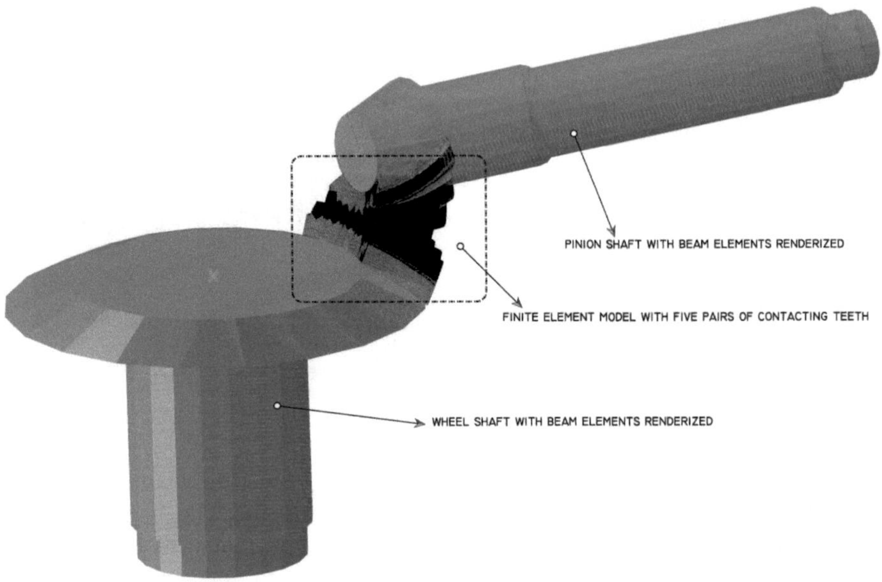

Fig. 4 Finite element model of a spiral bevel gear drive including gear supporting shafts with rendered beam elements according to shaft diameters in each section

of the spiral bevel gear model throughout the face width. The resulting boundary conditions allow for taking into account both shaft deflections and torsional deformation of gear tooth surfaces under torque transmission of the gear drive.

6. The beam elements defined on the portion of the supporting shaft under the spiral bevel gear rim are provided with circular sections which cover up to the bottom part of the rim, in order to model the whole body of the spiral bevel gear. Figure 4 schematizes the model of the supporting shaft sections and spiral bevel gear body sections for the purpose of considering torsional deformation of gear tooth surfaces accurately.

3 Analytical Determination of Relative Errors of Alignment in Spiral Bevel Gear Drives

In this section, an analytical procedure for determination of relative errors of alignment for spiral bevel gears will be described. Basically, this is based on mathematical concepts provided by analytic geometry. The different types of alignment errors which are considered in spiral bevel gear drives are: (i) ΔA_P as the axial displacement of the pinion, (ii) ΔA_W as the axial displacement of the wheel, (iii) $\Delta\Sigma$ as the change of the shaft angle Σ, and (iv) ΔE as the shortest distance between shafts. The above-mentioned alignment errors are illustrated, along with the nomenclature employed in this section, in Fig. 5.

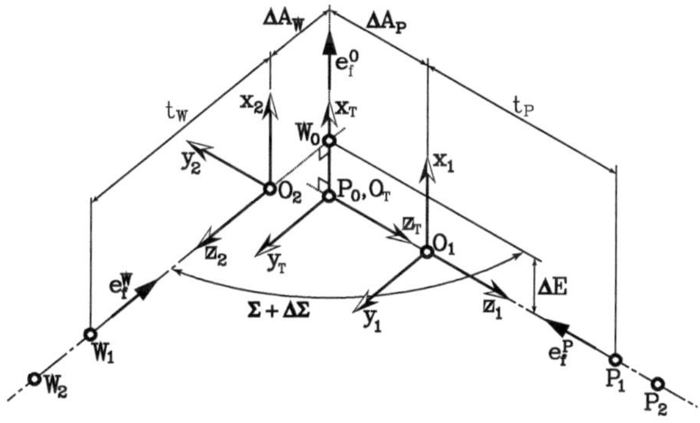

Fig. 5 Errors of alignment and coordinate system definition

3.1 Definition and Computation of Reference Points

Firstly, points P_1, P_2, W_1, and W_2, shown in Fig. 5, will be defined in a vectorial space $E \in \mathbb{R}^3$. Coordinate system S_T constitutes the fixed coordinate system for application of the TCA and local synthesis algorithms, the origin of which is located on the pinion axis of rotation and the direction of axis x_T oriented along the minimum distance between the axes of the pinion and the wheel when they are misaligned. Coordinate system S_f (not represented in Fig. 5) constitutes the fixed coordinate system rigidly connected to the housing and does not coincide with coordinate system S_T for the spiral bevel gear deformed position under load. Coordinate systems S_1 and S_2 represent fixed coordinate systems the origins of which are located on the pitch cone apices and axes z_1 and z_2 oriented along the pinion and wheel axes of rotation, respectively. Points P_1 and P_2 constitute the intersection points between the spiral bevel pinion axis of rotation and its corresponding front and back planes, respectively, whereas points W_1 and W_2 represent counterparts of points P_1 and P_2 for the spiral bevel wheel. Their position vectors, represented in the fixed coordinate system connected to the housing, S_f, for the shafts deformed position when transmitting nominal torque, which will be obtained by FEA, are given in Eqs. (1a), (1b), (2a), and (2b).

$$\mathbf{r}_f^{(P_1)} = \begin{bmatrix} x_{P_1} & y_{P_1} & z_{P_1} \end{bmatrix}^T \tag{1a}$$

$$\mathbf{r}_f^{(P_2)} = \begin{bmatrix} x_{P_2} & y_{P_2} & z_{P_2} \end{bmatrix}^T \tag{1b}$$

$$\mathbf{r}_f^{(W_1)} = \begin{bmatrix} x_{W_1} & y_{W_1} & z_{W_1} \end{bmatrix}^T \tag{2a}$$

$$\mathbf{r}_f^{(W_2)} = \begin{bmatrix} x_{W_2} & y_{W_2} & z_{W_2} \end{bmatrix}^T \tag{2b}$$

Next, unit vectors $\mathbf{e}_f^P$ and $\mathbf{e}_f^W$ of the supporting shafts of the spiral bevel pinion and wheel, respectively, in coordinate system S_f for their deformed position during nominal torque transmission are derived introducing Eqs. (1a), (1b), (2a), and (2b) into (3) and (4), respectively.

$$\mathbf{e}_f^P = \begin{bmatrix} x_{eP} \\ y_{eP} \\ z_{eP} \end{bmatrix} = \frac{\mathbf{r}_f^{(P_1)} - \mathbf{r}_f^{(P_2)}}{\left\| \mathbf{r}_f^{(P_1)} - \mathbf{r}_f^{(P_2)} \right\|} = \frac{1}{\sqrt{(x_{P_1} - x_{P_2})^2 + (y_{P_1} - y_{P_2})^2 + (z_{P_1} - z_{P_2})^2}} \begin{bmatrix} x_{P_1} - x_{P_2} \\ y_{P_1} - y_{P_2} \\ z_{P_1} - z_{P_2} \end{bmatrix} \quad (3)$$

$$\mathbf{e}_f^W = \begin{bmatrix} x_{eW} \\ y_{eW} \\ z_{eW} \end{bmatrix} = \frac{\mathbf{r}_f^{(W_1)} - \mathbf{r}_f^{(W_2)}}{\left\| \mathbf{r}_f^{(W_1)} - \mathbf{r}_f^{(W_2)} \right\|} = \frac{1}{\sqrt{(x_{W_1} - x_{W_2})^2 + (y_{W_1} - y_{W_2})^2 + (z_{W_1} - z_{W_2})^2}} \begin{bmatrix} x_{W_1} - x_{W_2} \\ y_{W_1} - y_{W_2} \\ z_{W_1} - z_{W_2} \end{bmatrix} \quad (4)$$

Taking into account the components of both position vectors $\mathbf{r}_f^{(P_1)}$ and $\mathbf{r}_f^{(W_1)}$, given by Eqs. (1a) and (2a), respectively, and unit vectors $\mathbf{e}_f^P$ and $\mathbf{e}_f^W$, given by Eqs. (3) and (4), respectively, it is possible to set out continuous equations corresponding to lines which represent the spiral bevel pinion and wheel supporting shafts in their deformed position, respectively. The aforementioned equations are shown in (5) and (6).

$$\frac{x - x_{P_1}}{x_{eP}} = \frac{y - y_{P_1}}{y_{eP}} = \frac{z - z_{P_1}}{z_{eP}} \quad (5)$$

$$\frac{x - x_{W_1}}{x_{eW}} = \frac{y - y_{W_1}}{y_{eW}} = \frac{z - z_{W_1}}{z_{eW}} \quad (6)$$

Additionally, points of intersection between the spiral bevel pinion and wheel supporting shafts and the perpendicular line common to both of them require definition, which will be denoted as P_0 and W_0, respectively. Their corresponding position vectors represented in coordinate system S_f are given by Eqs. (7) and (8).

$$\mathbf{r}_f^{(P_0)} = \begin{bmatrix} x_{P_0} & y_{P_0} & z_{P_0} \end{bmatrix}^T \quad (7)$$

$$\mathbf{r}_f^{(W_0)} = \begin{bmatrix} x_{W_0} & y_{W_0} & z_{W_0} \end{bmatrix}^T \quad (8)$$

Along these lines, unit direction vector $\mathbf{e}_f^0$ of perpendicular line common to the crossing supporting shafts of the spiral bevel pinion and the wheel in coordinate system S_f, respectively, can be derived taking into account components of unit direction vectors $\mathbf{e}_f^P$ and $\mathbf{e}_f^W$, given by Eqs. (3) and (4), respectively, and introducing them into Eq. (9). Vectors $\mathbf{i}_f$, $\mathbf{j}_f$ and $\mathbf{k}_f$ constitute the unit vectors of coordinate axes x_f, y_f and z_f, respectively.

$$\mathbf{e}_f^0 = \begin{bmatrix} x_{eo} \\ y_{eo} \\ z_{eo} \end{bmatrix} = \mathbf{e}_f^P \times \mathbf{e}_f^W = \begin{bmatrix} \mathbf{i}_f & \mathbf{j}_f & \mathbf{k}_f \\ x_{eP} & y_{eP} & z_{eP} \\ x_{eW} & y_{eW} & z_{eW} \end{bmatrix} = \begin{bmatrix} y_{eP}z_{eW} - y_{eW}z_{eP} \\ x_{eW}z_{eP} - x_{eP}z_{eW} \\ x_{eP}y_{eW} - x_{eW}y_{eP} \end{bmatrix} \tag{9}$$

In the manner of the previous case, considering the components of position vector $\mathbf{r}_f^{(P_0)}$, given by Eq. (7), and unit direction vector $\mathbf{e}_f^0$, given by Eq. (9), it is possible to derive the continuous equation corresponding to the aforementioned common perpendicular line, which is given by Eq. (10).

$$\frac{x - x_{P_0}}{x_{eo}} = \frac{y - y_{P_0}}{y_{eo}} = \frac{z - z_{P_0}}{z_{eo}} \tag{10}$$

Furthermore, the position vectors of the pitch cone apices corresponding to the spiral bevel pinion and wheel, which match origins O_1 and O_2 of coordinate systems S_1 and S_2, respectively, in their deformed position during power transmission are derived introducing Eqs. (1a), (2a), (3) and (4) into (11) and (12). Parameters t_P and t_W represent the distances from the spiral bevel pinion and wheel front planes to their corresponding pitch cone apices, respectively.

$$\mathbf{r}_f^{(O_1)} = \begin{bmatrix} x_{O_1} \\ y_{O_1} \\ z_{O_1} \end{bmatrix} = \mathbf{r}_f^{(P_1)} + t_P \mathbf{e}_f^P \tag{11}$$

$$\mathbf{r}_f^{(O_2)} = \begin{bmatrix} x_{O_2} \\ y_{O_2} \\ z_{O_2} \end{bmatrix} = \mathbf{r}_f^{(W_1)} + t_W \mathbf{e}_f^W \tag{12}$$

Finally, in order to derive the components of position vectors $\mathbf{r}_f^{(P_0)}$ and $\mathbf{r}_f^{(W_0)}$, given by Eqs. (7) and (8), it is required that we compute the intersection points between spiral bevel pinion and wheel crossing shafts, represented by Eqs. (5) and (6), respectively, and their corresponding common perpendicular line, given by Eq. (10). For that purpose, it is necessary to set out the system of linear equations given by Eq. (13).

$$\begin{gathered} \frac{x_{P_0} - x_{P_1}}{x_{eP}} = \frac{y_{P_0} - y_{P_1}}{y_{eP}} = \frac{z_{P_0} - z_{P_1}}{z_{eP}} \\ \frac{x_{W_0} - x_{W_1}}{x_{eW}} = \frac{y_{W_0} - y_{W_1}}{y_{eW}} = \frac{z_{W_0} - z_{W_1}}{z_{eW}} \\ \frac{x_{W_0} - x_{P_0}}{x_{eo}} = \frac{y_{W_0} - y_{P_0}}{y_{eo}} = \frac{z_{W_0} - z_{P_0}}{z_{eo}} \end{gathered} \tag{13}$$

The system of linear equations given by Eq. (13) can be expressed in matrix form through Eq. (14). The final solution of the aforementioned system of linear equations is given by (15).

$$\underbrace{\begin{pmatrix} y_{eP} & -x_{eP} & 0 & 0 & 0 & 0 \\ 0 & z_{eP} & -y_{eP} & 0 & 0 & 0 \\ 0 & 0 & 0 & y_{eW} & -x_{eW} & 0 \\ 0 & 0 & 0 & 0 & z_{eW} & -y_{eW} \\ -y_{e0} & x_{e0} & 0 & y_{e0} & -x_{e0} & 0 \\ 0 & -z_{e0} & y_{e0} & 0 & z_{e0} & -y_{e0} \end{pmatrix}}_{\mathrm{A}} \underbrace{\begin{pmatrix} x_{P_0} \\ y_{P_0} \\ z_{P_0} \\ x_{W_0} \\ y_{W_0} \\ z_{W_0} \end{pmatrix}}_{\mathrm{x}} = \underbrace{\begin{pmatrix} x_{P_1} y_{eP} - y_{P_1} x_{eP} \\ y_{P_1} z_{eP} - z_{P_1} y_{eP} \\ x_{W_1} y_{eW} - y_{W_1} x_{eW} \\ y_{W_1} z_{eW} - z_{W_1} y_{eW} \\ 0 \\ 0 \end{pmatrix}}_{\mathrm{b}} \tag{14}$$

$$\mathrm{x} = \mathrm{A}^{-1}\mathrm{b} \tag{15}$$

3.2 Computation of Relative Alignment Errors

Once the components of unit director vectors $\mathbf{e}_f^P$, $\mathbf{e}_f^W$, and $\mathbf{e}_f^0$, as well as position vectors $\mathbf{r}_f^{(P_0)}$ and $\mathbf{r}_f^{(W_0)}$, have been obtained, relative errors of alignment for a spiral bevel gear set can be derived.

Derivation of relative misalignment errors between spiral bevel gears is described below:

- Shortest distance between rotation axes of the spiral bevel pinion and wheel, ΔE:

$$\Delta E = \frac{\left|\left(\mathbf{r}_f^{(W_1)} - \mathbf{r}_f^{(P_1)}\right) \cdot \mathbf{e}_f^0\right|}{\left\|\mathbf{e}_f^0\right\|} = \frac{\left|(x_{W_1} - x_{P_1})x_{e0} + (y_{W_1} - y_{P_1})y_{e0} + (z_{W_1} - z_{P_1})z_{e0}\right|}{\sqrt{x_{e0}^2 + y_{e0}^2 + z_{e0}^2}} \tag{16}$$

- Change of the shaft angle, $\Delta\Sigma$:

$$\Delta\Sigma = \arcsin\left(\left\|\mathbf{e}_f^0\right\|\right) - \Sigma = \arcsin\left(\sqrt{x_{e0}^2 + y_{e0}^2 + z_{e0}^2}\right) - \Sigma \tag{17}$$

- Axial displacement of the spiral bevel pinion, ΔA_P:

$$\Delta A_P = \left\| \mathbf{r}_f^{(O_1)} - \mathbf{r}_f^{(P_0)} \right\| = \sqrt{(x_{O_1} - x_{P_0})^2 + (y_{O_1} - y_{P_0})^2 + (z_{O_1} - z_{P_0})^2} \quad (18)$$

- Axial displacement of the spiral bevel wheel, ΔA_W:

$$\Delta A_W = \left\| \mathbf{r}_f^{(O_2)} - \mathbf{r}_f^{(W_0)} \right\| = \sqrt{(x_{O_2} - x_{W_0})^2 + (y_{O_2} - y_{W_0})^2 + (z_{O_2} - z_{W_0})^2} \quad (19)$$

4 Brief Description of Computational Procedure for Compensation of Errors of Alignment

In order to obtain favorable conditions of meshing and contact of a gear drive under power transmission, a computational procedure of optimal design has been proposed. Basically, it is based on the following steps:

1. Application of FEA to the to-be-optimized spiral bevel gear set taking into account shaft geometry and bearing locations. As a result, and by application of the procedure described in the previous section, the relative errors of alignment between the pinion and wheel of the spiral bevel gear set are obtained.
2. Application of the local synthesis method, taking into account the obtained relative errors of alignment. The optimal microgeometry of the pinion tooth surfaces will be obtained.
3. Application of FEA to the spiral bevel gear set with the pinion designed in the previous step. The main purpose of this step is checking that the contact path remains inside the mating surface limits and both contact and bending stresses remain as uniform as possible throughout the cycle of meshing. The relative errors of alignment between pinion and wheel will be computed again, and Step 2 will be applied until the errors of alignment obtained by FEA are close enough to those considered in Step 2. Usually, several iterations between stages 2 and 3 have to be applied before the best design for the gear set is reached.

5 Numerical Example

In this section, an example of design of a face-milled generated spiral bevel gear drive is considered to illustrate the proposed procedure of determination of relative errors of alignment in spiral bevel gear drives and their compensation through

Table 1 Basic geometrical design data of a face-milled spiral bevel gear drive

Design feature	Pinion	Wheel
Number of teeth	16	40
Outer transverse module (mm)	4.9485	
Shaft angle, Σ (°)	90.0000	
Mean spiral angle (°)	35.0000	
Hand of spiral	LH	RH
Rotation direction	CCW	CW
Face width (mm)	31.0000	
Outer whole depth (mm)	8.9868	8.9868
Outer addendum (mm)	6.0389	2.0820
Outer dedendum (mm)	2.9480	6.9049
Outer working depth (mm)	8.1208	8.1208
Pitch angle (°)	21.8014	68.1986
Root cone angle (°)	20.6720	64.9232
Face cone angle (°)	25.0768	69.3280

changes in the microgeometry of the gear tooth surfaces. The basic geometrical design parameters of the spiral bevel gear drive studied in this section are represented in Table 1. Geometric parameters (diameters and lengths) of pinion and wheel supporting shafts with definition of the bearing location are shown in Fig. 6. Table 2 represents the machine-tool settings for pinion and wheel generation of the to-be-optimized spiral bevel gear set, together with those obtained for the optimized design proposed below. The machine-tool settings for the existing geometry are considered as known, whereas the machine-tool settings for the optimized geometry

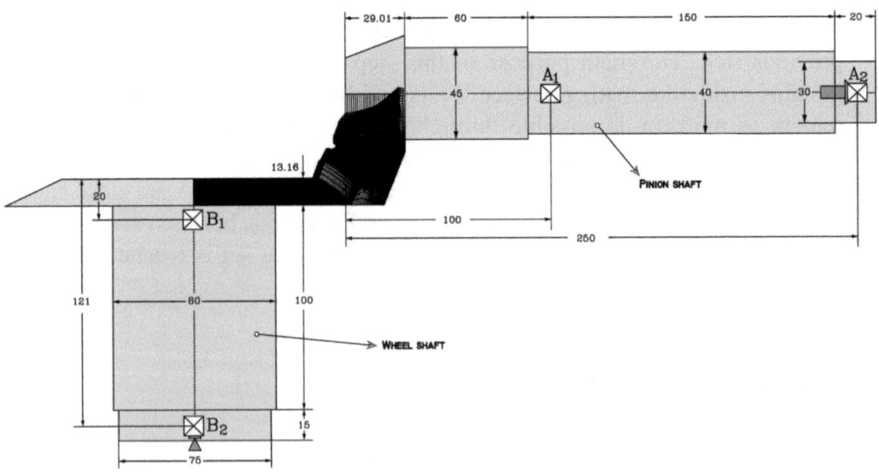

Fig. 6 Pinion and wheel supporting shaft dimensions and bearing location. Dimensions are given in mm

Table 2 Basic machine-tool settings and cutter geometry for generation of the spiral bevel gears for existing and optimized gear set

	Existing geo.			Optimized geo.		
	Pinion		Wheel	Pinion		Wheel
Tooth side	Cv	Cc	Both	Cv	Cc	Both
Mean cutter radius (mm)	N/A	N/A	76.2000	N/A	N/A	76.200
Cutter tip point radius (mm)	80.1924	71.0985	N/A	79.1771	72.0127	N/A
Point width (mm)	N/A	N/A	2.5000	N/A	N/A	2.5000
Blade pressure angle (°)	20.0000	20.0000	20.0000	20.0000	20.0000	20.0000
Blade edge radius (mm)	1.1906	1.1906	1.2371	1.2370	1.2370	1.2371
Machine center to back (mm)	2.4583	−2.5152	0.0000	5.9679	−4.1252	0.0000
Sliding base (mm)	−1.7142	0.0415	−0.8034	−2.9532	0.6099	−0.8034
Blank offset (mm)	−4.2722	2.1799	0.0000	−14.5996	8.0699	0.0000
Radial distance (mm)	85.3929	72.4535	78.2788	94.6080	67.5556	78.2788
Cradle angle (°)	51.3312	52.3630	52.8824	52.9458	51.1479	52.8824
Machine root angle (°)	20.6720	20.6720	64.9232	20.6720	20.6720	64.9232
Velocity ratio	2.9017	2.5071	1.0753	3.1877	2.3477	1.0753
Modified roll Coef. C	0.0155	−0.0159	0.0000	0.0261	−0.0233	0.0000
Modified roll Coef. D	−0.0473	0.0367	0.0000	−0.1386	0.0721	0.0000

are computed by using the expected errors of alignment and the local synthesis method.

In this work, ABAQUS® software [4] has been used for finite element analysis of the spiral bevel gear set. Gear teeth and rim volumes are meshed using three-dimensional solid element type C3D8I. Basically, type C3D8I represents hexahedral first order elements enhanced by incompatible deformation modes in order to improve their bending behavior. Supporting shafts are meshed with linear three-dimensional beam element type B31. Basically, type B31 is based on Timoshenko beam theory, and, consequently, transverse shear deformation is allowed. Spiral bevel wheel active surfaces are considered master surfaces, whereas those corresponding to the spiral bevel pinion are considered slave surfaces. Steel has been considered as material for the whole spiral bevel gear drive, including supporting shafts, with general properties of Young's module of 210 GPa and Poisson's ratio 0.30. A torque of 500 Nm has been applied to the spiral bevel pinion supporting shaft end where bearing A_2 is located (see Fig. 1) and two cycles of meshing are always studied in order to check out the repetitiveness of the obtained results, which means that boundary conditions are far enough and do not influence the results of finite element analysis.

Figure 7 shows the comparison of the evolution of maximum Mises stresses on the pinion and wheel active tooth surfaces of the considered spiral bevel gear drive under a torque of 500 Nm when (i) no shafts are considered and (ii) the pinion and wheel supporting shafts are considered. The influence of shaft deflections on the evolution of maximum Mises stresses in both pinion and wheel members of the

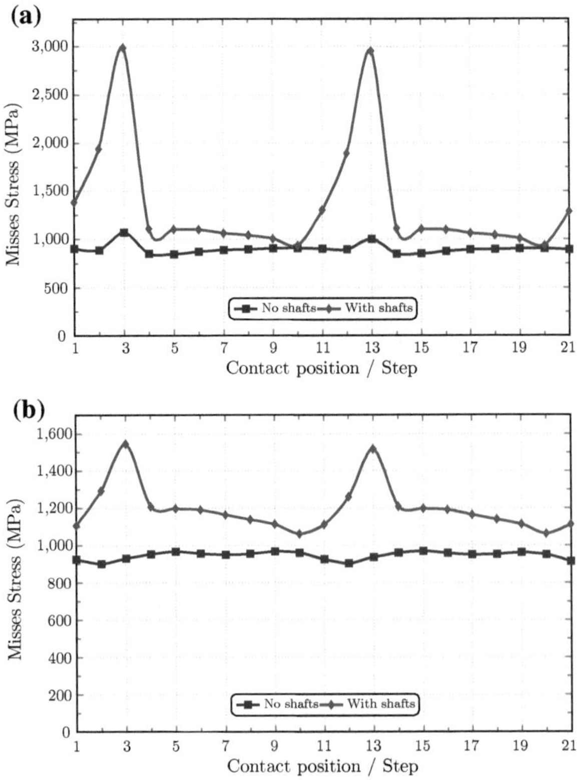

Fig. 7 Comparison of the evolution of maximum Mises stresses on **a** pinion, and **b** wheel of the given spiral bevel gear set when shafts are not considered versus the case where shafts are considered

spiral bevel gear set is very important, yielding severe edge contacts that may lead to a premature failure of the gear drive in the case where geometry of shafts will allow for flexible behavior due to their length or unfavorable bearing positions. In this analysis, the pinion convex side of the pinion tooth surfaces has been considered as driving and the concave side of the wheel tooth surfaces as driven.

Figure 8 shows the evolution of maximum Mises stresses on the pinion and wheel active tooth surfaces of the considered spiral bevel gear drive considering supporting shaft geometry for a torque applied to the pinion shaft end varying from 100 to 500 Nm. As can be expected, the higher the applied torque, the higher the maximum Mises stresses obtained. It can be observed that, for a torque between 200 and 300 Nm, edge contacts appear yielding a high increment of Mises stresses for some contact positions along the cycle of meshing.

Figure 9 shows the contact patterns and function of transmission errors when considering the relative errors of alignment caused by shaft deflections for different torques applied to the pinion shaft varying between 0 Nm (perfectly aligned gear drive) and 500 Nm. Those results reveal that, firstly, under no load, a longitudinally-oriented bearing contact and a predesigned parabolic function of transmission errors of 8 arcsec of maximum level are indeed obtained, and

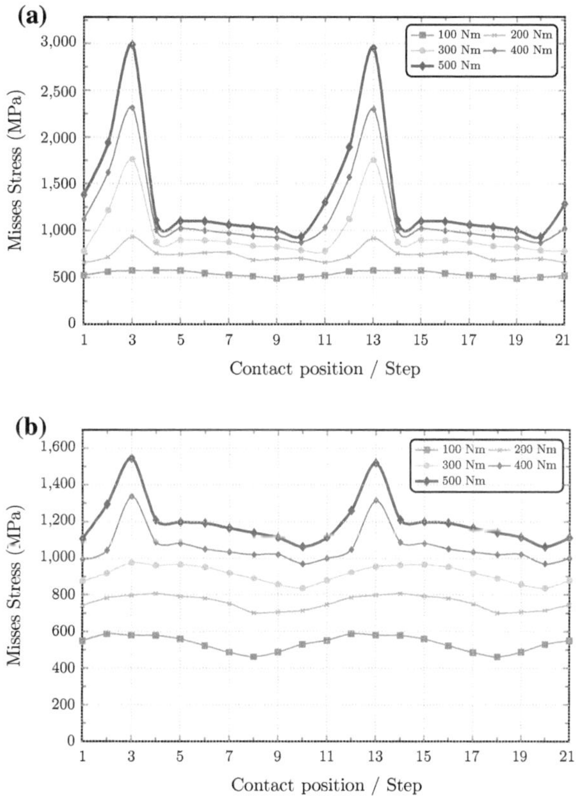

Fig. 8 Evolution of maximum Mises stresses on **a** pinion, and **b** wheel of the given spiral bevel gear set for a torque variation from 100 to 500 Nm applied to the pinion shaft

secondly, as the transmitted torque elevates, the contact pattern is shifted towards the heel of the pinion and wheel tooth surfaces and at the same time the maximum level of transmission errors is increased.

Table 3 shows the evolution of relative errors of alignment for each iteration of design that has been compensated through redesign of, in each case, the pinion member of the spiral bevel gear set. A longitudinally oriented path of contact is always considered.

Figure 10 shows the contact patterns and function of transmission errors when relative errors of alignment obtained in the fourth iteration of design have been compensated by pinion microgeometry redesign through application of the local synthesis method. Now, for low torque transmission, the contact patterns are shifted towards the toe edges of the active gear tooth surfaces and function of transmission errors yield high maximum values. However, when torque is increasing to its nominal value, contact patterns are shifted towards the center of the active pinion and wheel tooth surfaces and the function of transmission errors changes until it reaches its predesigned shape. In this case, the spiral bevel gear set will have an optimum performance at the nominal torque transmission yielding the best achievable contact pattern with low levels of noise and vibration.

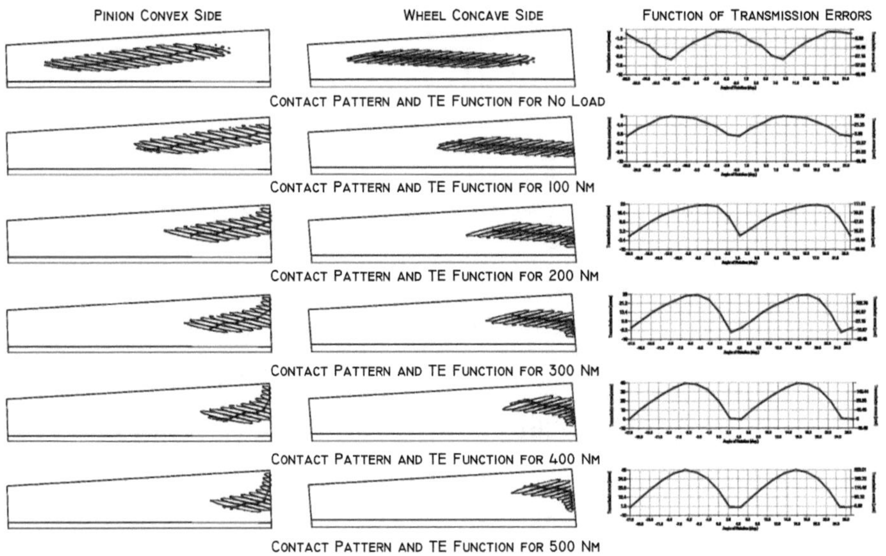

Fig. 9 Contact patterns and functions of transmission errors when the applied torque to the pinion varies from 0 to 500 Nm for the given spiral bevel gear set

Table 3 Evolution of relative errors of alignment compensated in each iteration of design

	Iteration 1	Iteration 2	Iteration 3	Iteration 4
Shortest distance between axes, ΔE (mm)	0.602069	0.705698	0.734774	0.743196
Axial displacement of the pinion, ΔA_P (mm)	−0.010494	−0.010158	−0.010108	−0.010093
Axial displacement of the wheel, ΔA_W (mm)	0.613510	0.630203	0.635877	0.637494
Change of the shaft angle, $\Delta\Sigma$ (°)	−0.205626	−0.212362	−0.214584	−0.215228

Figure 11 shows the evolution of maximum Mises stresses on the pinion and wheel active tooth surfaces of the optimized spiral bevel gear drive for a torque applied to the pinion shaft end varying from 100 to 500 Nm. Now, the higher the applied torque, the lower the maximum Mises stresses obtained, reaching a lower, smooth evolution for the nominal torque of 500 Nm.

Finally, Fig. 12 shows the contact stresses for the pinion of the given spiral bevel gear set for one position of meshing when shaft deflection is considered and for the pinion of the optimized spiral bevel gear drive when errors of alignment due to shaft deflections have been compensated by appropriate design of the pinion microgeometry. Figure 13 shows the contact stresses on the wheel for both the given and the redesigned geometry of the pinion of the spiral bevel gear set.

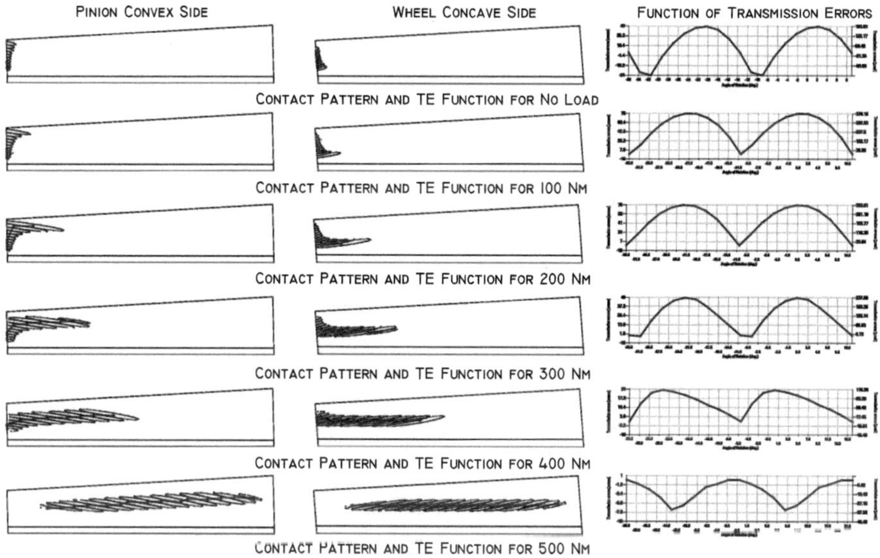

Fig. 10 Contact patterns and function of transmission errors when the applied torque to the pinion varies from 0 to 500 Nm for the optimized geometry

Fig. 11 Evolution of maximum Mises stresses on **a** pinion, and **b** wheel of the optimized spiral bevel gear set for a torque variation from 100 to 500 Nm applied to the pinion shaft

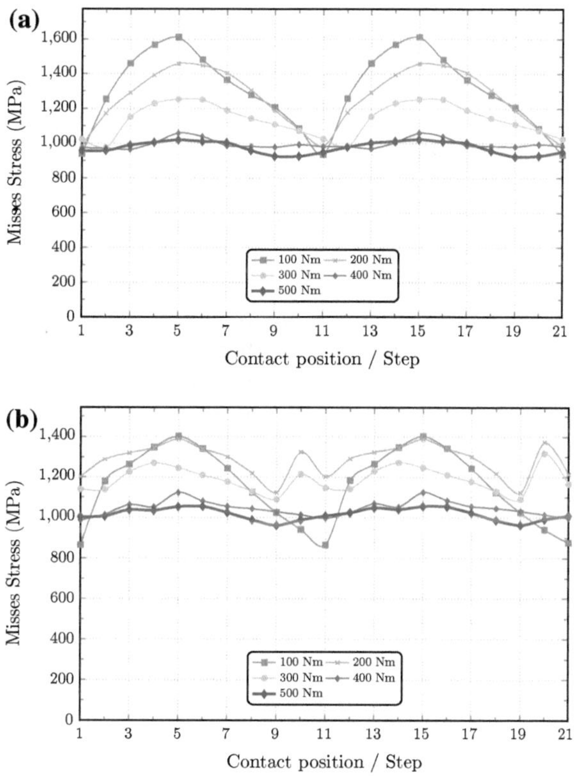

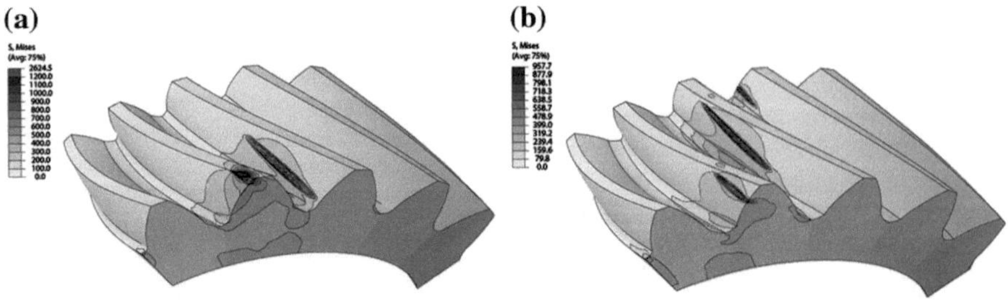

Fig. 12 Contact stresses on the pinion tooth surfaces for **a** existing design, and **b** optimized design with consideration of shaft deflections

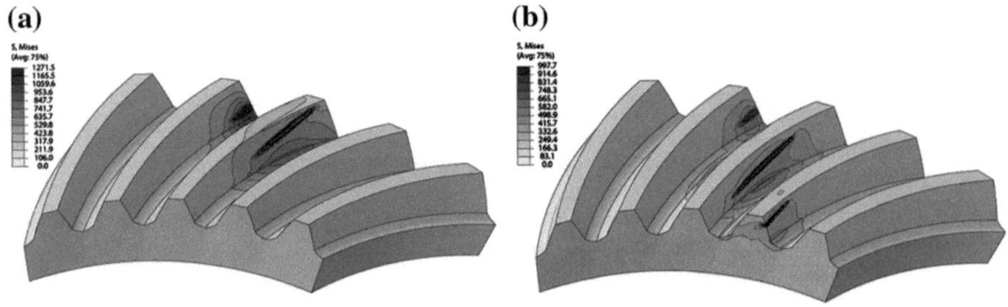

Fig. 13 Contact stresses on the wheel tooth surfaces for **a** existing design, and **b** optimized design with consideration of shaft deflections

6 Conclusions

An analytical procedure of determination of errors of alignment between spiral bevel gears has been proposed. It has been integrated as part of a design approach for spiral bevel gear drives based on the local synthesis method. The numerical results derived from the proposed procedure, together with the application of the finite element analysis, have been introduced as initial data in the local synthesis method, in order to redesign the pinion microgeometry to compensate supporting shaft deflections in the design procedure of spiral bevel gear drives. The effectiveness of the proposed approach has been demonstrated through a numerical example.

Acknowledgments The authors express their deep gratitude to the Spanish Ministry of Economy and Competitiveness (MINECO), for the financial support of research projects refs. DPI2013-47702-C2-1-P and IPT-2011-0711-020000 (financed jointly by FEDER funds), and for the FPI scholarship ref. BES-2011-045995.

References

1. American Gear Manufacturers Association: ANSI/AGMA ISO 23509-A08 (2008) Bevel and hypoid gear geometry
2. Argyris J, Fuentes A, Litvin FL (2002) Computerized integrated approach for design and stress analysis of spiral bevel gears. Comput Methods Appl Mech Eng 191(11–12):1057–1095
3. Chen YC, Liu CC (2011) Contact stress analysis of concave conical involute gear pairs with non-parallel axes. Finite Elem Anal Des 47(4):443–452
4. Dassault Systemes Simulia Corporation, Providence, Rhode Island 02909-2499 (US) (2010) ABAQUS/Standard User's Manual
5. Fan Q, Wilcox L (2011) New developments in tooth contact analysis (TCA) and loaded TCA for spiral bevel and hypoid gear drives. Gear Technol 24(3):26–35
6. Fuentes A, Litvin FL, Mullins BR, Woods R, Handschuh RF (2002) Design and stress analysis of low-noise adjusted bearing contact spiral bevel gears. J Mech Des, Trans ASME 124 (3):524–532
7. Gonzalez-Perez I, Roda-Casanova V, Fuentes A, Sanchez-Marin FT, Iserte JL (2012) A finite element model for consideration of the torsional effect on the bearing contact of gear drives. J Mech Des, Trans ASME 134(7):071007
8. Kolivand M, Kahraman A (2009) A load distribution model for hypoid gears using ease-off topography and shell theory. Mech Mach Theory 44(10):1848–1865
9. Li S (2007) Finite element analyses for contact strength and bending strength of a pair of spur gears with machining errors, assembly errors and tooth modifications. Mech Mach Theory 42 (1):88–114
10. Litvin FL, Fuentes A (2004) Gear geometry and applied theory, 2nd edn. Cambridge University Press, New York
11. Litvin FL, Fuentes A, Hayasaka K (2006) Design, manufacture, stress analysis and experimental tests of low-noise high endurance spiral bevel gears. Mech Mach Theory 41 (1):83–118
12. Mao K (2007) Gear tooth contact analysis and its application in the reduction of fatigue wear. Wear 262(11–12):1281–1288
13. Roda-Casanova V, Iserte-Vilar JL, Sanchez-Marin F, Fuentes-Aznar A, Gonzalez-Perez I (2011) Development and comparison of shaft-gear models for the computation of gear misalignments due to power transmission. In: Proceedings of the ASME design engineering technical conference, vol. 8, pp 279–287
14. Thomas J (1990) Gear design, manufacturing and inspection manual, Chap. 9. 15. society of automotive engineers (SAE)

Ease-Off and Application in Tooth Contact Analysis for Face-Milled and Face-Hobbed Spiral Bevel and Hypoid Gears

Q. Fan

Abstract The ease-off concept was introduced to describe the mismatch, a deviation between conjugation and non-conjugation, between two tooth surfaces from a pair of mating bevel gears in contact. However, a complete mathematical description of ease-off and the algorithm of computation were not found. The application of ease-off concept in tooth contact analysis (TCA) allows for a numerical determination of contact patterns and transmission errors of highly conformable contact or almost conjugate contact of tooth surfaces. The paper analytically describes a generalized theory of ease-off and its application in tooth contact analysis of both face-milled and face-hobbed spiral bevel and hypoid gears with complex tooth surface modifications. The implementation of the ease-off algorithm applied to the tooth contact analysis is illustrated with two examples of, respectively, a face-milled gear drive and a face-hobbed hypoid gear drive.

Keywords Ease-off · Tooth contact analysis (TCA) · Face-mill · Face-hob · Hypoid gears

1 Introduction

Spiral bevel and hypoid gears are widely used in vehicles, aircrafts, and industrial gearboxes, transmitting rotation between intersecting or crossing axes.

During the past several decades, advanced computerized methods of design and manufacture of spiral bevel and hypoid gears have been comprehensively presented by certain gear scientists. Baxter [1, 2] developed the vector method of describing the tooth surface geometries of hypoid gears and the theory of second order surfaces and its application in the synthesis of the contact characteristics of hypoid gears, which provided theoretical foundations for Gleason computerized bevel gear design methodology. Litvin [12, 13] systematically presented methods on computerized

Q. Fan (✉)
The Gleason Works, Rochester, New York, USA
e-mail: qfan@gleason.com

V. Goldfarb and N. Barmina (eds.), *Theory and Practice of Gearing and Transmissions*, Mechanisms and Machine Science 34,
DOI 10.1007/978-3-319-19740-1_15

analysis and synthesis of spiral bevel and hypoid gears. The local synthesis method integrated with tooth contact analysis was developed to optimize the contact patterns and transmission errors. Krenzer [10, 11] developed the corrective machine setting technique for bevel gears and other computerized methods for bevel gear design. Stadtfeld [16–21] developed a series of advanced bevel gear optimization design methods and processes such as the Universal Motion Concept (UMC) and its application in the synthesis of higher order crowning of tooth surfaces for hypoid gears that provide optimized contact characteristics and advanced tooling systems, as well as processes for manufacturing different types of gears on CNC hypoid gear machines. Fan [3–7] presented some advanced developments in computerized modeling and simulation of meshing, as well as synthesis of spiral bevel and hypoid gears. Computerized simulation and modeling of hypoid gears, including modeling of tooth surface generation, simulation of meshing, tooth contact analysis, stress analysis, and tooth contact analysis under load, have been the subjects of many other bevel gear researchers [8, 9, 14, 15].

The ease-off concept was introduced to describe the mismatch, a deviation between conjugation and non-conjugation due to tooth surface modification, between two tooth surfaces from a pair of mating bevel gears in contact. Ease-off analysis and tooth contact analysis are normally major modules of software packages for hypoid gear design. Ease-off topography can be used as a visual tool for analysis and synthesis of contact characteristics. Stadtfeld [16, 17] developed the tooth contact optimization method and the higher order tooth surface modification method called the Universal Motion Concept (UMC) by using ease-off topography. Fan [7] developed a method of optimizing the face cone element for spiral bevel and hypoid gears using an ease-off method. This paper analytically defines and describes a generalized theory of ease-off and its application in the development of an advanced TCA program for both face-milled and face-hobbed spiral bevel and hypoid gears with multiple features of tooth surface modifications.

2 Face Milling and Face Hobbing

In the gear industry, there are two categories of manufacturing processes for spiral bevel and hypoid gears, namely face milling and face hobbing. The major differences between face milling and face hobbing are: (1) in face hobbing, a timed continuous indexing is provided, while in face milling, the indexing is intermittently provided after cutting each tooth side or slot, which is also called single indexing. As shown in Fig. 1, in face milling, the inside and outside blades cut into the same slot at one time, while in face hobbing, the inside and outside blades cut into neighboring slots continuously; (2) the lengthwise tooth curve of face milled bevel gears is a circular arc with a curvature radius equal to the cutter radius, while the lengthwise tooth curve of face hobbed gears is a kinematic curve called the extended epicycloid, which is generated by the continuous indexing, a relative rolling motion between the cutter and the generating gear; (3) face hobbing gear

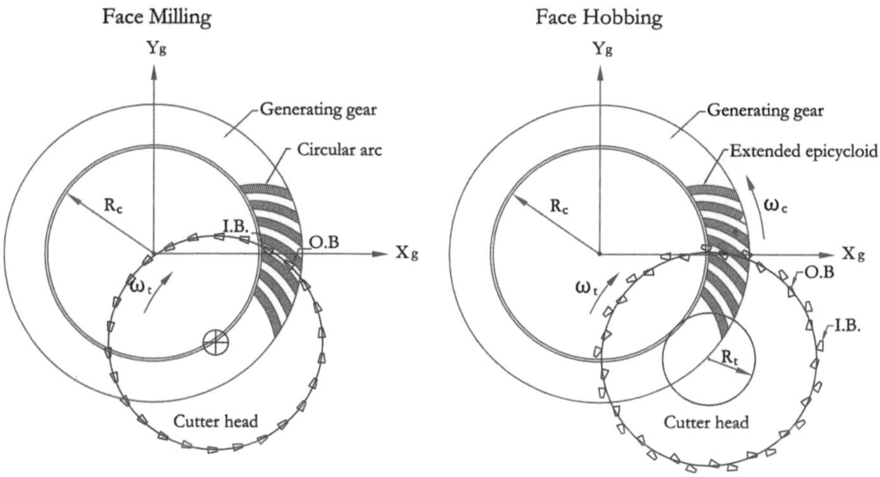

Fig. 1 Face milling and face hobbing

designs use the uniform tooth depth system, while most face milling gear designs use tapered tooth systems; (4) due to the continuous indexing, the grinding process cannot be applied to face hobbing. Therefore, for carburized and hardened gears for which the grinding process is needed for a hard finish, face milling has to be employed.

The pinion member of both face milling and face hobbing is always cut with the generated method through which the relative rolling motion between the generating gear represented by the cradle and pinion work is provided. However, the gear member can be cut with either the generated method or the non-generated (Formate®) method in which the relative rolling motion between the cradle and gear work is not provided, namely, the cradle is held at rest. The Formate® method offers higher productivity than the generated method because the generating roll is not applied in the Formate® method. However, the generated method offers more freedom for controlling the tooth surface geometries. That is why the pinion surface geometry is normally adjusted and optimized to match the gear member during the development process.

3 Definition of Ease-Off

Ease-off is the deviation from conjugation between two tooth surfaces from a pair of mating members in contact. If we consider a pair of conjugated tooth surfaces, the instant contact between the two surfaces is in line contact. The tooth surfaces of conjugated members are envelopes to each other. For instance, involute cylindrical gears are conjugated members if tooth surfaces are not modified or crowned.

However, spiral bevel and hypoid gears normally mesh in point contact due to the manufacturing process and the pre-designed modification of the tooth surfaces. The meshing of tooth surfaces in point contact in any instant is called non-conjugation. Conjugated tooth surfaces are sensitive to the errors of alignment, which may cause edge contact and concentrated contact stresses at the contact point.

Mathematically, we consider a pair of mating gear members, a pinion and a gear, whose real tooth surfaces are represented by their respective position vectors and unit normals as,

$$\begin{cases} \mathbf{r}_1 = \mathbf{r}_1(u_1, \theta_1) \\ \mathbf{n}_1 = \mathbf{n}_1(u_1, \theta_1) \end{cases}, \tag{1}$$

$$\begin{cases} \mathbf{r}_2 = \mathbf{r}_2(u_2, \theta_2) \\ \mathbf{n}_2 = \mathbf{n}_2(u_2, \theta_2) \end{cases} \tag{2}$$

where position vector $\mathbf{r}_1$, $\mathbf{n}_1$, $\mathbf{r}_2$, and $\mathbf{n}_2$ are represented in the coordinate system S_1 and S_2 that are rigidly connected to the pinion member and the gear member, respectively; u_1, θ_1, u_2 and θ_2 are associated pinion and gear tooth surface parameters. If one of the mating members, say the pinion, is given as a generating member to generate the gear member under the given gear ratio defined by their numbers of teeth, then the virtual gear tooth surface that is conjugate to the pinion can be determined as,

$$\begin{cases} \mathbf{r}_{2c} = \mathbf{M}_{21}(\phi_1)\mathbf{r}_1(u_1, \theta_1) \\ \mathbf{n}_{2c} = \mathbf{L}_{21}(\phi_1)\mathbf{n}_1(u_1, \theta_1) \\ f_{1c}(u_1, \theta_1, \phi_1) = \frac{\partial \mathbf{r}_{2c}}{\partial \phi_1} \cdot \mathbf{n}_{2c} = 0 \end{cases}. \tag{3}$$

Here, $\mathbf{r}_{2c}$ and $\mathbf{n}_{2c}$ are the position vector and the unit normal vector of the conjugate gear member and are represented in the coordinate system S_2; f_{1c} is the equation of meshing; $\mathbf{M}_{21}$ and $\mathbf{L}_{21}$ are the matrices transforming coordinates for the position vector and the unit normal from system S_1 to S_2; and ϕ_1 is the rotation angle of the pinion from its initial generating position. We notice that the conjugate gear tooth surface is related only to the pinion geometry and the relative motion of the mating members. Given ϕ_1 an instantaneous rotation angle, Eq. 3 represents a space curve, which is called the instantaneous generating line. The conjugate gear tooth surface is the envelope of the family of instantaneous generating lines represented in the coordinate system connected to the gear member. The instantaneous generating lines are also called potential contact lines in tooth contact analysis, since the real contact points are located on the potential contact lines. The contact pattern on the gear tooth surface can be determined using the potential contact lines represented in the gear coordinate system S_2.

Vice versa, if we take the gear member as the generating member and the pinion as the generated member under the same gear ratio, then the virtual pinion tooth surface that is conjugate to the gear can be determined as,

$$\begin{cases} \mathbf{r}_{1c} = \mathbf{M}_{12}(\phi_2)\mathbf{r}_2(u_2, \theta_2) \\ \mathbf{n}_{1c} = \mathbf{L}_{12}(\phi_2)\mathbf{n}_2(u_2, \theta_2) \\ f_{2c}(u_2, \theta_2, \phi_2) = \frac{\partial \mathbf{r}_{1c}}{\partial \phi_2} \cdot \mathbf{n}_{1c} = 0 \end{cases} . \tag{4}$$

Here, $\mathbf{r}_{1c}$ and $\mathbf{n}_{1c}$ are the position vector and the unit normal vector of the conjugate pinion member and are represented in the coordinate system S_1; f_{2c} is the equation of meshing; $\mathbf{M}_{12}$ and $\mathbf{L}_{12}$ are the matrices transforming coordinates for the position vector and the unit normal from system S_2 to S_1; and ϕ_2 is the rotation angle of the gear from its initial generating position. The conjugate pinion tooth surface is related only to the gear geometry and the relative motion of the mating members. In the same manner, given ϕ_2 as an instantaneous rotation angle, an instantaneous generating line is obtained. The conjugate pinion tooth surface is the envelope of the instantaneous generating lines represented in the coordinate system connected to the pinion member. The potential contact lines can also be represented in the pinion coordinate system S_1 for determination of the contact pattern on the pinion tooth surface. The relation of motion parameters ϕ_1 and ϕ_2 are related by the numbers of teeth N_1 of the pinion and N_2 of the gear as,

$$\frac{\phi_2}{\phi_1} = \frac{N_1}{N_2}. \tag{5}$$

Ease-off is defined as a deviation between the conjugate tooth surface and the real tooth surface or modified tooth surface in the direction of tooth surface normal, a difference between conjugation and non-conjugation. Ease-off is also called mismatch by some gear researchers. Analytically, it can be defined as,

$$\begin{cases} \Delta_1 = [\mathbf{r}_1(u_1, \theta_1) - \mathbf{r}_{1c}(u_2, \theta_2, \phi_2)] \cdot \mathbf{n}_1(u_1, \theta_1) \\ f_{1c}(u_2, \theta_2, \phi_2) = 0 \end{cases} \tag{6}$$

or,

$$\begin{cases} \Delta_2 = [\mathbf{r}_2(u_2, \theta_2) - \mathbf{r}_{2c}(u_1, \theta_1, \phi_1)] \cdot \mathbf{n}_2(u_2, \theta_2) \\ f_{2c}(u_1, \theta_1, \phi_1) = 0 \end{cases} . \tag{7}$$

Since ease-off Δ_1 and Δ_2 describe the difference between conjugation and non-conjugation of a pair of mating pinion and gear, the ease-off Δ_1 relative to the pinion member and the ease-off Δ_2 relative to the gear member are the same, namely, $\Delta_1 = \Delta_2 = \Delta$.

4 Ease-Off Topography

Ease-off can be numerically calculated using discretization of tooth flanks with grid points and presented as topographic elevation for visualization. For bevel gears, a grid of n rows and m columns is defined in the L-R axial plane of a bevel gear, as shown in Fig. 2. The grid should be extended to the flank boundary defined by the blank data and follow the root line in order to cover the whole tooth surface, including the fillet surface. Given any grid point by L and R values, the following system of nonlinear equations can be used to solve for the surface parameters u_i, θ_i, $\mathbf{r}_i$, and $\mathbf{n}_i$ subscript $i = 1, 2$,

$$\begin{cases} \sqrt{x_i^2(u_i,\theta_i) + y_i^2(u_i,\theta_i)} = R \\ z_i(u_i,\theta_i) = L \end{cases}. \tag{8}$$

Subsequently, using Eqs. 1–7, we can numerically calculate the ease-off Δ at each grid point. Ease-off topography can also be presented by discretized potential contact lines. Ease-off Δ is calculated at each point on the discretized potential contact lines. Figure 3 shows the ease-off topography represented by a regular grid (right side) and the potential contact lines (left side), respectively. For a conjugate gear pair, such as a pair of involute cylindrical gears without modification, the ease-off Δ is zero at all points and the ease-off topography is a "flat" surface on the base plane.

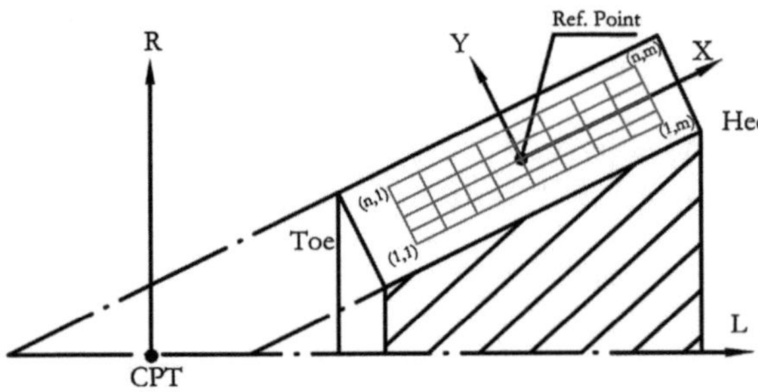

Fig. 2 Definition of a tooth flank grid

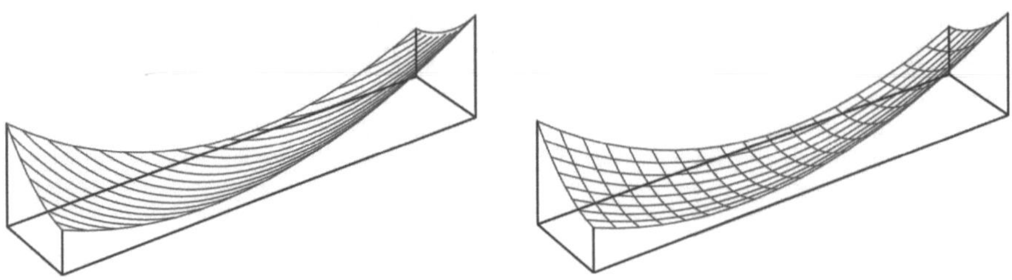

Fig. 3 Ease-off topography

5 Tooth Contact Analyses Using Ease-Off Concept

The ease-off concept can be used to analyze tooth contact characteristics, determining the contact patterns and transmission errors. An advanced version of the TCA program has been developed for the spiral bevel and hypoid gear drives, which consists of the following features and steps:

Step 1: Exact definition of the tool geometry by considering complex cutting edge modification elements such as Toprem and Flankrem, as shown in Fig. 4. Toprem is used either to relieve the tooth's lower dedendum surface right above the fillet line in order to avoid edge contact or interference or to prepare the root surfaces for hard finish process. Flankrem is used for either tooth tip relief or tooth tip chamfering. Flankrem and Toprem can be either straight or blended. Both elements can be either intersected or tangent to the main active profile part; when profiles are in tangency at connections, they are called "blended". A combination of different geometric tool features is considered and shown in Table 1.

Stadtfeld [18, 21] developed an advanced multi-face sharpening technology accompanied by the advanced cutter system for bevel gears. To consolidate these new tooling developments, the effective tool geometry is generally defined by pressure angle α, point radius R_b, rake angle λ, hook angle κ, and blade offset angle δ. We consider that the tooth surface of the generating gear is formed by the trace of a cutting edge of the tool whose geometry can be generally defined by the position vector and the unit tangent represented in the cutter head coordinate system S_t as,

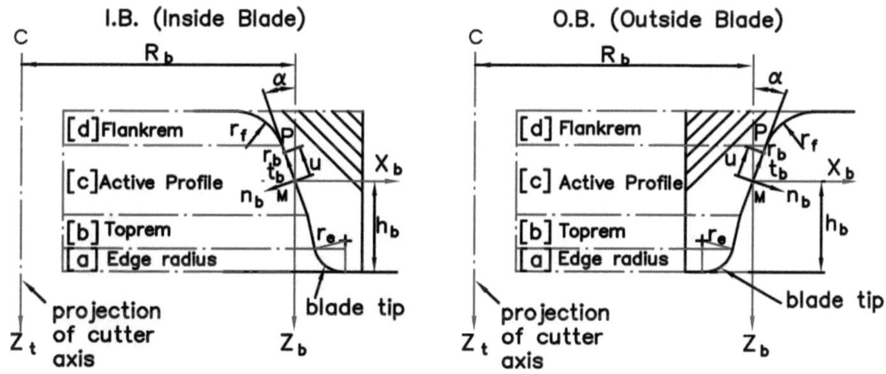

Fig. 4 Definition of tool geometry

Table 1 Geometric tool features

	Toprem		Flankrem	
	Straight	Curved	Straight	Curved
Straight active profile	Intersected	Blended	Intersected	Blended
Curved active profile	Intersected	Blended	Intersected	Blended

$$\mathbf{r}_t = \mathbf{M}_{tb}(\delta, \lambda, \kappa, R_b, \alpha)\mathbf{r}_b(u), \tag{9}$$

$$\mathbf{n}_t = \mathbf{M}_{tb}(\delta, \lambda, \kappa, R_b, \alpha)\mathbf{n}_b(u), \tag{10}$$

$$\mathbf{t}_t = \mathbf{M}_{tb}(\delta, \lambda, \kappa, R_b, \alpha)\mathbf{t}_b(u). \tag{11}$$

Here, position vector $\mathbf{r}_b(u)$, unit normal vector $\mathbf{n}_b(u)$ and unit tangent vector $\mathbf{t}_b(u)$ define any point on the blade edge with unified parameter u. Matrix $\mathbf{M}_{tb}$ represents the coordinate transformation from blade edge coordinate system S_b to system S_t that is connected to the cutter with origin at the machine center plane.

Step 2: Using the universal tooth surface generation model for modeling of the bevel gear generation process [3], we can represent the machine setting parameters in polynomials up to 6th orders as,

$$\begin{cases} R_a = R_{a0} + R_{a1}q + \ldots + R_{a4}q^4 + R_{a5}q^5 + R_{a6}q^6 \\ X_b = X_{b0} + X_{b1}q + \ldots + X_{b4}q^4 + X_{b5}q^5 + X_{b6}q^6 \\ s_r = s_{r0} + s_{r1}q + \ldots + s_{r4}q^4 + s_{r5}q^5 + s_{r4}q^6 \\ E_m = E_{m0} + E_{m1}q + \ldots + E_{m4}q^4 + E_{m5}q^5 + E_{m6}q^6 \\ X_p = X_{p0} + X_{p1}q + \ldots + X_{p4}q^4 + X_{p5}q^5 + X_{p6}q^6 \\ \gamma_m = \gamma_{m0} + \gamma_{m1}q + \ldots + \gamma_{m4}q^4 + \gamma_{m5}q^5 + \gamma_{m6}q^6 \\ j = j_0 + j_1q + \ldots + j_4q^4 + j_5q^5 + j_6q^6 \\ i = i_0 + i_1q + \ldots + i_4q^4 + i_5q^5 + i_6q^6 \end{cases}, \tag{12}$$

Here, q is the cradle rotational increment; R_a is the ratio of generating roll; X_b is the sliding base; s_r is the cutter radial setting; E_m is the offset; X_p is the work head setting; γ_m is the root angle; j is the swivel angle; and i is the cutter head tilt angle. The machine setting parameters represent corresponding kinematic elements of the hypoid generator illustrated in Fig. 5.

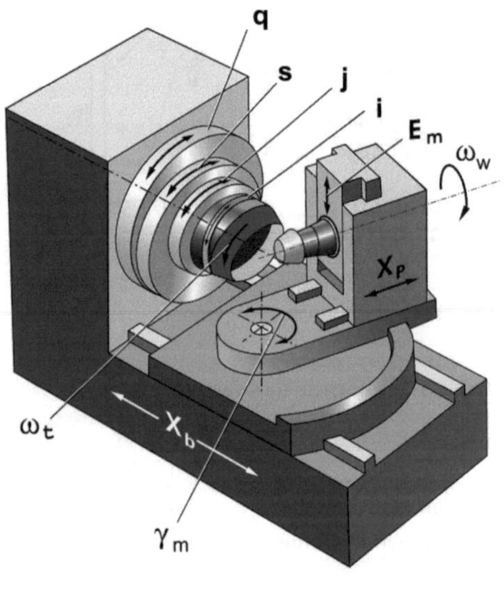

Fig. 5 A kinematic model of hypoid generators

The geometry of the tooth surfaces of a pair of mating pinion and gear can be generally represented by the position vector, unit normal and unit tangent in the coordinate systems S_1 and S_2 that are rigidly connected to the pinion and the gear, respectively, as follows,

$$\text{Pinion}: \begin{cases} \mathbf{r}_1 = \mathbf{r}_1(u_1, \theta_1, \varphi_1) \\ \mathbf{n}_1 = \mathbf{n}_1(u_1, \theta_1, \varphi_1) \\ \mathbf{t}_1 = \mathbf{t}_1(u_1, \theta_1, \varphi_1) \\ f_1(u_1, \theta_1, \varphi_1) = 0 \end{cases}, \tag{13}$$

$$\text{Generated gear}: \begin{cases} \mathbf{r}_2 = \mathbf{r}_2(u_2, \theta_2, \varphi_2) \\ \mathbf{n}_2 = \mathbf{n}_2(u_2, \theta_2, \varphi_2) \\ \mathbf{t}_2 = \mathbf{t}_2(u_2, \theta_2, \varphi_2) \\ f_2(u_2, \theta_2, \varphi_2) = 0 \end{cases}, \tag{14}$$

$$\text{Formate gear}: \begin{cases} \mathbf{r}_2 = \mathbf{r}_2(u_2, \theta_2) \\ \mathbf{n}_2 = \mathbf{n}_2(u_2, \theta_2) \\ \mathbf{t}_2 = \mathbf{t}_2(u_2, \theta_2) \end{cases}. \tag{15}$$

Step 3: Assembly of the pinion and gear members in their running position, as shown in Fig. 6, and representing the tooth surfaces of both members in a global coordinate system S_f that is fixed to the frame, including the assembling parameters, nominal offset and displacement $E_0 + \Delta E$, gear axial displacement ΔG, pinion axial displacement ΔP, and nominal shaft angle and displacement $\Sigma_0 + \Delta\Sigma$, as shown in Fig. 7. Given the initial contact position on the gear flank, the assembling displacements are determined using the improved algorithm described in [4]. Default displacement values of ΔE, ΔG, ΔP, and $\Delta\Sigma$ are zero.

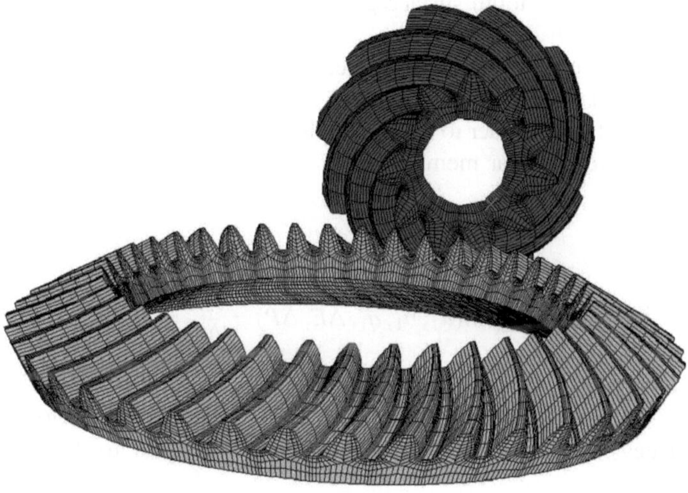

Fig. 6 A contact model of a mating pinion and gear

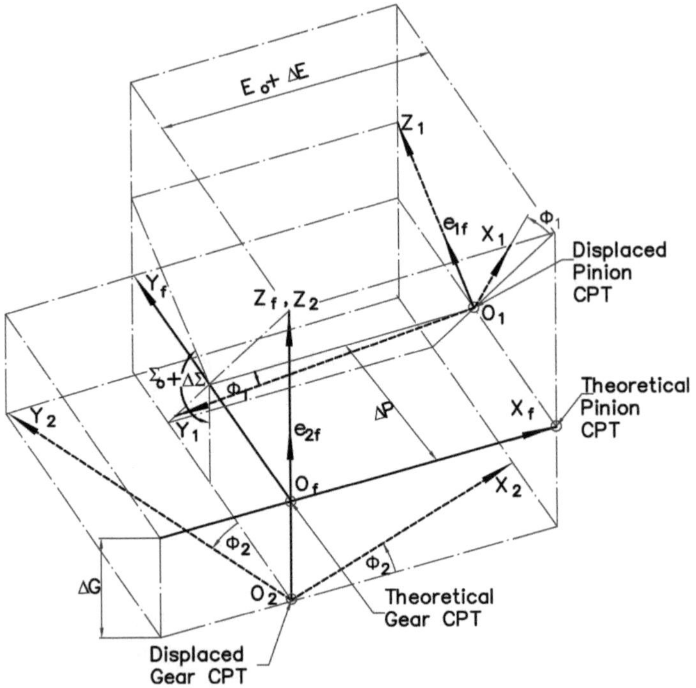

Fig. 7 Coordinate systems associated with the contact model in Fig. 6

Using the concept of potential contact lines, an advanced version of TCA has been developed for modeling and simulation of meshing for both face-milled and face-hobbed spiral bevel and hypoid gear drives. The algorithm of TCA is based on the identification of the tooth surface gaps along the potential contact lines and is determined by Eq. (9) or (10).

For a given contact position on the gear surface, the following contact formulation can be used to solve the parameters of the pinion surface and displacement values for ΔE and ΔP in order to determine the initial running position of the pinion member relative to the gear member:

$$\begin{cases} \mathbf{r}_{1f}(u_1, \theta_1, q_1, \Delta E, \Delta P) = \mathbf{r}_{2f} \\ f_1(u_1, \theta_1, q_1) = 0 \\ m(u_1, \theta_1, q_1, \Delta E, \Delta P) = \frac{N_2}{N_1} \end{cases} \tag{16}$$

where $\mathbf{r}_{1f}$ and $\mathbf{r}_{2f}$ are position vectors of the pinion and gear tooth surfaces and are represented in the coordinate system S_f; N_1 and N_2 are the number of teeth of the pinion and gear; m is the ratio of the moment of the gear to the pinion and is represented as,

$$m = \frac{(\mathbf{e}_{2f} \times \mathbf{r_{2f}}) \cdot \mathbf{n_{2f}}}{(\mathbf{e}_{1f} \times \mathbf{r_{1f}}) \cdot \mathbf{n_{1f}}}.$$

Equation 16 represents a condition of initial contact in five nonlinear scale equations that can be solved using an iteration method. The angular displacements of the pinion and the gear, ϕ_1 and ϕ_2 are explicitly represented in terms of functions of the surface normals as,

$$\phi_1 = \sin^{-1}\left\{\frac{\cos\delta_1[\mathbf{e}_{2f}\cdot\mathbf{e}_{1f}(\mathbf{e}_{1f}\cdot\mathbf{n}_{1f}) - \mathbf{e}_{2f}\cdot\mathbf{n}_{2f}]}{(\mathbf{e}_{1f}\times\mathbf{n}_{1f})\mathbf{e}_{2f}}\right\} - \delta_1, \tag{17}$$

$$\phi_2 = \sin^{-1}\left\{\frac{\cos\delta_2[\mathbf{e}_{1f}\cdot\mathbf{e}_{2f}(\mathbf{e}_{2f}\cdot\mathbf{n}_{2f}) - \mathbf{e}_{1f}\cdot\mathbf{n}_{1f}]}{(\mathbf{e}_{2f}\times\mathbf{n}_{2f})\mathbf{e}_{1f}}\right\} - \delta_2. \tag{18}$$

Here,

$$\delta_1 = \tan^{-1}\left\{\frac{[(\mathbf{n}_{1f}\times\mathbf{e}_{1f})\times\mathbf{e}_{1f}]\cdot\mathbf{e}_{2f}}{(\mathbf{n}_{1f}\times\mathbf{e}_{1f})\cdot\mathbf{e}_{2f}}\right\},$$

$$\delta_2 = \tan^{-1}\left\{\frac{[(\mathbf{n}_{2f}\times\mathbf{e}_{2f}\times\mathbf{e}_{2f}]\cdot\mathbf{e}_{1f}}{(\mathbf{n}_{2f}\times\mathbf{e}_{2f})\cdot\mathbf{e}_{1f}}\right\}$$

where $\mathbf{n}_{1f}$ and $\mathbf{n}_{2f}$ are unit normals of the pinion and gear tooth surface, which are calculated under $\phi_1 = 0$ and $\phi_2 = 0$; and $\mathbf{e}_{1f}$ and $\mathbf{e}_{2f}$ are unit vectors on the axes of the pinion and gear shown in Fig. 7. Equations 16 and 17 are derived based on the contact condition that, at the contact point, the surface normals are co-linear.

Step 4: Determination of the effective contact boundary, the overlapped area of mating tooth surfaces. The contact boundary on the gear member is determined using the conjugate image of the pinion tooth surface and vice versa.

Step 5: Calculation of the potential contact lines on both members using Eqs. 1–5 under the given displacement values of ΔE, ΔG, ΔP and $\Delta\Sigma$, and the ease-off value Δ at each point of potential contact lines within the effective boundary of contact shown in Fig. 8.

Step 6: Calculation of the contact pattern and transmission errors under the given surface separation. Ease-off Δ can also be interpreted as the surface "gap". The gaps under load will be closed and the contact pattern forms. The minimum gap along each potential contact line represents the transmission error.

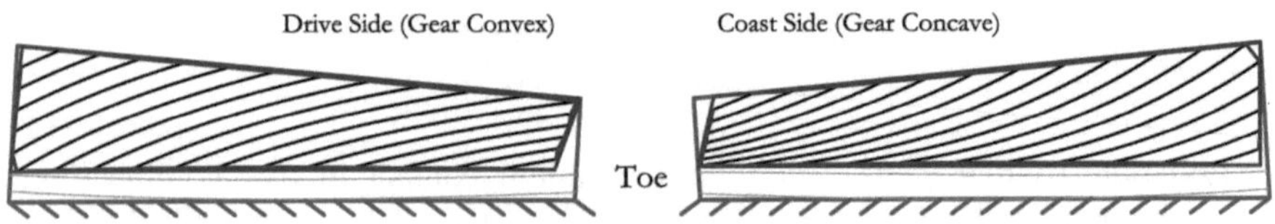

Fig. 8 Potential contact lines represented in gear flanks

Step 7: Calculation of the root line, Toprem line and Flankrem line on both pinion and gear surfaces to visualize the location of contact patterns on both tooth surfaces relative to the root and tip reliefs of the tooth surfaces.

Step 8: Calculation of the instantaneous backlash along the path of contact. The backlash at the transverse cross section of the mating members is calculated. At each contact point on the tooth, the thickness and space of the mating members are calculated and the backlash is calculated.

Step 9: Optimization of the pinion face cone to fit the gear root line and obtain maximum active working surface, as well as to avoid the root-and-tip interference.

In the blank design of spiral bevel and hypoid gears, the face cone is defined as an imaginary cone tangent to the tops of the teeth. Traditionally, the face cone element, or generatrix, is a straight line intersecting the face cone apex. On the other hand, the tooth root lines are traced by the blade tips and are normally curved lines due to the fact that (1) Formate gears have circular arc root lines; (2) the teeth are generated by a tilted circular cutter head, as shown in Fig. 9; (3) bevel gears that are generated with the UMC motions have the root lines of higher-order curves.

Fig. 9 Formation of gear tooth root lines

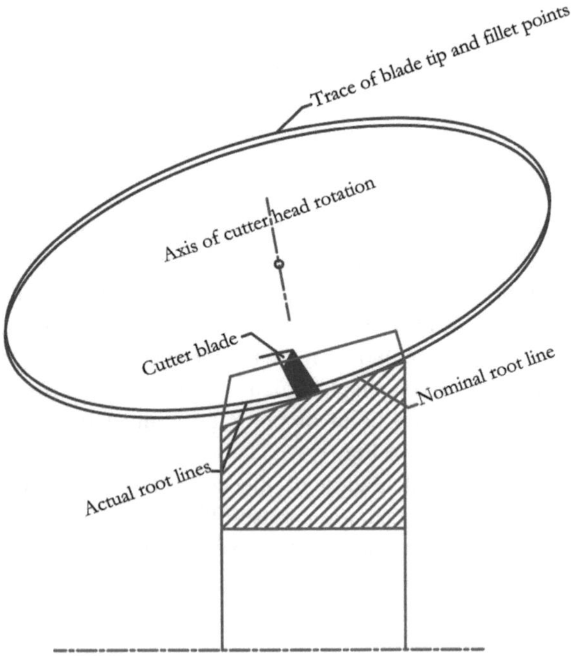

As a result, the tooth tip geometry generally does not fit the mating member's real root lines, providing an uneven tooth root-tip clearance. Since the pinion flank contour is transformed onto the gear flank as a conjugate image, the relative position of the pinion tip and gear root lines can be visualized and the root-tip interference may be observed if it exists. Figure 10 illustrates an example of a pinion tip with a gear root.

To avoid root-tip interference, the pinion teeth may be shortened, which results in reduced working area of the tooth surfaces and the contact ratio shown in Fig. 11.

Another solution is to optimize the pinion tip line that fits the corresponding gear root line so that maximum contact area can be obtained, as shown in Fig. 12.

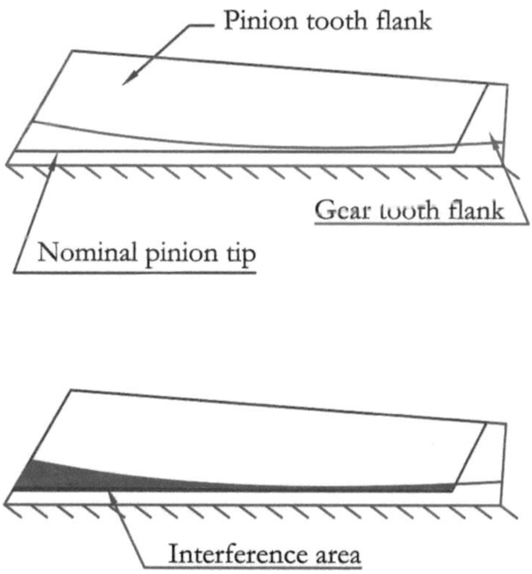

Fig. 10 Interference of pinion tip with gear root

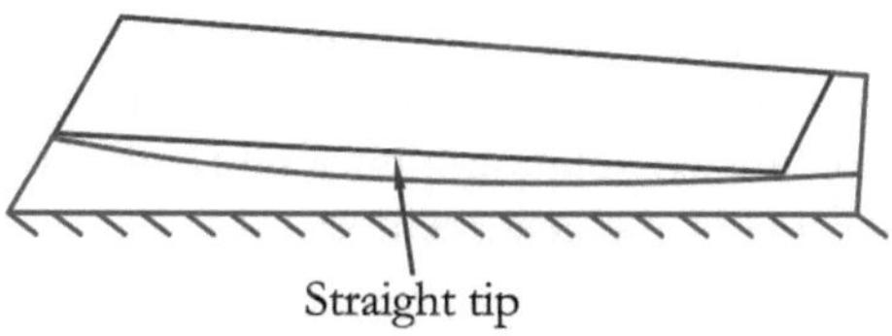

Fig. 11 Shortened pinion tooth for avoidance of interference

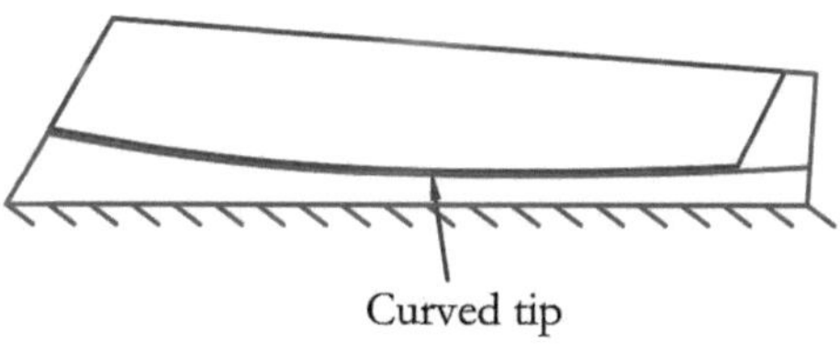

Fig. 12 Optimized pinion tip

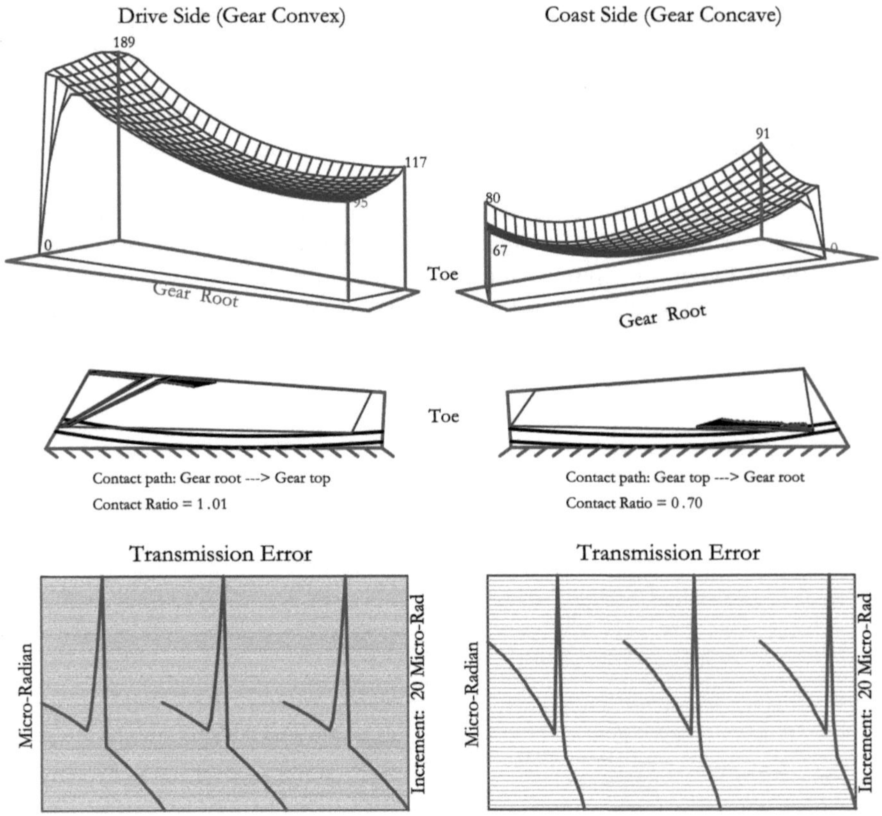

Fig. 13 TCA with pinion tip and gear root interference

An example of TCA with tip-root interference is shown in Fig. 13 where the pinion tip interferes with the real gear root at the heel. Figure 14 shows the solution of eliminating the interference by shortening the pinion teeth with a straight line. Figure 15 shows an optimized solution with modification of the pinion tip by a curve that fits the root line of the gear member.

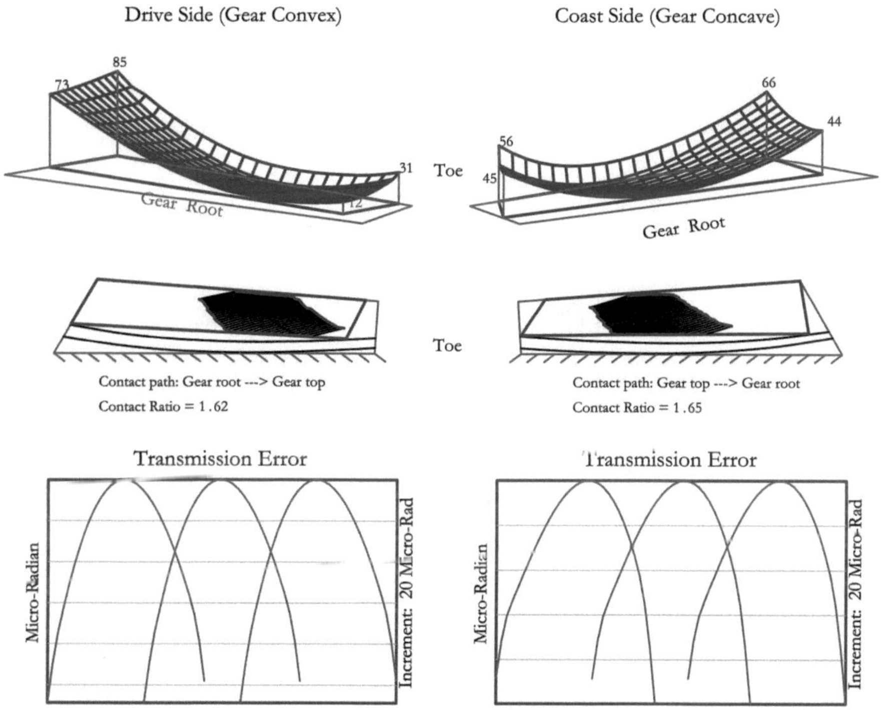

Fig. 14 TCA with shortened pinion tip

6 Numerical Examples

Two examples of TCA, a face-hobbed design and a face-milled design, are shown in Figs. 16, 17, 18, and 19. The basic design data of the two designs is shown in Table 2. In the face-hobbed design, Toprem and Flankrem are used for pinion and gear members. The fillet line, Toprem line and Flankrem line are shown on the tooth flanks. The ease-off topographies show the relief on the tip and root area. Figures 16 and 17 show the TCA results for the gear member and the pinion member, respectively.

Figures 18 and 19 show a highly conformable contact condition of a face-milled design. The ease-off topographies are “flat”, which represents an almost conjugate contact condition with large contact patterns and low transmission errors.

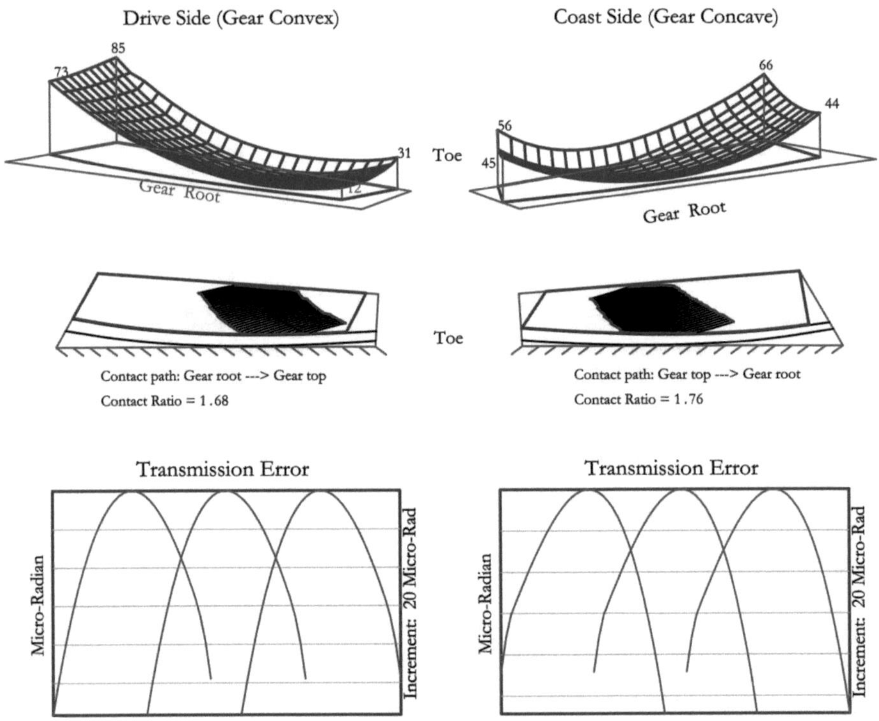

Fig. 15 TCA with optimized pinion tip/face cone

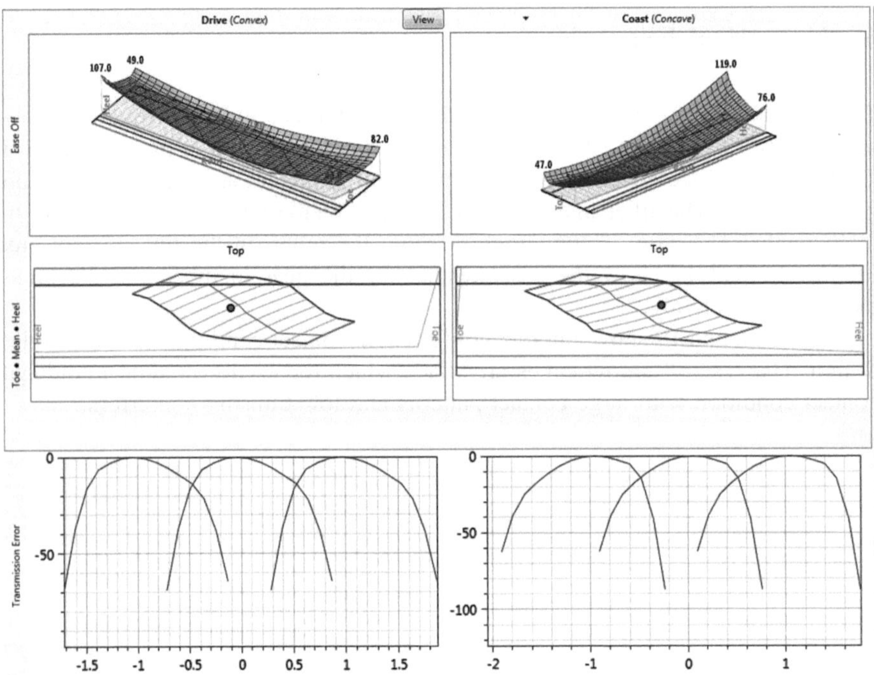

Fig. 16 TCA of the face-hobbed design—presented for the gear member

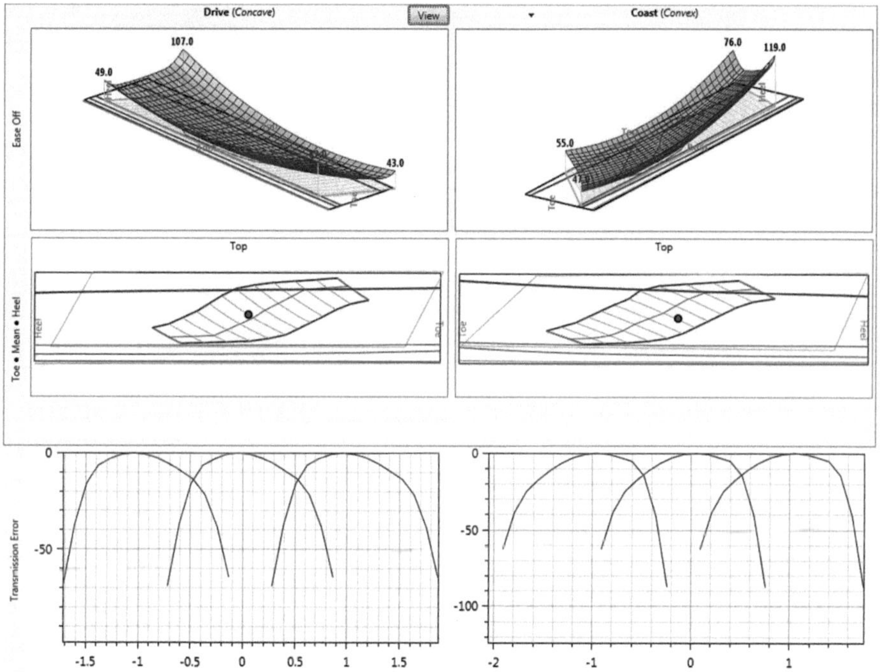

Fig. 17 TCA of the face-hobbed design—presented for the pinion member

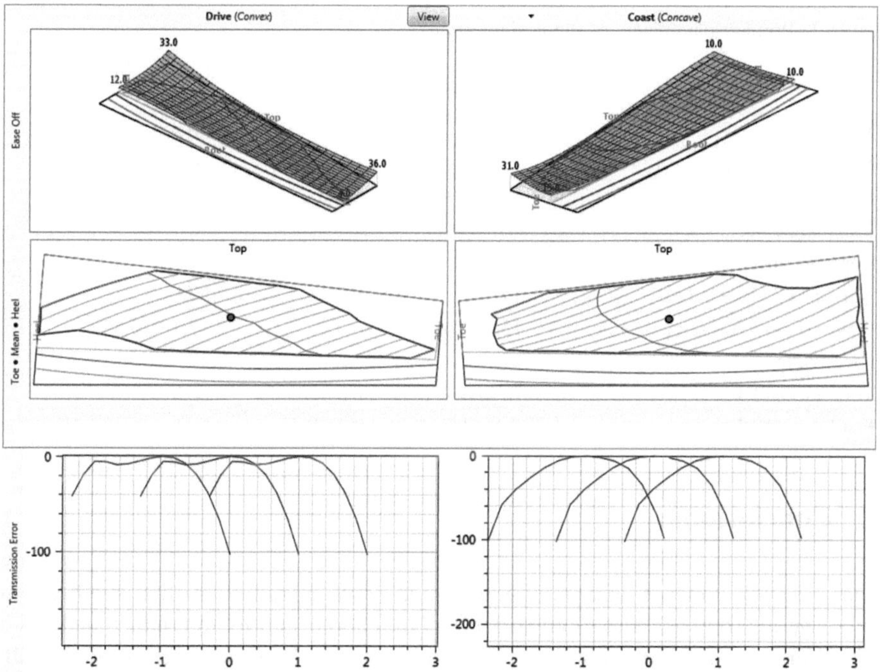

Fig. 18 TCA of the face-milled design—presented for the gear member

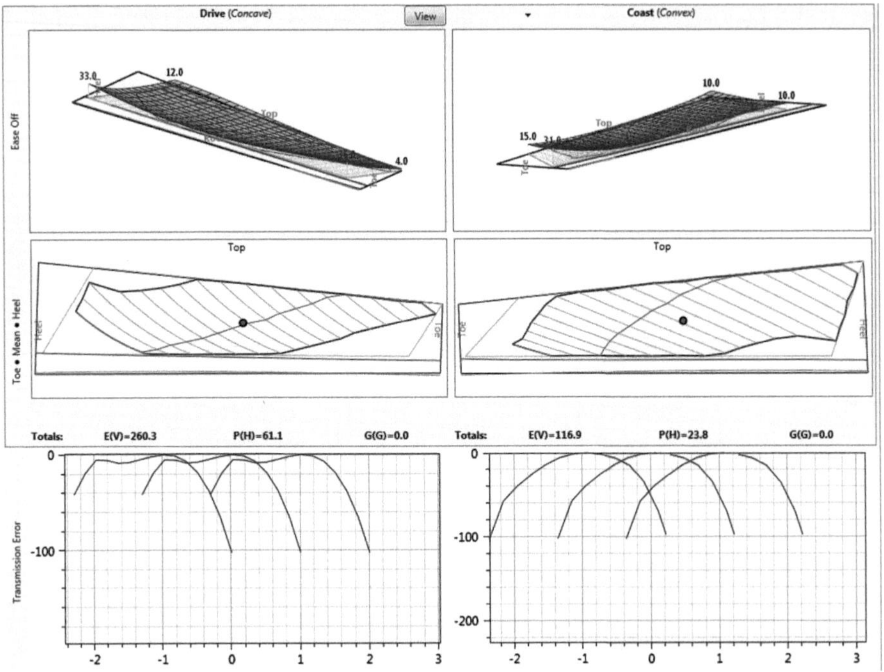

Fig. 19 TCA of the face-milled design—presented for the pinion member

Table 2 Primary parameters of the designs (mm)

	Face hobbing design		Face milling design	
Members	Pinion	Gear	Pinion	Gear
Number of teeth	11	39	9	41
Module	5.065		5	
Shaft angle (degree)	90		90	
Running offset	38.1		31.75	
Number of blade groups	17		N/A	
Pressure angle drive/coast side	15.44/24.56		17.13/20.87	
Face width	41	34	38	32
Toprem OB/IB	B.T./B.T.	I.T./I.T.	N/A	N/A
Flankrem OB/IB	B.F./B.F.	B.F./B.F.	N/A	N/A

7 Conclusions

The ease-off concept and methodology is very effective in the synthesis of micro-geometries for gear tooth surface modifications and tooth contact analysis in which the ease-off topography represented by potential contact lines is numerically computed and the tooth surface contact condition is scanned over the overlapped contact area of both members for identification of contact patterns and transmission errors, as well as possible interference. This paper comprehensively describes the ease-off concept and the computational algorithm, as well as its application. Based on the algorithm described, an advanced TCA program has been developed. Two examples, a face-hobbing design and a face-milling design, are presented and illustrated with TCA graphic output.

References

1. Baxter ML (1966) Exact determination of tooth surfaces for spiral bevel and hypoid gears, AGMA Semi-Annual Meeting, p 139
2. Baxter M L (1973) Second-order surface generation, J Ind Math Soc, vol 23, p 2
3. Fan Q (2006) Computerized modeling and simulation of spiral bevel and hypoid gears manufactured by gleason face hobbing process. ASME J Mech Des 128(6):1315–1327
4. Fan Q (2007) Enhanced algorithms of contact simulation for hypoid gear drives produced by face-milling and face-hobbing processes. ASME J Mech Des 129(1):31–37
5. Fan Q, DaFoe R, Swanger J (2008) Higher-order tooth flank form error correction for face-milled spiral bevel and hypoid gears. ASME J Mech Des 130(7):072601
6. Fan Q (2010) Tooth surface error correction for face-hobbed hypoid gears. ASME J Mech Des 132(1):011004
7. Fan Q (2011) Optimization of face cone element for spiral bevel and hypoid gears. ASME J Mech Des 133(9):091002
8. Fuentes A, Gonzalez-Perez I, Litvin FL, Hayasaka K, Yukishima K (2005) Design, manufacture, and evaluation of prototypes of low-noise high-endurance spiral bevel gear drives, ASME IDETC/CIE, PTG, DETC2005-84013. Long Beach, USA
9. Kolivand M, Kahraman A (2010) An ease-off based method for loaded tooth contact analysis of hypoid gears having local and global surface deviations. ASME J Mech Des 132(7):071004
10. Krenzer TJ (1984) Computer aided corrective machine settings for manufacturing bevel and hypoid gear sets. Fall Technical Meeting, Washington, D.C
11. Krenzer TJ (2007) The bevel gear, published by Ted Krenzer, New York
12. Litvin FL (1994) Gear geometry and applied theory, Prentice Hall, New Jersey
13. Litvin FL, Fan Q, Fuentes A, Handschuh R F (2001) Computerized design, generation, simulation of meshing and contact of face-milled formate^{TM}-cut spiral bevel gears, NASA Report, /CR-2001-210894, ARL-CR-467
14. Shih YP, Fong ZH, Lin GCY (2007) Mathematical model for a universal face hobbing hypoid gear generator. ASME J Mech Des 129(1):38–47
15. Simon V (2007) Load distribution in spiral bevel gears. ASME J Mech Des 129(2):201–209
16. Stadtfeld HJ (2000) Advanced bevel gear technology, The Gleason Works, New York
17. Stadtfeld HJ, Gaiser U (2000) The ultimate motion graph. ASME J Mech Des 122(3):317–322
18. Stadtfeld HJ (2003) The two-sided-ground bevel gear cutting tool. Gear Technology, May/Jun
19. Stadtfeld HJ (2008) gear encyclopedia, The Gleason Works, New York
20. Stadtfeld HJ (2010) Tribology aspects in angular transmission systems, Gear Technology, California, pp 46–52
21. Stadtfeld HJ (2012) Bevel gear cutting blade measurement, Gear Solutions, California

Methods and Results of Composite Gears Design

S. Shil'ko, V. Starzhinsky and E. Petrokovets

Abstract Due to expansion of the micro- and nanocomposites application for manufacturing machine components and friction units, the problem of calculation methods for determination of deformability, strength and wear resistance parameters of gear drives made of essentially inhomogeneous disperse-reinforced materials is addressed. The potentialities of analytical and numerical methods are analyzed. The original three-level (micro, meso- and macro) method for tribomechanical parameters optimization of the gears by controlling material reinforcing is presented. Through specific examples, the potentialities of polymer reinforcement for obtaining functional materials for gears which allow for an increase in damping capability, shape stability and life time of the gear driven by the criterion of wear and bending strength have been studied.

Keywords Composites · Gears · Calculation methods · Deformability · Wear resistance

1 Introduction

Assessment of the efficiency of gear drives using the criteria of strength, deformability, wear resistance, etc., is based on analysis of the stress strain state (SSS) in the mating parts (gears) and in the area of their contact interaction. As a rule, the engineering methods for calculating such machine components [1–3] are based on the compact analytic expressions obtained at solving classical contact problems of the theory of elasticity [4, 5] and approximation of the tooth profile. In many instances this approach provides acceptable accuracy.

However, in order to take into account the real geometrical shape of gears [6] and specific features of deformation of structurally inhomogeneous and physically

S. Shil'ko (✉) · V. Starzhinsky · E. Petrokovets
Metal-Polymer Research Institute of NAS of Belarus, Gomel, Belarus
e-mail: shilko_mpri@mail.ru

V. Goldfarb and N. Barmina (eds.), *Theory and Practice of Gearing and Transmissions*, Mechanisms and Machine Science 34,
DOI 10.1007/978-3-319-19740-1_16

nonlinear materials (e.g., polymer composites), it is necessary to apply the 3D discretization of the area being analyzed using finite and boundary elements in order to overcome limitations of the analytic methods (there is a vast bibliography on the subject, so it is not cited herein; comparison of analytical and numerical methods through the solving of test problems was performed, in particular, in [7]).

The urgency of researches in the field of mechanics and tribology of the composite gears is motivated, primarily, by the extensive use of disperse-filled polymers, as the methods of reinforcement of a polymer matrix with particles or short fibers, including the nanoparticles, is highly technological and yields considerable improvement in both strength and tribological characteristics [8]. For instance, analysis of SSS and wear of composite gears made of glass- and carboplastics in [9] shows a several-fold increase in the service life of gears due to the combined boosting of bending strength and wear resistance.

Calculation and optimization of gear drives with gear wheels made of polymeric composites are complicated due to the necessity of obtaining a considerable volume of the initial data, such as modules of elasticity, strength limits, and parameters of velocity sensitivity of the aforementioned materials to numerous potential compositions of materials. Their experimental determination is a costly and time-consuming procedure.

In this respect, the inverse problem of optimization of the material composition is of considerable practical importance (mainly, the concentration and dispersity of the reinforcing and antifriction filler) based on the specified geometry and loading conditions; this approach supplements and, in certain cases, substitutes for the usually practiced routine and labour-intensive tests.

This means that, in designing a gear drive, it shall be necessary to perform a series of calculations to take into account variations of the filler parameters until the specified strength, deformability and wear resistance criteria or their superpositions with weight factors are met. However, it must be admitted that an optimal variant of the composite gears can hardly be obtained using only analytical description or, vice versa, only numerical modeling.

The authors are of the opinion that the afore-mentioned problems may be effectively solved through multi-level modeling [10–12] allocating the following stages: micromechanical prediction of the composite properties, material model verification at the mesomechanical level based on experimental data and the stage of the structure calculation (macrolevel).

2 Calculation Methods for Composite Gears

2.1 Simplified Description of the Stress-Strain Properties of Composite Materials

As referred to in the introduction, application of the traditional analytical and numerical methods for calculating operational characteristics of gearing with wheels made of polymer composites is encumbered with the internal structure variety and complexity of description of interphase interaction of the mentioned inhomogeneous materials, finite deformations and nonlinear deformation of the polymer matrix, including its pronounced viscoelastic properties.

In a relatively small number of cases, the deformability, strength and wear resistance properties of polymer composite materials in terms of the volume content, elastic modules and strength limits of components (matrix and filler) can be obtained with the accuracy acceptable for engineering practice using classical methods of the theory of elasticity. Selection of the corresponding models is determined by the type of filler particles and their arrangement in the matrix material (reinforcement scheme).

Let us consider specific features of the simplified determination of mechanical properties of polymer composites with respect to predicting strength, deformability and wear resistance of the gears [9].

Reinforcement with the disperse filler. Introduced into the matrix of the disperse particles or arbitrarily oriented short fibers with the filler volume content V_f, the obtained inhomogeneous material can be considered as quasi-isotropic, which makes it possible approximately to calculate the elasticity modulus E_c, Poisson's ratio ν_c and strength limit σ_c using additivity law (rule of mixtures)

$$E_c = \alpha E_f V_f + E_m(1 - V_f); \tag{1}$$

$$\nu_c = \alpha \nu_f V_f + \nu_m(1 - V_f); \tag{2}$$

$$\sigma_c = \beta \sigma_f V_f + \sigma_m(1 - V_f). \tag{3}$$

Coefficients α and β in Eqs. (1)–(3) for the random fiber distribution in the matrix according to [13] are equal to 0.375, which is an upper limit of the true value obtained without taking into account the fiber strength dispersion and upon availability of an ideal adhesion bond between the fiber and the matrix. A more accurate assessment of the aforementioned coefficients can be made on the basis of the reference value of the composite real strength σ_c with the known filler volume content V_f

$$\alpha = \beta = \frac{\sigma_c - \sigma_m\left(1 - V_f\right)}{\sigma_f\, V_f}. \tag{4}$$

Assumption of the material's quasi-isotropy allows for the application of the Hertz solution of the contact problem for compressed elastic isotropic cylinders which is traditionally used for calculating the parameters of contact in gearing. Considering the mating metal wheel nondeformable as compared to the composite one, we obtain the dependencies to find the area of contact a and nominal contact pressure p_a

$$a = \sqrt{\frac{4\,F_t R\,(1 - \nu^2)}{\pi\, b_w E}}; \tag{5}$$

$$p_a = \frac{F_t}{2\,a\,b_w}, \tag{6}$$

where F_t is the circumferential force; R is the equivalent radius; b_w is the gear face width; and a is the half-width of the contact area.

Directional reinforcement. The filler orientation has an effect on the material's rigidity and strength, as well as on its wear resistance. Therefore, it is justifiable to use the gradient material with elastic matrix which provides for a reduction in gearing vibration activity and the reinforcing fibers oriented towards the tensile stresses in the surface layer which initiate the development of fatigue fractures.

By way of example, let us consider the structure of a gear formed by successive application of monolayers in the plane of which the reinforcing fibers are arranged in a random manner or with the fibers oriented parallel to one of the axes of symmetry. In such cases, there is a plane where all directions with respect to elastic properties are equivalent. Elastic properties of the mentioned transversely isotropic body are characterized by four independent constants, E_1, E_3, ν_1 and ν_3; while the remaining elastic constants are calculated by formulas, e.g.

$$G_{12} = E_1/(2(1 + \nu_1)). \tag{7}$$

Hill yield criterion can be used as a strength condition for materials with mild anisotropy [13], while for considerably anisotropic directionally reinforced materials, it should be necessary to apply more complex criteria.

It is justifiable to calculate stresses for assessing contact strength and wear rate of tribojoints (to which the gears are related) by using the numerical boundary element method [14] on the basis of correlation between stresses τ_i, p_i and displacements u_T, u_N at an arbitrary point j under elementary load on element i on the tooth surface

$$\begin{Bmatrix} u_{Tj} \\ u_{Nj} \end{Bmatrix} = \begin{Bmatrix} K_{ij}^{TT} \, K_{ij}^{TN} \\ K_{ij}^{NT} \, K_{ij}^{NN} \end{Bmatrix} \begin{Bmatrix} \tau_j \\ p_i \end{Bmatrix}. \tag{8}$$

For plane strain (2D) case, the generalized Hooke's law for an orthotropic body takes the following form [15, 16]

$$\begin{pmatrix} \varepsilon_x \\ \varepsilon_y \\ \gamma_{xy} \end{pmatrix} = \begin{pmatrix} S_{11} & S_{12} & 0 \\ S_{12} & S_{21} & 0 \\ 0 & 0 & S_{66} \end{pmatrix} \begin{pmatrix} \sigma_x \\ \sigma_y \\ \tau_{xy} \end{pmatrix}. \tag{9}$$

where ε_x, ε_y, γ_{xy} are the deformed state components; S_{11}, S_{12}, S_{21}, S_{66} are the compliance coefficients; and σ_x, σ_y, τ_{xy} are the stress state components.

For the transversely isotropic body at in-plane deformation, we shall have

$$\begin{aligned} S_{11} &= 1/E_1 - \nu_3^2/E_3; \\ S_{12} &= -\nu_1\nu/E_1 - \nu_3^2 - \nu_3^2/E_3. \end{aligned} \tag{10}$$

In the case of orthotropic material, the constants are calculated by formulas

$$\begin{aligned} S_{11} &= \frac{1-\nu_{13}\nu_{31}}{E_1}; \quad S_{12} = \frac{\nu_{12}+\nu_{23}\nu_{31}}{E_2}; \\ S_{22} &= \frac{1-\nu_{32}\nu_{23}}{E_2}; \quad S_{66} = \frac{1}{G_{12}}. \end{aligned} \tag{11}$$

Here, E_1, E_2, E_3 are Young moduli along the main elasticity directions; ν_{ij} are Poisson's ratios which determine the transverse deformation in direction j under the influence of stresses in direction i; and G_{12} is the shear modulus which characterizes the change in angle between the directions parallel to the main elasticity axes x and y.

The following relation exists between elastic constants $\nu_{12}/E_1 = \nu_{21}/E_2$. To make expression (8) more specific, it is possible to apply the potential theory by introducing the influence functions for orthotropic half-plane displacements under the action of distributed loads $p(\xi)$ and $t(\xi)$ as [4, 5]

$$\begin{aligned} K_{ij}^{TT} &= S_{11}(\gamma_1+\gamma_2)/\pi \int_{x_i-0.5h}^{x_i+0.5h} \tau(\xi)\ln\left|x_j-\xi\right| d\xi; \\ K_{ij}^{TN} &= (S_{11}\gamma_1\gamma_2 - S_{12}) \int_{x_i-0.5h}^{x_i+0.5h} p(\xi)\, d\xi; \\ K_{ij}^{NT} &= -(S_{22}/\gamma_1\gamma_2 - S_{12}) \int_{x_i-0.5h}^{x_i+0.5h} \tau(\xi)\, d\xi; \\ K_{ij}^{NN} &= (S_{22}/\pi)(1/\gamma_1 + 1/\gamma_2) \int_{x_i-0.5h}^{x_i+0.5h} \sigma(\xi)\ln_i|x-\xi|\, d\xi_j, \end{aligned} \tag{12}$$

where $\gamma_{1,2} = \sqrt{\frac{S_{66}+2S_{12} \pm \sqrt{\left(S_{66}+2S_{12}^2\right) - 4S_{11}S_{22}}}{2S_{11}}}$.

Henceforth, when expressions (12) are used in Eq. (8), calculation of the composite gear SSS is performed to assess the contact strength and wear rate (Sect. 2.3).

2.2 *Detailed Description of the Stress-Strain Properties of Composite Materials*

To analyze the strength and deformability of the disperse-filled materials of the gears, it is possible to use, alongside the aforementioned simplest rule of mixtures (1)–(3), more accurate dependencies of the mechanics of composite materials (described, e.g., in [13, 15–18]). They enable the determination of effective mechanical characteristics of the composite material over a wide range of elastic parameters of the polymer matrix, filler and interphase layer, as well as the filler volume content, form and particle size, including those at the nanodimensional level.

Two-phase (a), three-phase (b) and four-phase (c) structural models are applied which reflect, to a varying degree, the inhomogeneity of the material in question (Fig. 1). Application of the three-phase model assures acceptable accuracy of predicting effective properties of the material at the filler volume content up to 50 %, which covers practically all grades of the structural composites. It is important that, over a wide range of the filler particles mean radius (from 10 nm to 100 mcm), the interphase layer exerts a significant influence on the composite stiffness, thus the interphase parameters, such as its thickness, shall be determined as accurately as possible by applying the procedure of the material model identification on the basis of available experimental data.

It is assumed that the components of the dispersion-reinforced material (matrix, filler particles and interphase layer) are isotropic, linearly elastic, and adhesively bonded. The filler particles are modeled as ellipsoids in order to describe them as spherical particles, as well as short fibers. The composite as a whole is treated as a macroscopically isotropic and linearly elastic material, the characteristics of which can be determined by the Mori-Tanaka model and the method of "composite inclusion" [17].

For the criteria of the matrix and interphase layer yield, we have taken the critical points of the strain tensor rate. The following parameters are being calculated: the effective Young modulus, Poisson's ratio of the composite, and its yield limit under a uniaxial stress state. The value of the uniaxial stress which initiates the plastic flow in the matrix or interphase layer is taken as the yield limit of the composite.

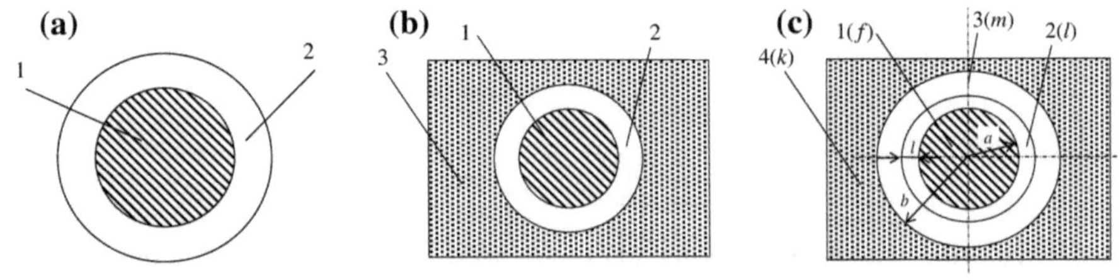

Fig. 1 Models of the disperse-filled material: *1* filler; *2* matrix; *3* interphase layer; *4* surrounding composite volume

Preparation of the initial data for calculations at the mesomechanical level presumes setting of the values of Young modulus E_m (MPa), Poisson's ratio ν_m and yield point σ_m^{cr} (MPa) of the matrix material; Young modulus E_f (MPa) and Poisson's ratio ν_f of the filler material; Young modulus E_l (MPa), Poisson's ratio ν_l and yield point σ_l^{cr} (MPa) of the interphase layer material; the volume content of the filler φ; anisometric parameter of the filler particles (the ellipsoidal minor semi-axis ratio to the major one) χ (ranges from 0 to 1) and the relative interphase layer volume (the layer volume ratio to that of the filler) q.

Then, the components of the elasticity moduli tensors of isotropic materials of matrix $\mathbf{C}_m$, filler $\mathbf{C}_f$ and interphase layer $\mathbf{C}_l$ are calculated.

$$C_{i11} = C_{i22} = C_{i33} = \frac{E_i(1-\nu_i)}{(1+\nu_i)(1-2\nu_i)};$$

$$C_{i12} = C_{i21} = C_{i13} = C_{i31} = C_{i23} = C_{i32} = \frac{E_i\nu_i}{(1+\nu_i)(1-2\nu_i)};$$

$$C_{i44} = C_{i55} = C_{i66} = \frac{E_i}{2(1-2\nu_i)}.$$

In this case, index i may take the values of m, f, l. The remaining components of the elasticity moduli tensors are equal to zero.

Proceeding from the Eshelby tensor components, it is possible to determine the volume content of the filler in the "composite inclusion" φ_f, formed by the filler particle and the interphase layer. Upon completion of orientation averaging the tensor components of the elasticity moduli of the "composite inclusion", the values of the effective volume modulus K, shear modulus G, Young modulus E and Poisson's ratio ν of the combined inclusion are calculated for chaotic orientation of the filler particles; these values are required for subsequent calculation of the gear SSS (Sect. 2.4)

$$K = \frac{1}{9}\left[C_{fl11} + 2\left(C_{fl22} + C_{fl23}\right) + 4C_{fl12}\right];$$

$$G = \frac{1}{30}\left[2C_{fl11} + 7C_{fl22} - 5C_{fl23} - 4C_{fl12} + 12C_{fl66}\right];$$

$$\nu = \frac{3K - 2G}{2(3K + G)}, \quad E = 2G(1+\nu).$$

Detailed description of the contact deformation taking into account physical nonlinearity of the matrix polymer material. Nonlinearity of the polymer material "stress-strain" curve, the typical appearance of which is shown in Fig. 2, can be described by the finite elements method as using stiffness matrix D with variable coefficients $\{\sigma\} = [D(\{\varepsilon\})] \cdot \{\varepsilon\}$. The equations which relate the strain rate and the stress are used. At the current iteration, the stress and strain vectors are determined; stress and strain rates are calculated.

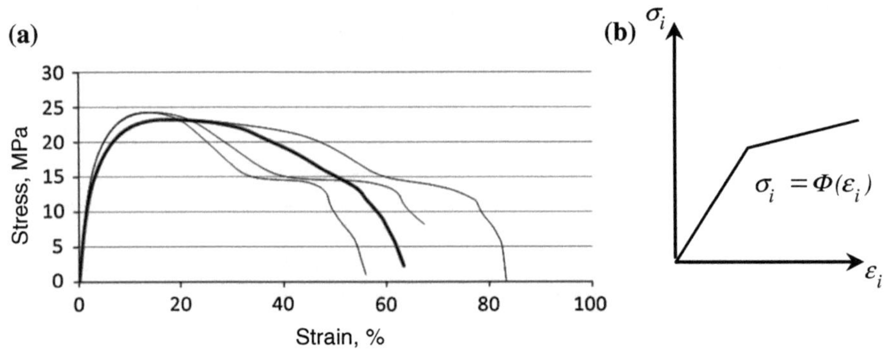

Fig. 2 Nonlinearity of elastic deformation (**a**) and its approximation (**b**)

Fig. 3 Convergence of the combined method

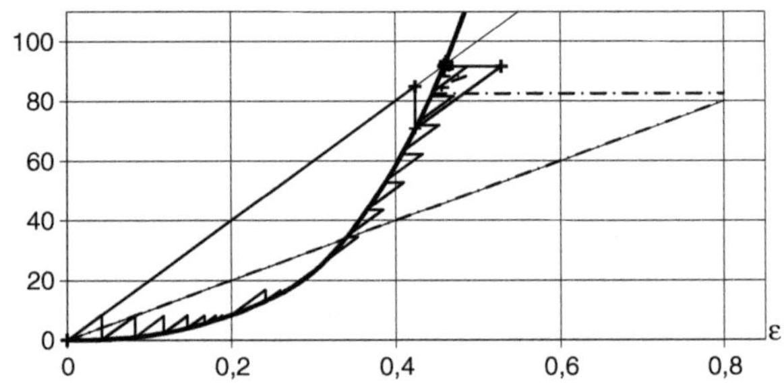

By using the known dependence $\sigma_i = \Phi(\varepsilon_i)$, ε_0 and σ_0 are determined, vectors of initial stresses and strains for next iteration $\varepsilon_0(\{\varepsilon\}) = const$ and $\sigma_0(\{\varepsilon\}) = const$ are calculated. It is justifiable to apply options which do not require restructuring of the stiffness matrix when using iterative methods for solving the problem.

To increase the rate of convergence, a combined method was proposed in [19] which takes into account the polymer material nonlinearity as $\{\sigma\} = [D] \cdot (\{\varepsilon\} + \varepsilon_0(\{\varepsilon\})) - \sigma_0(\{\varepsilon\})$. This enables us to perform the factorization of the stiffness matrix and reduce the time required to solve the problem.

From Fig. 3, it follows that the result remains unchanged for the analyzed options of the matrix coefficient values $[D]$ and method of applying the external load which proves the high stability of the solution.

2.3 Determination of Wear Resistance of the Gears Made of Composite Materials by Calculations

When the gear drives with wheels made of polymer materials are in operation, due to their high deformability, the wear is usually localized in the dedendum area, which increases the possibility of teeth shear fracture or breakage in the indicated weak section. In view of this, it shall be necessary to increase the gears' bending stiffness and strength, as well as the wear resistance of the structural plastics used.

The results of testing composite gears made of textolite, glass- and carbon-filled plastics [8] prove that reinforcement of the polymer matrix with high-strength and high modulus particles or fibers, including those at the nanodimensional level, provides for a many-fold increase in the gear life by the criterion of wear, especially in power transmissions.

Besides, there are known methods for combining improved strength and tribo-engineering characteristics based on use of the gear composite structure. In this case, the antifriction polymer layer with the working surface of the defined profile, e.g., the involute one, is formed on the strong and rigid gear workpiece (as a rule, manufactured of metal) [20].

Specific features of the wear calculations for composite gears made of reinforced polymer materials are shown below.

Wear parameter calculation. Let us consider a gear pair in which one wheel is made of composite material, including the matrix and the filler, with various elastic and strength characteristics.

To assess the reinforcement effectiveness, it is possible to apply the criterion of the teeth life time due to bending fracture in the area of maximal wear. The gearing life t_I is determined by the following dependence

$$t_I = \frac{h_{IP}}{2\,a\,\lambda\,n\,Z_S I_h}, \tag{13}$$

where h_{IP} is the allowable wear; a is the contact zone half-width; λ is the slip ratio; n is the rate of rotation; Z_s is the number of gear meshings; and I_h is the wear rate.

The slip ratio $\lambda = v_s/v_r$ increases when the point in question moves in a direction away from the pitch point. Therefore, it is possible to set (as a reference section) the dedendum area where the friction length $L = 2a\lambda$ is maximal and tooth strength depletion at bending is most probable.

Allowable wear h_{IP} for the section in question is determined by the difference between the non-worn tooth thickness S and tooth thickness S_w assigned according to the strength condition, as well as by the value of the tooth thinning due to the guaranteed backlash $j_{n\min}$

$$h_{IP} = S - S_W - j_{n\,\min}. \tag{14}$$

At determining S_w the following strength criteria may be used [21]:

(a) shear strength condition

$$\tau_S = \frac{1.9\,T}{m^2\,z\,\psi_m\,S_W} \le \tau_{SP}; \tag{15}$$

(b) bending strength condition

$$\sigma_{\mathrm{F}} = \frac{23\,T\,K}{z\,m\psi_m\,S_{\mathrm{W}}} \leq \sigma_{\mathrm{FP}}. \tag{16}$$

In formulas (15) and (16), T is the torque at the gear, N mm; m is the module, mm; z is the number of teeth; and $\psi_m = b_W/m$ is the face width coefficient.

A specific feature of the calculation is that the reinforcing phase affects all frictional and mechanical characteristics of the material and contact parameters, including wear rate I_h and the contact zone half-width in Eq. (13), allowable stresses τ_{SP} in (15), and σ_{FP} in (16).

The wear rate $I_h = dh/dS$ in Formula (13) is determined by calculation or experimentally on tribometers.

At calculation of the wear rate I_h, as a structurally sensitive characteristic, it shall be necessary to take into account the conditions of friction interaction in the toothed meshing and type of wear. Since it has been experimentally found that multiple deformations of surface layers, development of microcracks and periodical dissociation of particles occur under conditions of rolling with slipping, the fatigue wear mechanism was used at determination of the gear wear rate I_h.

To determine I_h, the dependence including complexes of the following main influencing parameters has been used: contact pressure p_a, elasticity modulus E, friction coefficient f, as well as strength limit σ_0 and friction fatigue index t [2]:

$$I_h = K_1\,0.5^{t-5/4}\,p_a^{t/5+1}\,E^{4t/5-1}\,\Delta^{2t/5}\left(\frac{k_p f}{\sigma_0}\right)^t, \tag{17}$$

where K_I is the coefficient which takes into account the form and height of individual asperities ($K_I = 0.2$); Δ is the integrated index of microgeometry of the metal counterbody (for polished surface $\Delta = 0.10$); and k_p is the coefficient which takes into account the fracture pattern in near-surface layers (for high-elastic materials, the value is taken to be equal to 3).

Approximate values of friction fatigue parameters of composite t and σ_0 can be found by using the bulk fatigue index x and strength limit σ_{-1} under cyclic load

$$t = x;\ \sigma_0 = \sigma_{-1}\,N_\sigma^{1/x}. \tag{18}$$

To make a more correct choice of t and σ_0 values, it shall be necessary to perform tribotesting with different filler contents. Thus, it should be necessary to experimentally determine the wear rate I_h, friction coefficient f and equilibrium roughness parameters, and then apply the equation for the wear rate (17), taking into account the elasticity and strength parameters to find the value of index t by the matching condition of the predicted I_c and experimental I_e wear rates.

Additional reserve to improve wear resistance of the reinforced gears is an increase of allowable wear h_{IP} if its value is assigned by the strength conditions.

By comparing, for example, the reinforced tooth thickness S^c_W, which meets the bending strength criterion, with the similar parameter of the gear tooth made of non-reinforced material S^m_W, the following is recorded in accordance with (16)

$$\frac{\left(S^c_W\right)^2}{\left(S^m_W\right)^2} = \frac{\sigma^m_{FP}}{\sigma^c_{FP}}. \tag{19}$$

Since (14) implies

$$h^c_{IP} = S^m_W + h^m_{IP} - S^c_W\,, \tag{20}$$

then, based on (19) and (20), we obtain

$$h^c_{IP} = S^m_W + h^m_{IP} - S^m_W\sqrt{\frac{\sigma^c_{FP}}{\sigma^m_{FP}}}. \tag{21}$$

Let us introduce an additional condition of the permissible wear h_{IP} proportionality to the tooth thickness in non-worn condition

$$h_{IP} = K_h\left(S^m_W + h^m_{IP}\right), \tag{22}$$

where K_h is the proportionality factor.

Considering that h^m_{IP} is the specified value, on the basis of (20) and (21), we obtain the following expression for comparative analysis of the change in the permissible wear of the composite gear (to calculate absolute values of the permissible wear, it shall be necessary to determine the non-worn tooth thickness S_c in dependence (16) according to the known formulas of the theory of involute meshing)

$$h^c_{IP} = h^m_{IP}\left(1 + \frac{1-K_h}{K_h}\left(1 - \sqrt{\frac{\sigma^m_{FP}}{\sigma^c_{FP}}}\right)\right). \tag{23}$$

Thus, calculation of the composite gear wear resistance includes the following stages:

(1) determination of friction and mechanical characteristics of the composite for the given coefficient of filling V_f;
(2) solution of the problem on teeth contact (determination of the contact zone a, contact pressure p_a and tangential stress τ_a);
(3) calculation of the wear rate I_h by the fatigue theory;
(4) life time evaluation by formula (13).

At determining the wear rate of the gears made of material with oriented fibers which is characterized by the aforementioned set of elastic constants, it shall be necessary to use the generalized expressions (17) in the form proposed in [9]

$$I_h \cong \left[p_a^{1+\beta t-t}\tau_a^t\right]\left[\theta_0^{1+\beta t-t}\sigma_0^{-t}\right], \tag{24}$$

where $\theta_0 = \frac{2(1-\nu^2)}{\pi E}$.

A specific example of calculation of the composite gear wear resistance using the obtained relations is described in Sect. 3.3 below.

2.4 Three-Level Method of Designing Gears Made of Disperse-Reinforced Composites

Figure 4 shows the scheme for applying the three-level method of designing the structural elements made of composites (the basics of which have been described in [10–12]) to the analysis of strength, deformability and wear resistance of the gears made of disperse-filled materials.

At the first scale level, an intrinsic (repeated) fragment of the composite structure is schematized by a set of related and minimal-in-size structural elements—periodicity cells. The choice of the micromechanical model presupposes adequate and, at the same time, simple mathematical description of the real structure and deformational behavior of the material. The defining relations for the selected structural

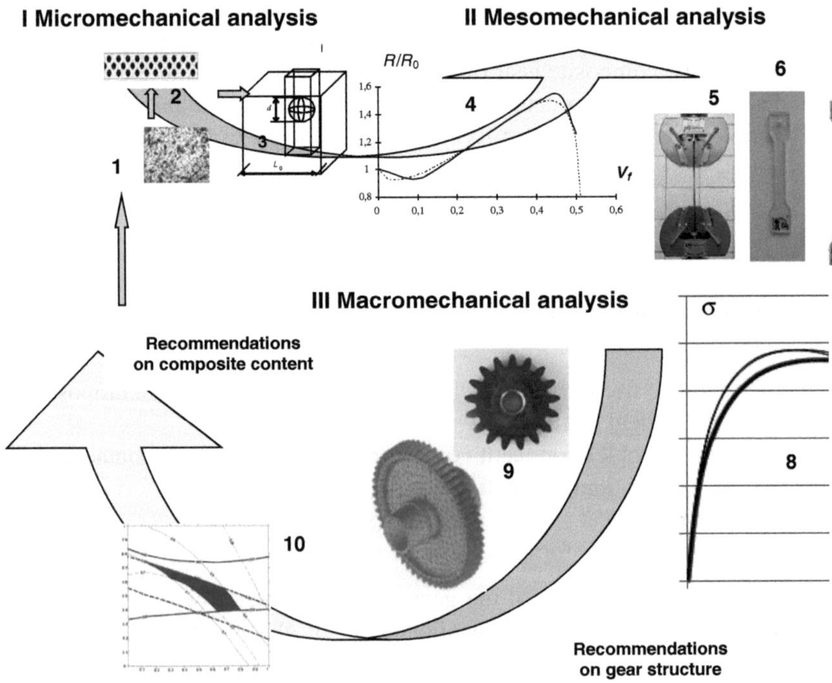

Fig. 4 Scheme of three-level strength calculation for the gears

element are recorded, which take into account the properties of separate components and the composite structure (dimensions and shape of reinforcing inclusions, the interphase layer thickness, the density of microdefects, etc.). The reference deformation or strength characteristic is calculated for the representative composite volume under varying parameters of the filler (volume content, particles form, etc.) (stage 4).

Effective mechanical characteristics determined at the 1st (micromechanical) level and verified at the 2nd (mesomechanical) level serve further as initial data for macromechanical (3rd level) calculation of the gear and the gear drive as a whole.

Figure 4 shows the main stages of the study:

1. microstructural analysis of the composite;
2. structure schematization;
3. structure element (periodicity cell) formation;
4. determination of dependence between the composite strain and strength characteristics and the filler parameters at applying the micromechanical model;
5. development of the mechanical tests scheme and selection of the representative volume of the composite;
6. fabrication of the experimental specimen of the composite representative volume;
7. calculation of the strain-stress state of the composite (specimen) representative volume;
8. verification of the mesomechanical model of the representative volume;
9. solid state modeling and numerical analysis of the gear strain-stress state;
10. optimization of the material composition and gear geometry by a group of quality criteria.

As an example, let us consider the calculation of the gear drive formed by a metal pinion and a gear wheel made of disperse-reinforced composite (polyamide for the matrix, glass fiber for the filler).

The initial data for calculation of the composite gear taking into account the deformable interphase layer by the method [10–12] include the Young modulus of the matrix material $E_m = 2$ GPa; Poisson's ratio of the matrix material $\nu_m = 0.45$; the yield limit of the matrix material $\sigma_m = 50$ MPa; the Young modulus of the filler material $E_f = 70$ GPa; Poisson's ratio of the filler material $\nu_f = 0.2$; the anisometry parameter χ (the relation between the ellipsoid small axis modeling the filler particle and its large axis) set equal to 0.1; the filler volume content $V_f = 0.05 \div 0.35$; the Young modulus of the interphase layer material $E_I = 1$ GPa; Poisson's ratio of the interphase layer material $\nu_I = 0.4$; the yield limit of the interphase layer material $\sigma_I = 100$ MPa; and the relative volume of the interphase layer q (the relation between the layer volume and the filler volume) varied within $0.001 \div 0.1$. The results of the calculation of the composite elasticity modulus and Poisson's ratio are listed in Table 1.

After the micromechanical model has been identified, analysis of the strain-stress state of the spur gear with module $m = 3.5$ mm (the geometrical parameters of which are shown in Fig. 5) was performed.

Table 1 Elastic modules of the composite material for various volume content of the filler for anisometry parameter $q = 0.1$

V_f	0.05	0.10	0.15	0.20	0.25	0.30	0.35	0.70
E_c (GPa)	2.4	2.8	3.3	3.8	4.4	5.0	5.6	13.9
ν_c	0.445	0.439	0.433	0.426	0.421	0.415	0.409	0.369

Profile point	Point coordinates, mm	
	x	y
M	0.0	18.542
E2	-0.369	18.538
CE2-E1	-0.361	18.157
RE2-E1	0.381	
E1	-0.666	18.386
CE1-2	8.581	11.466
RE1-2	11.549	
2	-1.510	17.084
C2-3	4.805	13.568
R2-3	7.228	
3	-2.216	15.284
C3-4	3.401	13.911
R3-4	5.782	
4	-2.350	14.519
C4-L	-0.897	14.366
R4-L	1.461	
L	-2.338	14.129
L1	-2.333	14.096
CL1-F	-3.716	13.867
RL1-F	1.402	
F	-3.353	12.513

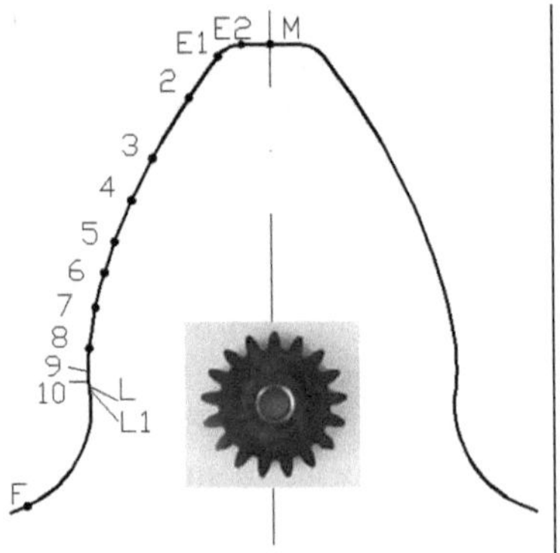

Fig. 5 Model of spur tooth with coordinates of profile points

At the finite element modeling of the tooth meshing, the contact of repeated fragment of the gear and rigid circular cylinder of the equivalent radius $R = 2.3$ mm simulating a metal pinion was studied for two characteristic meshing phases (Fig. 6). Specific load F/b_w = 10 N/cm was applied along the normal to the tooth involute profile and, on the assumption of its plane strain state, maximum equivalent and contact stresses were determined which were then compared with the limiting values through criteria of bending and contact strength, respectively.

By solving the problem of the indicated mating bodies' contact in the tooth top zone of a composite gear, we obtained the contact pressure and equivalent stress distributions which determine the onset of plastic deformation when the material in the surface layer reaches the yield limit. The maximum values of these stresses for different volume content of the filler for the relative interphase layer volume $q = 0.1$ are shown in Table 2.

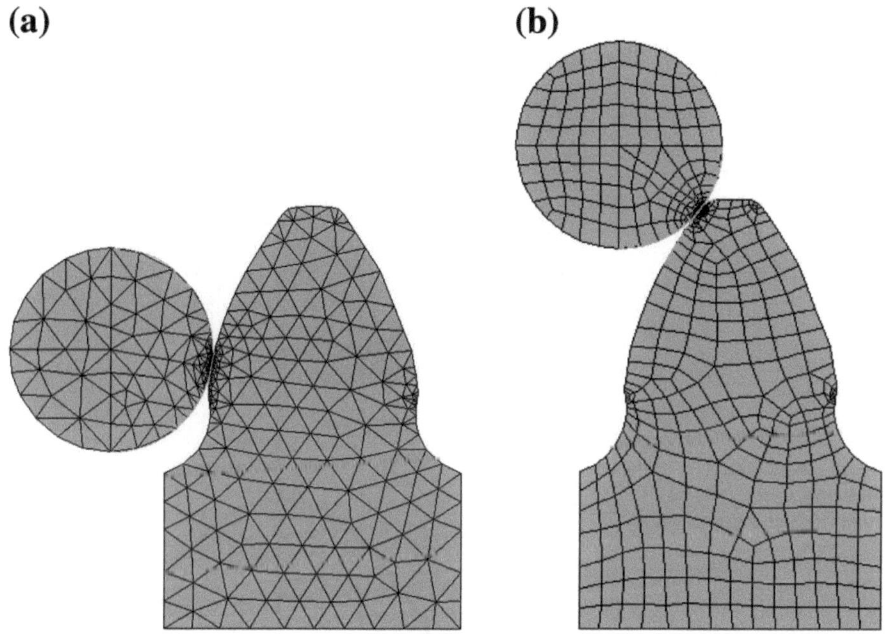

Fig. 6 FEM simulation of a tooth meshing at contact in the pitch point (**a**) and on tooth *top* (**b**)

Table 2 Results of the finite element analysis of the stress state in the contact region and subsurface layer (q = 0.1)

V_f	0.05	0.10	0.15	0.20	0.25	0.30	0.35	0.70
p (MPa)	5.87	6.24	6.55	6.81	7.06	7.37	7.58	9.29
σ_{eqv} (MPa)	52.4	56.2	59.1	61.8	64.4	67.4	69.6	87.3

By using the calculated elastic characteristics, we obtained the contact pressure and equivalent stress distributions which determine the onset of plastic deformations when the material in the surface layer reaches the yield limit.

At parameter q = 0.001, increase of the filler volume content up to the limiting (determined by the mixing technology) value V_f = 0.70 causes a larger (almost threefold) increase in the maximum contact pressure and an even greater increase (more than fivefold) in the equivalent stress.

Figure 7 shows that equivalent stress concentration occurs in the vicinity of the contact zone.

The distribution of pressures in the concentrated contact in question using coordinates related to the contact half-width x/a is shown in Fig. 8. The calculated values may be used for the contact strength estimation and wear prediction.

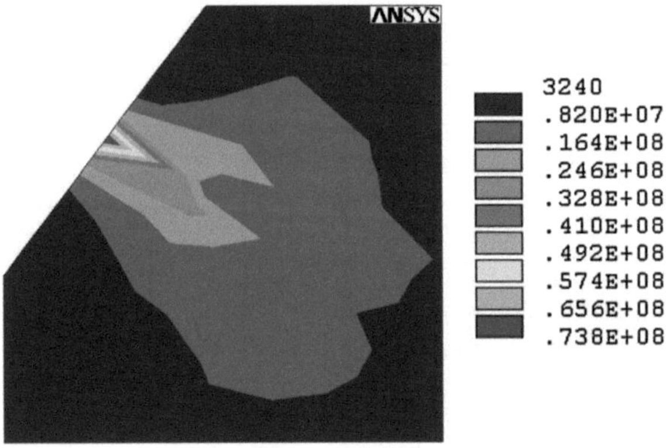

Fig. 7 Distribution of equivalent stresses, p_a, in the gear subsurface layer

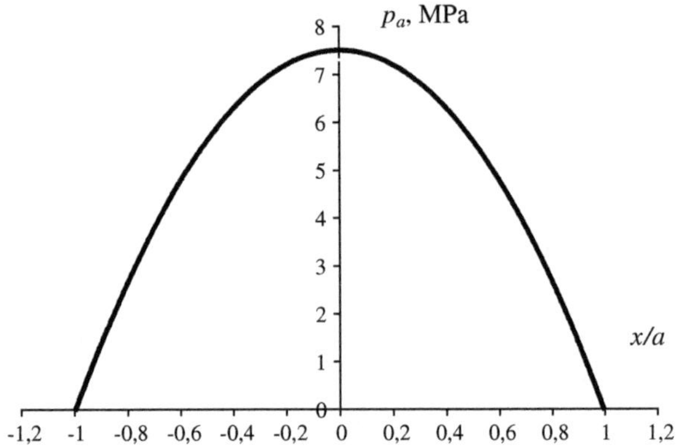

Fig. 8 Distribution of contact pressures at the joint of composite and metal gears

3 Consideration of Specific Features of Polymer Materials in Assessing Deformability of Gears

3.1 Analysis of the Influence of a Material's Poisson's Ratio on Gear Deformability

The polymer gears display moderate vibration activity, which is largely attributed to a low elasticity modulus (less than 10 GPa) of the antifriction and, at the same time, structural plastics from which they are made (carbon and glass-filled polyamides, textolite, etc.). On the other hand, elevated bending and contact compliance, which is intrinsic for the polymer gear teeth, is a source of kinematic errors in the meshing. Unlike with metallic gears, the elastic characteristics of which are adjusted by geometrical correction or thermal treatment within a relatively narrow interval of values, when developing composite gears, a designer should employ wide potentialities of the optimization of deformation characteristics of the composite gears by varying the structure of the material, e.g., reinforcing filler content, while the teeth geometry remains unchanged.

Thus, the methods of composite materials science make it possible to obtain the materials with abnormally low (negative) values of Poisson's ratio $\nu < 0$ called "auxetics" [22]. These materials are significantly resistant to shear strains; this property may be used to increase the shape stability of the polymer gears when working as a part of power transmission systems. Therefore, to adjust the elastic properties, it is justifiable to use the three-level (micro-, meso- and macromechanical) method for designing the composite machine parts and structural elements the general scheme of which with respect to gears is described in Sect. 2.4.

It is of interest to define the dependence of the contact and bending stiffness of a tooth and to find the actual contact ratio of the composite gearing versus Poisson's ratio within a theoretically admissible interval of values of the given parameter for an isotropic material.

Using the example of a metal-polymer gear (module $m = 3.5$ mm) discussed in Sect. 2.3, corresponding calculations have been performed. It is noteworthy that variations in the sign of Poisson's ratio significantly affect the distribution of horizontal u_x and vertical u_y elastic displacements (Figs. 9 and 10). To isolate the components of the contact and bending stiffness of the tooth, we have performed calculations for two variants of fixing the gear fragment in question (Fig. 6). In the

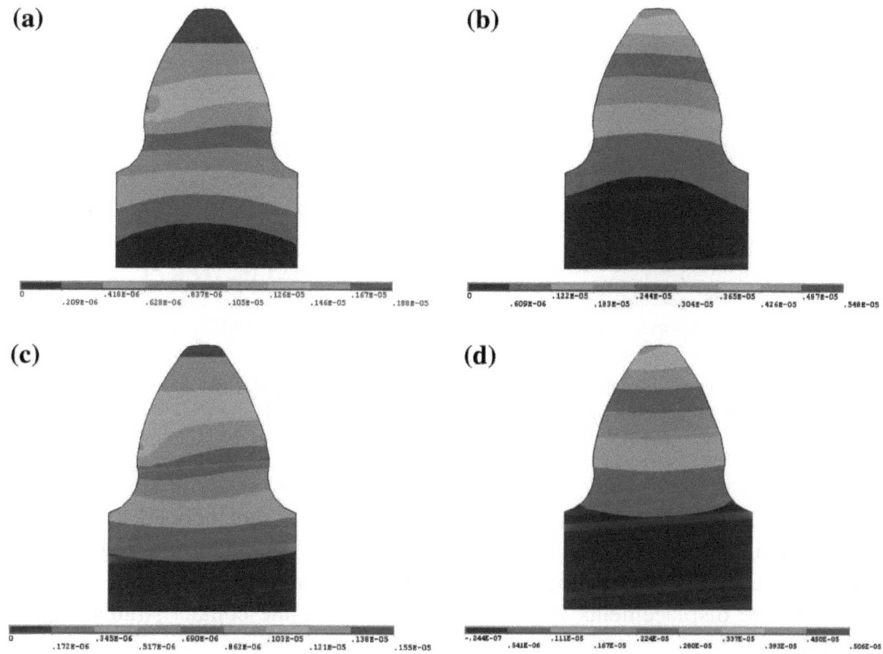

Fig. 9 Distribution of displacements u_x: contacting at the pitch point (**a**) and on tooth tip land (**b**) when Poisson's ratio is $\nu = 0.4$; contacting at the pitch point (**c**) and on tooth tip land (**d**) when Poisson's ratio is $\nu = -0.4$

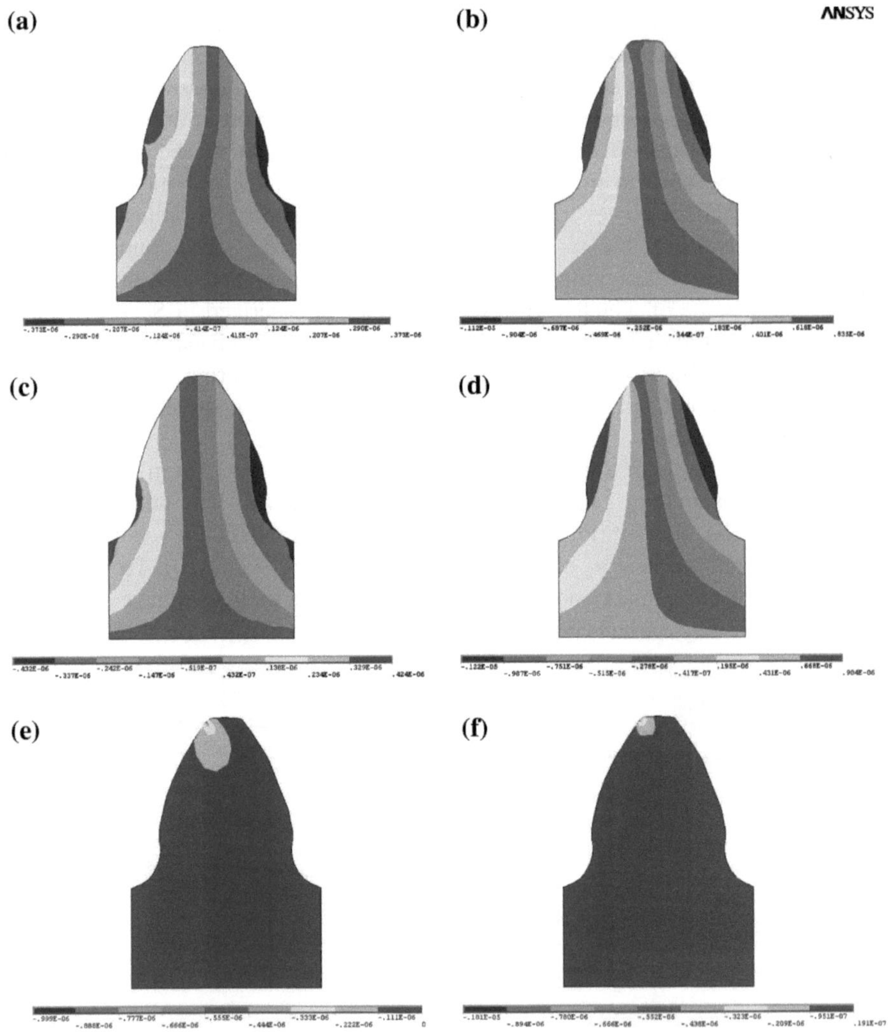

Fig. 10 Distribution of displacements u_y: contacting at the pitch point (**a**) and on tooth tip land (**b**) at $v = 0.4$; contacting at the pitch point (**c**) and on tooth tip land (**d**) at $v = -0.4$; on tooth tip land at $v = 0.4$ (**e**) and $v = -0.4$ (**f**) at additional fixing of tooth over its *right-side* contour

case of the lower edge fixing, displacements u_x and u_y are defined by both the bending and contact deformations (Figs. 9 and 10c). The additional fixing of the tooth at the right-side contour was needed to minimize the bending factor, wherefore the displacements were induced mainly by the contact strains (Figs. 9 and 10d).

Table 3 lists the maximal values of displacements, bending displacement coefficient Δ_1 along x axis and the ratio $\Delta_2 = \Delta_1/u_x$ for the top contact under varied ν.

Table 3 Contact and bending stiffness values of tooth depending on Poisson's ratio ν

ν	0.426	0.426[a]	0.1	0.1[a]	−0.1	−0.1[a]	−0.4	−0.4[a]	−0.9	−0.9[a]
u_x (μm)	5.48	1.25	6.29	1.50	6.13	1.49	5.06	1.26	1.11	0.29
u_y (μm)	0.84	−0.99	1.06	−1.21	1.06	−1.20	0.90	−1.01	0.20	−0.23
Δ_1	4.23	–	4.79	–	4.64	–	3.75	–	0.82	–
Δ_2	0.77	–	0.76	–	0.76	–	0.74	–	0.74	–

Fixing over the right tooth contour is marked by [a]

The reduction of the bending (to a larger extent) and contact displacements with decreasing Poisson's ratio (especially at $\nu < -0.4$) is the evidence of the considerable increment of tooth rigidity (Figs. 11 and 12).

In practice, it is important to estimate the effect of teeth compliance on the transverse contact ratio. The actual (real) transverse contact ratio $\varepsilon_{\alpha r}$, which is determined taking into account other elastic displacements of the teeth,

$$\varepsilon_{\alpha r} = \varepsilon_\alpha + \Delta\varepsilon_\alpha \tag{25}$$

consists of the theoretical transverse contact ratio ε_α, realized in the case of the absolutely rigid teeth and increment of the overlap factor $\Delta\varepsilon_\alpha$ induced by the elastic displacements [23]:

$$\varepsilon_\alpha = \frac{z_1}{2\pi}\varphi_1; \tag{26}$$

$$\Delta\varepsilon_\alpha = \frac{z_1}{2\pi}(\varphi_1' + \varphi_2''), \tag{27}$$

where φ_1 is the transverse angle on the pinion disregarding the effect of displacements; φ_1' is a supplementary transverse angle on the pinion at engagement; and φ_1'' is the same at disengagement.

The approximate relations for defining supplementary overlap angles are presented in [24]

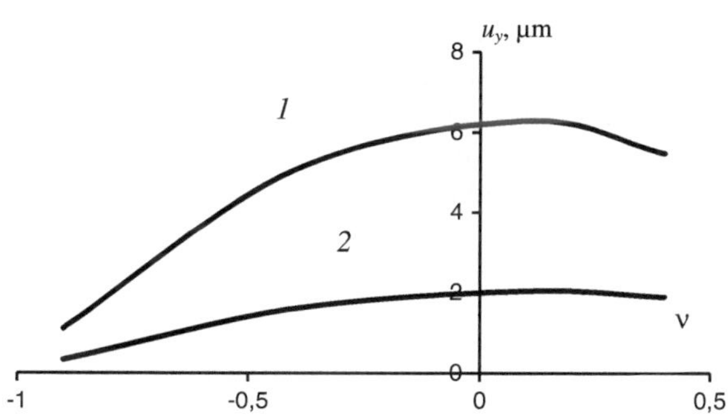

Fig. 11 Dependence of displacements along y axis versus Poisson's ratio ν: *1* at tooth tip land; *2* at pitch point

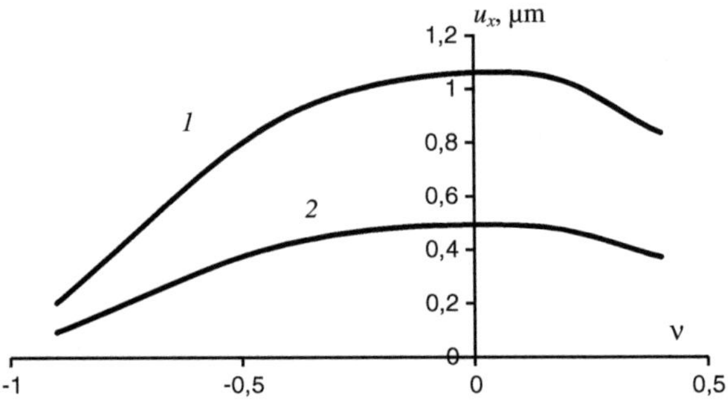

Fig. 12 Dependence of displacements along *x* axis *versus* Poisson's ratio *ν*: *1* at tooth tip land; *2* at pitch point

$$\varphi_1' = \sqrt{2\delta_{V2}/\rho_{a1}(1+\rho_{a1}/\rho_{p2})};$$
$$\varphi_2' = \sqrt{2\delta_{U2}/\rho_{a2}(1+\rho_{a2}/\rho_{p1})} - \delta_{U2}/r_{b1}, \tag{28}$$

where δ_{V2}, δ_{U2} are teeth displacements at points V_2 and U_2; and $\rho_{a1(2)}$, $\rho_{p1(2)}$ are the curvature radii of the teeth at points $a_{1(2)}, p_{1(2)}$. If the gear ratio is $u = 1$,we shall have

$$\varphi_1' = \sqrt{2\delta_V/\rho_a(1+\rho_a/\rho_p)};$$
$$\varphi_2' = \sqrt{2\delta_U/\rho_a(1+\rho_a/\rho_p)} - \delta_U/r_b, \tag{29}$$

where ρ_a and ρ_p are the profile curvature radii at the starting point (on tooth tip land) and end (on dedendum) of the effective line of action, correspondingly; δ_U and δ_V are tooth displacements at the points of start (*U*) and end (*V*) in the one-tooth mesh zone, respectively.

To analyze the effect of Poisson's ratio *ν* on the supplementary transverse contact ratio $\Delta\varepsilon_\alpha$, we may use the dimensionless displacement factors of the teeth $\chi = \delta E b_w / F_t$ at points *U* and *V* which can be found from the nomograms [20] for the gears with a standard basic rack.

Figure 13 illustrates an increment of the overlap factor $\Delta\varepsilon_\alpha$ calculated by formulas (27) and (29) and its variations with reducing Poisson's ratio similar to the dependencies of displacements $u_x(\nu)$ and $u_y(\nu)$ (Figs. 11 and 12) found by the finite-element method.

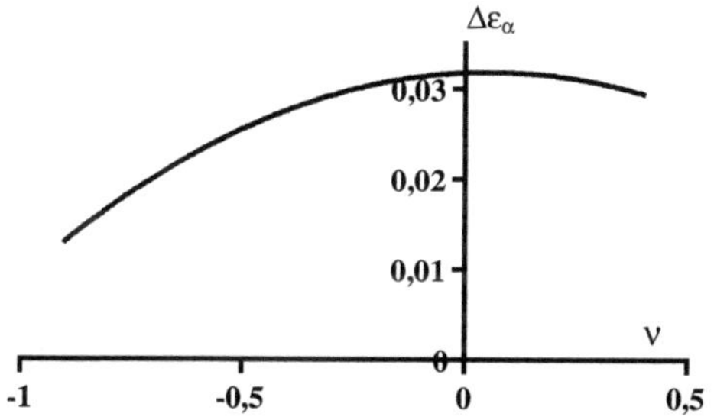

Fig. 13 Dependence of transverse contact ratio increment versus Poisson's ratio ν

3.2 *Analysis of the Influence of a Material's Viscoelasticity on Composite Gear Deformability*

Rheonomous properties of the polymer materials that manifest themselves as creeping and relaxation exert significant influence on the deformability and strength of structural elements. Creep deformations reduce shape stability and worsen the product performance characteristics by being, for example, a source of kinematic errors in the gearing. At the same time, the viscoelastic effects play a significant role by facilitating vibration damping and, in certain cases, favorable redistribution of stresses and an increase in the load-bearing capacity as the result of the structure's "adjustment" to external loads.

In view of this, description of the rheonomous properties of the applied materials with subsequent analysis of the strain-stress state of the structure in viscoelastic setup is the indispensable stage of the polymer structure strength calculations. It should be noted that analytic solutions of the theory of viscoelasticity are applicable only for bodies of canonical forms. This makes it difficult to use them in calculations of polymer gears of complex configuration. On the other hand, the finite-element method broadly used for solving applied problems in elastic and elastoplastic setup is rarely used at the moment for modeling viscoelastic behavior of real structures and machine parts.

The aim of the present study is to develop strength calculation methods for polymer elements of structures of complex geometrical shape through the example of the gears and taking into account the material viscoelasticity.

Study method. A number of viscoelastic models have been proposed so far by domestic and foreign scientists to describe the creeping and relaxation processes of polymer materials, e.g., [25]. Since the widespread software products for finite-element analysis, e.g., ANSYS [26], make use of Prony's linear viscoelasticity model, let us write the defining relations to determine the rheological constants of the material based on the results of standard mechanical tests [27] as

$$\sigma_{ij}(t) = \int_0^t 2G(t-\tau)\frac{e_{ij}(\tau)}{d\tau}d\tau + \delta_{ij}\int_0^t K(t-\tau)\frac{d\varepsilon(\tau)}{d\tau}d\tau. \quad (30)$$

Here, $G(\xi) = G_\infty + \sum_{i=1}^{n_G} G_i e^{-\frac{\xi}{\lambda_i^G}}$; $K(\xi) = K_\infty + \sum_{i=1}^{n_K} K_i e^{-\frac{\xi}{\lambda_i^K}}$; $G(0) = G_\infty + \sum_{i=1}^{n_G} G_i = \mu$; $K(0) = K_\infty + \sum_{i=1}^{n_K} K_i = K$; G and K are shear and volumetric moduli, respectively; and t is the time. Taking into account the known correlations for elastic constants [15]

$$E = \frac{\mu(3\lambda + 2\mu)}{\lambda + \mu};\ \nu = \frac{\lambda}{2(\lambda+\mu)};\ \mu = \frac{E}{2(1+\nu)}$$

and assumption of the constancy of the material's Poisson's ratio $\nu = \text{const}$, we shall have the following dependencies

$$ER_E(t-\tau) = 2(1+\nu)G'_\tau(t-\tau);$$
$$K'_\tau(t-\tau) = \frac{2(1+\nu)}{3(1-2\nu)}G'_\tau(t-\tau).$$

Then, Eq. (30) can be written in the following form:

$$\sigma_{ij}(t) = \frac{E}{1+\nu}\left[\frac{\nu}{1-2\nu}\varepsilon(t)\delta_{ij} + \varepsilon_{ij}(t)\right] - \frac{E}{1+\nu}\int_0^t R_E(t-\tau)\left[\frac{\nu}{1-2\nu}\varepsilon(\tau)\delta_{ij} + \varepsilon_{ij}(\tau)\right]d\tau.$$

We obtain an analytical force-versus-time dependence in the relaxation test

$$P(t) = E\varepsilon_0 S_0\left(1 - \frac{2(1+\nu)}{E}\left(\sum_{i=1}^{n_G} G_i\left(1 - e^{-\frac{t}{\lambda_i^G}}\right)\right)\right),$$

where S_0 is the specimen cross-section area.

For example, to determine the viscoelastic parameters of polymer material Polyamide 6, mechanical tests of standard cylindrical specimen for relaxation under compression were performed on an Instron 5567 machine. The obtained dependencies of the normalized force versus time for various strain levels ε at the Young modulus of the material E = 1650 MPa are shown in Fig. 14.

It is noteworthy that stabilization of the compression load (for the given material) occurs rather quickly, which allows for reducing the testing time (in this case—to 3 h), i.e., to obtain the required data from the short-term stress relaxation testing.

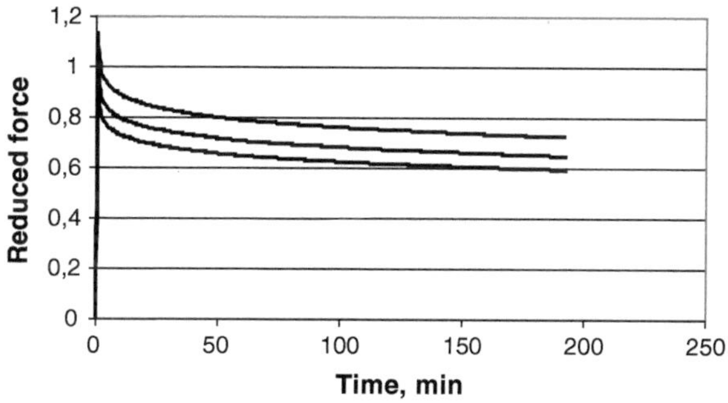

Fig. 14 Dependence of the normalized force $\left(\frac{P(t)}{E\varepsilon S_0}\right)$ versus time: lower curve corresponds to $\varepsilon = -0.051$, middle—to $\varepsilon = -0.043$, upper one—to $\varepsilon = -0.034$ (difference is not more than 20 %)

The following expression was written for square deviation (disparity) Δ:

$$\Delta = \sum_{i=1}^{4} \left(P^i_{eqv} - 1 + \frac{2(1+\nu)}{E} \left(\sum_{i=1}^{n_G} G_i \left(1 - e^{-\frac{t}{\lambda_i^G}} \right) \right) \right).$$

Henceforth, the power characteristic under deformation $\varepsilon = -0.043$ as 4 pairs of time and normalized force values (Table 4) was used for calculation of Prony's model parameters.

By minimizing the disparity, we obtain constants G_∞, G_1, λ_1 depending on the Young modulus on the assumption of the known Poisson's ratio (hereinafter used as $\nu = 0.4$).

Identification results. At identification of the Prony's model ($n_G = 1$) for Polyamide-6 material at $E = 1650$ MPa, $\nu = 0.4$ the following nuclei of volumetric and shear relaxation (respectively) were obtained:

$$K(\xi) = 1783.8 + 962.3 \times e^{-\frac{\xi}{4.3}};$$

$$G(\xi) = 382.8 + 206.5 \times e^{-\frac{\xi}{4.3}}.$$

In this case, $K_\infty = 1783.8$ MPa, $K_1 = 962.3$ MPa, $\tau_1^K = 4.3$ min, $G_\infty = 382.8$ MPa, $G_1 = 206.5$ MPa, and $\tau_1^G = 4.3$ min.

According to the ANSYS program description [26], the following additional parameters have been introduced: $G_0 = G_\infty + G_1$, $K_0 = K_\infty + K_1$, $\alpha_1^G = \frac{G_1}{G_0}$, $\alpha_1^K =$

Table 4 Values of the normalized force for 4 points of time

t_i (min)	2.6	16.1	50.1	111.6
$\frac{P}{E\varepsilon_0 S_0}$, $E = 1650$ MPa	0.86	0.77	0.72	0.68

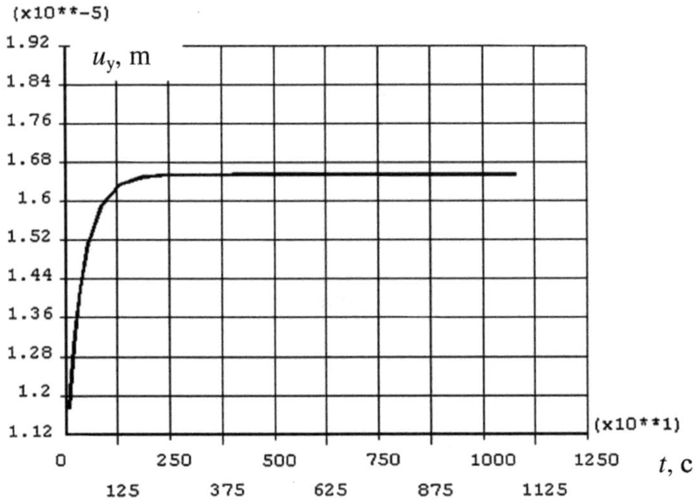

Fig. 15 Dependence of displacements along y axis versus time

$\frac{K_1}{K_0}$ which take the following values for the available experimental data: $G_0 = 589.3$ MPa, $K_0 = 2750$ MPa, $\alpha_1^G = 0.345$, $\alpha_1^K = 0.345$. Elastic parameters should also be added: $E = 1650$ MPa, $\nu = 0.4$.

Analysis of gear deformability. To analyze viscoelastic effects during operation of a cylindrical metal-polymer gear, we studied, as in the examples above (see Sects. 2.3, 3.1), the mating of the unit tooth of a polymer gear with module $m = 3.5$ mm and the rigid circular cylinder of the equivalent radius simulating a metal pinion.

The calculation data also include the previously determined characteristics of the gear material (Polyamide-6): Young modulus $E = 1650$ MPa, Poisson's ratio $\nu = 0.4$, friction coefficient $f = 0.3$; rheological constants $\alpha_1^G = 0.35$, $\tau_1^G = 4.3$ min. The results of calculation of displacements in time are shown in Fig. 15.

It is noteworthy that the rheonomous properties of the polymer material noticeably manifest themselves in the initial period when the major increase in viscoelastic displacements occurs.

3.3 Example of Calculation of the Composite Gears on Wear Resistance

The wear resistance calculation described in Sect. 2.2 has been performed to assess the effectiveness of reinforcement with glass and carbon fibers of the polyamide gearing wheels of the starting motor drive. The following gearing parameters were set: module $m = 3.5$ mm; number of teeth $z = 15$; relative radius of curvature in the wearing zone $R = 10.87$ mm; face width $b_w = 13.5$ mm; torque $T = 4040$ N mm; rate of rotation $n = 2850\ \text{min}^{-1}$; slip ratio $\lambda = 0.2$; allowable wear for the basic option (without a filler) $h^{\text{m}}_{\text{IP}} = 1$ mm.

According to the reference data, the matrix elasticity and strength characteristics (Polyamide PA-6) are as follows: $E_m = 1.5$ GPa; $\nu_m = 0.4$; $\sigma_m = 70$ MPa. The corresponding parameters of the filler are: (a) glass fiber $E_f = 70$ GPA; $\nu_f = 0.2$; $\sigma_f = 3500$ MPa; and (b) carbon fiber $E_f = 250$ GPa; $\nu_f = 0.2$; $\sigma_f = 3500$ MPa. Coefficients α and β in (1)–(3), which have been determined on the basis of strength characteristics and the Young modulus of glass-filled polyamide with the fiber volume content $V_f = 0.15$ and $\sigma_c = 180$ MPa, $E_c \cong 5$ GPa, were equal to 0.375 and 0.23, respectively. The coefficient of sliding friction of the materials in question was selected based on the calculated pressure according to experimental dependencies *f(p)*. The polyamide strength limit and friction fatigue index are $\sigma_0^m = 180$ MPa and $t = 2$.

The results of calculation of elasticity and strength parameters of glass- and carbon-filled plastics, as well as contact pressures in the engagement under varied volume content of the fiber, are shown in Table 5. One can see a considerable increase in elasticity modulus as the result of reinforcement: for example, at $V_f = 0.3$, a sixfold increase is observed for glass-filled plastic and an almost 20-fold increase for carbon-filled plastic; the strength limit increases almost 4.5 times. At the same time, nominal contact pressure for the first composite increases by 2.5 times and for the second one, by 4 times. Nevertheless, we can assume that the contact loading remains lower than the limit value due to reduction in the matrix creeping and considerable increase in strength σ_c.

Assessment of the friction fatigue parameters of composite σ_c^0 was performed according to the similar parameter for the matrix and the composite tensile strength.

$$\sigma_c^0 = \sigma_m^0(\sigma_c/\sigma_m).$$

Table 5 Dependence of parameters E, σ_c and contact pressure p_a versus degree of reinforcement V_f

V_f	Young modulus E (GPa)		Strength limit σ_c (MPa)	Contact pressure p_a (MPa)	
	Glass-filled plastic	Carbon-filled plastic		Glass-filled plastic	Carbon-filled plastic
0.00	1.50	1.50	70	16	16
0.05	2.78	6.11	107	22	32
0.10	4.05	10.72	144	26	45
0.15	5.32[a]	15.34	180[a]	30	51
0.20	6.60	19.95	217	34	59
0.25	7.87	24.57	254	37	65
0.30	9.15	29.18	291	40	71
0.35	10.43	33.78	327	42	76

[a]Characteristics of glass-filled polyamide

Due to a lack of data on the influence of the degree of reinforcement V_f on the friction fatigue index of the composite t^c, this parameter was assumed to be constant and corresponding to the matrix material: $t^c = t^m$.

Coefficient k_p, which takes into account the fracture pattern in surface layers, was taken to be equal to 3 for high-elastic materials; the value of a microgeometric complex for a polished surface is $\Delta = 0.10$; the coefficient which takes into account the form and height of individual asperities $K_1 = 0.2$.

Strength characteristics σ_f of both types of fillers were the same, which allowed us to find out the wear resistance dependence on the elasticity modulus and friction coefficient.

Based on the calculated contact pressures, the friction coefficients obtained for glass- and carbon-filled plastics were set. The friction coefficient for the first material is relatively stable and, thus, was assumed to be constant under varied V_f: $f_f = f_m = 0.4 = const.$ For carbon-filled plastic, since the friction coefficient change was set within the range from $f = f_m = 0.4$ (polyamide matrix) to 0.3 at $V_f = 0.05$; at $V_f = 0.10$ and higher, we assumed that $f = 0.2$. This was motivated by the fact that the carbon-filled plastic friction coefficient decreases rather quickly at the initial stage of reinforcement (under relatively low pressures p_a = 0.4–0.6 MPa). At increase of the degree of filling, the friction coefficient stabilizes. Decrease of f under pressure rise is also observed for considerably higher loads (up to 90 MPa), which are close to the p_a values presented in Table 5.

Change in the filler content (both for glass- and carbon-filled plastic) under constant pressure results in considerable (several fold) reduction of the wear rate which is most notable at introduction of the filler up to 10 vol.%. Reduction of the tooth contact area at reinforcement stipulates the change in pressure which offsets the mentioned effect: at $V_f = 0.3$, the wear rate at filling with carbon fiber I_h^y decreases in total by 60 %, while in the case of glass fiber (I_h^y), by 45 %. In these circumstances, the wear rate of carboplastic is slightly higher than that of the material reinforced with glass fiber, in particular, at a low degree of filling. This can be explained by the increased elasticity modulus and contact pressure in the first case. However, higher stiffness of carboplastic results in reduction of the friction distance and, eventually, wear resistance of the gear made of this material proves to be 6.4 times higher than the initial one (4 times higher for glass plastic).

The collateral effect of the increase in service life (at $V_f = 0.05$, 5 times, at a higher degree of filling, by an order of magnitude) is assured by the increase in allowable wear h_{IP} owing to the increase in tooth bending strength.

The data obtained by testing composites with antifriction components prove the possibility of getting the material for which, at optimal content of the filler, maximal bending strength and minimal friction coefficient are attained at the same time. Wear resistance calculation for glass plastic using approximation for the given variant of change $f = 0.17 + 15(V_f - V_{fo})^2$ with a coordinate of the friction coefficient minimum $V_{fo} = 0.12$ demonstrated an increase in wear resistance by an order of magnitude.

A specific feature of the wear of filled polymers is the formation in the friction zone of the filler "exposure" areas which exert dual influence on the wear. On the one hand, the filler coming out into the friction zone facilitates more uniform distribution of load, while, on the other hand, it stipulates localization of temperatures on the contact spots and catastrophic wear of the binder in the surrounding areas.

In this case, zone formation with various physical effects occurs: these zones are homogeneous as far as load transmission is concerned and inhomogeneous with respect to surface temperature. Since the competitive action of strengthening and heat generation mechanisms under external friction occurs, it is obvious that maximum effectiveness of reinforcement is reached when using the carbon-filled-plastic.

The performed complex analysis of the reinforcing filler influence shows that the increase in wear resistance of the gear made of composite is assured by reduction of the friction distance, increase of the allowable wear and, to a lesser extent, reduction of the wear rate. As a result, the gears reinforced with carbon fibers have higher wear resistance (at the same matrix) as compared to the glass-filled materials.

4 Conclusions

Specific features of calculating the strength, deformability and wear resistance of gears made of polymer composites are addressed. Methods describing the effects of the reinforcement of polymer materials with disperse and fibrous filler, as well as viscoelastic properties of polymer composites, are proposed.

An original three-level (micro-, meso- and macro) method which allows for determining (with high accuracy and minimum computational cost) the strain-stress state, deformability, strength and wear resistance of composite gears, as well as optimum composition of the material on the basis of requirements for the gearing, is described.

Calculations of elastic and viscoelastic displacements, as well as wear resistance of spur cylindrical gear, have been performed for approbation of the methods. In particular, it has been shown that the deformability of polymer gears varied over a wide range by changing Poisson's ratio of the material and that considerable increase in the bending and contact stiffness values of teeth occurs as Poisson's ratio reduces in the area of its negative values.

References

1. Frolov KV (2008) Contemporary tribology: totals and perspectives. Publishing House, LKI (in Russian)
2. Kragel'sky IV, Dobychin MN, Kombalov VS (1977) Osnovy raschetov na trenie i iznos (Basis of calculations on friction and wear). Publishing House, Mashinostroenie, Moscow (in Russian)

3. Hebda M, Chichinadze AV (Ed) Reference book on triboengineering, vol 3. Publishing House, Mashinostroenie, Moscow (1989) (in Russian)
4. Goryacheva IG (2001) Mechanics of friction interaction. Publishing House, Nauka (in Russian)
5. Johnson KL (1987) Contact mechanics. Cambridge University Press, Cambridge; Publishing House, Mir (1989)
6. Goldfarb VI, Lunin SV, Trubachov ES (2004) Direct digital simulation for gears, vol 1. Izhevsk
7. Shil'ko SV, Petrokovets EM, Semenova TV (2001) Estimation of contact stiffness of coatings: comparison of calculation methods. Part 1. J Frict Wear 22:6–12
8. Sytar VI, Kuzyaev IM, Burya AI et al (2004) Optimization of the triboengineering characteristics of a phenylon-based composition. J Frict Wear 25:219–222
9. Shil'ko SV, Starzhinsky VE (1993) Prediction of wear resistance of gearing with wheels made of reinforced composites. J Frict Wear 14:7–13
10. Shil'ko SV, Starzhinsky VE, Petrokovets EM, Chernous DA (2013) Two-level calculation method for tribojoints made of disperse-reinforced composites. Part 1. J Frict Wear 34:65–69
11. Shil'ko SV, Starzhinsky VE, Petrokovets EM, Chernous DA (2014) Two-level calculation method for tribojoints made of disperse-reinforced composites. Part 2. J Frict Wear 35:47–55
12. Starzhinsky VE, Shalobaev EV, Shil'ko SV et al (2012) Elements of device drives. In: Pleskachevskii YM (ed) Calculation, design, technologies. Publishing House, Belaruskaya Navuka (in Russian)
13. Fudzii T, Dzako M (1982) Fracture mechanics of composite materials. Mir, Publishing House (in Russian)
14. Crouch SL, Starfield AM (1987) Boundary element methods in solid mechanics. Publishing House: George Allen and Unwin, London
15. Christensen R (1979) Mechanics of composite materials. Wiley, New York; Publishing House, Mir (1982)
16. Kravchuk AS, Maiboroda VP, Urzhumtsev YuS (1985) Mechanics of polymer and composite materials. Nauka, Publishing House (in Russian)
17. Shil'ko SV, Chernous DA, Panin SV (2012) Mesomechanical analysis of polymer composites reinforced by short fibers with taking into account an interphase layer. Mech Compos Mater 48:171–178
18. Shil'ko SV, Chernous DA, Panin SV (2013) The modified Takayanagi model of deformation for dispersed-filled composites. Part. 2. Determination of elastic moduli and yield limit with taking into account the interface layer. J Compos Mater Struct 19:181–195 (in Russian)
19. Shil'ko SV et al (2002) Simulation of contact interaction in MEMS conjugations. In: Bulletin Gomel State Technological University named by P.O. Suchoy, vol 3, pp 31–38 (in Russian)
20. Belyi VA, Starzhinsky VE, Scherbakov SV (1981) Metal-polymer gear drives. Publishing House, Nauka i Technika (in Russian)
21. Blagodarnyi VM (1985) Wear and strength computation of fine-module gear drives. Publishing House, Mashinostroenie (in Russian)
22. Konyok DA, Wojciechowski KW, Pleskachevsky YM, Shil'ko SV (2004) Materials with negative Poisson's ratio (The review). J Compos Mater Struct 10:35–69 (in Russian)
23. Muller R (1972) Maschinenteile aus Kunststoff. Ausbau 8:491–501
24. Shandalov KS (1968) Influence of teeth flexibility on contact ratio. In: Kolchin NI (ed) Gear and worm drives. Publishing House, Mashinostroenie, pp 90–101 (in Russian)
25. Koltunov MA (1976) Creep and relaxation. Publishing House, Vysshaya Shkola (in Russian)
26. ANSYS Revision 10–14
27. Gavrilenko SL, Shil'ko SV (2014) Identification of Prony linear viscoelastic model on the results of test at relaxation under compression. In: Composite book, Theoretical and applied mechanics, vol 29, pp 219–223 (in Russian)

Selection of Overlap Ratios in Helical Gear with Low Vibration Excitation

O. Kosarev

Abstract A mathematical model of excitation of oscillations in tooth contact of a helical gear is refined where oscillations are excited by a pitch error. It is shown that choosing the magnitudes of transverse and overlap contact ratios in a helical gear provides minimization of vibration excitation caused by simultaneous action of three forcing factors: variable rigidity of meshing, impulse loading of teeth and pitch error. Examples of calculation of exciting forces, amplitude-frequency characteristics of oscillations and tooth loads are given.

Keywords Vibration excitation · Tooth mesh · Helical gear · Transverse and overlap contact ratios · Time-varying mesh rigidity · Pitch error · Exciting forces · Amplitude-frequency characteristics · Tooth stresses · Impact loading

Decreasing the vibration and dynamic loading state of helical (herringbone) gears remains an urgent problem for modern mechanisms and machines.

Processes of excitation of oscillations in helical involute gears with tooth mesh frequency f and methods of decreasing their vibration excitation are considered.

Forcing factors or sources of excitation of oscillations with tooth mesh frequency and its harmonic components are as follows: variable rigidity of meshing, impulse (discrete) teeth loading, pitch error, profile error, helix angle error, skewing. There is also excitation, causing oscillations with side spectral frequencies $f \pm f_o$, where f_o is a rotational frequency [1, 2], but it is not considered here.

Among the pointed factors, there are two forcing factors that are independent of the accuracy of manufacture and assembly of gears: variable rigidity of meshing and impulse loading of teeth. They practically always exist and depend on the transverse contact ratio ε_α and the overlap ratio ε_β. In practice, contact ratio selection is performed by a design engineer on the basis of experience or experiments. At present, there is still no theory for choosing the contact ratios according to criteria of quietness of helical gears.

O. Kosarev (✉)
Blagonravov Institute of Machine Sciences, Russian Academy of Sciences,
Malyi Khariton'evskii per. 4, Moscow 101990, Russia
e-mail: oikosarev@yandex.ru

V. Goldfarb and N. Barmina (eds.), *Theory and Practice of Gearing and Transmissions*, Mechanisms and Machine Science 34,
DOI 10.1007/978-3-319-19740-1_17

Excitation of oscillations in gearing relates to a kinematic type and is defined by the product $C(t)\Delta(t)$, where $C(t)$ is the rigidity of meshing of a pair of teeth, $\Delta(t)$ is an error, and t is time.

Both functions $C(t)$ and $\Delta(t)$, in the general case, are variable in time. In case of excitation of oscillations by variable rigidity of meshing and impulse loading of teeth, the constant magnitude—static load P_{st} (or static deformation) serves as an error. In other cases, different errors are represented as $\Delta(t)$: pitch error, profile error, skewing in meshing and deviation of the helix angle. Functions of these errors are variable while the contact line moves along the field of meshing (along the tooth flank).

Every function appearing in the product $C(t)\Delta(t)$ is the forcing factor, and they are related by mutual influence. For example, in a case of great pitch errors that exceed static deformation, teeth entering into contact can knock in dynamics the earlier loaded teeth out of contact, thereby influencing contact ratios and correspondingly changing the total rigidity of meshing. Because of pitch error and "deformational error", teeth entering into contact mate at the off-design point out of the contact area with different linear velocities. This fact leads to edge contact of teeth that increases contact ratios. "Deformational error" is the total deformation of all pairs of loaded teeth at the moment that the next pair of teeth enters into contact.

All pointed forcing factors act simultaneously and excite oscillations with one and the same tooth mesh frequency. It is impossible to separate these factors in a physical experiment. It can be done only by numerical simulation [1, 2]. Variable rigidity of meshing $C(t)$ takes part in excitation together with all forcing factors, according to the formula $C(t)\Delta(t)$.

The influence of contact ratios on three forcing factors (variable rigidity of meshing, pitch error and impulse loading) is considered in this article. The term "exciting force" is conditionally applied below instead of the expression $C(t)\Delta(t)$ that has the force dimension.

Let us detail the applied notions of the forcing factors. The function of meshing rigidity of a pair of helical teeth (in contrast to spur teeth) is variable in time as is shown in Fig. 1. The magnitude of the gear pitch error is defined as $\Delta = p_{b2} - p_{b1}$, where p_{b2} and p_{b1} are gear pitches of the driven and driving wheels. The function $\Delta(t)$ has a form of inclined line [1, 2]. This error leads to a variable instant gear ratio and, together with "deformational error", to edge contact of teeth.

A factor of impulse (discrete) loading of teeth is conditioned by the fact that the torque in a gear is imparted by teeth that consequently enter into contact during a short period of time. The initial parameters of this factor are static load and initial conditions of oscillation at the moment of application (removal) of this load. It is impossible to inscribe this factor in the form of a separate formula. This factor is taken into account in software when defining the dynamic loads on teeth in equations of motion of an oscillating system.

In order to provide the evaluation of exciting forces, we will represent them in a frequency domain in the form of a Fourier series but without considering edge contact areas, for reasons of simplification. Functions of rigidity of a pair of teeth $C_k(t)$ and gear pitch error $\Delta(t)$ of the previous mathematical model are shown in [2]. In the earlier model, the error $\Delta(t)$ depended on ε_α and ε_β.

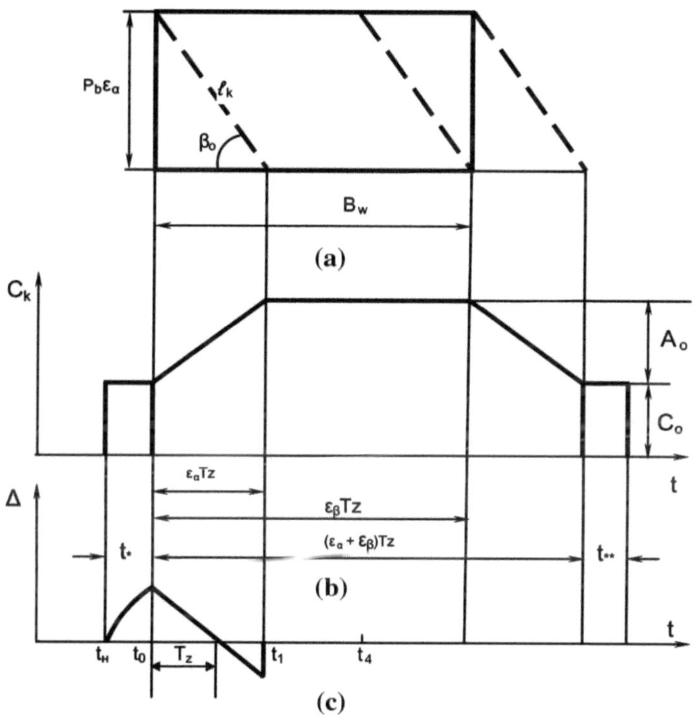

Fig. 1 To calculate exciting force FΔ: **a** contact area, **b** rigidity of tooth pair Ck(t), **c** function of pitch error Δ(t)

The present work proposes a new physical-mathematical model of the gear pitch error $\Delta(t)$. This model differs from the previous one by the fact that the time of function $\Delta(t)$ operation is limited here by the time of motion of a contact line along the involute profile in a face section between addendum and dedendum of the tooth. This time is defined by the coefficient ε_α, as in a spur gear. Further motion of the inclined contact line takes place along a spiral flank of a helical tooth defined by the overlap ratio ε_β. This motion is not considered in the new model, since it is not correlated with an involute profile error.

Figure 1 shows: (a) contact area, (b) rigidity of meshing of a pair of helical teeth $C_k(t)$, and (c) gear pitch error $\Delta(t)$. The function of rigidity of the kth pair of teeth C_k (the contact line l_k on the contact area corresponds to this function, Fig. 1a varies according to the trapezoid law during the time $t = \varepsilon_\Sigma T_z$, where summarized contact ratio $\varepsilon_\Sigma = \varepsilon_\alpha + \varepsilon_\beta$, T_z is a period of tooth mesh frequency, t_*, t_{**} is the time of edge contact, and C_0 and $(C_0 + A_0)$ are magnitudes of rigidities at the entrance (exit) and in the middle part of the contact area. The calculation results given below were obtained for the case of $C_0 = A_0$.

Analytical expressions of exciting forces in frequency-response area are obtained in the following way. Time functions of rigidity and the corresponding error are inscribed for every kth pair of teeth. According to the formula of kinematic excitation, the product of these functions is taken. The product is expended in the Fourier series for the period of the gearwheel turning $T_0 = zT_z$. Then, it is summarized according to a number of gearwheel teeth z and a constant component is subtracted from it.

For a helical gear, the exciting force due to variable rigidity is determined by formula [2]

$$F_c = 2C_0 w_{st} \sum_m \left[\frac{\sin \pi n(\varepsilon_\alpha + \varepsilon_\beta)}{\pi m} + \frac{A_0}{C_0} \frac{\sin(\pi n \varepsilon_\alpha)\sin(\pi n \varepsilon_\beta)}{\varepsilon_\alpha \pi^2 m^2} \right] \times \cos n\left[\omega t - \pi(\varepsilon_\alpha + \varepsilon_\beta)\right]. \quad (1)$$

For a new type of helical gear, the formula of the exciting force due to pitch error is obtained as a result of the above-stated operations with the product (Fig. 1)

$$F = (C_0 + A_0 \frac{t}{T_z})\, \Delta_0\, (1 - \frac{t}{T_z}) \quad \text{for} \quad (t_0 < t < t_1).$$

After expending this function in a Fourier series, we have

$$F_\Delta = \frac{a_0}{2} + \frac{2C_0\Delta_0}{T_0} \left\{ \left[\frac{1}{n\omega} - \frac{t_1}{n\omega}\left(\frac{1}{T_z} - \frac{A}{C_0 t_1}\right) - \frac{A_0}{C_0 t_1 T_z}\left(\frac{t_1^2}{n\omega} - \frac{2}{n^3\omega^3}\right)\right] \sin n\omega(t_1 - t) \right.$$
$$- \left[\left(\frac{1}{T_z} - \frac{A_0}{C_0 t_1}\right)\frac{1}{n^2\omega^2} + \frac{A_0}{C_0 t_1 T_z}\frac{2t_1}{n^2\omega^2}\right] \cos n\omega\,(t_1 - t)\Big]$$
$$\left. + \left[\frac{1}{n\omega} + \frac{A_0}{C_0 t_1 T_z}\frac{2}{n^3\omega^3}\right] \sin n\omega t + \left(\frac{1}{T_z} - \frac{A_0}{C_0 t_1}\right)\frac{1}{n^2\omega^2}\cos n\omega t \right\}$$
$$\frac{a_0}{2} = \frac{C_0\Delta_0}{T_0}\left[t_1 - \left(\frac{1}{T_z} - \frac{A_0}{C_0 t_1}\right)\frac{t_1^2}{2} - \frac{A_0}{C_0 t_1 T_z}\frac{t_1^3}{3}\right]$$

After summation over the tooth number and subtracting the constant component, this formula is reduced to the form

$$F_\Delta = R_1 \sin n\,\omega t + R_2 \cos n\,\omega t \quad (2)$$

where n is the harmonic component, and $\omega = 2\pi f$, R_1 and R_2 are amplitudes (their expressions are not mentioned here because of their inconvenience).

Calculation of time-dependent forced oscillations is performed by a method of numerical simulation of oscillations. The method includes: (a) model of vibration excitation; (b) dynamic model of a gear; (c) set of modifications of equations of motion for the period of meshing; (d) logic algorithm of controlling the choice of equation modifications for areas of the current meshing period; and (e) software [1, 2].

The dynamic model of a helical gear (that can be a single-rim or a herringbone gear, Fig. 2) consists of two completely rigid bodies—gearwheels 1 and 2.

Fig. 2 Dynamic model of herringbone gear. Dynamic model of helical gears

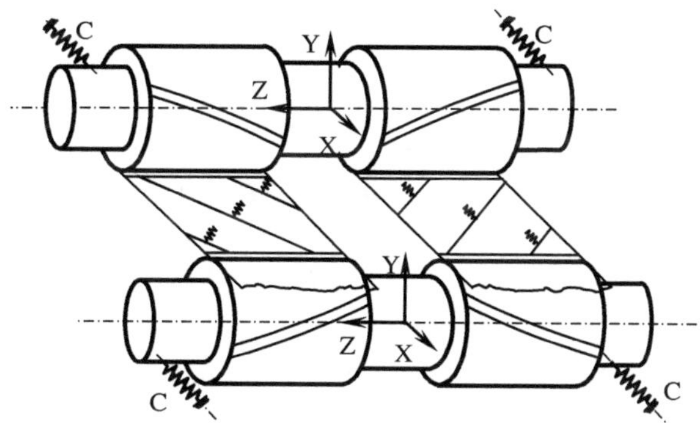

Gearwheels are mounted on elastic supports with rigidities C_{xi} and C_{zi} ($i = 1, 2$) in face and axial directions and bound to each other by springs C_k, simulating rigidities of meshing of the kth pairs of teeth. The pointed springs C_k are located in a meshing plane. Each of them corresponds to a definite contact line and moves together with it along the contact area during rotation of the gearwheels. Each spring in space is oriented along the line of action of the normal force P_κ in meshing. The gear is loaded by a static torque.

The following designations are accepted: $C_z = A_0\varepsilon_\beta + C_0\varepsilon_\Sigma$ is the mean value of the resultant rigidity of meshing; K_z is the damping coefficient in meshing; $C_{x1}, C_{x2}, K_{x1}, K_{x2}$ are coefficients of rigidities and damping of gearwheel supports in the face direction; $C_{z1}, C_{z2}, K_{z1}, K_{z2}$ are coefficients of rigidities and damping of gearwheel supports in the axial direction; X_i, Y_i, Z_i $i = 1,2$ are coordinate systems of gearwheels 1 and 2. Directions of coordinate axes are: Z_i along the gearwheels axes of rotation, X_i parallel to the contact area and contact, Y_i perpendicular to the contact area, $\alpha = \cos(\vec{x}, \vec{e})$, $\beta = \cos(\vec{y}, e), \gamma = \cos(\vec{z}, \vec{e})$ are guiding cosines of springs, l, m, n are unit moments, and $\alpha = \cos\beta_0, \gamma = \sin\beta_0, \beta_0$ is the helix angle on a base cylinder.

Periodicity of repetitions of tooth re-mating processes allows for composing and solving equations of gearwheel oscillations within each period of tooth mesh frequency under the correlation of final and initial conditions of vibration displacement and vibration velocity. A logic algorithm is developed that traces current periods of meshing, choice of type (modification) of equations, signs of loads on teeth P_k and removing of negative loads from equations.

The system of differential equations of motion of a helical gear model has the following form:

$$m_1 \underset{1}{\ddot{x}} + \alpha\Sigma P_k + C_{x1}x_1 + K_{x1}\dot{x}_1 = 0,$$

$$m_1 \underset{1}{\ddot{z}} + \gamma\Sigma P_k + C_{z1}z_1 - K_{z1}\dot{z}_1 = 0,$$

$$I_1 \ddot{\underset{1}{\varphi}} + \alpha \Sigma P_k r_1 = P_{st} r_1,$$

$$m_2 x_2 - \alpha \Sigma P_k + C_{x2} x_2 + K_{x2} \dot{x}_2 = 0,$$

$$m_2 \ddot{\underset{2}{z}} - \gamma \Sigma P_k + C_{z2} z_2 + K_{z2} \dot{z}_2 = 0,$$

$$I_2 \ddot{\underset{2}{\varphi}} - \alpha \Sigma P_k r_2 = -P_{st} r_2,$$

where $\Sigma P_k = P_k + P_{k\text{-}1} + P_{k\text{-}2} + P_{k\text{-}3}$ is the sum of forces in the meshing of the kth pairs of teeth; C_k, $C_{k\text{-}1}$, $C_{k\text{-}2}$, $C_{k\text{-}3}$ are their rigidities; and x_1, z_1, φ_1, x_2, z_2, φ_2 are displacements of centers of inertia and angles of rotation of gearwheels 1 and 2.

After introducing the dimensionless parameters related correspondingly to static deformation $w_{st} = P_{st}/C_z$ or static load, time $\tau = t/T_z$ and reducing the initial equation system to a canonical form of the Runge-Kutta equations [2] we have:

$$\dot{z}_1 = \frac{C_z}{m_1 f_z^2}\left[-\alpha S_0 \bar{P}_k - \alpha S_1 \bar{P}_{k-1} - \alpha S_2 \bar{P}_{k-2} - \alpha S_3 \bar{P}_{k-3} - \frac{C_{x1}}{C_z} Y_1 - \frac{K_{x1}}{C_z} f_z z_1\right],$$

$$\dot{z}_2 = \frac{C_z}{m_1 f_z^2}\left[-\gamma S_0 \bar{P}_k - \gamma S_1 \bar{P}_{k-1} - \gamma S_2 \bar{P}_{k-2} - \gamma S_3 \bar{P}_{k-3} - \frac{C_{z1}}{C_z} Y_1 - \frac{K_{z1}}{C_z} f_z z_2\right],$$

$$\dot{z}_3 = \frac{C_z r_1^2}{I_1 f_z^2}\left[1 - \alpha S_0 \bar{P}_k - \alpha S_1 \bar{P}_{k-1} - \alpha S_2 \bar{P}_{k-2} - \alpha S_3 \bar{P}_{k-3}\right],$$

$$\dot{z}_4 = \frac{C_z}{m_2 f_z^2}\left[\alpha S_0 \bar{P}_k + \alpha S_1 \bar{P}_{k-1} + \alpha S_2 \bar{P}_{k-2} + \alpha S_3 \bar{P}_{k-3} - \frac{C_{x2}}{C_z} Y_4 - \frac{K_{x2}}{C_z} f_z z_4\right], \quad (3)$$

$$\dot{z}_5 = \frac{C_z}{m_2 f_z^2}\left[\gamma S_0 \bar{P}_k + \gamma S_1 \bar{P}_{k-1} + \gamma S_2 \bar{P}_{k-2} + \gamma S_3 \bar{P}_{k-3} - \frac{C_{z2}}{C_{z\,1}} - \frac{K_{z2}}{C_z} f_z z_5\right],$$

$$\dot{z}_6 = \frac{C_z r_2^2}{I_2 f_z^2}\left[-1 + \alpha S_0 \bar{P}_k + \alpha S_1 \bar{P}_{k-1} + \alpha S_2 \bar{P}_{k-2} + \alpha S_3 \bar{P}_{k-3}\right].$$

The tooth loads are defined with regard to the new model of pitch error

$$\bar{P}_k = \bar{C}_k S + \frac{K_z}{\bar{C}_z} f \dot{S} + H_1\left[\bar{C}_k (S_s + \bar{\Delta}) f_\tau - \bar{C}_k S_s - \frac{K_z}{C_z} f \dot{S}_s (1 - f_\tau)\right]$$
$$+ H_2 \bar{C}_k \bar{\Delta}[1 - (\tau - k)],$$

$$\bar{P}_{k-1} = \bar{C}_{k-1} S + \frac{K_z}{C_z} f \bar{S} - \bar{C}_{k-1} \bar{\Delta}(\tau - k),$$

$$\bar{P}_{k-2} = \bar{C}_{k-2}S + \frac{K_z}{C_z} f\,\dot{S} - H_3\left[\frac{1}{2}\bar{C}_{k-2}\left(\frac{a'}{w_{st}}\right)\left(\frac{2\pi}{z_p}\right)^2(\tau-\tau_3)^2\right],$$

$$\bar{P}_{k-3} = \bar{C}_{k-3}S + \left(\frac{K_z}{C_z}\right) f\,\dot{S} - \frac{1}{2}\bar{C}_{k-3}\left(\frac{a'}{w_{st}}\right)\left(\frac{2\pi}{z_p}\right)^2(\tau-\tau_3+1)^2,$$

where $S = \alpha y_1 + \gamma y_2 + \alpha y_3 - \alpha y_4 - \gamma y_5 - \alpha y_6$; y_j are coordinates of linear and reduced angular shifts of gearwheels; $z_j = dy_j/d\tau$; $j = 1 - 6$, S_s is the initial magnitude of S; S_0, S_1, S_2, S_3, H_1, H_2, H_3 are coefficients of equation modification; $\tilde{C}_k, \tilde{C}_{k-1}, \tilde{C}_{k-2}$ are dimensionless rigidities of tooth pairs; f_τ is the function of the edge contact; z_p is the pinion tooth number; and $\tau_3 = k + \varepsilon_\Sigma - 2$.

Fortran-based programs are developed for solving the Eqs. (3), evaluating amplitude-frequency characteristics of gearwheel oscillations, oscillation amplitudes spectrum and calculation of total (dynamic plus static) tooth loads. For reasons of simplification, the gear with $\varepsilon_\Sigma \approx 3$ is considered, though it has no crucial influence on the final results.

In a helical gear, there are two contact ratios ε_α and ε_β for minimization of two exciting forces F_c and F_Δ. But according to the new physical model, the excitation of oscillations caused by pitch error depends only on the coefficient ε_α. Thus, the coefficient ε_α may be chosen according to the condition of minimizing the force F_Δ (2).

Dependencies of modules of amplitudes of exciting forces F_Δ from the coefficient ε_α are shown in Fig. 3, where 1 is a spur gear, 2 is a helical gear according to the old model, and 3 is the helical gear according to the new model (2). Magnitudes of forces are dimensionless in fractions of $2C_0\Delta$ for $C_0 = 5 \times 10^5$ (N/mm), $\Delta = 10$ µm. Analysis of the presented data shows that curve 2 (obtained for $\varepsilon_\beta = 1.5$) qualitatively differs from curve 3, which illustrates the considerable differences of

Fig. 3 Amplitude of exciting force FΔ versus transverse contact ratio εα

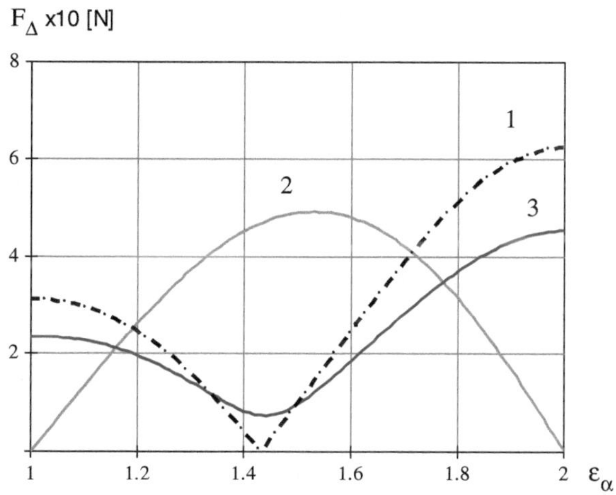

Fig. 4 Amplitude of exciting force Fc versus contact ratio εβ

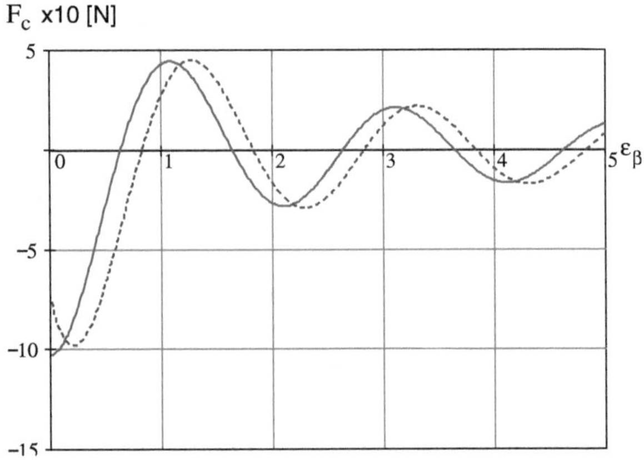

the new model. Curves 1 and 3 have a similar form. Their difference is related only to the fact that, in a helical gear, the motion of the contact line along the involute profile of teeth is accompanied by teeth rigidity variation, and in a spur gear, this rigidity is constant. The new model reflects more precisely the physical nature of pitch error in a helical gear. When ε_α approaches integral numbers, the force F_Δ increases. According to curve 3, we choose the coefficient $\varepsilon_\alpha = 1.43$.

According to formula (1), at integral values of coefficients ε_α and ε, exciting force $F_c = 0$. But to minimize both forces F_Δ and F_c simultaneously, we assume $\varepsilon_\alpha \approx 1.43$, and that magnitude ε_β will be defined by formula (1).

Figure 4 shows the dependence of the amplitude of exciting force F_c on coefficient ε_β for $\varepsilon_\alpha = 1.43$ as the continuous line. Magnitude F_c is dimensionless in fractions of $2C_0w_{st}$ for $C_0 = 5 \times 10^5$ (N/mm), $w_{st} = 10$ μm. This dependence has a form of "damped cosine wave" and defines several values of coefficients ε_β where $F_c = 0$. According to this relation, we get $\varepsilon_\beta = 1.63$. At other values ε_α, the result is analogous, but zero values of function are shifted along axis ε_β (dashed line for $\varepsilon_\alpha = 1.2$). Hence, by selecting the appropriate values of contact ratios ε_α and ε_β, one can get minimum excitation at simultaneous action of exciting forces F_Δ и F_c. However, formulas (1) and (2) are approximate because they don't consider: edge contact, excitation by impulse loading of teeth, disjunction of teeth and interrelation of forcing factors.

The physical essence of minimizing the factor of impulse loading of teeth consists in providing the constancy of the sum of loads on tooth pairs entering and exiting the meshing. The constancy is preserved when one force is increased and another one is decreased, but their joined action doesn't cause oscillations in a system. Note that a total constancy of the sum of all loads on teeth can be theoretically obtained by profile-longitudinal modification of helical teeth [3].

For simultaneous consideration of the action of all three forcing factors, the calculations were performed through a method of numerical simulation of oscillations, taking into consideration edge contact, impulse loading of teeth and real dynamics of the tooth meshing process.

The resulting calculations establish that the factor of impulse loading is minimal at contact ratios $\varepsilon_\alpha = 1.3$ and $\varepsilon_\beta = 1.695$, at an error $\Delta = 0$ and static load $P_{st} = 5000\ N$. In this case, the amplitude of oscillations with tooth mesh frequency $A \approx 0$ (order 10^{-4}), i.e., oscillations are not practically excited. Here and further on in the text, the values Δ and A are relative in fractions of static deformation.

Figure 5a shows loads on teeth within minimization of the impulse loading of teeth for a quasi-static mode $f = 100$ Hz.

Loads on teeth $P_k - P_{k\text{-}3}$ are dimensionless, each representing the relation of a total load (as a sum of dynamic and static components) related to P_{st}. A change of loads during three periods of meshing $3T_z$ is shown (64 targets k correspond to one period T_z). One may notice that load variation is repeated within every period T_z. At one period T_z, tooth pairs are designated: 1 is a pair entering the meshing P_k, 4 is a pair exiting from the meshing $P_{k\text{-}4}$, and 2 and 3 are intermediate pairs $P_{k\text{-}2}$ and $P_{k\text{-}3}$ correspondingly. The total load on tooth pairs 1 and 4 is almost constant. If we

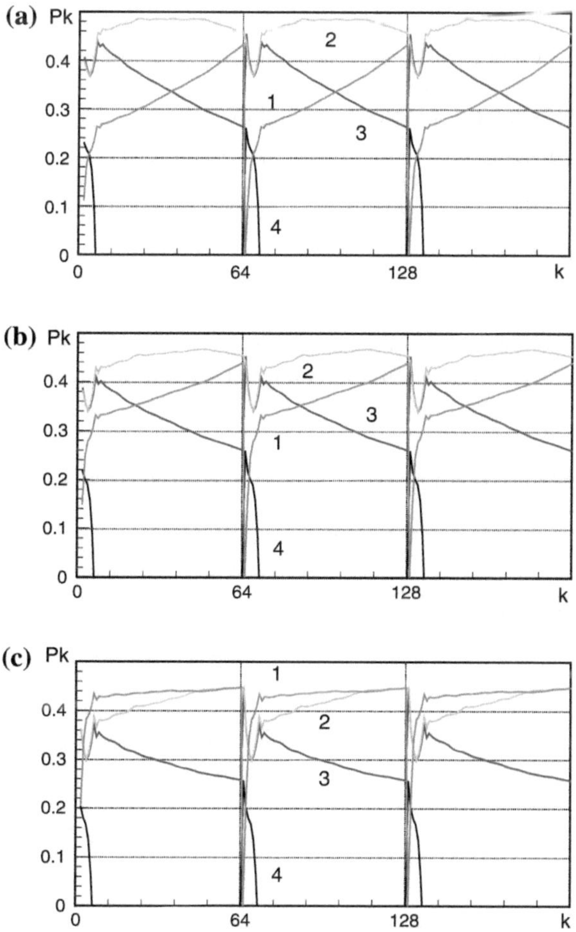

Fig. 5 Tooth stress. Pitch error: **a** Δ = 0, **b** Δ = 0.4, **c** Δ = 1

follow one and the same tooth pair being in contact according to ε_Σ, the numbers correspond to it: 1 for the first period, 2 for the second, and 3 for the third. In this case, the total time of meshing for each pair exceeds $3T_z$ because of edge contact of teeth at the entrance (pair 1) and exit from the mesh (pair 4). In Fig. 5b, c, loads on the same teeth are shown, but with errors $\Delta = 0.4$ and $\Delta = 1$ correspondingly. Amplitudes of oscillations with increase of error Δ are increased correspondingly to $A = 0.027$ and $A = 0.067$. This is connected to the negative influence of error Δ on conditions of minimizing the impulse loading.

Figure 6 presents amplitude-frequency characteristics of oscillations according to parameter S in a range of frequencies $f = 0 \div 10$ kHz. Oscillations are excited by all forcing factors at two variants of parameters: (1) $\varepsilon_\alpha = 1.2$ and $\varepsilon_\beta = 1.2$; and (2) $\varepsilon_\alpha = 1.3$ and $\varepsilon_\beta = 1.695$, for $\Delta = 0.4$. In the first variant, the relative amplitude of oscillations A at resonance $f = 3600$ Hz is $A = 0.712$, and in the second variant, the amplitude $A = 0.077$ at resonance $f = 4200$ Hz. Amplitudes of oscillations resulting from the performed choice of contact ratios (even with error $\Delta = 0.4$) are decreased considerably, almost 10 times (20 dB). This decrease occurs at all frequencies in the range of tooth mesh frequency but at one static load.

Figure 7 shows loads on teeth P_k, and curves 1 and 2 at resonance modes correspondingly for these two variants of parameters. We see that, in the second variant, the load differs slightly from a static one. It should be noted that, after choosing the contact ratios $\varepsilon_\alpha = 1.3$ and $\varepsilon_\beta = 1.695$, an excitation due to pitch errors still remains. The influence of errors Δ on amplitude-frequency characteristics of oscillations at these contact ratios is shown in Fig. 8, where amplitude-frequency characteristics correspond to: 1—$\Delta = 0.2$; 2—$\Delta = 0.4$; 3—$\Delta = 0.8$; 4—$\Delta = 1.6$. At $\Delta = 0$, the resonance is absent and the amplitude of oscillations is $A = 2 \times 10^{-3}$. When error Δ is increased, amplitudes A are increased in direct proportion. Hence, the important conclusion follows: such a combination of parameters ε_α, ε_β and P_{st} exists where oscillations are not excited in the absence of manufacture errors.

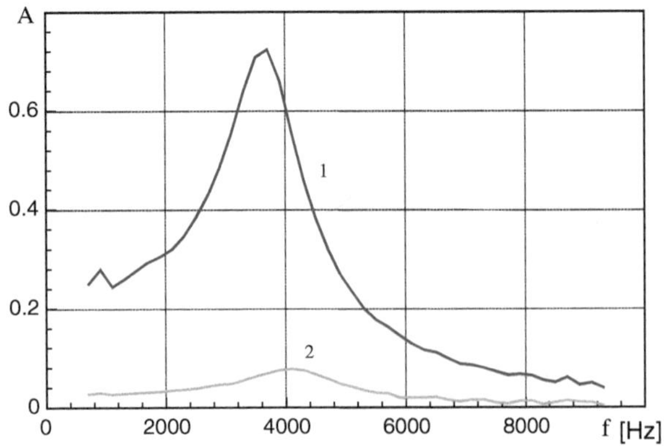

Fig. 6 Frequency response of oscillations with mesh-frequency f of a helical gear with contact ratio: *1* εα = 1.2, εβ = 1.2; *2* εα = 1.3 , εβ = 1.695

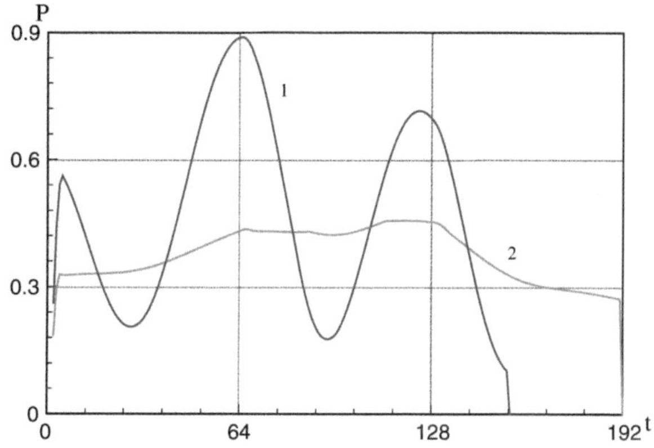

Fig. 7 Tooth stress in resonance modes of a helical gear with contact ratio: *1* εα = 1.2, εβ = 1.2; *2* εα = 1.3, εβ = 1.695

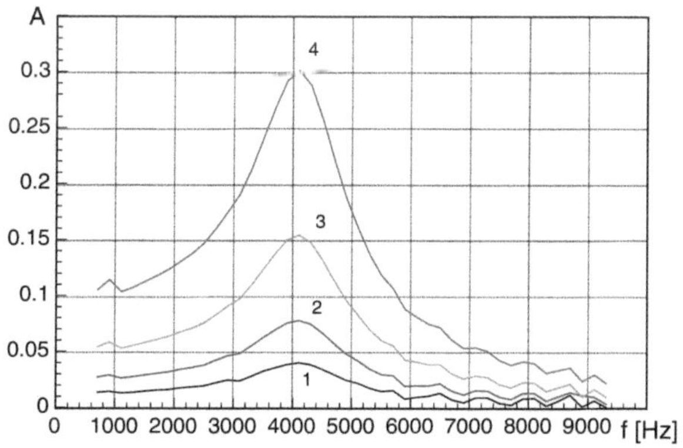

Fig. 8 Frequency response of oscillations with mesh-frequency f of a helical gear with different pitch errors: *1* Δ = 0.2; *2* Δ = 0.4; *3* Δ = 0.8; *4* Δ = 1.6

The higher efficiency of a helical gear (as compared to a spur one) in terms of decreasing the vibration and dynamic loading is explained by the ability to control two contact ratios. Choice of the coefficient ε_α leads to minimization of the excitation due to the pitch error, and choice of the coefficient ε_β leads to minimization of the excitation due to variable rigidity. Both these coefficients are finally specified by numerical simulation of oscillations. This method allows for minimizing three factors of excitation simultaneously: variable rigidity of meshing, impulse loading of teeth and pitch errors in conditions close to real dynamic process of teeth re-meshing. Such a choice of contact ratios allows for minimizing the vibration excitation in a helical (herringbone) gear.

References

1. Kosarev OI (1995) Model of vibration excitation in helical meshing. Vestnik mashinostroeniya N11:3–10
2. Kosarev OI, Bedniy IA, Mamonova MG (2011) The herringbone gear vibrations decrease in its source. Vestnik mashinostroeniya N11:19–24
3. Kosarev OI (2009) Modification of helical gear teeth purposely to reduce the tooth mesh frequency oscillations. Vestnik mashinostroeniya. N5:7–9

Terminology and Design of Asymmetrical Gears for Aircraft

A. Novikov, V. Golovanov, D. Dorofeyev and V. Dorofeyev

Abstract Main calculation principle of gear wheel geometry with asymmetrical tooth profile is presented. The paper presents calculation of bending stresses by means of YFS coefficient which takes into account the tooth shape and stress concentration. Graphic method of YFS coefficients calculation is described. Graphic method of stiffness calculation of teeth with asymmetrical profile and results of vibration testing are given.

Keywords Gears · Asymmetrical tooth profile · Bending stresses · Stiffness · Vibration

1 Introduction

Spur gears with asymmetrical teeth have been developed as a result of the search to increase the carrying capacity of gears. In truth, the contact and bending resistance of such gears is rising, meaning the new construct is very useful in certain cases, e.g., for aircraft gearboxes, to name only one of many examples. The theory of the geometrical calculation of these gears was developed more completely by Profs. Vulgakov [1], Litvin, Lian, Kapelevich [2, 3], Novikov, Paikin, Dorofeyev, Ananiev, Arnaudov, Dorofeyev [4–7], Alipiyev [8] and other scientists who have developed various design techniques for asymmetric gears.

A. Novikov (✉) · V. Golovanov · D. Dorofeyev · V. Dorofeyev
CIAM—Central Institute of Aviation Motors, Moscow, Russia
e-mail: vld@ciam.ru

V. Golovanov
e-mail: vvg@ciam.ru

D. Dorofeyev
e-mail: dl@list.ru

V. Dorofeyev
e-mail: vld@ciam.ru

V. Goldfarb and N. Barmina (eds.), *Theory and Practice of Gearing and Transmissions*, Mechanisms and Machine Science 34,
DOI 10.1007/978-3-319-19740-1_18

Advantages of asymmetrical gears are:

1. Bending stresses are diminished by 14–28 %.
2. Contact stresses are diminished by 18 %.
3. The basic standards: ISO 6336 [9] or national standards DIN 3990 [10], GOST 21354 [11], BDS [12] with a small number of additions are applicable to transmission with asymmetrical gears.
4. Gears with asymmetrical teeth can be produced by available machine tools.

2 Terminology

Further presentation demands a specified terminology. It is important to note that when dealing with gears with asymmetrical teeth, it is necessary to use double terminology.

2.1 Design Terminology

α_n—Basic Side—Defines the width of the teeth on the operating Pitch circle.
α_{nN}—Adjacent Side—Defines the width of the teeth at the addendum.

The edge of a basic rack profile of the basic side is assigned. The edge of a basic rack profile of the adjacent side is calculated by the width of the teeth on the addendum circle.

2.2 Manufacturing Terminology

Flat Side—The side with the greater angle of basic rack
Abrupt or Bluff Side—The side with the smaller angle of basic rack

3 Key Rule of Asymmetrical Teeth Designing

On the operating portion of the teeth, there is a point D, which is constant at variation of the initial contour angle (Fig. 1).

Fig. 1 Terminology which is used at designing and manufacture

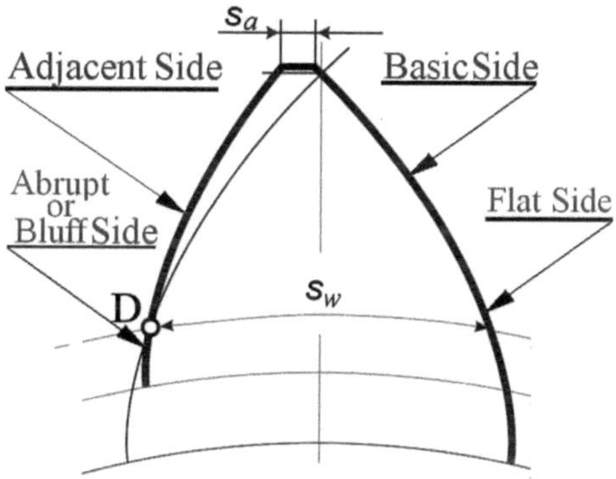

4 Selection of Optimal Geometry

Calculation of factor YFS of teeth for various edges of the basic and adjacent sides is executed. The data are presented in Table 1.

5 The Factor Y_{FS} Considering the Teeth Form and Stress Concentration at a Root Fillet of Teeth

Calculation of the factor Y_{FS} can be executed through a finite element method (Figs. 2 and 3). But there is a faster, more exact way, a specially designed method of complex integral equations [13, 14]. Diagrams for the definition of factor YFS

Table 1 Results of factor YFS calculation

α_{nN}/α_n	Basic Side							
Adjacent Side	$h_a^*=1.0$; $c^*=0.25$ $r_F^*=0.2$							
	15°	20°	25°	30°	33°	35°	40°	45°
10°	4.3	4.12	3.91	3.68	3.6	3.52	3.4	3.54
15°	4.1	3.93	3.74	3.54	3.47	3.4	3.4	3.71
20°	3.96	3.78	3.6	3.45	**3.29**	3.37	3.5	
25°	3.8	3.65	3.49	3.4	3.42	3.44		
30°	3.7	3.55	3.45	3.44	3.51	3.57		
33°	3.65	3.48	3.51	3.56	3.62			
35°	3.6	3.54	3.56	3.64				
40°	3.7	3.76	3.7					

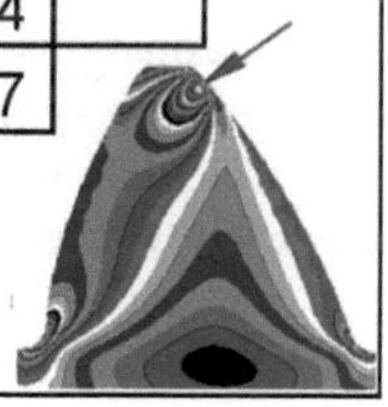

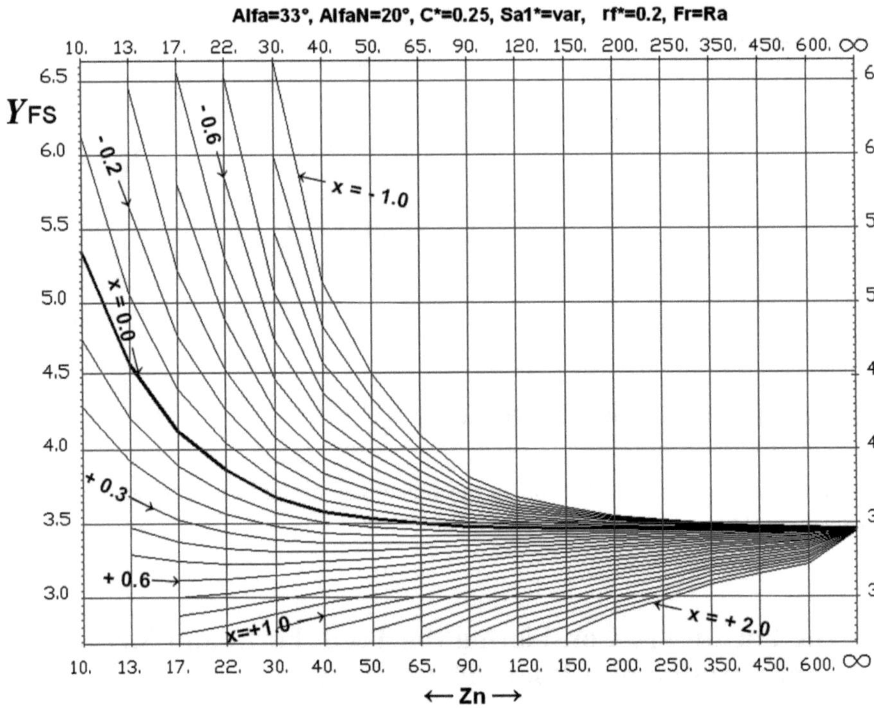

Fig. 2 Tooth shape factor for asymmetrical teeth with parameters corresponding to Table 1

for four basic racks are displayed in Figs. 4, 5, 6 and 7. Calculation of factor YFS for separate basic racks is executed under the following conditions:

For the gears produced with a basic rack, $\alpha_n = 33°$ and $\alpha_{nN} = 20°$.

The force is applied to a tip. It is designated in Figs. by symbols $F_r = R_a$.

The machining method is a generating method that uses a hob without a protuberance.

The tooth thickness is diminished by magnitude ds, equal to half of the backlash accepted as being equal to 0.1 mm.

6 Specific Stiffness C′

Calculation of the factor c′ is also executed through a method involving complex integral equations, at force application at the circle Rv = z/2 + 1. In Fig. 3, it is designated by symbols Fr = Rv. The method of stiffness calculation is described in [14]. Calculations were executed for contact stresses $\alpha_H = 1000$ N/mm^2, with the subsequent recalculation at specific stiffness, which has dimension N/mm/μm.

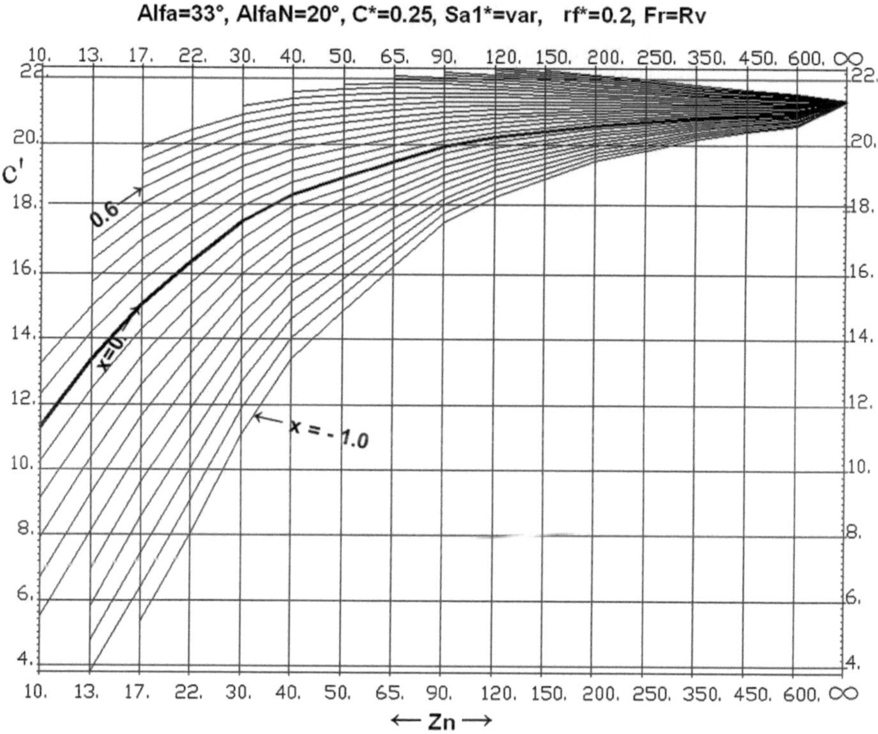

Fig. 3 Factor c′ of asymmetrical teeth with parameters corresponding to the optimal contour table

The results shown in Fig. 3 apply to gears of disk shape. Calculations are executed with account of a backlash.

Diagrams for the definition of factor c′ for three basic racks are given in Fig. 3.

Factors c′ are described for a gear ratio u = 1. If u ≠ 1, the factor c′ is defined by the formula

$$c' = 2\frac{c'(z1, x1) \cdot c'(z2, x2)}{c'(z1, x1) + c'(z2, x2)}$$

7 Example of Calculation of a Gear with Asymmetrical Teeth

Calculations of asymmetrical gears are executed by means of the program AEROFLANK. Profile errors are equal to zero (Figs. 8, 9, 10, 11 and 12).

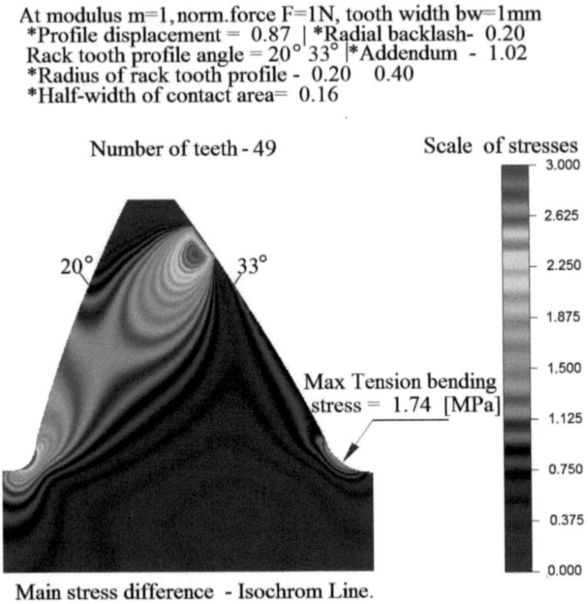

Fig. 4 Asymmetrical gears. Geometrical parameters and main stress difference

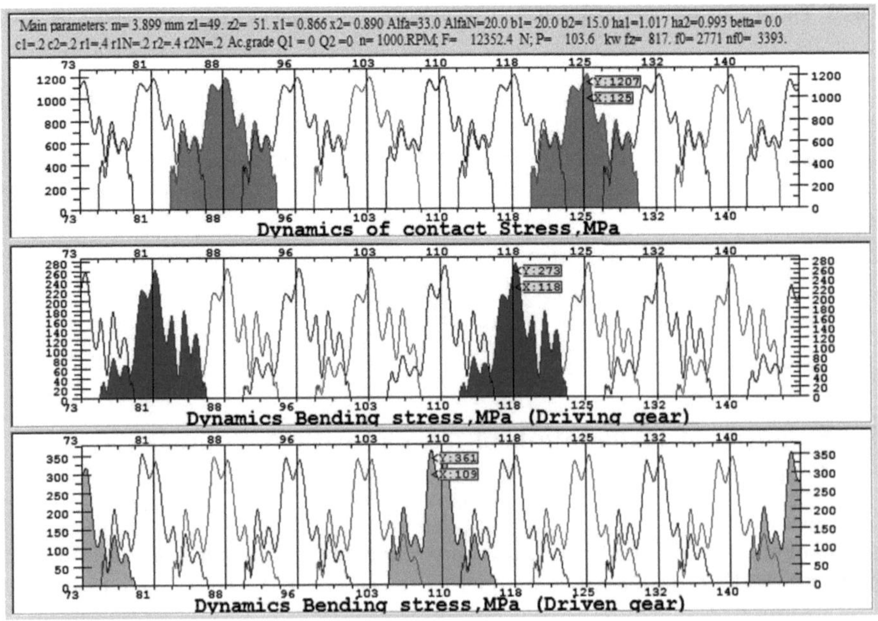

Fig. 5 Asymmetrical gears. Dynamics of contact stress; Dynamics Bending stresses at driving and driven gears

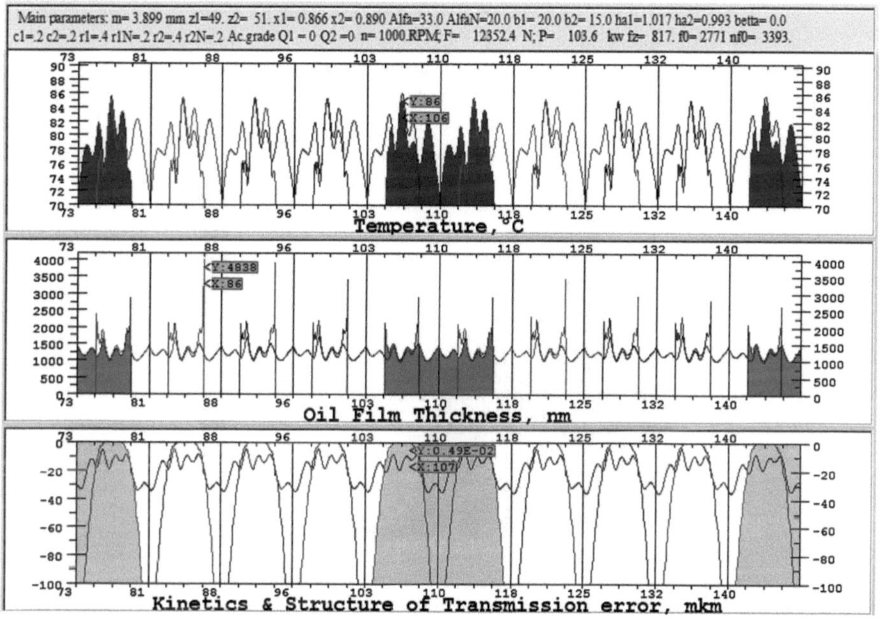

Fig. 6 Asymmetrical gears. Temperature; Oil Film Thickness; Kinetics and Structure of Transmission error

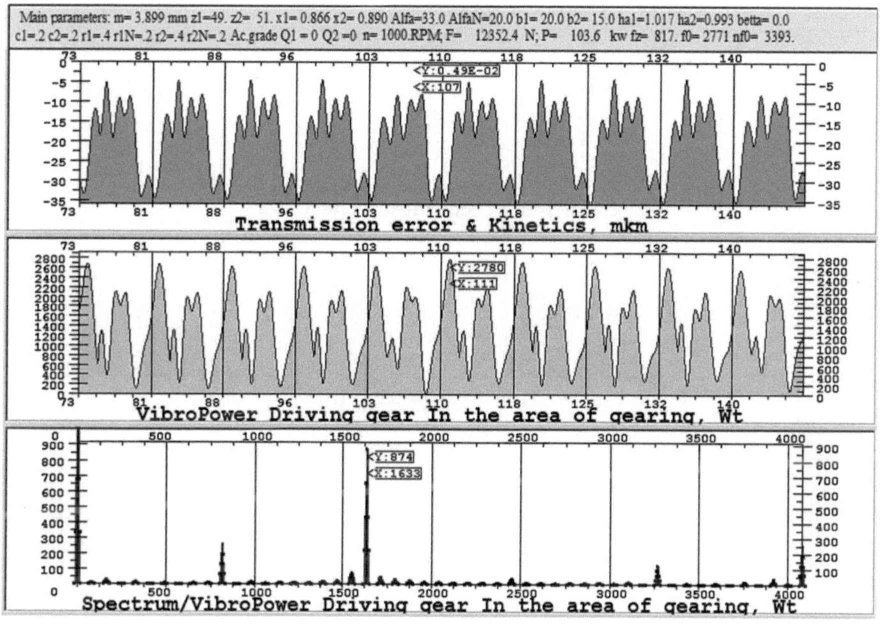

Fig. 7 Asymmetrical gears. Transmission error and Kinetics; VibroPower; Spectrum/VibroPower

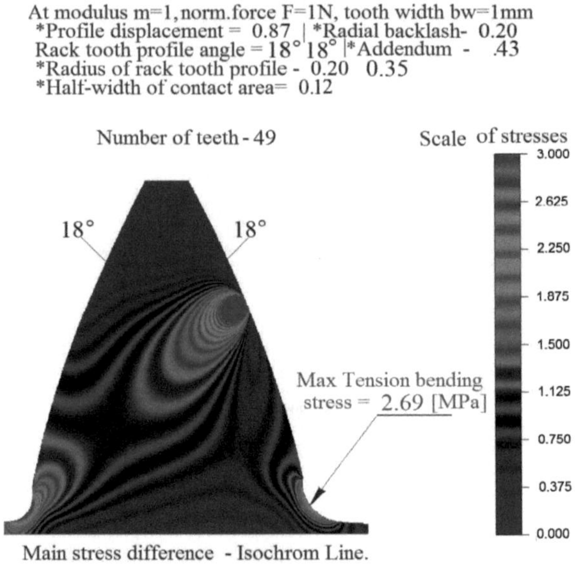

Fig. 8 HCR gears. Geometrical parameters and main stress difference

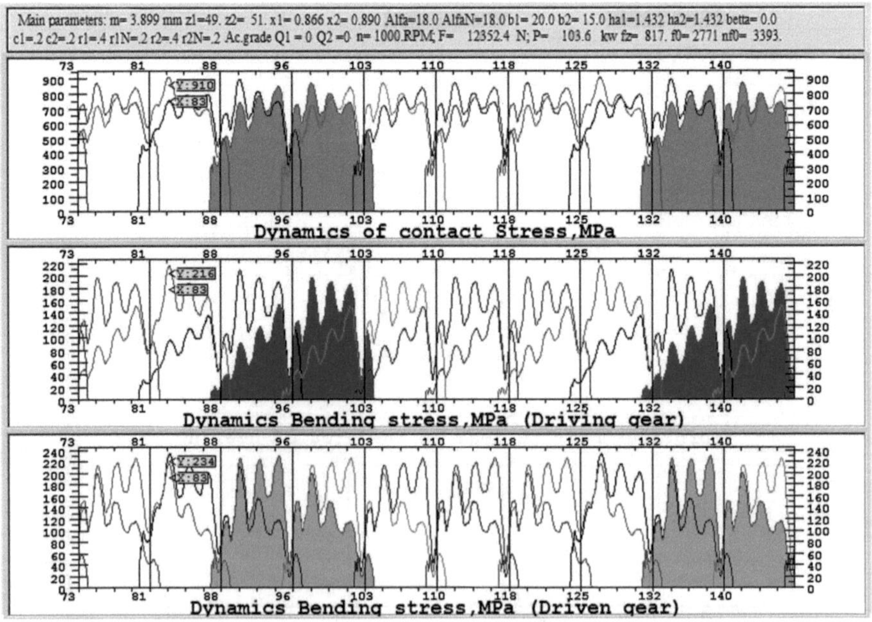

Fig. 9 HCR gears. Dynamics of contact stress; Dynamics Bending Stresses at driving and driven gears

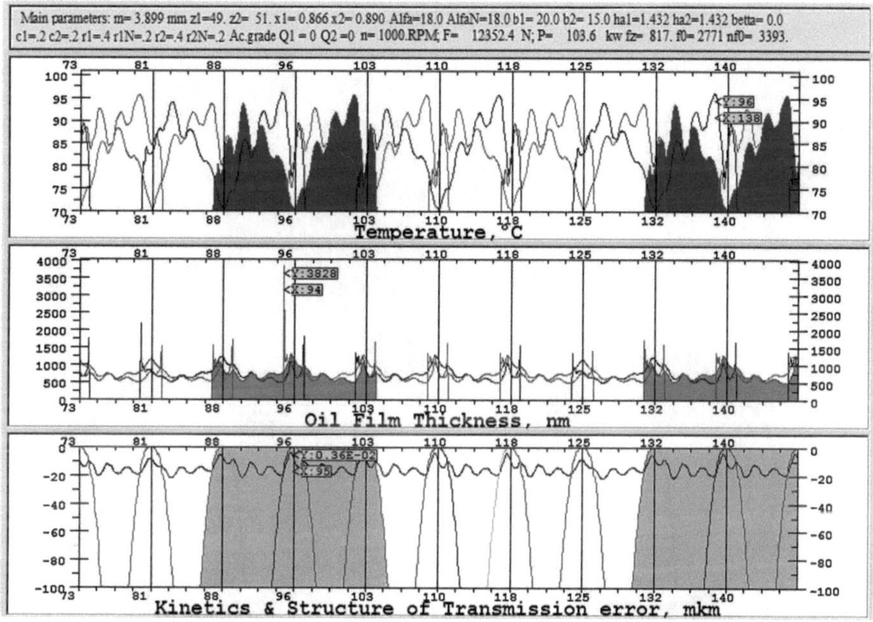

Fig. 10 HCR gears. Temperature; Oil Film Thickness; Kinetics and Structure of Transmission error

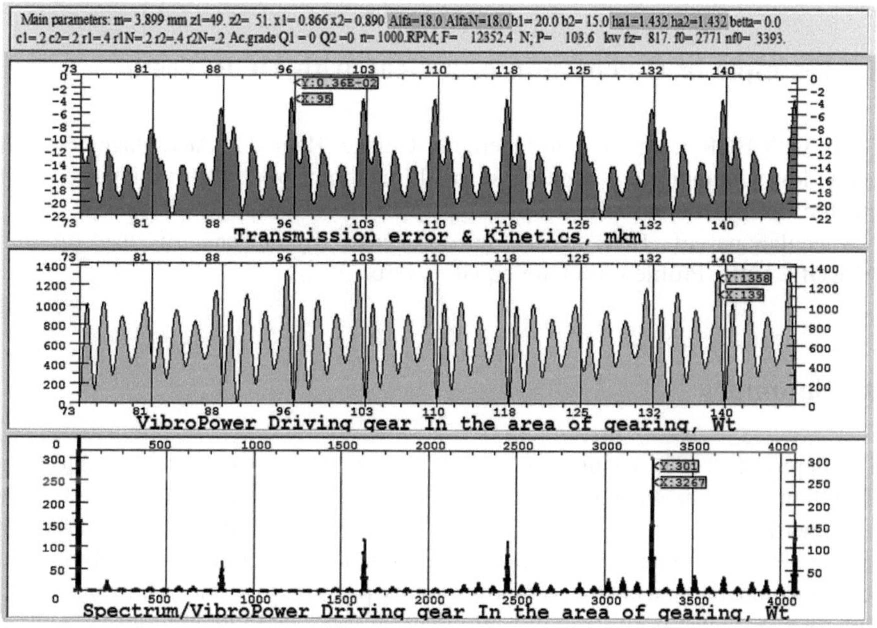

Fig. 11 HCR gears. Transmission error and Kinetics; VibroPower; Spectrum/VibroPower

Fig. 12 Asymmetrical gears—some of the TV7-117S turboprop engine gearbox components and assemblies

8 Example of Calculation of a Gear with HCR Teeth

Gears with HCR teeth have an Operating Contact Ratio >2. Advantages of HCR gears: low vibration, low bending stress. Limitation: low oil film thickness, manufacture from heat-resisting steels of type VCS-10 (Tables 2, 3, 4 and 5).

Calculations of HCR gears are executed by means of the program AEROFLANK. Profile errors are equal to zero.

9 Summary

The approach to the asymmetric and HCR gear design, demonstrating the basic definition of gear teeth and mesh parameter, has been developed. Application of the asymmetric gears in the TV7-117S, TV3-117VMA-SBM1, NK-93 turboprop engine gearbox resulted in extremely low weight to output torque ratio, about 50 % less in comparison with the gearboxes of its predecessors, the AI-20 and AI-24 turboprop engines. It also significantly reduced noise and vibration levels, and cut down the duration and expense of operational development.

Table 2 Basic gear geometry parameters

Gear		Drive	Coast
Number of gears		49	51
Center distance, mm		201	
Operating module, mm		3.899	
Basic rack profile, °	Drive flat flank	33	
	Coast bluff flank	20	
Radius of rack profile, mm	Drive flat flank	0.4 (module)	
	Coast bluff flank	0.2 (module)	
Drive flank operating contact ratio		1.2	
RPM (drive)		1000	
Output torque, Nm (coast)		1122	

Table 3 Basic gear HCR geometry parameters

Gear		Drive	Coast
Number of gears		49	51
Center distance, mm		201	
Operating module, mm		3.899	
Basic rack profile, °	Drive flat flank	18	
	Coast bluff flank	18	
Radius of rack profile, mm	Drive flat flank	0.35 (module)	
	Coast bluff flank	0.2 (module)	
Drive flank operating contact ratio		2.05	
RPM (drive)		1000	
Output torque, Nm (coast)		1122	

Table 4 Application of gears with asymmetrical and HCR teeth in turboprop engine gearbox

Criterion (active profile)	Asymmetrical	HCR
Contact stress, MPa	1207	910
Bending stress, MPa (driving gear)	273	216
Bending stress, MPa (driving gear)	361	234
Temperature, °C	86	96
Oil film thickness, μm	1.0	0.5
VibroPower, Wt	2780	1358

Table 5 Application of gears with asymmetrical and HCR teeth in turboprop engine gearbox

The aircraft engine	Assemblies	Type gears
TV3-117VMA-SBM1	Rear reduction gear	HCR
NK-93	Gearbox between fan and turbine	HCR
TV7-117S	Gearbox between air screw and turbine	Asymmetrical

The new design and technological approaches that have seen realization in the TV7-117S, TV3-117VMA-SBM1, NK-93 engine gearbox have been recommended for the development of gearboxes of advanced aviation engines.

References

1. Vulgakov EB (1995) Theory of involute gears. Mashinostroenie, Moscow, 320 pp
2. Litvin FL, Lian Q, Kapelevich AL (2000) Asymmetric modified gear drives: reduction of noise, localization of contact, simulation of meshing and stress analysis. Comput Methods Appl Mech Eng 188:363–390
3. Novikov AS, Paikin AG, Dorofeyev VL, Ananiev VM, Kapelevich AL (2008) Application of gears with asymmetric teeth in turboprop engine gearbox. Gear Technology, January/February, pp 60–65
4. Dorofeev V, Arnaudov K, Dorofeyev D (2011) Increase of strength of toothed gearings with help the gears with asymmetric teeth/eight international congress. Machines, Technologies, Materials, s 83–87, 19–21 September 2011
5. Dorofeev V, Arnaudov K, Dorofeyev D (2011) Appointment of parameters of an initial contour of gearing wheels with asymmetrical teeth. In: The bulletin of national technical university HPI. The collection of proceedings. Thematic release, Problems of a mechanical drive. NTU HPI, Kharkov. 2011, №29. –стр. 53–60
6. Vulgakov EB, Dorofeyev VL (2002) Computer designing of involute gearings in generalizing parameters. In: Conversion in engineering industry, pp 148–154
7. Dorofeyev VL (1995) Direct synthesis of the aviation Involute gearings, Moscow, 22 pp
8. Alipiev OL (2010) Asymmetric involute-lantern meshing formed by identical spur gears with a small number of teeth. In: International conference on gears. Garching near Munich, Germany, 4–6 Oct 2010, pp 925–940
9. ISO 6336 (2006) Calculation of load capacity of spur and helical gears, International Standardization Organization
10. DIN 3990 (1987) Tragfahigkeitsberechnung von Stirnradern, Teil 1 bis 5, Beuth Verlag, Dezember
11. GOST 21354-87 (1987) Cylindrical involute gears of external gearing. Strength calculation, Russia Standardization Organization
12. BDS 17108-89 (1990) Cylindrical involute gears of external gearing. Strength calculation, Sofia. Sofia. Committee on quality
13. Dorofeyev VL (2010) Problems of internal gear pair design and methods of their solution. In: International conference on gears. Garching near Munich, Germany, 4–6 Oct 2010, pp 1443–1446
14. Dorofeyev VL (2003) Principle of analytical separation of elastic transitions on contact and bending and its application for calculation of teeth gearings. In: International conference Power Transmissions'03, Bulgaria

Quality Control of Spur Gears on the Basis of Simulating Their Production Processes

M. Kane

Abstract The paper presents the main results of investigations carried out by the author on variation of teeth and base surface accuracy (matching bore and rim faces) of spur gears in the main operations of their machining. The results of the influence of the base surface accuracy and cutting conditions on the teeth accuracy in gear-milling by a worm hob and gear-shaping, as well as the methods of choosing gear-machining manufacturing route and the requirements of their quality in the intermediate operations of their machining providing the necessary quality of completed gears, are described. Information about СТБ 1251-2001 (State Standard of the Republic of Belarus) that has necessary techniques and recommendations to solve this problem is also given.

Keywords Quality · Spur gears · Technology · Accuracy · Processes simulation · Statistical methods

1 Introduction

One of the most effective methods for obtaining the high quality of machine parts machining is to design the production process based on the regularities of operation-by-operation parameter variations of machine part quality and their forming in main operations, taking into account the machining conditions (quality and design of the workpiece, equipment and tooling conditions, etc.). Unfortunately, this approach is now rarely used in the practice of mechanical engineering. One of the reasons for this is the widespread opinion that the quality of most machine parts is mainly formed in the finishing operation of their machining, and the lack of mathematical models for most parts production processes. This opinion is wrong and was so proved long ago by the authors of the technological heredity theory, Professors A.M. Dalskiy, A.I. Yascheritsin, A.A. Matalin, and their followers.

M. Kane (✉)
Belorussian National Technical University, Minsk, Belarus
e-mail: kane_08@mail.ru

V. Goldfarb and N. Barmina (eds.), *Theory and Practice of Gearing and Transmissions*, Mechanisms and Machine Science 34,
DOI 10.1007/978-3-319-19740-1_19

2 Experimental Investigations of Spur Gear Production Processes

For mathematical modeling of variation and formation of spur gear quality parameters during their production, we carried out extensive experimental investigations on variations of spur gear teeth and base surface accuracy parameters in the main operations of gear machining. We also carried out investigations on the influence of the base surface accuracy and cutting conditions on tooth accuracy when gear hobbing and shaping.

Within the scope of the given problem, we carried out the experimental investigations into operation-by-operation variations of 11 teeth accuracy parameters and 5 base surface accuracy parameters of cylindrical gears for tractors, trucks and cutting-machines (m = 2–5 mm, z = 23–51, gear material—steels 45, 40X, 25XГТ, 20XH3A, 18XГТ,[1] and others), which were performed after 8 operations of gear teeth machining and 11 operations of base surface machining in manufacturing and laboratory environment. We considered such cylindrical gear production processes as shaving with rigid and expanding arbor, carburization and nitro-carburization in the clutchless machines (chemical thermal treatment), tooth generation, gear honing, gear grinding with tapered and worm grinding wheels; drilling, core drilling, single and double broaching, honing, burnishing, shot-blasting treatment and chemical thermal treatment of matching bore; rough, semi-finish and finish turning, chemical thermal treatment and grinding of rim faces. The investigations were executed with a large amount of material (65 gear batches were processed and measured in the manufacturing environment, and 15 batches were processed and measured in the laboratory environment; the quantity of each batch was 50–100 pieces) and covered about 90 % of the standard sizes of spur gears being made in the automotive and machine-building industries.

3 Theoretical Principles of Modeling the Production Process for Spur Gears

In processing the experimental data, it was shown that it was reasonable to use correlated-regressive analysis (CRA) for modeling the operation-by-operation variation of spur gear quality parameters in the machining operations under investigation by measuring results of one sufficient volume sample. Analysis of quality parameter distribution laws (192 distributions were investigated) and of stationary state and ergodicity of processes under consideration for selected quality parameters, as well as phase analysis of gear kinematic accuracy parameters in

[1]The steel's chemical composition and properties are in accordance with ГОСТ 1050–74 and ГОСТ 4543–71. ГОСТ—State Standard of the Russian Federation.

different kinds of gear-machining, were all carried out. These investigations confirmed the reasonability of CRA application for modeling the gear production processes in question.

Tables 1, 2 and 3 show some results of the investigations performed that characterize the influence of the technological heredity at variation of gear quality parameters in their machining, as well as the scope for modeling operation-by-operation interrelations of quality parameters for gear production processes.

Table 1 Parameters of operation-by-operation interrelations of spur gear tooth accuracy in their machining

Tooth machining operation	Tooth accuracy parameter	r_{xy}	B (%)	F	ε_{av} (%)	K_v
1.	2.	3.	4.	5.	6.	7.
Shaving	F'_{ir}	0.54	31.3	0.84	26.3	1.1–1.4
		–	–	–	–	–
	F_{pr}	0.67	53.2	0.63	17.2	1.03–1.2
		0.74	65.0	0.55	16.1	1.1–1.35
	F''_{ir}	0.44	11.4	0.95	24.5	1.25–3.7
		(0.35)	(9.9)	(0.94)	(15.8)	(1.5–2.9)
		0.72	5.5	0.35	9.1	1.1–1.8
	F_{rr}	0.28	11.9	0.98	25.6	1.4–3.6
		0.56	24.4	0.73	16.4	1.3–2.4
	F_{vwr}	0.22	5.3	0.97	27.6	0.3–1.3
		(0.12)	(1.5)	(1.01)	(30.1)	(0.8–1.4)
		0.29	6.0	0.97	23.5	0.6–0.9
	f'_{ir}	0.07	2.5	1.01	15.3	1.7–3.4
		–	–	–	–	–
	f_{ptr}	0.31	9.3	0.91	24.2	1.3–2.1
		0.41	11.2	0.93	15.4	1.8–2.5
	f''_{ir}	0.45	7.5	0.78	20.4	1.2–3.6
		(0.28)	(6.7)	(0.98)	(24.5)	1.2–2.4
		0.52	15.0	0.59	19.2	(1.3–3.8)
	f_{fr}	0.12	1.9	1.02	17.8	1.9–3.2
		0.32	7.0	0.94	13.1	2.1–4.4
	f_{pbr}	0.25	4.4	0.96	14.5	1.3–1.9
		0.20	9.8	0.93	20.0	1.4–2.1
	$F_{\beta r}$	0.48	8.8	0.92	23.1	1.4–3.1
		(0.31)	(7.8)	(0.98)	(28.7)	(1.6–4.5)
		0.55	25.0	0.76	22.0	1.8–4.8
Chemical thermal treatment	F''_{ir}	0.38–0.90	48–81	0.20–0.98	10–27	0.74–0.89
		0.31–0.85	45–79	0.30–1.0	12–31	0.70–0.88

(continued)

Table 1 (continued)

Tooth machining operation	Tooth accuracy parameter	r_{xy}	B (%)	F	ε_{av} (%)	K_v
1.	2.	3.	4.	5.	6.	7.
	F_{rr}		46–78	0.30–0.81	18–23	0.67–0.91
		0.31–0.72	45–76	0.33–0.90	20–26	0.65–0.89
Chemical thermal treatment	F_{vwr}	0.40–0.81	47–75	0.48–0.77	16–22	0.83–0.97
		0.35–0.76	46–74	0.51–0.82	18–25	0.77–0.96
	f''_{ir}	0.35–0.60	51–82	0.67–0.93	16–35	0.58–0.95
		0.30–0.56	50–81	0.70–0.95	18–39	0.53–0.93
	$F_{\beta r}$	0.46–0.85	54–92	0.29–0.84	10–34	0.58–0.95
		0.35–0.76	61–90	0.32–0.93	12–36	0.53–0.93
Gear burnishing	F''_{ir}	0.55–0.94	68–89	0.39–0.73	15–20	1.2–1.4
	f''_{ir}	0.54–0.86	60–88	0.66–0.75	13–30	1.1–1.5
	$F_{\beta r}$	0.61–0.95	69–85	0.37–0.68	15–25	1.0–1.15
Gear honing	F''_{ir}	0.50–0.95	69–88	0.15–0.92	9.6–25.0	1.05–1.25
	f''_{ir}	0.40–0.78	59–77	0.58–0.90	15–35	1.1–1.35
	$F_{\beta r}$	0.54–0.85	67–80	0.31–0.75	10–27	1.05–1.2
Gear grinding	F''_{ir}	0.19–0.26	0–2	–	17–23	4–6.3
		0.1–0.2	0–2		9–30	3.2–6.1
	f''_{ir}	0.11–0.13	0–1	–	6–30	5.2–9.6
		0.02–0.1	1–2		22–44	2.9–4.2
	$F_{\beta r}$	0.1–0.32	1–10	–	6–40	1.9–7.0
		0.14–0.29	1–2		12–36	2.1–4.1

Notes 1. One value in the column corresponds to an arithmetical mean of the results obtained; two values in the column correspond to minimum and maximum values for all experiments. *2*. Letter designations of the tooth accuracy parameters are accepted in accordance with standard ГОСТ 1643-81. *3*. For shaving, the values of parameters given in the numerator of the fraction were obtained in the manufacturing environment, those in the denominator were obtained in the laboratory environment, and those in brackets, in gear shaving with the expanding arbor. *4*. For chemical thermal treatment, the parameters given in the numerator were obtained in nitro-carburizing, those in the denominator, in carburizing. *5*. For gear grinding, the parameters given in the numerator were obtained in gear shaving by two-sided tapered grinding wheel, those in the denominator, by worm grinding wheel. *6*. Kv is a coefficient of the machining accuracy variation for the given accuracy parameter in the operation in question (see formula (5))

It was found that the technological heredity takes place for most gear quality parameters and their machining operations ($r_{xy} > 0.273$, B > 10 %) where r_{xy} is a coefficient of the pair correlation between the values of the given quality parameters before (x) and after (y) the operation under consideration; B is a part of y dispersion extended from the previous operation.

The analysis of the results obtained along with previous investigations permitted supposing that the $\bar{y} = f(\bar{x})$ interrelations can be described by the polynomials of the first and the second power

Table 2 Parameters of operation-by-operation interrelations of spur gear matching bores accuracy in their machining

Matching bore machining operations	Matching bore accuracy parameter	r_{xy}	B (%)	F	ε_{av} (%)	K_v
Workpiece drilling	D_b	0.49–0.81	26.6–53.7	0.4–0.85	0.21–0.27	3.5–5.1
	N_r	0.28–0.59	11.5–35.7	0.76–1.0	0.3–3.7	1.7–1.9
	N_c	0.3–0.5	15–40	0.8–0.91	2.0–14.5	2.1–2.8
Core–drilling	D_b	0.49–0.74	37.7–55.8	0.43–0.76	0.5–2.5	3.8–5.9
	N_r	0.27–0.61	9.7–34.8	0.83–0.97	2.5–7.2	1.9–2.2
	N_c	0.28–0.41	10–32	0.7–0.84	3.1–11.6	2.5–3.6
Single broaching	D_b	0.36–0.63	26.5–48.7	0.7–0.9	2.4–7.3	3.7–5.6
	N_r	0.18–0.32	3.6–7.4	0.9–1.0	10.1–30.2	3.5–8.6
	N_c	0.16–0.33	10.7–15.6	0.9–1.1	18.5–40.4	3.6–5.5
Finish broaching	D_b	0.33–0.55	23.2–47.5	0.7–0.9	2.4–8.8	1.3–1.7
	N_r	0.16–0.25	4.1–12.8	0.96–0.99	12.1–25.9	1.1–2.1
	N_c	0.12–0.24	4.7–19.4	0.9–1.02	17.2–38.9	1.1–1.9
Fine boring	D_b	0.53–0.71	41.0–47.0	0.59–0.75	10.0–17.0	2.3–2.7
	N_r	0.35–0.67	37.0–48.0	0.61–0.82	8.0–14.0	1.1–1.6
	N_c	0.34–0.57	46.0–32.0	0.60–0.86	11.0–15.1	1.3–1.9
Honing	D_b	0.49–0.68	35.0–38.1	0.61–0.83	11.0–16.2	2.1–2.8
	N_r	0.3–0.5	30.0–39.2	0.73–0.91	10.1–15.2	1.5–2.1
	N_c	0.3–0.45	22.0–40.3	0.70–0.89	12.2–18.1	1.4–2.0
Chemical thermal treatment	D_b	0.32–0.70 / 0.25–062	16.3–30.1 / 18.1–32.5	0.65–1.2 / 0.8–12.6	0.3–8.3 / 1.5–12.6	0.07–0.39 / 0.06–0.41
Shot blasting	D_b	0.36–0.86	25.2–51.3	0.37–0.8	1.2–3.3	1.0–1.07
Burnishing	D_b	0.3–0.8	40.8–58.8	0.33–0.85	1.7–7.1	0.85–1.08

Notes 1. Accepted designations: D_b—bore diameter; N_r—bore non-roundness factor; N_c—bore non-cylindrical shape factor. *2.* Fine boring and honing of the bore were done after finishing boring. *3.* Minimum and maximum values of interrelation parameters for the experiments carried out are given in Table 4. For chemical thermal treatment, the parameters given in the numerator were obtained in nitro-carburizing, those in the denominator, in carburizing

$$\bar{y} = a + b\bar{x} \tag{1}$$

$$\bar{y} = c + d\bar{x} + e\bar{x}^2 \tag{2}$$

where x, y are the values of different gear quality parameters before and after a given operation; a, b, c, d, e are constant coefficients.

In order to choose the kind of connection that best describes the relations under investigation, along with the calculation of the coefficients in Eqs. (1) and (2), a number of parameters was also determined which characterize an adequacy of each theoretical relation in question to the empirical data. The residual sum of squares S,

Table 3 Parameters of operation-by-operation interrelations of spur gear rim faces accuracy in their machining

Face machining operation	Face accuracy parameters	r_{xy}	B (%)	F	ε_{av} (%)	K_v
Rough turning	E_f	0.5–0.9	21.1–55.3	0.4–1.1	5.3–16.7	2.5–6.2
	E_l	0.4–0.8	20.3–52.1	0.5–1.2	4.7–17.2	2.3–5.9
Semi–rough turning	E_f	0.4–0.8	18.5–47.6	0.5–1.2		1.7–4.7
	E_l	0.35–0.7	17.7–46.1	0.6–1.3		1.5–4.5
Finishing turning	E_f	0.35–0.89	18.7–47.3	0.19–0.62	3.9–12.3	1.5–3.8
	E_l	0.25–0.68	25.6–62.8	0.71–0.91	1.5–3.7	1.3–3.5
Chemical thermal treatment	E_f	0.3–0.7	21.1–59.1	0.4–0.9	3.3–15.1	0.8–0.9
		0.31–0.8	20.4–58.5	0.5–0.1.2	4.5–20.1	0.7–0.8
	E_l	0.3–0.6	18.9–55.3	0.5–1.3	4.1–14.7	0.85–0.95
		0.32–0.5	16.1–51.2	0.6–1.4	5.0–16.3	0.75–0.9
Grinding	E_f	0.33–0.67	15.5–40.2	0.6–1.1	6.1–22.3	2.2–4.1
	E_l	0.31–0.63	14.1–37.6	0.7–1.2	6.6–25.1	2.1–3.8

Notes 1. Accepted designations: E_f—runout of the mounting (supporting) face of the gearwheel blank when tooth-cutting; E_l—parallelism of faces of gearwheel blank when tooth-cutting. 2. For chemical thermal treatment, the parameters given in the numerator were obtained in nitro-carburizing, those in the denominator, in carburizing

F—Fisher's criterion (F-test) of an adequacy of the model, and the average relative error ε_{av} of the coupling equation were accepted as the model adequacy parameters. The reliability of the coefficients in Eqs. (1) and (2) was estimated by the Student's test. The analysis of the results obtained made it possible to find that, for all the production processes and gear quality parameters in question with the reliability $P \geq 95$ %, it is possible to describe the processes of operation-by-operation variations of these quality parameters by the polynomial of the first power (1) ($F < 1.65$, $\varepsilon_{av} < 30$ %).

4 Main Results of Production Process Analysis for Spur Gears

The presence of technological heredity in spur gear machining and the capability of modeling accuracy parameter variations of spur gears in their machining ensure the effective control of these processes, the required accuracy being effected by choosing the rational route of machining and gear quality requirements in intermediate operations, taking into account the main regularities of these gear parameter variations in their machining.

Since the deformation ratio of spur gears in their chemical thermal treatment depends both on the kind of chemical thermal treatment and gear design, we carried out, together with I.L. Barshai, the investigation into the influence of these factors on variability degree of gear teeth accuracy parameters in carburizing and nitro-carburizing. The results of the investigation conducted are given in chart form and are used in the СТБ 1251-2000 [1].

5 Design Techniques for the Manufacture of Spur Gears that Provide Their Quality

Based on investigations conducted, the techniques for choosing the manufacturing routes of cylindrical gear surfaces meeting the requirements for accuracy in machining in intermediate operations have been offered.

The presence of technological heredity of cylindrical gear accuracy parameters during the main operations of their machining gives every reason to assume that the gear production process may be admissible from a technical point of view when it allows for each of the quality parameters standardized in the draft to fulfill the condition:

$$K_{v.o} \leq \prod^{m} K_{v.i} \tag{3}$$

where Kv, o is the overall coefficient of accuracy parameter variation during the technological process; Kv,i is the coefficient of the given parameter variation during the i-operation; and m is the number of technological process operations.

$$K_{v.o} = \frac{\delta_w}{\delta_a}, \tag{4}$$

where δ_w is the given accuracy parameter tolerance in the workpiece (for base gear surfaces) or after the first operation of teeth forming; δ_a is the tolerance of the given parameter specified in the completed gear drawing.

$$K_{v.i} = \frac{\delta_{i-1}}{\delta_i} \tag{5}$$

where δ_{i-1} and δ_i are the tolerances on the given accuracy parameter in the previous (i − 1) and given (i) operations that can be obtained by the machining methods under consideration.

The values of the coefficient Kv.i for accuracy parameters and the processes of spur gear production were obtained experimentally and are given in Tables 1, 2 and 3.

The procedure of determining the requirements of the accuracy parameters in the intermediate operations takes into account the nature of parameter variation in the given operation and the random nature of this variation. Figure 1 [2] shows the calculation pattern of requirements of the quality parameter accuracy in the previous operation (Xadm.min, Xadm.max) based on the requirements of its parameter

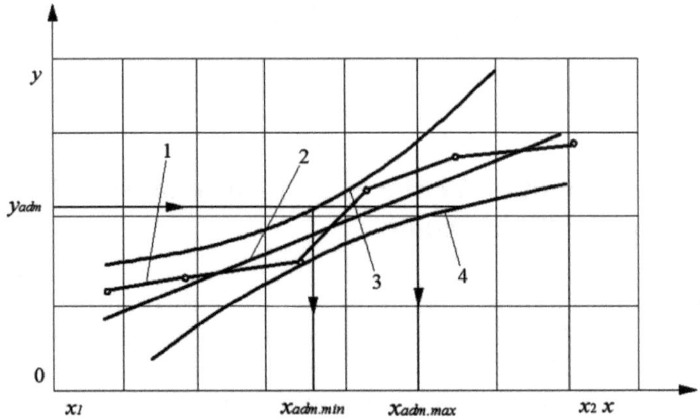

Fig. 1 The fundamental calculation pattern of the allowable accuracy parameter values for the part in the previous operation of its machining

accuracy (y) after the given operation. Here, 1 is the empirical regression line, 2 is the theoretical regression line, and 3 and 4 are confidence limits for line 2.

Using the given procedure and results of our investigations into the variation of spur gear accuracy parameters in their machining, recommendations were obtained regarding how to choose the requirements for the parameters of teeth and base surface accuracy before all their considered machining operations [1, 3]. Knowing the manufacturing route of gear surface that was relations (3)–(5) and the data given in Tables 1, 2 and 3, one can determine the requirements for surface accuracy (teeth, matching bore or rim face) in all its machining operations, taking into account the equipment condition and gear design [1, 3].

The earlier investigations showed that inaccuracies in workpiece base surfaces before gear-cutting cause up to 80 % of some teeth accuracy parameter dispersion in gear-milling. Therefore, the correct choice of requirements for the accuracy of gear workpiece base surfaces may be an effective means of obtaining the required teeth accuracy in gear-cutting. At present time, there is no procedure for such a choice that takes into account the influence of the base surface inaccuracies on the main teeth accuracy parameter. We carried out the experimental investigation into the influence of 3 base surface accuracy parameters in teeth-cutting (matching bore and rim face) on 14 teeth accuracy parameters in gear-milling and gear-shaping. The investigations were conducted in the manufacturing and laboratory environment. The parameters of spur gears were in the same limits as in the above-mentioned investigation. We determined the degree of closeness and interrelation of teeth accuracy parameters with base surface accuracy parameters (diameter clearance ΔD_0 in the mating of the gear with the machine arbor, abutting face runout in gear-cutting E_f, radial runout of the tooth rim E_f) [4].

It was shown that the interrelations between the investigated factors can be described by the relation (6) with sufficient accuracy ($\varepsilon_{av} < 20\ \%$)

$$y = a_0 + a_1 E_f + a_2 \Delta D_0 \quad (6)$$

where y is one of the measured tooth rim accuracy parameters; a_0, a_1, a_2 are constant coefficients of the equation that depends on the properties of the technological system and the relation characteristic.

The investigations permitted the development of the optimization procedure for the requirements of gear base surface accuracy and tooling in gear-cutting operations based on the condition of obtaining the necessary gear quality in a given operation [4].

The calculations of the optimal requirements of base surface accuracy for all investigated dimension types of gear, and kinds and conditions of gear-cutting were performed with the given procedure.

Since all the experimental material covered the substantial range (band) of gear dimensions, accuracy of workpiece bases and tooth rim and other machining conditions, it became possible to determine the maximum allowed interrelations of parameters E_f and ΔD_0 in gear-milling for the conditions considered with gear geometric dimensions and the required accuracy grade of tooth rim after gear-milling and gear-shaping [3].

We also conducted an investigation into the influence of the cutting speed V, m/min, and the feed S0, mm/rev, in cylindrical gear-milling on tooth accuracy parameters F_{rr}, F_{ir}'', f_{ir}'', $F_{\beta r}$. It was shown that interrelations between investigated factors can be described by the next relation with a sufficient accuracy ($\varepsilon_{av} < 20\ \%$)

$$y = b_0 + b_1 V + b_2 S_0, \quad (7)$$

here y is the tooth accuracy parameter.

The feed S_0 influences teeth accuracy in gear-milling to a degree 11–28 times greater than the cutting speed V.

The determined interrelations between the factors investigated are given in Table 4. These interrelations didn't differ substantially for gear materials considered (steels 20XH3A, and 25XГT) and gear hobbers (models 5B312 and AБCB12).

By the given relations, we obtained the recommendations in setting gear-cutting conditions ensuring the required accuracy of gear teeth [3, 5].

The designing procedure of spur gear manufacturing processes that ensure the required gear quality and appropriate recommendations are based on the results of

Table 4 The interrelations of gear tooth accuracy parameters with gear-milling conditions

Accuracy parameter	Kind of relation
F_{rr}	$F_{rr} = 0.042 + 0.002\ V + 0.011 S_0$
F_{ir}''	$F_{ir}'' = 0.081 + 0.0015\ V + 0.01 S_0$
f_{ir}''	$f_{ir}'' = -0.021 + 0.001\ V + 0.0045 S_0$
$F_{\beta r}$	$F_{\beta r} = 0.054 + 0.002\ V + 0.0088 S_0$

Fig. 2 Sequence of designing the cylindrical gear manufacturing process by means of СТБ 1251-2000

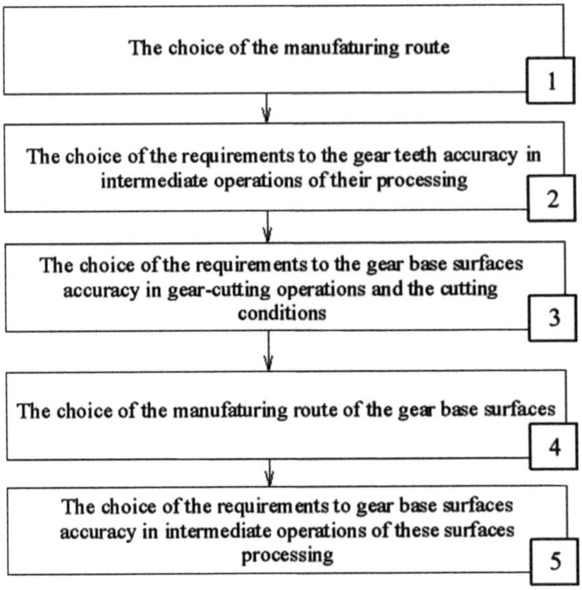

conducted investigations and are used in СТБ 1251-2000 [1]. The developed standard has been valid in the Republic of Belarus since 01.01.2001. The requirements of this standard are recommended.

The standard applies to the spur involute bore-type gears and shaft gears with the module m = 3.0…6.3 mm, reference diameter to 400 mm with 5–12 accuracy grades in the drawing of the completed part according to the standard ГОСТ 1643-81. The standard doesn't regulate the embodiment of the completed gears or the requirements regarding their quality.

The present document is essentially the first experience in standardizing production (process) engineering overall, and (in particular) spur gear production and design processes. The tasks of this document are to simplify and improve the development of the technological processes of spur gear manufacture, and to improve the reliability of these processes in obtaining the required gear accuracy.

The sequence of designing the spur gear manufacturing process by means of СТБ 1251-2000 is shown in Fig. 2.

The above enumerated tasks are interrelated. The results of the previous task solution are the basic data for the next task solution.

In adjusting the designed technological process, one has to determine relations 1, 4, 5, 6 and 7 in the machining location kit of gears ($n \geq 50$) in different operations and specify the manufacturing route, the requirements for gear quality in different operations and gear-cutting conditions, taking into account the actual machining conditions. The offered procedures can be used as well.

Practical application of the presented procedures and recommendations shows that they allow for improving the accuracy of the completed gears by 20–40 %, reducing flaw losses by 30–50 %, and bringing down the cost of the gear machining intermediate operations by 10–15 %.

6 Summary

The processes of the base workpiece surface accuracy influence of spur gears and gear-cutting conditions on their tooth accuracy, variations of tooth accuracy parameter values in operations of their manufacturing, hardening, and finishing can be mathematically modeled and described with sufficient accuracy (ε_{av} = 0.5… 20 %).

The capability of spur gear quality control in designing their manufacture processes by regulation of such technological factors as requirements of workpiece base surface accuracy before gear-cutting, cutting conditions in gear-cutting, and reasonable requirements for gear quality in intermediate operations of their production was shown.

The procedures and recommendations offered for controlling spur gear quality in designing their production processes are practically implemented and given in СТБ 1251-2000. They allow for improving the accuracy of the completed gears by 20–40 %, reducing flaw losses by 30–50 %, and bringing down the cost of gear machining intermediate operations by 10–15 %.

References

1. СТБ 1251-2000 (2001) Spur gears. Procedures of designing the manufacturing processes. Gosstandart RB, Minsk, 50 pp
2. Kane MM (1996) Choosing the reasonable spur gears accuracy in various operations of their machining. Vestnik Mashinostroeniya, #8, pp 3–8
3. Starzhinskiy VE, Kane MM et al (2007) Production engineering and methods of providing the quality of gearwheels and drives. Professiya, Saint-Petersburg, 830 pp
4. Kane MM, Yakovlev GM, Aleshkevich IL (1974) Probabilistic-statistical analysis of the influence of spur gears workpiece base surfaces on the gear-machining accuracy. Vestnik Mashinostroeniya, #10, pp 46–50
5. Kane MM, Ivanov BV, Zagorskaya NB (2014) Development of the optimization procedure of cutting conditions in the example of spur gear-cutting. STIN, #1, Moscow, pp 17–20

New Concept of the Process of Designing Gearboxes and Gear Systems

V. Goldfarb, O. Malina and E. Trubachev

Abstract The paper describes classical and newly developed approaches to design of gears and gearboxes. Advantages and drawbacks of these approaches are considered. Examples of applying the advanced mathematical apparatus and techniques of computer-aided design of gears by example of spiroid gears and gearboxes are given. Conclusions of progressive ways of applying the proposed mathematical apparatus and advanced techniques of gear systems design are made.

Keywords Mathematical modelling · Design of gears and gearboxes · Computer-aided design

1 Introduction

One of the most important and crucial elements of the process of increasing the quality of the produced unit is enhancement of the structure of its design, with the creation and application of new approaches, and advanced methodology of structural design that allows not only for optimizing the produced units but significantly increasing variability within their development, thus finding new areas of effective application. In mechanical engineering and, in particular, its classical branch—gearbox engineering and creation of the methodology of gearbox development was formed over a rather long time period, coming to life as guidance papers, reference books, textbooks and other scientific and practical publications [1–7 and oth.], which are being actively and fruitfully applied now in gearbox development. However, time and advanced possibilities are demanding the development of approaches to the design process for produced units, and creation of a new formalization of this process.

V. Goldfarb (✉) · O. Malina · E. Trubachev
Kalashnikov ISTU, Izhevsk, Russia
e-mail: veniamingoldfarb@yahoo.com

V. Goldfarb and N. Barmina (eds.), *Theory and Practice of Gearing and Transmissions*, Mechanisms and Machine Science 34,
DOI 10.1007/978-3-319-19740-1_20

Though ideas described in this work are invariant with respect to a specific type of gearbox, the main object discussed further will be a gearbox with a spiroid gear.

2 Classical Approach to the Design of Gearboxes

The design process of the overwhelming majority of mechanical engineering production can be reduced to a number of typical (even standardized, to some extent) procedures. In general, they are:

- formation and specification of the design assignment (we introduce this part of the work to the design process deliberately, though, probably de jure, it precedes the design process itself when assigning the initial data. But de facto, the design assignment is formed and, moreover, specified only during the designing process);
- development of a draft design project, consisting of:
- generating the variants of layout decision;
- designing the main elements of the structural unit determining their dimensions and the possibility of their mutual arrangement;
- layout variant drawing, allowing the designer to specify the geometrical parameters, which is why it is accompanied by a great number of detailed analyses of structural elements;
- comparative analysis of layout variants, making the design decision;
- development of the detailed project.

The largest part of the works is related to the second stage—the draft design process. It is very poorly formalized. Decisions here depend mainly on the specific subject area, and experience accumulated both by the designer himself and in the specific area in general. These decisions are often contextual (they are made only upon a certain problem arising) and heuristic.

The above-detailed notion concerns the design of even comparatively simple objects—gearboxes. The classic sequence of gearbox design is known according to a great number of textbooks and reference books, and is as follows:

- analysis of the chosen type of gear (gears for a multistage gearbox);
- formation of the layout scheme of the gearbox;
- determining the arrangement and type of bearing supports;
- calculation of shafts;
- analysis and design of casing and other parts;
- manufacturing design (determining the setting parameters of equipment for complex manufacturing operations, special tooling and measurement devices).

This sequence operates reliably, allowing for the creation of rather serviceable production units. However, at least two arising problems can be singled out here. The first problem is: in most practical cases (except for special ones), the design process is carried out not according to definite values of loading, kinematic and mounting parameters of a gearbox, but for a certain range of such values. Since corresponding designed loading, lifetime and layout restrictions are different here, for each variant of the layout decision, this problem causes a whole bulk of chains of parallel calculations providing different evaluations and conclusions.

The second problem is largely the consequence of the first (though it appears independently): a great number of possible variants complicate analysis, and the presence of a great number of layout restrictions is overcome, in most cases, heuristically (for example, by simple "progressing" this or that unit).

Though both described problems are subjective (success in overcoming them depends on the skills and thoroughness of the designer), their demonstrations are too evident and numerous not to notice them and not to seek ways of reliably overcoming them. To illustrate this statement, Fig. 1a shows examples of extremely unsuccessful layouts of gearboxes widely applied until recently in Russian mechanical engineering as compared with better analogs that have appeared lately (Fig. 1b). Let us skip the detailed analysis of specifying separate units and elements of gearboxes; we acknowledge a significant failure in terms of mass and overall dimensions.

The evident advanced means of solving the pointed problems is automatizing the design process.

The methodological obstacle to overcoming the pointed problems is the above-mentioned determinacy of the design process. As was stated before, arising layout problems are solved according to their appearance within thorough refinement of variants of solutions, and their non-optimal or even unrealizable character is not revealed at once. Refinement of variants which seemed to be preferable for a designer due to these or those subjective reasons (for example, based on his/her

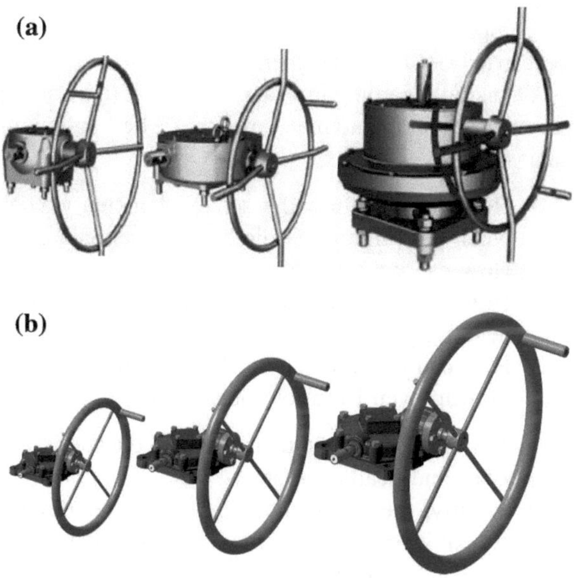

Fig. 1 Non-optimal (**a**) and close to optimal (**b**) results of gearbox development

own experience) often lead to worsening the resultant variant when overcoming certain layout or manufacturing problems. This process often eliminates variants from consideration which have been rejected since they seemed to be unfavorable. The detailed consideration of the whole variety of solutions (combinations of solutions) is accompanied by computational difficulties. Therefore, simply following the typical (standard) logics of engineering ("manual") design within computer-aided design improves the result in a way, but does not ultimately solve the problem of increasing the quality of the design decisions and shortening the time of development of competitive production units.

3 New Approach to the Process of Gearbox Design

Another approach to implementation of the design process is based on results of investigation into human creative activity. The philosophy of creativeness states that the process of creating anything new involves both human consciousness operating according to cause-and-effect algorithms generated during life and fixating interconnections of elements of the knowledge domain, and human sub-consciousness operating according to direct-search algorithms with the possibility of evaluating the sustainability of combinations obtained after combinatorial search. According to the proposed theory, the creative and talented individual is a personality whose sub-consciousness is able to reach consciousness at the moment of realization of the sought combination and whose consciousness is able to perceive and "logically" analyze the obtained solution.

Though the pointed theory of creative activity is not new, it was impossible to implement it in practice to obtain new results for design engineering activity due to its complexity and the incredible number of direct search procedures which should be carried out by a designer.

Decision regarding this task turned out to be non-trivial, even with application of advanced computational means. The following questions still remained unanswered:

(a) How can we form the set of elementary knowledge for further application when performing the direct-search procedure?
(b) How can we save this set in a compact way?
(c) How can we widen this set and supplement it with new knowledge without negatively affecting its correctness?
(d) How can we optimize the direct-search procedure, since it leads to information blowup even for a low power set?

Let us give answers to the stated questions regarding the design process of gearboxes and gearbox systems.

Evidently, the set of modules for further direct search, namely the set of elementary knowledge of gearbox component structural elements and their design features, can be obtained by analysis of existing design solutions.

Applying the principles of modularity and hierarchy as the main description principles, the principle of decomposition as the resolving principle, the principle of sufficiency as the restriction principle, and the multilevel principle as the representation principle, one can obtain the necessary and sufficient set of modules to solve the design task and the comprehensible model of the object structure in terms customary for an expert designer.

The model of an object can be expressed by multiple Cartesian products of the set of modules of the object except for non-implementable variants [8]:

$$F_{im\,v} = \prod_{i=1}^{K_1} (M) \backslash (F_z \cup F_{id\,v}), \tag{1}$$

where M is the set of modules of the object class, K_1 is the power of the set M, F_z is the set of non-implementable (forbidden) variants of the object, and $F_{id\,v}$ is the set of identical variants of the object.

The process of obtaining the pointed set of modules can be presented by the following set of stages:

(1) description by decomposition into structural elements (structure-forming modules) of existing layouts of gearboxes;
(2) description of structural elements;
(3) generalization of obtained knowledge.

Decomposition of the object into structure-forming modules can be carried out through a number of methods. The most preferable is the description of the object in terms customary to the designer: produced item, unit, sub-unit, assembly unit, sub-assembly unit, part, surface. Thus, when describing the gearbox structure, the graph is obtained with the following structure. The upper level comprises one apex —a gearbox (it represents the set of apexes of the first level V_1); the second level comprises gearbox units: worm unit, gearwheel unit, casing unit, all of them representing the set of apexes of the second level V_2; the third (the set of apexes V_3) involves assembly units: assembled elements like bearing supports and casing; the fourth (V_4) contains parts: bearings, caps, seals, barrels; the fifth (V_5) involves elements of parts of the fourth level: mounting, base, operating surfaces; the sixth (V_6) comprises geometrical primitives for imaging the surfaces of parts. The example of decomposition of the gearbox structure is schematically shown in Fig. 2.

Description of the object structure by decomposition is performed from the top downwards, beginning from resolving the apex-object into its components and continuing by resolving the obtained components into their substructure components and until the lowest level except for standard functional elements, which, as was shown above, do not demand resolution into components.

The pointed description can not be considered complete, though it gives representation of the gearbox structure. It does not mark the features of a variant of structural elements, or their specification providing the structural completeness of description.

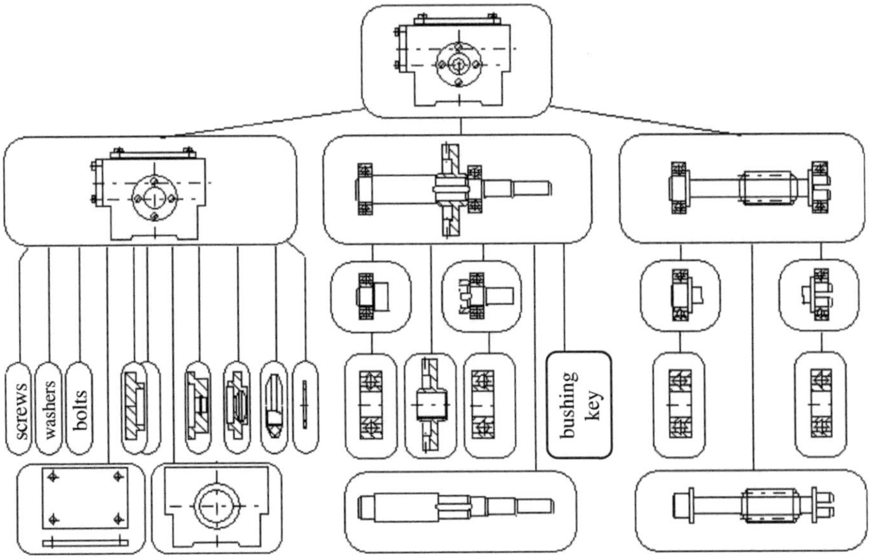

Fig. 2 Decomposition of the gearbox structure to functional elements

In order to achieve structural completeness, each structure-forming element is characterized, that is, described by a group of features marking, for instance, the mutual arrangement of elements, production technique and operating features, geometrical parameters, shape, material, etc. Characterization of elements is carried out from the top downwards, so that the characteristics of the part could not be mistakenly given to the unit comprising the pointed part (Fig. 3).

As the result of the pointed description, we get the tree-type graph model comprising apexes, which are the set of structural elements, and their characteristics and ribs, which are the links pointing at a hierarchy of structural elements and affiliation of features.

Despite the apparent simplicity, implementation of decomposition for certain design decisions required additional formalization to save the unified hierarchical approach. Introduction of the following concepts should be referred to elements of the pointed formalization.

1. Attribute—phantom. The phantom in the model is the functional apex designating a structural element but which is not per se the structural element; it simply denotes another purpose of the apex existing in this model. For example, in a double-stage gearbox, the output shaft of the first stage and the input shaft of the second stage prove to be one and the same part, however, due to peculiarities of hierarchical description, it will be represented in the model by two apexes, one of which is a phantom (Fig. 4).
2. Attribute—“standard”. The “standard” attribute is assigned to structural elements of the layout, with characterization implying only a reference to

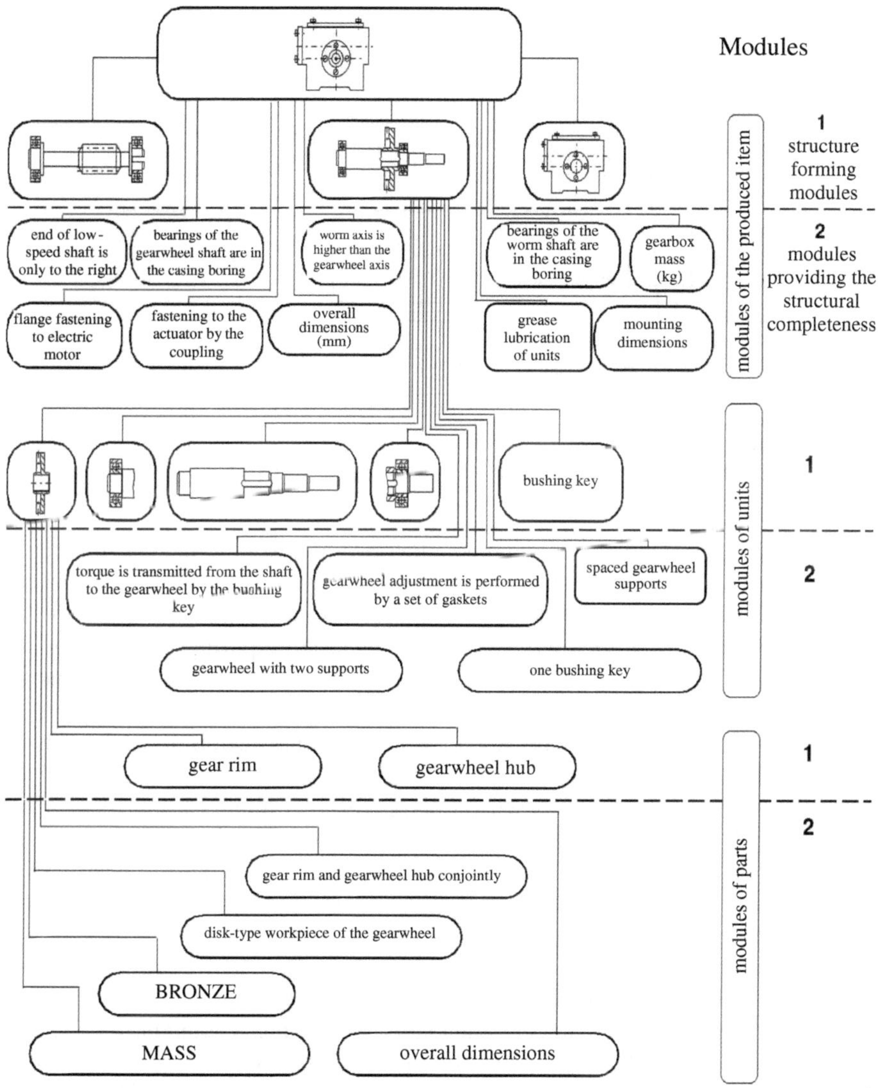

Fig. 3 Example of characterization of gearbox elements

corresponding standards, thus preventing their further decomposition and characterization of the module. This attribute can be assigned to bearings, bronze grades, threads, etc.

3. Attribute—"twin", denoting the presence of several single-type elements in the structure. Applying this attribute exempts the expert from the necessity of describing the similar element once again.

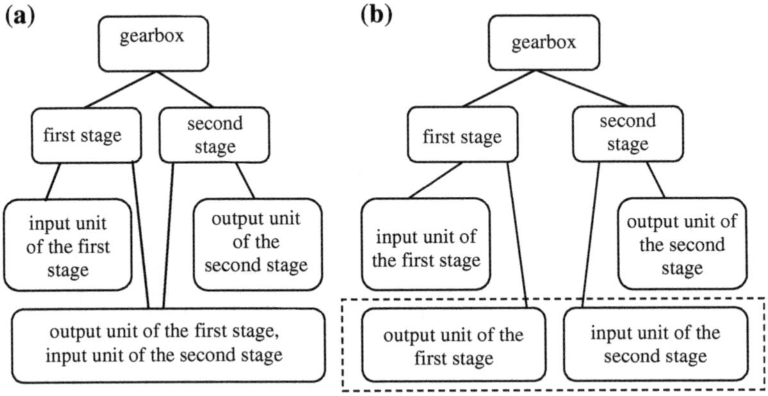

Fig. 4 **a** Actual representation of the gearbox structure; **b** hierarchic representation of the gearbox structure where the apex "input unit of the second stage" is a phantom

Application of the pointed approach for development of models for certain design decisions and further analysis of a rather large number of pointed models allowed for the following conclusion: no model can be used as the model for the whole class of objects, therefore the following tasks requiring the solution is a generalization of the obtained models and forming the set of those peculiarities which distinguish one design solution from the other.

An example of the model possessing complete information about its class of objects would be the model proposed in the work "Search design" by Polovinkin [9]. This model is developed based on comparison of different models of object designs and is represented by an "AND-OR" tree where structural elements and characteristics forming the structure of the design element are united by the "AND" apex, and structural elements or characteristics alternative with respect to each other are united by the "OR" apex.

The above-described model of the class of objects has a drawback. The presence of functional elements united in the "OR" apex implies designing the new variants of the object by means of bypassing the tree-type structure. Taking into account the multiple branching of the structure of complex objects, such an approach leads to a high work content of the synthesis process. To overcome this drawback, a new model of the class of objects has been proposed where only apexes-characteristics can be united by the "OR" apex.

Let us consider the graph method of developing such a generalized model.

The pointed method of creating the generalized model is based on the composition of graphs of certain variants of the object. The generalized model can be obtained as an "AND-OR" tree by semantic imposing of the graphs of certain variants of the object one onto the other, namely, implementing the following steps [10]:

1. Intersection of graphs of trees of certain variants

$$G_0(V_0, E_0) = \bigcap_{i=1}^{N} G_i(V_i, E_i),$$

where N is the number of initially analyzed variants of structure.
2. Analysis of the set V_0 to single out the obligatory component V_0':

- if there is the route from the root apex of the resultant graph v_0 denoting the object as a whole to the apex v_i where $i = \overline{1 \ldots K}$ $((K = |V_0 - 1|))$, then $V_{0j+1}^{'} = V_{0j}^{'} \cup \{v_i\}$, where $V_{0j}^{'}$ and $V_{0j+1}^{'}$ are two consequent states of the set of pseudo-obligatory apexes, and $V_{01}^{'} = \{v_0\}$.

Therefore, the sub-graph with the set of apexes being the component of coherence containing the root apex represents the pseudo-obligatory part of the structure of any variant of design for this class of production item, which is the skeleton of the generalized tree-graph.
3. To obtain the obligatory component of the graph, the set of obtained apexes is provided to the expert for analysis and to answer the question: Does any known structure of objects of this class have the analyzed structural unit in its composition? The positive answer allows for assigning the apex to the set of obligatory apexes, thus allowing us finally to obtain $G_0^{'}(V_0^{'}, E_0^{'}) \subseteq G_0(V_0, E_0)$. The negative answer implies that the considered apex is the imaginary alternative of the first kind. Appearance of alternative apexes of the first kind is the consequence of the initial sampling of objects. Alternatives of the first kind are eliminated from the component of coherence containing the root apex and are designated by the sign NILL-alternative.
4. Supplementing the generalized tree graph with the alternative component, implying the formation of the remaining alternative apexes, is commenced with consideration of the root apex and is the alternative process, each iteration implying execution of the following procedures when moving along the generalized graph from one level to the other:

- for those initial graphs with numbers fixed at the considered apex of the generalized tree graph (if the apex is obligatory or alternative of the first kind, then for all initial graphs), the sub-graph is drawn $G_{lik}(V_{lik}, E_{lik})$ with the set of apexes comprising the considered lth apex of the ith level v_{li} of the generalized tree graph and all adjacent apexes of the (i + 1)th level of the initial graph $G_k(V_k, E_k)$;
- sets $V_{lik}^{s} = V_{lik} \backslash V_0^{'} \backslash V_s$ are formed where $V_s = \{v_{li}\}$ consisting of unobligatory apexes;
- after comparative analysis of the obtained sets V_{lik}^{s} and their consecutive overview, the apexes opposite, according to the content, are singled out. If at least a pair of such apexes is found, they are united by the "OR" apex and relocated to the generalized tree graph to the (i + 1)th level, joined to the

apex v_{li} and related to the set $V''_{li0} \subset V''_0$ where V''_0 is the set of all apexes united by the "OR" apex. Each apex is marked here by the number of graphs designating the set of apexes to which this apex belongs. If the number of alternative apexes does not reveal numbers of all analyzed graphs or the parent apex is marked by the NILL sign, then the set of such alternatives is called incomplete, and in order to represent the model correctly, it is supplemented with the NILL-alternative, the value of which is $\overline{\bigvee_i (v_{il+1})}$ where v_{il+1} are values of alternatives of the incomplete set;

- the remaining apexes $(\bigcup_{k=1}^{N} V_{lik} \backslash V'_0 \backslash V''_{li0})$ are included in the set V'''_0, designated as the set of imaginary alternatives. Here, imaginary alternatives of the first kind, along with numbers of all graphs under consideration, are marked by the NILL sign, unlike alternatives of the second kind designated only by numbers of corresponding graphs. The generalized tree graph is considered to be generated when analysis for all apexes is executed (Fig. 5 a–c).

The generated generalized model is characterized by compactness of data storage for a great number of variants of the object structure and can be applied to obtain new variants. However, such a model has a drawback of redundancy of the stored information, since only those modules of the object are of interest when creating the production item which can distinguish one variant of the object from the other. To overcome the redundancy, formula (1) should be transformed as follows [11]:

$$F_{im\,v} = M_{obm} \times \prod_{i=1}^{K_2} (M_{opm}) \backslash (F_z \cup F_{idm}), \tag{2}$$

where M_{obm} is the set of obligatory modules of the object, M_{opm} is the set of optional modules of the object, and K_2 is the power of the set M_{opm} However, the set of optional modules M_{opm} is complete only when the following sets are exhaustive: the set of apexes of the generalized graph corresponding to modules of the structure, the set of groups of alternative characteristics of functional apexes, and the set of values of groups of alternative characteristics. It is possible only when the initial sampling involves all original (differing by at least one module) structural solutions, which is impossible when developing real CAD systems.

Therefore, the subsequent stage of developing the model of the object class is overcoming the redundancy of the generalized model and solving the task of its expansion. Development of the expanding classifier is proposed to be applied as the means of solving the pointed task.

The classifier represents the list of classification features containing questions, their answers specifying the features that distinguish one layout from the other.

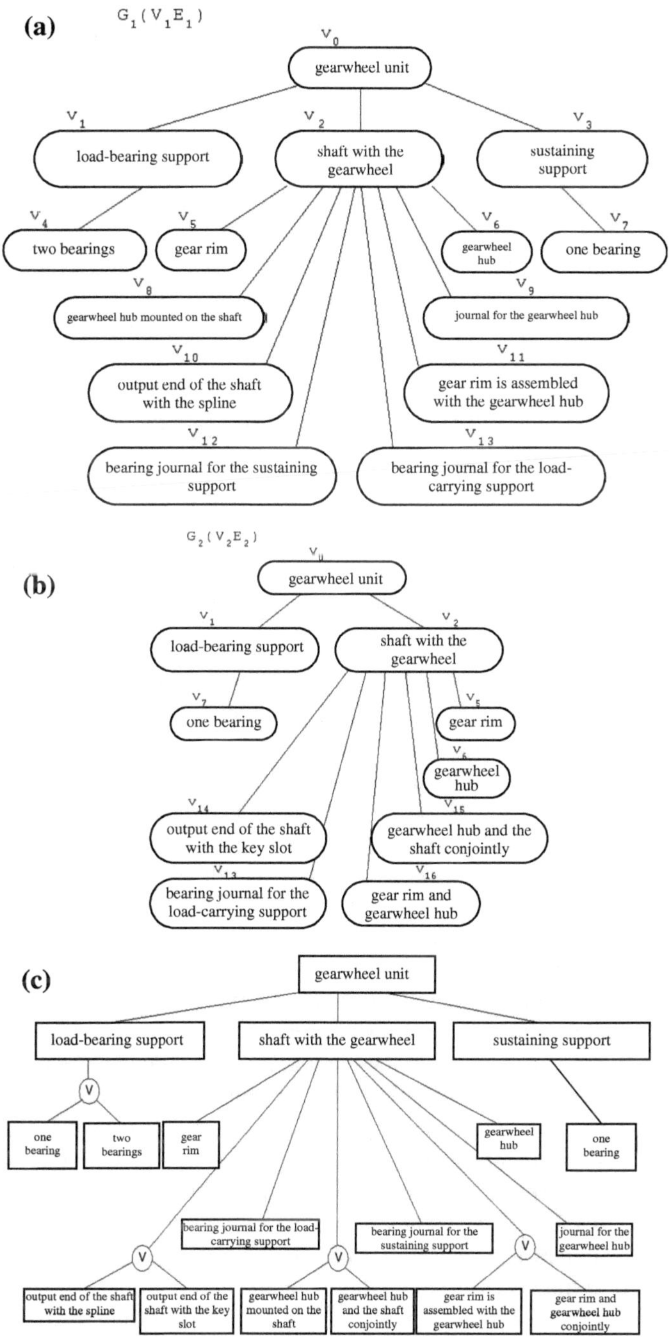

Fig. 5 Example of forming the generalized tree graph

Development of the classifier is executed by analyzing the generalized model.

1. If the considered apex is structure forming but not obligatory, the feature "presence of [name of the apex]" is stated, which should be referred to its parent apex. For example, the optional apex "flywheel" as a component of the functional element "unit of the input shaft of the spiroid gearbox" generates the feature "the presence of the flywheel". The feature stated in such a way always consists of two results: "yes" and "no".
2. For a group of alternative characteristics, the feature is stated by singling out the category—unchangeable part of characteristics. For example, the feature "Diameter" will be stated from values of characteristics: "Diameter 20", "Diameter 25", and "Diameter 30", and values of this feature will be: 20, 25, and 30. It is necessary to supplement the set of results of each of the stated features by the value NILL-F if the parent apex is alternative and by the value NILL-P if the pointed characteristics are not mentioned at the apex of one of the design variants.

Another advantage of the developed classifier is convenience of its expansion—supplementing with new knowledge.

Classifier expansion can be executed through several methods:

1. Expansion of the set of feature results. For this purpose, the expert is proposed to enumerate all known answers to the question stated at the feature in addition to already existing results. For example, for the apex "casing", the feature "shape" is stated with the results "parallelepiped" and "cylinder", and it can be supplemented with the result "parallelepiped + cylinder" when expanding the classifier. However, the set of results of some features represent the range of values with a definite pitch of discretization. For instance, the feature "loading torque" referring to the functional element "output shaft of the spiroid gearbox" implies the set of results as ranging from 500 Nm up to 32,000 with the pitch 500 Nm. In this case, enumeration by the expert of possible results can turn out to be unreasonably time-consuming. It is more reasonable to correlate this feature with information about limits of the range of possible values and discretization pitch. Such correlation is proposed to execute as the "Continuous" attribute. Being correlated with the feature, this attribute keeps and provides data on the range and pitch of discretization of values of feature results. The set of results can also be supplemented by the value NILL-P if the expert gives a negative answer to the question "Is one of the set of feature results always present in the layout if the apex to which this feature is referred is there?".
2. Expansion of the set of features referred to a certain functional apex. For this purpose, the expert is proposed to evaluate the current set of features of the apex with regard to the necessity of stating its name and pointing out possible results.
3. Correction of the generalized model and classifier by adding the functional apex. If it is necessary to add a functional apex, the expert must provide a description of its whole sub-graph, including features and their results. If the added apex is optional, the feature "presence of [added apex]" should be stated for its apex.

4. Expansion of the set of reference-classifiers. Evidently, adding a new reference-classifier increases the number of possible characteristics thus expanding the set of variants of the object described by the classifier. The added references should correspond to the principle of limiting the area of analysis, that is, should not contain the description of functional elements which are not applied within the investigated class of objects.
5. The result of classifier development is the possibility of modifying the model of class of objects to the following form:

$$F_{imv} = M_{obm} \times \prod_{i=1}^{N} (p_i) \backslash F_z, \tag{3}$$

where N is the power of the set of features, p_i is the *ith* feature, and F_z is the set of non-implementable (forbidden) variants of the object.

Analysis of the set of forbidden variants showed that their certain sub-sets are non-implementable for one and the same reason. To optimize the data storage, it was decided to keep not the forbidden variants but rather the reasons for their non-implementable character, which have been called "forbidden figures". This concept was first introduced as the component of the theory of characterization analysis by the academician Gorbatov [11].

Forbidden variants and, therefore, forbidden figures may have different natures [11].

A relatively forbidden figure is a combination of modules that is non-implementable due to initial restrictions imposed by the user to the model. Absolutely forbidden figures are combinations of modules that are always non-implementable independently on initial assigned restrictions.

Besides, the set of forbidden figures can be represented as $Z = Z_E \cup Z_f$ where Z_E is the set of empirical, absolutely forbidden figures obtained by expert survey and based only on expert knowledge, and Z_f is the set of functional, absolutely forbidden figures caused by the presence of definite calculation relations within the subject area, like $y = f(x)$ allowing for evaluation of one or several other parameters on the basis of certain parameters of the synthesized object.

Generation of the set Z_E of empirical forbidden figures takes place gradually and represents the interactive process. Each step of this process implies generation of intermediate sets of results and expert evaluation of these sets in order to single out and further eliminate variants which contain forbidden figures or which are the forbidden figures themselves. This is stipulated by the fact that such forbidden figures may have the power from 2 to n where n is the number of features participating in describing the layout of the object. At that, none of them must involve other forbidden figures of less power, that is, by the moment of analyzing the set of results with the power k, all forbidden figures of the power less than k should be determined.

Generation of the intermediate set at the ith step takes place by Cartesian product of the intermediate set obtained at the $(i - 1)$ step after deleting the forbidden variants of the initial set representing the integration of all features participating in description of the object. The intermediate set of the first step is nothing but the initial set M_{Π}.

Evidently, the number of such Cartesian products must correspond to the number of considered features, since the forbidden figure may have (theoretically) the power equal to the power of the set describing one of the variants of the complete set.

Obviously, generating the set of empirical forbidden figures is a rather time-consuming process. These sets of great power generated by features with a continuous range of values are extremely complicated in their evaluation. That is why investigation of the subject area to single out new functional and logic relations is the most vital for implementing the proposed ideology of the process of structural synthesis.

Having generated the classifier and the set of forbidden figures, we filled the model of the class of objects described by the formula (3) with data.

According to this formula, the set of possible design solutions can be obtained in two stages: by multiple products of the set of features and further deleting the set of forbidden variants that contain forbidden figures.

However, application of the pointed model as the model of the process of design solution synthesis is not possible due to the problem of "curse of dimensionality" because of extremely high demands on the memory space to store intermediate and final variants of the object and because of high computational load when synthesizing the variants of the object and deleting the forbidden variants.

The model of the process of synthesis can be obtained through transformation of the model of class of objects consisting of optimizing the necessary volume space and computational resources.

In order to optimize the process of synthesis, the following methods were proposed:

- Interchange of synthesis and analysis [12]. This method implies division of the synthesis process into elementary steps, each of them comprising the stage of synthesis as additional multiplying by the next feature with obtaining the intermediate results and the stage of analysis implying the truncation of the set of intermediate variants by deleting the variants containing forbidden figures.
- Ordering the features to be multiplied within synthesis [13]. In connection with the previous method, this one is the powerful tool of controlling the rate of increasing the number of intermediate variants. For example, if the object is described by features $A = \{a_1, a_2\}, B = \{b_1, \ldots, b_{100}\}, C = \{c_1, \ldots, c_5\}$, their multiplying in the order A-B-C will result in the higher average number of intermediate variants after multiplying by the next feature than in the order A-C-B.

Moreover, ordering the features allows for approaching the greater number of forbidden features at the beginning of synthesis, thus subjecting a lesser number of intermediate variants to analysis.

- Analysis of functional forbidden figures and their generating functional dependencies. This method implies classifying the functional relations, evaluation of their computational complexity and obtaining new relations with less computational complexity. It ensures minimization of computational complexity when analyzing the intermediate variants for forbidden figures.
- Prevention of generating the forbidden variants instead of analysis of the set of variants for forbidden figures [14]. As for empirical forbidden figures, this method implies application of markers indicating that the intermediate variant can be the argument of the forbidden figure. Synthesis of the next set of intermediate variants is performed by means of a matrix with columns containing the current intermediate variants and lines containing results of the next feature. Therefore, forbidden variants are not generated within synthesis. This ensures minimization of the number of intermediate variants and eliminates the necessity of additional analysis of the set of intermediate variants for forbidden figures.

In order to eliminate the analysis of functionally forbidden figures, the method is applied as direct generation of allowed variants. Thus, when the intermediate set of variants of the power N is multiplied by the feature P, which is the result of a function of certain previously multiplied features, in order to obtain the intermediate set of allowed variants of the power (N + 1), for each of initial variants, it is enough to calculate the value (or values) of the function having substituted the corresponding results-arguments into this function and having checked affiliation of these values to the set of possible results of the additionally multiplied feature. For example, when the feature E is additionally multiplied for the initial variant $(a_i, b_j, \ldots, d_k)$, the result of calculating the function $f(a_i, d_k)$ is the values x, y, z, with x corresponding here to the possible result e_1, and z corresponding to the possible result e_3, whereupon the new set of intermediate variants will include variants $(a_i, b_j, \ldots, d_k, e_1)$ and $(a_i, b_j, \ldots, d_k, e_3)$

Analysis for forbidden figures will further imply such a number of additional activities at the stage of synthesis which allows for preventing the generation of forbidden figures.

- Application of binary logics to code the variant of the object [15]. This method implies writing the variant of the object as the sequence of zeros and units denoting the presence or absence of a result of some feature in the variant of the object. Optimization effect of applying the binary logics appears due to peculiarities of data storage in the computer. Such an approach is supposed to decrease the memory space necessary to keep variants, since binary logics allow for applying the minimum memory space for data storage.

The above-described methods of solving the tasks of optimizing the process of synthesis give the greatest effect within their combination. The common trait of the first three methods is the influence of the order of feature multiplying on the efficiency of these methods. Let us consider the factors influencing the order of multiplying the features within the synthesis.

The first factor is the number of results of each feature, since the number of intermediate variants increases proportionally to it when multiplying by the corresponding feature. It should be taken into account that imposing the design assignment results in reducing the power of features, since their values are specified by truncating the set of its results. The coefficient of such a truncation can be determined as the ratio of the number of forbidden results of the feature to the number of all results of the feature:

$K_p = \frac{F_P}{M_P}$ where F_P is the power of forbidden results of the feature P, and M_P is the power of the set of all results of the feature P.

The second factor of ordering the features to be multiplied within synthesis is the rate of occurrence of results of the feature in forbidden figures. The higher the number of forbidden figures where the values of the feature are present, the more effective the truncation of the intermediate set.

The third factor is the power of forbidden figures, since checking the forbidden figures with the power 3 does not make any sense after multiplication of 2 features.

The fourth factor is application of functional relations available in the subject area to truncate the set. Functional relation possesses, as a rule, a clearly defined set of initial data and a set of results, therefore the desired sequence of features to be multiplied is evident: initial data are multiplied first and results are calculated at the end without producing the forbidden variants. Obviously, the fourth factor is of crucial influence on optimization of the process of synthesis if the value of the function result is not assigned before calculation. It is possible to overcome the pointed problem through synthesis of the family of functions, when the initial functional relation $P_n = f_1(P_1, P_2, \ldots, P_{n-1})$ is the basis for generating the set of inverse functions with respect to each argument $F = \{P_n = f_1(P_1, P_2, \ldots, P_{n-1}),\ P_1 = f_2(P_2, \ldots, P_n),\ P_2 = f_3(P_1, P_3, \ldots, P_n),\ P_{n-1} = f_n(P_1, P_2, \ldots, P_{n-2}, P_n)\}$

Investigation of the function reversibility allowed for proposing the following classifications.

According to reversibility, the functions are divided into:

- reversible—the result of conversion is the function of the power series, which is calculated by substitution of values of arguments;
- conditionally reversible—imitation of solving the task $x = f^{-1}(y)$ is achieved through iterative methods of solving the linear equations implying the multiple calculation of the initial function $y = f(x)$. These methods are, for instance, the method of half divisions, chord method, Newton's method, and Tchebyshev's method. Application of such methods is possible only when the function is continuous within the definition domain;
- irreversible—imitation of solving the task $x = f^{-1}(y)$ is achieved only by calculating the function $y = f(x)$ for all possible values x;

- interval reversible—they have different types of reversibility at different intervals of the definition domain.

  According to the type of function obtained after conversion, reversible functions can be divided into:
- analytically reversible functions—the inverse function can be obtained by means of relatively simple identical transformations;
- numerically reversible—for functions of one variable which are irreversible analytically but differentiable continuously, the inverse function can be obtained as the power series.

According to the pointed classifications of initial functions, it was accepted to classify the obtained inverse functions with regard to the calculation type as:

1. Analytical, that is, analytical functions.
2. Numerical, represented by the power series.
3. Iterative, their calculation being based on application of iterative methods.
4. Direct-search, their calculation being executed through calculating the values of the initial function for all possible values of the argument with respect to which the inverse task $x = f^{-1}(y)$ should be solved. At first sight, such a "function" does not provide the optimization effect.

The considered methods and factors of optimization, and investigations devoted to conversions of functional relations, allowed for accepting the following procedure of feature multiplying in order to solve the tasks of minimizing the number of intermediate variants within synthesis:

At the first stage, features are multiplied participating in families of functions, that is, interrelated with a functional dependence.

At the second stage, features are multiplied with results participating in the remaining empirical forbidden figures.

Finally, features are multiplied with results that do not participate in forbidden figures.

The above-considered methodology has been partially applied when creating the layout of spiroid gearboxes for pipeline valves [16].

4 Conclusions

The mathematical model of the design process proposed in this work differs principally from the approach implemented within classical design procedures, since:

1. The mathematical apparatus of the layout synthesis involves the direct-search algorithm rather than the well-known relations (functional and empirical) of the subject area. Knowledge of the subject area stated as graphs, tables and formulas are only restrictions imposed on the process of synthesis and cutting the forbidden variants.

2. In description of a class of objects, there is no division of characteristics into "parameters" of structural elements and "criteria" of evaluating the results of the design process. Both parameters and criteria are integrated into the set of features.
3. The set of features participating in direct search is ordered with respect to the number of initial assignment data, therefore, the procedure of performing the analysis may differ a lot for various statements of tasks.
4. The proposed mathematical model allows for creating the system of designing where the set of input and output data is not strictly determined. Assigning the initial data, the designer specifies the value of any sub-set of features of the classifier describing the class of objects and the system completes the definition of the values of all the remaining features of the classifier.
5. Assigning the function of searching the best solution as the initial data, the system applies this function as the additional restriction and will provide the best variant out of the set of possible variants according to the assigned requirements.

Implementation of the pointed model as the model of the process of computer-aided design provides the researchers in the field of gears and gearboxes with the tasks of detecting and fixing functional, heuristic relations, and accumulating the empirics, since the knowledge obtained again is easily built into the system of computer-aided design developed according to the new approach, optimizes its functioning essentially and does not require its rebuilding.

References

1. Anuryev VN (2001) Reference book of mechanical engineering designer, vol 3. Mashinostroyeniye, Moscow (in Russian)
2. Standard GOST P 50891-96. General-purpose gearboxes. General technical conditions (in Russian)
3. Dietrikh Ya (1981) Design and development. System approach. Mir, Moscow, 456 pp (in Russian)
4. Davydov BL, Skorodumov BA, Bubyr YuV (1963) Gearboxes. Layouts, analysis, and testing. Mashgiz, Kiev, 474 pp (in Russian)
5. Kurmaz LV (2007) Designing of machine units and parts. Vysshaya shkola, Moscow, 455 pp (in Russian)
6. Lelikov OP (2007) Fundamentals of analysis and design of machine units and parts. Mashinostroyeniye, Moscow, 464 pp (in Russian)
7. Orlov PI (1988) Fundamentals of design. Reference guidance manual. Moscow, Mashinostroyeniye, 544 pp (in Russian)
8. Malina OV (2002) Theory and practice of automatizing the structural synthesis of objects and processes with application of methods of characterization analysis. D.Sc. in Engineering Thesis, Izhevsk, 392 pp (in Russian)
9. Polovinkin AI (1991) Theory of design of new machinery: laws of machinery and their applications. Informelectro, Moscow, 104 pp (in Russian)

10. Malina OV, Urzhumov NA (2004) Approach to optimization of the process of structural synthesis in systems of computer-aided design. In: Proceedings of the scientific technical forum with international participation "High technologies—2004". ISTU, Izhevsk, pp 95–102 (in Russian)
11. Malina OV, Valeyev OF (2013) Model of the process of structural synthesis of objects developed within discrete structures and features of its implementation. Vestnik of Kalashnikov ISTU, N2(57), pp 24–26 (in Russian)
12. Malina OV, Valeyev OF, Zarifullina EG (2013) Analysis of approaches to minimization of computer resources within the process of computer-aided structural synthesis of mechanical engineering medium complexity parts. Vestnik of Kalashnikov ISTU, N2(57), pp 22–24 (in Russian)
13. Malina OV, Valeyev OF (2013) Mathematical support of computer-aided design of gearbox systems. In: Theory and practice of gearing: Proceedings of International Symposium, ISTU, Izhevsk, pp 426–431 (in Russian)
14. Malina OV, Lukin EV, Valeyev OF (2012) Development of the model of optimizing the process of designing the low-speed heavy-loaded spiroid gearboxes for pipeline valves (PV). In the world of scientific discoveries, N1(25) (Mathematics, Mechanics, Computer Science), pp 134–152 (in Russian)
15. Malina OV, Valeyev OF (2013) Approaches to minimization of computer resources necessary to implement the process of structural synthesis of objects developed within discrete structures. Intelligent systems in production, N1(18), pp 29–34 (in Russian)
16. Goldfarb VI, Glavatskikh DV, Trubachev ES, Kuznetsov AS et al (2011) Spiroid gearboxes of pipeline valves. Veche, Moscow, 222 pp (in Russian)

Design of Shaping Machine and Tooling Systems for Gear Manufacturing

A. Krivosheya, Ju. Danilchenko, M. Storchak and S. Pasternak

Abstract The development and manufacturing of gear pairs is determined by a system combining the kinematic basis of the relative motions between workpiece and tool with the necessary production technology. This system unites the two subsystems of theory and technology. Via a multitude of parameters, gears can be assigned to different classes according to the range of existence. The paper describes a mathematical model for determining the range of existence for the general kinematic scheme of gear shaping. This model is based on an analysis of the existing kinematic schemes of theoretical and real shaping. Based on a morphological approach, individual kinematic shaping schemes are classified, and their mathematical models are developed. The shaping schemes, as well as the necessary translational and rotational motion matrices, are exemplarily presented for a gear manufacturing machine using the shaping principle. The modern approaches to the principles of designing equipment and instrumentation systems for gear manufacturing are presented. Methodical basics for selecting the optimal machine configuration depending on the technical and economic requirements for the machining and form of a gear's tooth profile are stated. The system for theoretical and technological optimization synthesis of instrumentation systems for gear manufacturing is presented.

Keywords Computer aided engineering · Gear · Kinematic shaping · Machine design · Neural network

---

A. Krivosheya
V. Bakul Institue for Superhard Materials, NAS, Kiev, Ukraine
e-mail: krivosheyatolja@ukr.net

Ju. Danilchenko
National Technical University of Ukraine "KPI", Kiev, Ukraine
e-mail: yumd@i.ua

M. Storchak (✉) · S. Pasternak
Institute for Machine Tool, University of Stuttgart, Stuttgart, Germany
e-mail: michael.storchak@ifw.uni-stuttgart.de

S. Pasternak
e-mail: sergii.pasternak@ifw.uni-stuttgart.de

V. Goldfarb and N. Barmina (eds.), *Theory and Practice of Gearing and Transmissions*, Mechanisms and Machine Science 34,
DOI 10.1007/978-3-319-19740-1_21

1 Introduction

Modern machine building uses gear pairs and multiple keyings of complicated form and structure. This includes two-parametric gear pairs and those with sinusoidal, involute and Novikov-Wildhaber profiles, as well as spline shafts and external splines such as, e.g., those with polygonal surfaces [1, 2, 10, 18]. The increasingly complicated form and structure of gear pairs, as well as the rise in demand regarding their functional, qualitative and operating characteristics, require a new approach for the planning and production technique of such components and assemblies.

The planning and production of gear pairs is determined by the gear manufacturing system generating the gear shape. It consists of two mutual subsystems: the theoretical and the production technological subsystem. The theoretical synthesis solves the task of the optimum theoretical shaping of gear pairs. The production technological synthesis, however, solves the task of the optimum real succession of shaping, guaranteeing retention of the required form, quality and productivity in each machining operation.

The theoretical shaping with a reference form body or reference form profile constitutes the main characteristic in the planning of gear pairs, wherein lies the difference compared to other components of machine building. Although there are other methods for theoretically arranging complete gear tooth surfaces with a constant profile along the flank line [17], the theoretical shaping with a reference form body is virtually the only method for producing gear pairs with complicated structure, as well as their modification [9, 11].

During the theoretical shaping process, the reference form body executes a particular movement relative to the immovable, shaping gear. This movement, which is described in the coordinate system of the gear, is called a kinematic scheme of the theoretical shaping [13, 14, 16]. In this way, for example, the whole surface of the cylindrical involute tooth system is determined by a theoretical shaping with a standardized reference form profile. This serves as a counter former to the reference profile.

The real shaping of the gear tooth surface takes place step by step. Each successive, technological operation contains different tools and kinematic schemes. Analogous to the theoretical shaping, the kinematic scheme describes a relative motion of the tool or the reference body in the coordinate system of the gear.

During shaping on a machine tool, a fundamental difference exists with regard to kinematic schemes. The theoretical and real shaping differs from the kinematic scheme of the absolute motion of the gear pairs. Consequently, a gear tooth surface can be shaped with different gear cutting tools in the case of different kinematic schemes. On the other hand, a kinematic scheme of the real shaping can be realized with different kinematic schemes of the gear pairs' absolute motions.

In the accuracy theory of machine tools, this is defined as a synthesis of the layout of a machine tool with similar shaping options [15]. Hence, there is a defined number of kinematic schemes from theoretical and real shaping of the gear tooth systems. The number of kinematic schemes of absolute motions is, however, greater in the case of real shaping. In order to decide on an optimum kinematic scheme for the theoretical and real shaping and the absolute motions of the gear tooth systems in the case of real shaping, a seek area or an existence range of such kinematic schemes is to be determined.

An analysis of the kinematic schemes of shaping and absolute motion during shaping contributes to optimization of the process of theoretical and real shaping. This also contributes to the development of new gear pair types, gear cutting tools and gear manufacturing machines, as well as new gear cutting methods.

The scientific foundations acquired [4, 5, 7, 8] contain a classification of the kinematic shaping schemes for different cutting processes. The classification constitutes only a subset of the possible kinematic schemes here. The performed analysis shows that no summary of the kinematic shaping schemes exists in the gear cutting theory. In addition, there is neither systematization nor classification of these schemes and their mathematical models.

2 Mathematical Model of the Kinematic Scheme

In the face of algorithmization and programming, the apparatus of affine transformation has proved to be most suitable for describing the kinematic shaping schemes and surface shaping in machining mathematically [6, 12]. Here, a movement was represented as a particular affine transformation or operators to be mapped were represented as motion operators $\left(\overline{\varphi}_i^w, \bar{\ell}_i^w\right)$. Furthermore, it was demonstrated that a composite transformation equation of the affine space transforms into a composite equation of an arbitrarily multiparametric motion of a material body, including a shaping member. To simplify matters, it was suggested here that a rotation around an axis and any rectilinear motion or transfer into each reference point be entered. At the same time, the secondary coordinate systems are to be entered so that two axes of this coordinate system remain parallel in pairs. If a shaping member or tool are then determined in the first coordinate system and a member to be shaped or a workpiece are determined in the n + 1 coordinate system, the composite kinematic scheme of absolute motion of the shaping member in the n + 1 coordinate system represents a composite kinematic shaping scheme of different classes, kinds and types. Such an arrangement of the composite kinematic shaping scheme is characterized by the three classification features:

- Number of rotational and translational motions
- Reciprocal arrangement of axes
- Nature of the elemental movements and their functional connection

The well-known classification of the kinematic shaping schemes is only arranged according to the first feature, i.e., the number of affine transformations [12]. The number of affine transformations is three. Three rotations and three linear movements can be carried out here. The idea is, however, to justify and determine the necessary and sufficient number of affine transformations in the kinematic scheme, as well as to establish the particular motions in each reference point and the particular relative arrangement of the neighbouring reference points. This is carried out in order to determine an existence range of the composite kinematic shaping schemes concerning the gear tooth systems of different classes, types and kinds, as well as to work out a unified mathematical model, systematization and detailed classification of the kinematic shaping schemes. A solid body has six degrees of freedom. If all possible mobilities of the shaping member and member to be shaped are taken into account, six affine transformations are sufficient in the composite kinematic scheme of the gear tooth system. This determines all existing kinematic shaping schemes of different classes, types and kinds. Complicated spherical gears comprise up to five one-parametric rotations [18].

The Denavit-Hartenberg method is used for unifying the coordinate transformations of neighbouring coordinate systems [3]. In accordance with this method, a rectangular coordinate system is assigned to each *i*th kinematic pair. The z_i-axis is arranged along the pair axis, and the x_i-axis is arranged perpendicular to the z_i and z_{i+1} axes. The y_i-axis is chosen to result in a rectangular coordinate system. A relative arrangement of two neighbouring coordinate systems, $S_i(X_i, Y_i, Z_i)$ and $S_{i+1}(X_{i+1}, Y_{i+1}, Z_{i+1})$, corresponds to the general case according to Denavit-Hartenberg. The z_i and z_{i+1} axes, as well as the x_i and x_{i+1} axes, cross in the general case. In special cases, axes of neighbouring coordinate systems can be parallel to each other, correspond and intersect, for example, the z_i and z_{i+1} axes as well as the x_i and x_{i+1} axes.

The general mathematical model of a kinematic shaping scheme presented in Fig. 1 can be described with 4th-order matrices as follows:

$$\begin{cases} m_{rm_{(n+1)}\,T/P} = m_{v_n^w C_n^w} \times m_{\varphi_n^w \ell_n^w} \times \ldots \times m_{v_i^w C_i^w} \times m_{\varphi_i^w \ell_i^w} \times \ldots \\ \qquad \ldots \times m_{v_1^w C_1^w} \times m_{\varphi_1^w \ell_1^w} \times m_{r_{T1}} \\ \varphi_{iA}^{u} < \varphi_i^{u} < \varphi_{iB}^{u} \\ \varphi_{iA}^{v} < \varphi_i^{v} < \varphi_{iB}^{v} \\ \varphi_{iA}^{w} < \varphi_i^{w} < \varphi_{iB}^{w} \\ \left.\begin{array}{l} \varphi_i = f_{\varphi_i}(\varphi_1^w) \\ l_i = f_{l_i}(\varphi_1^w) \end{array}\right\} \text{coupling equation} \end{cases} \tag{1}$$

$m_{rm_{(n+1)}\,T/P}$—matrix of moving equation of shaper link (T) that is specified in (1). coordinate system relative to fixed part link (P), $m_{v_i^w C_i^w}$—matrix of location, i.e., matrix of coordinate transformation by conversion from *i*th to i + 1-th reference point, $m_{\varphi_i^w \ell_i^w}$—matrix of shaper link moving in the *i*th reference point, $m_{r_{T1}}$—matrix equation of shaper link surface in the (1). reference point, φ_i^u, φ_i^v—independent parameters of shaper link surface, φ_1^w, l_i—independent parameters of rotation and displacement shaper link, i—number of reference point, A, B—starting and finite indices.

The last two equations represent a coupling equation. By the special entry of the movements into each reference point and the special orientation of the coordinate

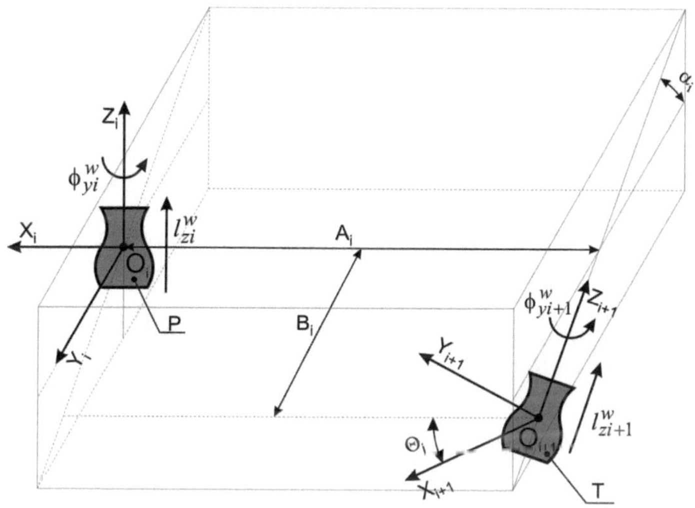

Fig. 1 Composite kinematic scheme of the relative arrangement of members

axes of neighbouring reference points, the motion and position matrices can be deduced as follows:

- Motion matrices

$$\begin{cases} m_{\varphi^w_{xi}\ell^w_i} = m_{\varphi^w_{xi}} \times m_{\ell^w_{xi}} \times m_{\ell^w_{yi}} \times m_{\ell^w_{zi}} \\ m_{\varphi^w_{yi}\ell^w_i} = m_{\varphi^w_{yi}} \times m_{\ell^w_{xi}} \times m_{\ell^w_{yi}} \times m_{\ell^w_{zi}} \\ m_{\varphi^w_{zi}\ell^w_i} = m_{\varphi^w_{zi}} \times m_{\ell^w_{xi}} \times m_{\ell^w_{yi}} \times m_{\ell^w_{zi}} \end{cases} \quad (2)$$

  The operators φ^w_x, φ^w_y and φ^w_z are positive, if an anticlockwise rotation occurs when considering the corresponding tip of the axis.
- Position matrices

$$\begin{cases} m_{v^w_{xi}C^w_i} = m_{v^w_{xi}} \times m_{C^w_{xi}} \times m_{C^w_{yi}} \times m_{C^w_{zi}} \\ m_{v^w_{yi}C^w_i} = m_{v^w_{yi}} \times m_{C^w_{xi}} \times m_{C^w_{yi}} \times m_{C^w_{zi}} \\ m_{v^w_{zi}C^w_i} = m_{v^w_{zi}} \times m_{C^w_{xi}} \times m_{C^w_{yi}} \times m_{C^w_{zi}} \end{cases} \quad (3)$$

The operators ℓ^w_x, ℓ^w_y and ℓ^w_z must be positive if a translation of the corresponding axis in a positive direction occurs by means of them. If a coordinate system is rotated anticlockwise around an axis and coincides with a subsequent coordinate system, then the operators v^w_x, v^w_y and v^w_z take on positive values. In the new coordinate system, the operators c^w_x, c^w_y and c^w_z contain the coordinates of the origin of a previous coordinate system here. The individual matrices in (2) and (3) represent well-known transformation and rotation matrices [11]. The parameters A_i, B_i, Θ_i and α_i can be converted explicitly into the parameters v_x^w, v_y^w, v_z^w, c_x^w, c_y^w, c_z^w of the coordinate transformation from (2) and (3) as follows:

(a) Conversion $S_i \rightarrow S_{i+1}$: (b) Conversion $S_{i+1} \rightarrow S_i$:

$$\begin{array}{ll} v_{xi}^w = \alpha_i & v_{xi+1}^w = -\alpha_i \\ v_{zi}^w = \Theta_i & v_{zi+1}^w = \Theta_i \\ c_{xi}^w = A_i \times \cos \Theta_i & c_{xi+1}^w = -A_i \\ c_{yi}^w = A_i \times \sin \Theta_i & c_{yi+1}^w = B_i \times \sin \alpha_i \\ c_{zi}^w = B_i & c_{zi+1}^w = -B_i \times \cos \alpha_i \end{array} \quad (4)$$

Correspondingly, the matrices of the coordinate transformations can be represented in the general case as follows:

(a) Conversion $S_i \rightarrow S_{i+1}$: $\mathrm{m}_{v_i^w c_i^w} = \mathrm{m}_{v_{zi}^w} \times \mathrm{m}_{v_{xi}^w c_i^w} \times \mathrm{m}_{c_{zi}^w} \times \mathrm{m}_{c_{yi}^w} \times \mathrm{m}_{c_{xi}^w}$

(b) Conversion $S_{i+1} \rightarrow S_i$: $\mathrm{m}_{v_{i+1}^w c_{i+1}^w} = \mathrm{m}_{v_{xi+1}^w} \times \mathrm{m}_{v_{zi+1}^w} \times \mathrm{m}_{c_{zi+1}^w} \times \mathrm{m}_{c_{yi+1}^w} \times \mathrm{m}_{c_{xi+1}^w}$

An important feature of the shaping systems of machine tools has to be taken into account for a particular presetting of the motions at each reference point: Two successive members of the system do not have more than one degree of freedom for relative motion [15]. This means that a member can either rotate around an axis which is connected to the coordinate system of the neighbouring member or move linearly along a fixed axis. The member can remain immovable as well. This feature is necessary since there must be a constructively stable and unchanging bearing system for the movement of each member. This guarantees the desired accuracy, productivity and reliability of the machine tool on the whole.

For a particular presetting of the motions, there must be the rotation around the z_i-axis and a linear movement along the same axis. This axis is determined along the axis of the kinematic pair, in accordance with the Denavit-Hartenberg method. In view of the up to six degrees of freedom of a shaping member and the presetting that conducts only one single movement at each reference point, it is possible to describe a composite kinematic shaping scheme, as illustrated in Fig. 2. The motion

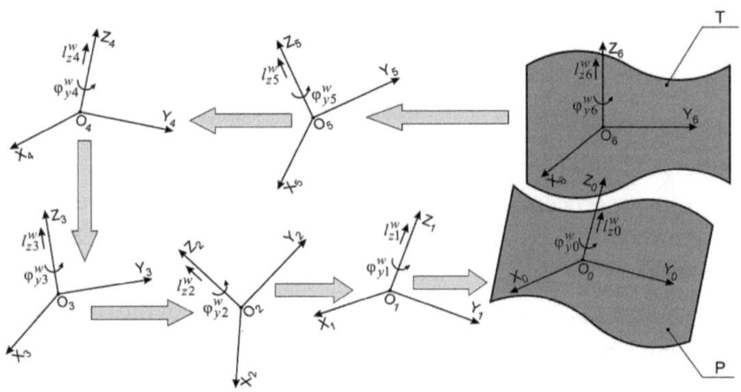

Fig. 2 Schematic representation of the existence range of a composite kinematic shaping scheme

can be translational or rotational, $\bar{\ell}^{w}_{zi}$ or $\bar{\phi}^{w}_{zi}$. The presetting of the coordinate system and the motions correspond to a concrete machining process here. The arrows indicate the direction of the individual affine transformations.

The reference points are to be numbered starting with the reference point at which the gear tooth system's member to be shaped is set in advance [15]. The index "0" is assigned to the immovable reference point here. If no coordinate transformation is carried out, the neighbouring reference points correspond and have the same index.

The composite kinematic shaping scheme of the gear tooth system can be restricted to seven matrices for a particular presetting of the motions and the arrangement of the coordinate systems of neighbouring reference points (Eq. 5):

$$\begin{cases} m_{rm0\,T/P} = m_{v_1^w c_1^w} \times m_{\varphi_{z1}^w \ell_{z1}^w} \times m_{v_2^w c_2^w} \times m_{\varphi_{z2}^w \ell_{z2}^w} \times m_{v_3^w c_3^w} \times \\ \qquad \times m_{\varphi_{z3}^w \ell_{z3}^w} \times m_{v_4^w c_4^w} \times m_{\varphi_{z4}^w \ell_{z4}^w} \times m_{v_5^w c_5^w} \times m_{\varphi_{z5}^w \ell_{z5}^w} \times \\ \qquad \times m_{v_6^w c_6^w} \times m_{\varphi_{z6}^w \ell_{z6}^w} \times m_{ru6} \\ \varphi_{1A}^u \le \varphi_1^u \le \varphi_{1A}^u \\ \varphi_{1B}^v \le \varphi_1^v \le \varphi_{1B}^v \\ \varphi_{z6A}^w \le \varphi_{z6}^w \le \varphi_{z6B}^w \\ \varphi_{zi}^w = f_{\varphi_{zi}^w}\left(\varphi_{z6}^w\right) \\ \ell_{zi}^w = f_{\ell_{zi}^w}\left(\varphi_{z6}^w\right) \end{cases} \quad (5)$$

According to the Denavit-Hartenberg convention [3], these matrices describe the general case and consist of two motion matrices at each reference point and five matrices of the coordinate transformation. The matrices of the unperformed motions at the reference points and coordinate transformations are either represented as $m_{m_{qi}^w} = diag(1,1,1,1)$ or not considered.

3 Classification of Kinematic Schemes

The existence range of the kinematic shaping scheme of the gear tooth system can be illustrated in the form of a graph, considering its first two classification features (Fig. 3). In accordance with the graph, the existence range of the kinematic shaping

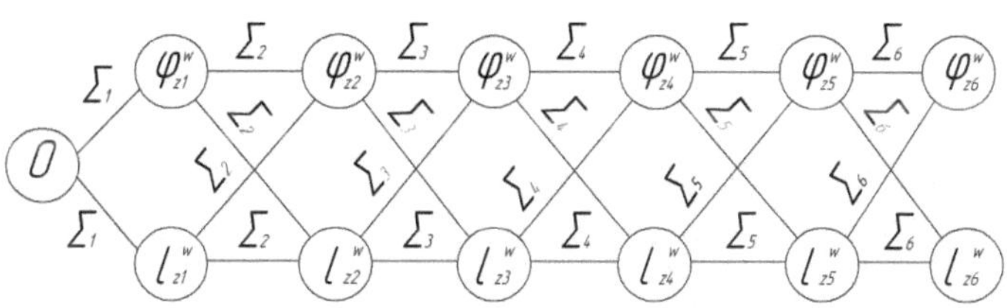

Fig. 3 Schematic classification graph of the existence range of the kinematic shaping scheme of the gear tooth system

schemes comprises a substantial number of variants but only a portion of them is important in practice. In the form of a table, all possible variants can be presented as particular graphs of the kinematic shaping scheme together with their mathematical model. A class of the kinematic scheme is determined by the general number of rotational and translational motions of a shaping member. A subclass is, however, established by matrices of the relative arrangement of the reference points. Correspondingly, there are 32 subclasses in each class. Table 1 lists examples of the

Table 1 Classification of kinematic schemes

Class of scheme	Subclass of scheme	Graphs of individual kinematic shaping schemes	Mathematical model of individual kinematic shaping scheme
1	1	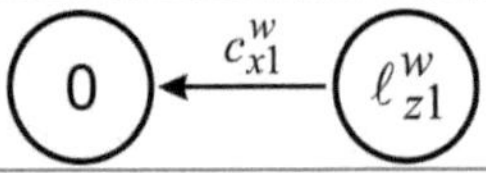	$m_{rm0\ T/P} = m_{c^w_{x1}} \cdot m_{\ell^w_{z1}} \cdot m_{rT1}$
	2	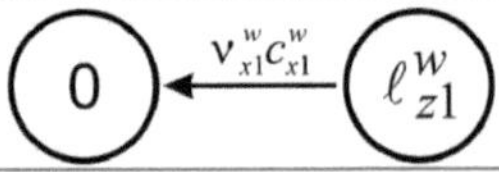	$m_{rm0\ T/P} = m_{v^w_{x1}c^w_{x1}} \cdot m_{\ell^w_{z1}} \cdot$ $\cdot m_{rT1}$
	3	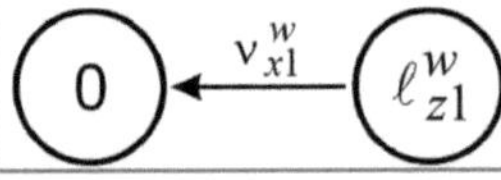	$m_{rm0\ T/P} = m_{v^w_{x1}} \cdot m_{\ell^w_{z1}} \cdot m_{rT1}$
2	1	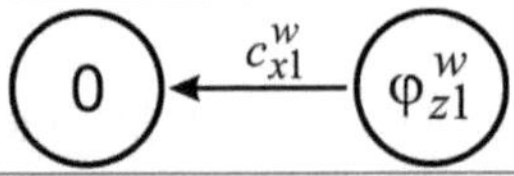	$m_{rm0\ T/P} = m_{c^w_{x1}} \cdot m_{\varphi^w_{z1}} \cdot m_{rT1}$
	2	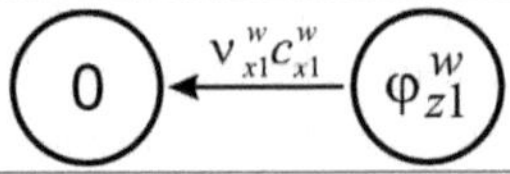	$m_{rm0\ T/P} = m_{v^w_{x1}c^w_{x1}} \cdot m_{\varphi^w_{z1}} \cdot$ $\cdot m_{rT1}$
	3	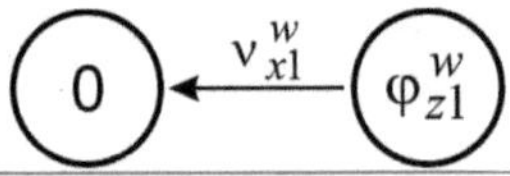	$m_{rm0\ T/P} = m_{v^w_{x1}} \cdot m_{\varphi^w_{z1}} \cdot m_{rT1}$
3	1	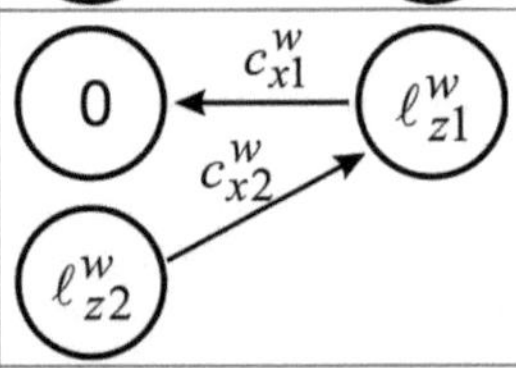	$m_{rm0\ T/P} = m_{c^w_{x1}} \cdot m_{\ell^w_{z1}} \cdot m_{c^w_{x2}} \cdot$ $\cdot m_{\ell^w_{z2}} \cdot m_{rT2}$
	2	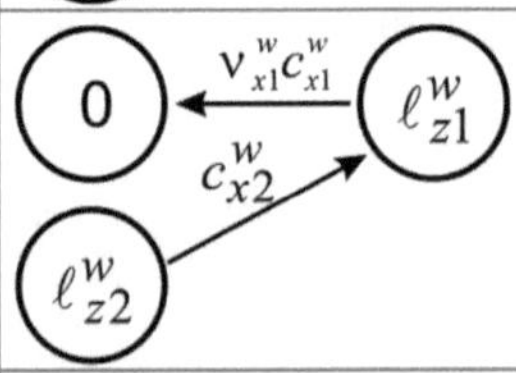	$m_{rm0\ T/P} = m_{v^w_{x1}c^w_{x1}} \cdot m_{\ell^w_{z1}} \cdot$ $\cdot m_{c^w_{x2}} \cdot m_{\ell^w_{z2}} m_{rT2}$
	3	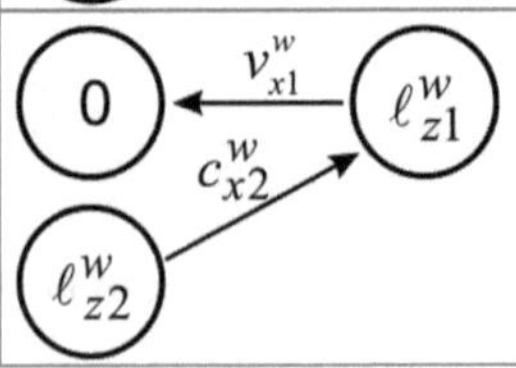	$m_{rm0\ T/P} = m_{v^w_{x1}} \cdot M_{l^w_{z1}} \cdot m_{c^w_{x2}} \cdot$ $\cdot m_{l^w_{z2}} \cdot m_{rT2}$

first three subclasses. A presetting method of arbitrary particular kinematic shaping schemes and their mathematical models is apparent from these examples.

The absolute motion of the gear manufacturing members can be realized from each individual kinematic shaping scheme of the gear tooth system presented in Table 1 by means of a defined set of kinematic schemes.

4 Synthesis of the Machine Tool Layout

A particular coordinate code of the machine tool is assigned to each kinematic scheme in the shaping [15]. In addition, each coordinate code comprises a set of realizable layouts of the machine tool. In this way, a preset relative motion is limited by those motions that are absolutely realizable. The shaping options of a machine tool are characterized by all the machining schemes that are realizable on the machine. The layout of the machine tool is shown with an example. The machining is realized here according to the following shaping scheme:

The two neighbouring reference points of the areas $X_1Y_1Z_1$ and $X_2Y_2Z_2$ coincide or correspond here. Accordingly, the mathematical model of the kinematic shaping scheme can be deduced as follows:

$$m_{rm0\,T/P} = m_{\phi^w_{z0}} \times m_{c^w_{y1}} \times m_{l^w_{z1}} \times m_{\phi^w_{z2}} m_{rT2} \tag{6}$$

With regard to the matrix code k assumed in [15], a coordinate code of the machine tool corresponding to Eq. (6) can be described as follows:

$$K_1 = 6236, \tag{7}$$

where k is 1, 2, …, 6. 1, 2, 3 are the codes of the translational motions along the x, y, z axes. 4, 5, 6 are the codes of the rotational motions along the x, y, z axes.

The following codes of the machine tool layout correspond to this coordinate code: $K_{C1} = 62360$; $K_{C2} = 62306$; $K_{C3} = 62036$; $K_{C4} = 60236$; $K_{C5} = 06236$. An immovable member is marked with “0”, for example, the bed.

It has to be noted here that the layout code K_{C3} corresponds to the conventional layout of a gear shaping machine (Fig. 4). If possible combinations of the matrix multiplication of Eq. (6) are used, a number of alternative coordinate codes is obtained which do not change the shaping scheme of the machine tool: $K_2 = 6326$; $K_3 = 6263$; $K_4 = 3626$.

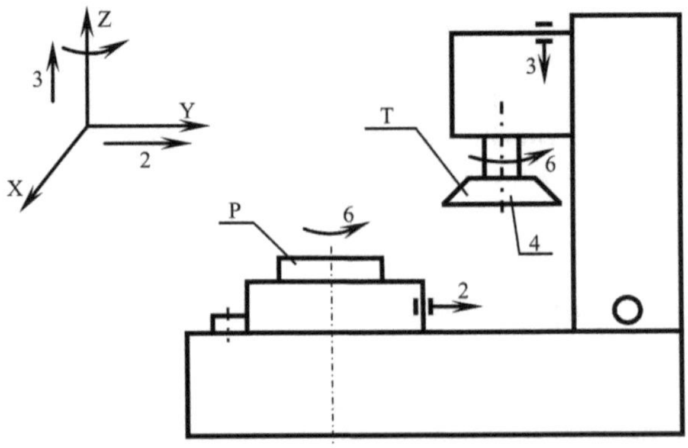

Fig. 4 Layout of the gear shaping machine

Each alternative coordinate code corresponds to a separate layout code, compiled analogously to the above-mentioned example. In accordance with [15], the common ground of the codes can be depicted as a matrix of possible layouts of machine tools, as presented in the following:

$$A = \begin{vmatrix} 62360 & 62306 & 62036 & 60236 & 06236 \\ 63260 & 63206 & 63026 & 60326 & 06326 \\ 62630 & 62603 & 62063 & 60263 & 06263 \\ 36260 & 36206 & 36026 & 30623 & 03623 \end{vmatrix}$$

The columns of matrix A correspond to the coordinate codes from K_1 to K_4, and the rows correspond to the motion variants between the shaping members of the machine tool (on the right of "0" and their members to be shaped (on the left of "0").

In practical use, the coordinate code of the machine shaping system should be given in the global coordinate system. The motion parameters in the kinematic shaping scheme are determined in local coordinate systems here. Correspondingly, each mathematical model of the individual kinematic shaping scheme has to be brought into line with the mathematical model of the machine's shaping system. This was achieved by a transformation to the major axes. Possibly similar transformations, such as $B = T^{-1} \cdot A \cdot T$, are taken into account here. For example, the following orthogonal coordinate transformations correspond to these conditions:

$$v_y^w = \frac{\pi}{2}, \quad m_{v_y^w} \times m_{l_z^w} \Rightarrow m_{l_x^w} \; m_{v_y^w} \times m_{\varphi_z^w} \Rightarrow m_{\varphi_x^w}$$
$$v_x^w = \frac{\pi}{2}, \quad m_{v_x^w} \times m_{l_z^w} \Rightarrow m_{l_y^w}, \; m_{v_x^w} \times m_{\varphi_z^w} \Rightarrow m_{\varphi_y^w}$$

Regarding the layout, this means that the coupled movable members move on the reciprocally perpendicular surfaces.

5 Optimized Synthesis Layouts Shaping Machine Tools

The input data on the machine tools' initial design stages are mathematical models shaping surfaces (1), while the corresponding movements' schemas and ways to implement these movements determine the layout of the machine. The analysis and synthesis problem of possible machine layouts is based on a mathematical description of communication forms of the treated surface with the shape of the tool and movement of the units forming system (FS) of the machine [5]. After conversion of the mathematical models of shaping surfaces (1) for problem analysis and synthesis of the possible machine layouts, the nominal form of the treated surface is represented as a trajectory of the relative motion of the forming points of the tool in the coordinate system of the workpiece.

In this case, the mathematical model FS of the machine or shaping function $\mathbf{r}_0$ is expressed as [15]:

$$\mathbf{r}_0 = A_{0,l} \cdot \mathbf{r}_l = A_{0,l} \cdot A_l \cdot \mathbf{e}^4 = \prod_{i=1}^{l} A_{i-1,i}^{k_i}(q_i) \cdot \prod_{j=l+1}^{l+p} A_{j-1,j}^{k_j}(q_j) \cdot \mathbf{e}^4, \quad l + p = n, \quad (8)$$

where $A_{0,l}$ is the matrix transformation coordinate of the FS machine; $\mathbf{r}_i$ is the radius-vector of the shaping points of the tool in the tool coordinate system (mathematical model of the tool); A_I is the transformation matrix of the tool starting point; and the coordinate transformation matrix for adjacent units FS machine and successive transformations starting point of the tool; and the motion parameters of the *i*th level FS machine and the *j*th conversion tool starting point; $A_{i-1,i}^{k_i}(q_i)$ and $A_{j-1,j}^{k_j}(q_j)$ are the coordinate transformation matrix for adjacent units FS machine and successive transformations of the tool starting point; and the motion parameters of the *i*th level FS machine and the *j*th conversion tool starting point; q_i and q_j are the motion parameters of the *i*th FS machine level and the *j*th conversion of the starting point tool; k_i and k_j are the coordinate transformation matrix codes, $k = 1, \ldots, 6$: 1, 2, 3 are a translational motion along the axes X, Y, Z, 4, 5, 6 are the rotational movement around the axes X, Y, Z; l and p are the number of mobile units (nodes) FS machine and changes of the tool starting point; and $\mathbf{e}^4 = (0, 0, 0, 1)^T$ is the radius vector of the tool coordinate beginning.

Equation (8) is the equation of the machined surface in the shaping form [15] with the number of parameters determined by the number of moving parts of the FS machine and the shape tool, and so it is reasonable to deduce the equality:

$$\mathbf{r}_0(u_0, \upsilon_0) = \mathbf{r}_0(q_1, q_2, \ldots, q_i, \ldots, q_n), \quad (9)$$

where u_0 and υ_0 are the curvilinear coordinates of the machined surface.

In accordance with (8), the shaping function $\mathbf{r}_0$ is uniquely determined by forming coordinate code k. Code k is an ordered sequence of codes k_i and k_j coordinate transformation matrices:

$$k = k_c \cdot k_u = \sum_{i=1}^{l} k_i \cdot \sum_{i=1}^{p} k_i, \tag{10}$$

where k_c and k_u are the coordinate codes of the machine and tool, respectively.

Changing of the sequence (4) order is consistent with the properties of the matrix $A_{i-1,i}^{k_i}(q_i)$ and $A_{j-1,j}^{k_j}(q_j)$. This is the basis for the synthesis of new layouts of the machine with the same shaping possibilities: (1) based on the synthesis of new coordinate machine code for a given tool shape $(k_u = const)$, [15]; (2) on the basis of the parameter redistribution of the machine and tool and the synthesis of new coordinate machine tool code [19].

In turn, a shaping function (8) in the form of (9) allows us to solve the problem of technological capabilities expansion of a given machine layout $(k_c = const)$ by forming thereon the same surface with the tool of another form $(k_u = var)$ [20]. This is achieved by the dependence (9) determination of the motion law of the processing unit PS by machining of a given shape surface, provided that the movements of other parts of the PS and tool shape are known [20].

The selection of suitable machine tools represents a significant part of production preparation. To guarantee the required accuracy, stiffness, reliability, performance, etc., machine tools need to fulfil a number of requirements, which substantially depend on the configuration of their structure. It is therefore necessary to determine the configuration of the machine tool structure in such a way that it meets the technical requirements to the highest degree [30, 31, 38].

The configuration of the machine tool structure is generally selected by highly qualified design engineers, who rely on their experience and intuition. Using intuition and experience, however, does not guarantee that the optimal configuration will be selected. This is mainly because a general algorithm for the selection of a machine tool structure's configuration does not exist, and because only known solutions are considered, but not all possible configurations 0.

To determine and further analyze the configuration of different machine tool structures, various scientific methods have been developed [21, 23–25, 32, 37, 38]. Special attention is given to the method of selecting of the optimal configuration of a machine tool's structure according to structural characteristics. The algorithm of this method consists of three phases [30, 31, 36, 38]:

- determination of the total number of possible configurations;
- structural extraction of subsets which fulfil the needed requirements;
- final selection of the optimal configuration of a machine tool's structure through comparison of their technical and economic characteristics.

Several studies have shown that the method of synthesis of a structure's configurations on the basis of the relations between the movements of tool, workpiece and machine tool units is the most suitable method for determining the total number of possible configurations by shape generation [23, 26, 32]. The subsets are structurally extracted according to the formulated conditions for configuration of a machine tool's structure, which determine the necessary spatial location and alignment of the machine tool units [30, 31, 36, 38]. In the final selection of the best configuration, methods of the cost-utility analysis are applied, the algorithm of which comprises compiling of important criteria, elaboration of a rating scale and evaluation by convolution of all criteria [21].

The manual performance of all steps mentioned above is very time-consuming and slows down the development of machine tools in general. This is unacceptable for modern market conditions and reduces competitiveness. Therefore, the automation of this process is necessary. Today, the process is implemented with special software for the determination of optimal configurations [30, 31, 36]. A particular feature of this software is that it was developed on the basis of Boolean logic. Such a design causes inflexibility in solving new tasks and optimizing the configurations of existing structures. To add a new condition into the algorithm for determining the optimal configuration, for example, the program code or the database of the entire software needs to be changed. This certainly entails an enormous amount of time, cost and effort. This disadvantage can be overcome by using the Fuzzy logic, e.g., of neural networks, which are characterized by high flexibility and their learning ability [27, 33, 34]. This learning ability creates the possibility of adapting the program or database without the software correction. This considerably reduces the time for solving new tasks and optimizing existing configurations.

This paper describes a possible path towards the automation of the first and second phases (see above) of the selection of optimal configurations of machine tool structure by means of neural networks.

6 Determining Possible Configurations of a Machine Tool's Structure

The total number of possible configurations of a machine tool's structure's can be determined according to the method described in [15, 23, 32]. This method is based on the relations between the movements of the workpiece, the tool and, hence, all machine tool units in the shape generation process. To simplify the machine tool design, complex relative movements of the tool and workpiece are combined from the elementary movements of the machine tool units: from three translational movements along the axes x, y, z and three rotary movements around the same axes. Accordingly, the relative movements of the tool and workpiece can be expressed as a coordinate code [15, 23, 32]:

$$k = k_1 \cdot k_2 \ldots k_i, \tag{11}$$

where k_i $(i = 1, \ldots, \mathbf{n})$ is the movement of the *i*th machine tool unit relative to the (*i*−1)th unit; 1, 2, 3 describe the codes of the translational movements along the axes x, y, z and 4, 5, 6, the codes of the rotary movements around the axes x, y, z; and **n** is the number of the mobile machine tool units.

The coordinate code is generated according to the kinematic scheme of the shape generation process. It is formulated with the following rules [15]:

1. The coordinate code should start with the code of the workpiece movement.
2. In the coordinate code, the codes of the rotary movements come first, followed by the codes of the translational movements.
3. Priority is given to the relative movements and then to the guided movements.
4. The codes of the movements are recorded according to the increase in their number.

Taking these rules into account, the coordinate code (1) for gear machining with profile milling cutters, Fig. 5, for example, is as follows:

$$k = 61234. \tag{12}$$

In order to receive the code of a complete configuration of a machine tool's structure from the coordinate code (1), which describes a set of mobile machine tool units, it is necessary to add an immobile basis unit—a machine bed. Symbolically, this is realized by adding the digit "0" to the coordinate code on any position. The thus received coordinate code is called the code of the configuration of the machine tool's structure. From the coordinate code (2), for example, the following structure's configuration's codes can be derived by adding "0":

$$\begin{aligned} K_{C_1} &= 612340;\ K_{C_2} = 612304;\ K_{C_3} = 612034; \\ K_{C_4} &= 610234;\ K_{C_5} = 601234;\ K_{C_6} = 062314. \end{aligned} \tag{13}$$

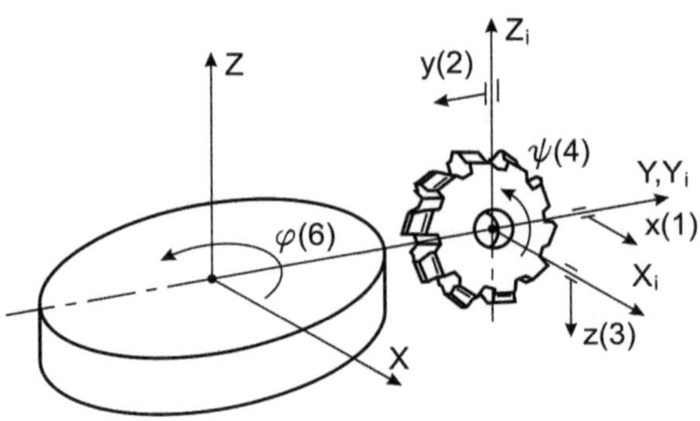

Fig. 5 Kinematic diagram of gear machining with profile milling cutters

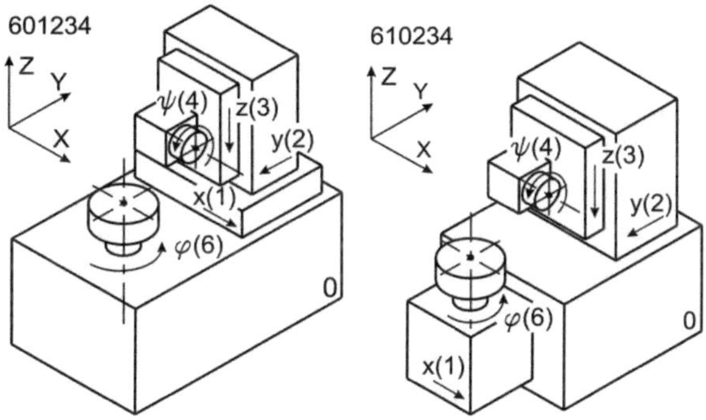

Fig. 6 Examples of configurations of a gear milling machine's structure

Figure 6 shows the configurations of the two machine tool structures that correspond to the codes 601234 and 610234.

The changing of the position of the machine bed and of the corresponding code is not the only possibility for determining alternative configurations of a machine tool's structure, which have the same shape generation abilities. New configurations can also be obtained through permutations, aggregations or disjunctions of machine tool units or digits in the code *k* of the structure's configuration. Only the following four transformations or a combination of these will not change the shape generation's scheme [15]:

- Permutation (rearrangement) of two neighbouring units, which perform translational movements along different axes: 12 = 21, 23 = 32, 31 = 13.
- Permutation (rearrangement) of two neighbouring units, the first of which performs the translational movement along the axis and the second of which performs the rotary movement around the same axis (or the unit, which represents a cylindrical pair): 14 = 41, 25 = 52, 36 = 63.
- Aggregation (combination) of several units into one, which performs the same movement relative to the same axis: 11 = 1, 22 = 2, 33 = 3, 44 = 4, 55 = 5, 66 = 6.
- Disjunction (separation) of one unit into several units, which perform the same movement: 1 = 11; 2 = 22; 3 = 33; 4 = 44; 5 = 55; 6 = 66.

Another option for obtaining alternative configurations of a machine tool's structure is a change in the spatial alignment of its coordinate system. Symbolically, this is realized by the circular permutation of the digits in the code of the structure's configuration according to two schemes: $1 \rightarrow 2 \rightarrow 3 \rightarrow 1$ and $4 \rightarrow 5 \rightarrow 6 \rightarrow 4$.

The total number of configurations with the same shape generation abilities, which can be obtained by using the three methods described above, is determined with the following equation [15]:

$$N = (n + 1) \cdot N_p \cdot N_f \tag{14}$$

Table 2 Structure's configurations fulfilling the first condition (1st class)

361240	361204	361024	360124	306124	036124
362140	362104	362014	360214	306214	036214
362410	362401	362041	360241	306241	036241
612340	612304	612034	610234	601234	061234
613240	613204	613024	610324	601324	061324
621340	621304	621034	620134	602134	062134
623140	623104	623014	620314	602314	062314
623410	623401	623041	620341	602341	062341
631240	631204	631024	630124	603124	063124
632140	632104	632014	630214	603214	063214
632410	632401	632041	630241	603241	063241

where n is the number of mobile machine tool units; N_f is the number of configurations which can be obtained by the four permissible transformations; and N_p is the number of configurations, which can be obtained by spatial rotations of the machine tool according to the change of its coordinate system.

The total number of possible configurations of the structure of a hobbing machine with coordinate code (2), calculated according to formula (4), equals 198. All possible configurations, except those additional configurations that consider the spatial rotations of the machine tool, are shown in Table 2.

The determination of the set of possible configurations of the structure of machine tools on the basis of their kinematic schemes of shape generation is automated with a C++ program, which was developed for this purpose. When the program is used, the coordinate code of the machine tool or of its kinematic scheme of shape generation should be entered. The result of the calculation is delivered as a text file with a complete list of all possible machine tool designs. Simultaneously, the program creates an input file for the neural network.

7 Structural Extraction of a Subset of Configurations

The configurations of a machine tool's structure that fulfil the required conditions are structurally extracted by means of neural networks. The significant difference and advantage of neural networks, compared to conventional methods, is their *learning ability* [28]. Neural networks are trained or instructed by means of examples that are combined into learning sets. The learning sets consist of input and corresponding output data. In the course of learning, the learning sets are analyzed and possible dependencies in their structure are determined. Due to these dependencies, rules are created, according to which the further classification can be carried out [22, 35].

The learning set of the neural network is shown in the form of a connectivity matrix T:

$$T = \begin{bmatrix} k_{1,1} & k_{1,2} & \dots & k_{1,j-1} & k_{1,j} \\ k_{2,1} & \ddots & & & k_{2,j} \\ \vdots & & \ddots & & \vdots \\ k_{i-1,1} & & & \ddots & k_{i-1,j} \\ k_{i,1} & k_{i,2} & \dots & k_{i,j-1} & k_{i,j} \\ c_1 & c_2 & \dots & c_{j-1} & c_j \end{bmatrix}, \tag{15}$$

which consists of the ith number of input vectors (input data of the learning algorithm):

$$\mathbf{p}_i = \begin{bmatrix} k_{i,1} & k_{i,2} & \dots & k_{i,j-1} & k_{i,j} \end{bmatrix}, \tag{16}$$

and of the target vector $\mathbf{T_C}$ (output data of the learning algorithm):

$$\mathbf{T_C} = \begin{bmatrix} \mathrm{c}_1 & \mathrm{c}_2 & \dots & \mathrm{c}_{i-1} & \mathrm{c}_i \end{bmatrix}. \tag{17}$$

The elements $k_{i,j}$ of the input vectors (16) are generated according to the structural conditions of the extraction of the configurations of the machine tool's structure. These conditions are formulated and determined in the form of structure equations. These structure equations are similar to the code (3) of the structure's configuration and have the same number of digits for a defined machine tool. Each of the structure equations, however, does not only correspond to an individual configuration, but to a configurations' set that fulfils a condition of the extraction [38].

The combination of structure equations is explained with the example of a gear milling machine with coordinate code (12). The machine is designed for the machining of medium size workpieces. In this case, the extraction conditions are formulated as follows:

1. The rotation axis of the workpiece should be vertical, since the structure's configurations with horizontal and tilted rotation axis are suitable for the machining of long parts with small diameter 0. This can be illustrated with the following equation, which describes a configuration's subset with the number "6" at the beginning of the code (the machine tool unit rotating around the vertical axis z):

$$\underset{\mathbb{R}}{\forall} B \underset{\mathbb{R}}{\exists} I\left(\phi_Z = \frac{\pi}{2}\right) : \{M_1 \Rightarrow 677777 \vee 767777 \vee 776777\}, \tag{18}$$

where $\mathbb{R}$ is the space of definition of all configurations of the machine tool's structure; B is a set of all existing configurations of the machine tool's structure and I, a subset of all configurations that need to be determined; ϕ_Z is an inclination angle of the rotation axis of the workpiece to the horizontal level; and M_1 describes the subset that fulfils the first extraction condition.

A random digit of the structure's configuration's code, i.e., from 0 to 6, can be on the position of number "7". This position is only required to ensure that all elements of the neural network's learning vector have the same number of characters. This guarantees a correct learning process and further program functions.

2. Tool and workpiece need to be mobile. The configurations with an immobile tool are very complicated and inefficient in production. The configurations with an immobile workpiece are only suitable for machining of very large parts [38]. For this condition, the position of the digit "0" or the machine bed's position is important; it can be in any position but the first and last. Therefore, the second condition can be described with the following logical equation:

$$\underset{\mathbb{R}}{\forall B}\ \underset{\mathbb{R}}{\exists} I(F \wedge W \Rightarrow \text{var}) : \{M_2 \Rightarrow 707777 \vee 770777 \vee 777077 \vee 777707\}, \tag{19}$$

Here, F and W are the motion parameters of the tool and workpiece; and M_2 is a subset that fulfils the second extraction condition.

3. To increase the stiffness, configurations in which a linear moving unit is attached to the rotating unit (moving-out quill) should be avoided [39]. Important elements for the neural network's learning set are the positions of the digits "0", "4" and "6" in the code of the structure's configuration, which exclude the possible combination of the digits "36" and "41". Accordingly, the logical equation for this condition is as follows:

$$\underset{\mathbb{R}}{\forall B}\ \underset{\mathbb{R}}{\exists} I(P = \emptyset) : \{M_3 \Rightarrow 677774 \vee 067774 \vee 677740\}. \tag{20}$$

Here, P represents a motion parameter of the quill; $\emptyset$ describes an empty set; and M_3 is a subset that fulfils the third extraction condition.

4. In order to avoid the weight effect of mobile units on the accuracy of the machine tool, horizontal mobile units should be attached to immobile units [39]. Accordingly, there should be a combination of the numbers "102" or "201" in the code of the structure's configuration. Then, the logical equation for the fourth condition can be described as follows:

$$\underset{\mathbb{R}}{\forall B}\ \underset{\mathbb{R}}{\exists} I(U_M \Leftrightarrow U_S) : \{M_4 \Rightarrow 710277 \vee 720177 \vee 771027 \vee 772017\}. \tag{21}$$

Here, U_m represents a condition parameter of mobile units; U_S is a condition parameter of immobile units; and M_4 is a subset that fulfils the fourth extraction condition.

5. The last extraction condition is obtained by the combination of the previously formulated four conditions (8–11):

$$\underset{\mathbb{R}}{\forall} B\ \underset{\mathbb{R}}{\exists} I : \{M_5 \Rightarrow M_1 \wedge M_2 \wedge M_3 \wedge M_4\}, \tag{22}$$

Here, M_5 is a subset that fulfils the fifth extraction condition.

If the configuration of the machine tool's structure fulfils the fifth condition, then it is optimal.

Additional conditions can be added to and/or excluded from the formulated extraction conditions. In the first case, the subset of the optimal configurations increases, and in the second case, it decreases. The decision as to which conditions should be added or excluded is made by the user or engineer according to the current state of technology.

Based on the formulated conditions or logical Eqs. (8–12), the input vectors (6) are generated. For the first extraction condition (8), for example, the following vectors of the learning set are received:

$$\mathbf{p}_1 = [6\ 7\ 7\ 7\ 7\ 7]; \quad \mathbf{p}_2 = [7\ 6\ 7\ 7\ 7\ 7]; \quad \mathbf{p}_3 = [7\ 7\ 6\ 7\ 7\ 7]. \tag{23}$$

By assigning the vectors (13) to a class c_i in the target vector $\mathbf{T_C}$ (7), it is ensured that they are bound to a certain extraction condition, i.e., for the first condition, $c_i = 1$, for the second condition, $c_i = 2$, etc. Accordingly, vectors $\mathbf{p_1}$, $\mathbf{p_2}$ and $\mathbf{p_3}$ are assigned to the same class $c_1 = c_2 = c_3 = 1$. Analogous to that, all other input vectors $\mathbf{p}$ are assigned to the classes in the target vector $\mathbf{T_C}$, according to the logical Eqs. (9–12). Then, the connectivity matrix T (5) is formed.

A radial-basis-function (RBF) network, modelled in Matlab, was used for learning. Such RBF networks can effectively solve the tasks of generalization and classification of vectors, since they have a large number of neurons, compared to the standard networks with direct transmission of signals and reverse spreading of errors [35]. Figure 7 shows a structure scheme of the modelled RBF network for the extraction of the optimal configuration of the machine tool's structure. The network consists of two layers: an RBF layer with an activating function $radbas\,(n) = e^{-n^2}$ and S_1 neurons and a linear output layer with an activating function $purelin\,(n) = n$ and S_2 neurons.

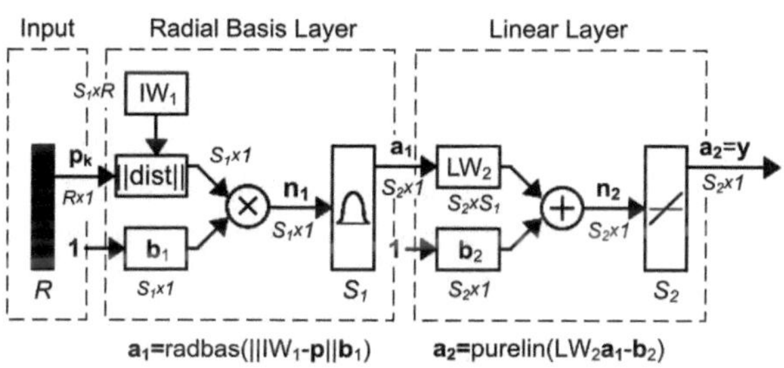

Fig. 7 Structural diagram of the neural network for the extraction of the structure's optimal configuration

During the learning with the connectivity matrix *T*, weight matrices are formed for the first IW_1 and second layer LW_2 [29, 35]. In the first layer, a distance between the new input vector $\mathbf{p_k}$ and the vectors of the learning set is determined. In this process, the data of the vector $\mathbf{p_k}$ is transmitted to the block ||dist||. In this block, the distance between the vector $\mathbf{p_k}$ and the weight matrix IW_1 is determined. The output data of the block ||dist|| is multiplied element by element with a displacement vector $\mathbf{b_1}$. Their product forms the input data of the activating function ***radbas***. If the output of the activating function—vector $\mathbf{a_1}$—is a number close to "1", the new vector $\mathbf{p_k}$ is the closest to the vector of the learning set 0. In the second layer, it is determined to which class the vector $\mathbf{p_k}$ should be assigned. The output vector $\mathbf{a_1}$ is multiplied with the weight matrix LW_2. Their product is added to the displacement vector $\mathbf{b_2}$. The sum is the input data of the activating function ***purelin***. The output of this function—vector $\mathbf{a_2}$—takes the value of the class to which the new vector $\mathbf{p_k}$ is assigned by the neural network [22].

The displacement vectors $\mathbf{b_1}$ and $\mathbf{b_2}$ provide a possibility for correcting the sensitivity of the neurons in corresponding layers. The user can change the values of the displacement vectors' elements [23]. The input vectors $\mathbf{p_k}$ of the neural network include all 198 possible configurations of the machine tool's structure, which correspond to the coordinate code (2), and are created analogously to the vectors of the learning set (13).

Tables 2, 3, 4, 5, 6 show a correlation of the possible configurations of the structure according to the extraction conditions formulated above. This correlation was received through manual calculation and combination. The configurations of the gear milling machine with the coordinate code (2), which are presented in Table 6, fulfil the fifth extraction condition (12). Therefore, they are the most suitable for the machining of medium size workpieces. One of these optimal machine tool configurations, which corresponds the code 631024, is shown in Fig. 8 on the left side.

Table 7 represents the classification of designs received by means of the RBF network. Tables 2 and 7 are identical; however, the classes to which the neural network assigned the codes of the structure's configuration are presented in Table 7

Table 3 Configurations of the structure fulfilling the second condition (2nd class)

361204	361024	360124	306124	
362104	362014	360214	306214	
362401	362041	360241	306241	
612304	612034	610234	601234	
613204	613024	610324	601324	
621304	621034	620134	602134	
623104	623014	620314	602314	
623401	623041	620341	602341	
631204	631024	630124	603124	
632104	632014	630214	603214	
632401	632041	630241	603241	

Table 4 Configurations of the structure fulfilling the third condition (3rd class)

612304	612304	612034	610234	601234	061234
613204	613204	613024	610324	601324	061324
621304	621304	621034	620134	602134	062134
623104	623104	623014	620314	602314	062314
631204	631204	631024	630124	603124	063124
632104	632104	632014	630214	603214	063214

Table 5 Configurations of the structure fulfilling the fourth condition (4th class)

		361024			
		362014			
			610234		
			620134		
		631024			
		632014			

Table 6 Configurations of the structure fulfilling all four conditions (5th class)

			610234		
			620134		
		631024			
		632014			

instead of the codes of the structure's configuration. As we can see from Tables 2, 6 and 7, the program has assigned to the 5-th class the same configurations, which were defined as optimal by manual calculation. This means that the instructed neuronal network correctly classifies and determines the optimal configurations of the structure in accordance with the extraction conditions.

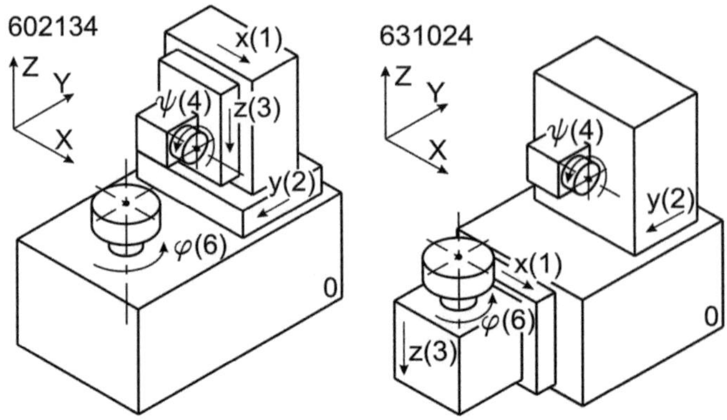

Fig. 8 Examples of the identified optimal configurations of the machine tool structure

Table 7 Classification of the structure's configurations by instructed neural network

1	2	4	2	2	1
1	2	4	2	2	1
1	2	2	2	2	1
3	3	3	**5**	3	3
3	3	3	3	3	3
3	3	3	**5**	3	3
3	3	3	3	3	3
1	2	2	2	2	1
3	3	**5**	3	3	3
3	3	**5**	3	3	3
1	2	2	2	2	1

To test the operating mode of the developed algorithm and program further, the optimal configurations of the structure were extracted for a hobbing machine with the same coordinate code (2), but in a case of the machining of large gear wheels. The learning data of the neural network will differ from the previous learning data in three conditions:

1. The second extraction condition is re-formulated as follows: The workpiece can make only one movement or be immobile, because the displacement of the large masses causes enormous energy consumption and reduces the efficiency [38]. Hence, the logical equation is as follows:

$$\underset{\mathbb{R}}{\forall} B \underset{\mathbb{R}}{\exists} I(W = 1 \vee const) : \left\{ M_2' \Rightarrow 077777 \vee 707777 \right\} \quad (24)$$

2. The fourth extraction condition is modified as follows: In order to avoid the weight effect of the units on the accuracy of the machine tool, the horizontal

mobile unit, which performs the largest displacements, should be attached to the immobile units [38]. Therefore:

$$\underset{\mathbb{R}}{\forall} B \underset{\mathbb{R}}{\exists} I(U_{M\max} \Leftrightarrow U_S) : \left\{ M_4' \Rightarrow \begin{matrix} 702777 \vee 770277 \vee 777027 \\ \vee 720777 \vee 772077 \vee 777207 \end{matrix} \right\}. \quad (25)$$

3. The fifth condition is formulated analogous to Eq. (12), but with the logical Eqs. (8, 10, 14, 15):

$$\underset{\mathbb{R}}{\forall} B \underset{\mathbb{R}}{\exists} I : \left\{ M_5 \Rightarrow M_1 \wedge M_2' \wedge M_3' \wedge M_4' \right\}. \quad (26)$$

Tables 2, 4, 8, 9 and 10 show the results obtained manually, and Table 11 correspondingly shows the results received by the neural network. The optimal machine tool configuration of the hobbing machine for large gear wheels, which corresponds with the code 602134, is shown in Fig. 8 on the right side. If the results are compared with each other, it can be concluded that the newly instructed neural network makes no mistakes in the classification and determination of the optimal configurations of the structure, despite the fact that the extraction conditions were modified.

Table 8 Configurations of the structure fulfilling the second condition (2nd class)

				306124	036124
				306214	036214
				306241	036241
				601234	061234
				601324	061324
				602134	062134
				602314	062314
				602341	062341
				603124	063124
				603214	063214
				603241	063241

Table 9 Configurations of the structure fulfilling the fourth condition (4th class)

361204	361204	361024			
		362014	360214		
		362041	360241		
		612034	610234		
613204	613204	613024			
			620134	602134	
			620314	602314	
			620341	602341	
631204	631204	631024			
		632014	630214		
		632041	630241		

Table 10 Configurations of the structure fulfilling all four conditions (5th class)

				602134	
				602314	

Table 11 Classification of the structure's configurations by the newly instructed neural network

4	4	4	2	2	2
1	1	4	4	2	2
1	1	4	4	2	2
3	3	4	4	3	3
4	4	4	3	3	3
3	3	3	4	5	3
3	3	3	4	5	3
1	1	2	4	4	2
4	4	4	3	3	3
3	3	4	3	3	3
1	1	4	2	2	2

8 Conclusion and Prospect

The performed systematization and classification represent a basis for the analysis and synthesis of different machining schemes of gear tooth systems. The kinematic shaping scheme of the gear tooth system is classified according to classes, types and kinds. In addition, the schemes are classified on the basis of the absolute motions of the gear manufacturing members during shaping, as well as their mathematical models. The analysis and synthesis of different machining schemes of the gear tooth system are effected according to criteria of accuracy, productivity and quality, as well as other criteria.

The algorithm described above allows for automated development of machine tools in the phase of selecting of the optimal configuration of their structure by means of neural networks. The results of the operating tests of the developed neural network correlate with the manual calculations if the structural extraction conditions are correctly formulated. The learning ability of the neural network guarantees high flexibility of the algorithm in solving new tasks and also in optimization of existing configurations of a machine tool's structure. No direct changes are required in the program or in its data base if the extraction conditions are changed.

As a further development of the algorithm and program, the structure equations should be generalized for all types of machine tools. These equations should be entered into a database of learning sets. This will considerably reduce the calculation time.

References

1. Anikin JV (1975) Sinusoidal engagement. Fundamentals of Geometry and Kinematics. VGU, Voronezh
2. Demenego A, Vecchiato D, Litvin FL, Nervegna N, Manco S (2002) Design and simulation of meshing of a cycloidal pump. Mech Mach Theory 37:311–332
3. Denavit J, Hartenberg RS (1955) A kinematic notation for lower-pair mechanisms based on matrices. ASME J Appl Mech 77:215–221
4. Erichov ML (1972) Principles of systematization, analysis methods and issue of synthesis. Dissertation, Dr. Sc. techn, Habarovsk
5. Granovski GI, Granovski VG (1985) Cutting of metal. Higher school, Moskow
6. Katsumi Nomizu (1994) Takeshi sasaki: affine differential geometry. Cambridge University Press, Cambridge, XIV
7. Kolchin NI (1949) Analytical calculation of plane and three-dimensional engagement. Mashgis, Moscow-Leningrad
8. Konovalov GG (1961) Fundamentals of new method of metal cutting. ASB, Minsk
9. Krivosheya AV, Petasyuk OU, Melnik VE, Korinets AV (2004) Procedure of setting and mathematical description of the initial form-generating profiles. J Superhard Mater 1:73–85
10. Litvin FL, Fuentes A, Gonzalez-Perez I, Carnevali L, Sep TM (2002) New version of novikov-wildhaber helical gears: computerized design, simulation of meshing and stress analysis Comput. Methods Appl Mech Eng 191:5707–5740
11. Litvin FL, Fuentes A (2004) Gear geometry and applied theory, 2nd edn. Cambridge University Press, Cambridge
12. Perepeliza BA, Kondusova EB, Filimonov EV (2002) Features of cutting tools 3D modeling on basis of space multiparameter reflection. Cutting and tool in technological systems NTU « KhPI », vol 62, pp 99–103
13. Radzevich SP (2007) Kinematic geometry of surface machining, CRC Press Inc, Boca Raton
14. Rodin PR (1981) Shaping fundamentals of surface by cutting. Vishcha Shkola, Kiev
15. Reshetov DN, Portman VT (1988) Accuracy of machine tools. New York, ASME-Press, p 304
16. Sheveleva GI (1999) Shaping and contact theory of moving bodies. "Stanki", Moscow
17. Vulgakov EB, Dorofeev VL (2002) Computer design of involute gearing by generalizing parameters. Convers Manuf 6:148–154
18. Wildenhof P (2007) Voraussetzungen und Möglichkeiten für die Realisierung einer Gleitflächenverzahnung. Konstruktion 59(9):107–113
19. Danilchenko Yu M, Krivosheya AV, Karska AA, Storchak MG, Pasternak SI (2011) Initial generation of machine tools coordinate codes modern technology in mechanical engineering, vol. 6, Kharkov: NTU "HPI", pp 23–28
20. Heisel U, Pasternak S, Storchak M, Schaal M, Danilchenko Yu (2010) Modellieren des Verzahnens mit Scheibenwerkzeugen ZWF Zeitschrift für wirtschaftlichen Fabrikbetrieb. Nr. 7–8:649–654
21. Averjanov OI (1987) The modular principle of machine tools with CNC building. Machine building, Moscow, 232 p

22. Dayhoff J (1990) Neural networks. Van Nostrand Reihold, New York
23. Inasaki I, Kishinami K, Sakamoto S, Takeuchi Y, Tanaka F (1997) Shape generation theory of machine tools. Yokendo Press
24. Ito Y (2008) Modular design for machine tools. McGraw Hill, New York, 504 p
25. Ito Y, Shinno H (1990) Structural description of machine tools. Trans JSME Ser C 46 (405):562–571
26. Iwata K, Sugimura N, Peng L (1990) A study of the fundamental design of machine structure for machining. Trans JSME Ser C 56(523):803–809
27. Jain AK, Mao J, Mohiuddin KM (1996) Artificial neural networks: a tutorial. Computer 29(3):31–44
28. Kudinov AV (2001) Features of neural network modeling tools. Machine and Tool 1, pp 13–18
29. Medvedev VS, Potemkin VG (2002) Neural networks. MATLAB 6. DIALOG, Moscow, 496 p
30. Moriwaki T (2008) Multi-functional machine tool. Annals of the CIRP 57(1):736–749
31. Moriwaki T, Nunobiki M (1992) Object-oriented design support system for machine tools. Trans JSME Ser C 58(546):655–660
32. Portman V, Inasaki I, Sakakura M, Iwatate M (1998) Form-shaping system of machine tools: theory and applications. Annals of the CIRP 47(1):329–332
33. Patan K (2008) Artificial neural networks for the modeling and fault diagnosis of technical processes. Springer, Berlin, Heidelberg, 206 p
34. Rey GD, Wender KF (2011) Neuronale Netze: Eine Einführung in die Grundlagen, Anwendungen und Datenauswertung. 2. Auflage, Huber, Bern, 205 S
35. Rutkowskaja D, Pilinskij M, Rutkowskaja L (2006) Neural networks, genetic algorithms and fuzzy systems. Telekom, Moscow, 452 p
36. Shinno H, Ito Y (1987) Computer aided concept design for structural configuration of machine tools: variant design using directed graph. Trans ASME J Mech Trans Autom Des 109: 372–376
37. Spicer P, Koren Y, Shpitalni M, Yip-Hoi D (2002) Design principles for machining system configurations. Annals CIRP 51(1):275–280
38. Vragov Yu D (1978) Analysis of machine tools configurations. (Fundamentals of componetik). Machine building, Moscow, 208 p
39. Weck M, Brecher C (2006) Werkzeugmaschinen. In 6 Bänder, Springer, Berlin Heidelberg

MIX
Papier aus verantwortungsvollen Quellen
Paper from responsible sources
FSC® C105338

If you have any concerns about our products,
you can contact us on
ProductSafety@springernature.com

In case Publisher is established outside the EU,
the EU authorized representative is:
Springer Nature Customer Service Center GmbH
Europaplatz 3, 69115 Heidelberg, Germany

Printed by Libri Plureos GmbH
in Hamburg, Germany